Dorothea Juliana Wallich

PATHWAYS TO THE
UNIVERSAL TINCTURE

DOROTHEA JULIANA WALLICH

Pathways to the Universal Tincture

COLLECTED ALCHEMICAL WRITINGS

Edited by Alexander Kraft, PhD

TRANSLATED BY ALEXANDER KRAFT
With Michael A. Putman and Aaron Cheak

RUBEDO
2025

Copyright Rubedo Press 2025

Pathways to the Universal Tincture: Collected Alchemical Writings
By Dorothea Juliana Wallich (1657–1725)
Edited by Alexander Kraft
Translated by Alexander Kraft,
Michael A. Putman, and Aaron Cheak

The works collected and translated in this volume were originally published in Leipzig by Joh. Heinichens Wittwe as: *Das Mineralische Gluten, doppelter Schlangen-Stab, Mercurius Philosophorum, langer und kurzer Weg zur Universal-Tinctur* (1705); *Der philosophische Perl-Baum. Das Gewächs der drei Principien zu Deutlicher Erklärung des Steins der Weisen* (1705); and *Schlüssel zu dem Cabinet der geheimen Schatz-Kammer der Natur zur Suche und Findung des Steins der Weisen, durch Fragen und Antwort gestellt* (1706).

Published by
RUBEDO PRESS
Auckland, New Zealand, 2025

ISBN: 978-0-9951245-1-6 (paperback)
ISBN: 978-0-9951245-3-0 (hardback)

The moral rights of the authors have been asserted.
All rights reserved. No part of this work may be reproduced without written permission from the publisher.
Brief passages may be cited by way of criticism, scholarship, or review, as long as full acknowledgement is given.

Design and Typography
by Aaron Cheak

SCRIBE SANGUINE
QUIA SANGUIS SPIRITUS

Table of Contents

Introduction
A Short Biography of Dorothea Juliana Wallich
Alexander Kraft, PhD
9

The Mineral Gluten
17

Das mineralische Gluten, doppelter Schlangen-Stab, Mercurius Philosophorum, langer und kurzer Weg zur Universal-Tinctur

The Mineral Gluten, Double Serpent Staff, Mercurius Philosophorum, Long and Short Way to the Universal Tincture

The Philosophical Pearl Tree
157

Der philosophische Perl-Baum. Das Gewächs der drei Principien zu deutlicher Erklärung des Steins der Weisen

The Philosophical Pearl Tree. The Growth of the Three Principles for Clear Explication of the Stone of the Wise

Key to the Cabinet of
Nature's Secret Treasure Chamber
369

Schlüssel zu dem Cabinet der geheimen Schatz-Kammer der Natur zur Suche und Findung des Steins der Weisen, durch Fragen und Antwort gestellt

Key to the Cabinet of Nature's Secret Treasure Chamber for Seeking and Finding the Stone of the Wise, Presented Through a Series of Questions and Answers

Annotations
679

Bibliography
697

ALCHEMICAL SYMBOLS
USED IN THIS BOOK

☉	*Sol*, sun; gold, Au
☽	*Luna*, moon; silver, Ag
♂	Mars; iron, Fe
♀	Venus; copper, Cu
♃	Jupiter; tin, Sn
♄	Saturn; lead, Pb
☿	*Mercurius*; mercury, Hg
⊖	Salt; sodium chloride, NaCl
⊕	Saltpeter, nitrate, potassium nitrate, KNO_3
⊕	*Vitriol*, sulfate, SO_4^{2-}
O	Alum, hydrated double sulfate salt of aluminum and potassium, $KAl(SO_4)_2 \cdot 12H_2O$
O	Circle of the cabbalistic figure
⊕	Verdigris, copper acetate, $Cu(CH_3CO_2)_2$
♄	Sulfur, S
♂	Arsenic, As_2O_3
♁	Antimony: in fact antimony trisulfide, Sb_2S_3
✶	Salmiac, ammonium chloride, NH_4Cl
♃	Cream of tartar; potassium bitartrate
♌	Leo; lion
♇	*Aqua Regia*, royal water, mixture of nitric acid, HNO_3 with hydrochloric acid, HCl
♇	*Aquafort*, strong water, concentrated nitric acid, HNO_3 (two different symbols are used in the books)
□	Square of the cabbalistic figure
△	Triangle of the cabbalistic figure (alchemical symbol for fire but not used in these books).

INTRODUCTION

A Short Biography of Dorothea Juliana Wallich

ALEXANDER KRAFT

DOROTHEA JULIANA WALLICH was born as Dorothea Juliana Fischer in 1657 in the central German town of Weimar in Thuringia. Her baptism is recorded for 8 August 1657[1] in the parish register of the town church of Weimar. Her father was Heinrich Fischer (1611-1665) First Tax Collector of the Duke of Saxe-Weimar, her mother Anna-Catharina Lippach (?–1664), daughter of the well known Protestant theologian David Lippach (1580–1653). On 1 March 1674, at the age of sixteen years and seven months Dorothea Juliana Fischer married the thirty-four year old Johann Wallich (1639–1711), Court Secretary of the Duke of Saxe-Weimar. The wealthy couple lived in Weimar and had no children.

In most sources from her lifetime and in some letters written by herself, her name was given as Dorothea Juliana Wallichin. But many letters she only signed with D.J.W., and in her three books this was printed as D.I.W. The suffix, -*in*, in her family name indicated that she was a woman. This female form was still common in early modern Germany. Today, this distinction in the family name between male or female persons is not in use anymore.

Perhaps around 1685, Wallich was in her late twenties, she became interested in chymistry including practical laboratory work. Wallich

1 This date is according to Julian calendar, which was still in use in the Protestant part of Germany until 1700. According to the modern Gregorian calendar her baptism took place on August 18, 1657.

worked a lot in this field in the next twenty years and became a well known expert of chymistry in the area of Thuringia and Saxony. She and her husband owned a small mine in the Schneeberg area of the Saxon ore mountains, called Dorotheenzeche. It is reported that she sucessfully extracted more silver from cobalt ores than other chymists were able to do during this time. At this time, Wallich was well acquainted with the court physician Georg Ernst Stahl (1659–1734) in Weimar. Stahl is well-known in the history of chemistry as the developer of the phlogiston theory. Around 1693, the Wallichs left Weimar and moved permanently to Schneeberg. Johann Wallich had also given up his good position as court secretary in Weimar. From then on, they focused exclusively on mining and alchemy.

During her many experiments, Dorothea Juliana Wallich produced a rose-colored salt and a rose-colored aqueous solution by treatment of a certain ore with nitric acid and sodium chloride. Today, we know that she had produced unpure cobalt(II) chloride salt and its aqueous solution. Both change their color with changes in temperature. At lower temperature, the salt and solution are rose-colored. At higher temperature, they change to a deep blue color. Due to the impurities, this was followed by a green coloration at even higher temperature. These color changes are highly reversible.

Dorothea Juliana Wallich thought she had found the stage of *cauda pavonis* (peacock's tail) in the great work (*opus magnum*), that is, the making of the philosophers' stone. In the *cauda pavonis* stage of the *opus magnum* an array of iridescent colors appear, either all colors at the same time or one color after the other. Her experimental discovery of the thermochromism of cobalt chloride and her assumption that this was a successful step towards the production of the philosophers' stone gave Wallich the idea of writing three alchemical books. These books are characterized by a particularly rich and creative use of alchemical code names and allegorical descriptions of processes. Although this often reads very well, it does not make it too easy for everyone to understand the text.

Since Wallich had produced her thermochromic tinctures and salts from only one specific ore or *minera*, she was convinced that this outwardly quite inconspicuous ore was the only one that was suitable to prepare the universal tincture. Today we know that this ore belongs to the so-called bismuth-cobalt-nickel formation. In this type of ore, bismuth is present in native form, while cobalt and nickel mostly occur in the form of arsenic compounds, with cobalt dominating. Veins with this bismuth-cobalt-nickel ore also often contain small amounts of silver.

This *minera* is a kind of main character in Wallich's three books. From various perspectives and viewpoints, Wallich attemped to explain why this ore has this significance and how the universal tincture can be made from it. Large passages in her first book describe how one can get to the universal tincture from her *minera* by more or less traditional sequences of alchemical procedures. However, some particular recipes are also presented at the end of book one.

In her second book, Wallich combined alchemy with the holy scriptures, adopting some of Jacob Böhme's ideas and formulations. But here too she repeatedly turned to her *minera*, which she tried to show are referred to in various passages in the Old and New Testament. Just as there is only one God in heaven, there is only one *minera* that can be used for the preparation of the philosophers' stone.

In her third book, Wallich addressed around two-hundred hypothetical questions. She provided information on many allegorical alchemical representations and sayings, but also on certain alchemical recipes and *Decknamen*. But here too, her *minera* plays a major role. Among other things, she described where it can be found and how it can be recognized. Wallich also reported various chemical reactions of this ore, including the preparation of thermochromic tinctures and salts. The universal tincture produced with it is also said to act as a medicine against all diseases. So her primary concern was not to become rich, but to help others, especially the sick and needy. Yet the transformation of base metals into gold remained an important goal.

She also pursued this goal in her further practical work. From a serious chymist who worked in the mining industry, she changed to become an alchemist, who now pursued the aim of converting base metals into gold by use of her universal tincture or by one of her particular processes. While her husband stayed in the mining town of Schneeberg, Wallich traveled through Germany and tried to find potent business partners for her alchemical work. Since she was not fully capable of doing business as a woman, she enlisted the help of a young admirer, Johann Ernst Heubel (1678–1740).

In Leipzig, Wallich worked for some time in the apothecary laboratory of Heinrich Linck (1638–1717), and it is also reported that she worked for the lords of Schönburg in Western Saxony. Not much is known about these attempts to make gold. More information is available about her work as a kind of court alchemist for three German princes between 1706 and 1710: these were count Anton Günther II of Schwarzburg-Sondershausen-Arnstadt (1653–1716, count from 1666), Johann Wilhelm Elector Palatine (1658–1716, duke from 1679), and duke Ernst Ludwig I of Saxe-Coburg-Meiningen (1672–1724, duke from 1706). In three more or less quite similar contracts with these three clients, Wallich promised to produce gold by chymical operations from a certain amount of silver. She wanted to transmute about ten percent of the silver into gold. In all three cases, she had first to prove that her chymical process did work properly. After this proof, her contracting partner would invest several thousand *thaler* to transfer this chymical process to a large scale operation. The profit of this large scale operation would be divided equally between Wallich and the corresponding duke. She wanted to use part of the money she would receive from the work's profits to build and maintain an orphanage. In Arnstadt, Düsseldorf, and Coburg, Wallich tried to produce gold from silver according to her particular processes. She was never successful in proving that her processes worked with the requested three small samples. She lost nearly all her money and her reputation in these years.

Although she was not successful during her time as court alchemist, she had proven that a woman could also work in the alchemical laboratories of great princes and successfully stand up to the men in this field. This fact, together with her three alchemical books, and the discovery of the thermochromism of cobalt chloride, makes her the most important female alchemist of the Early Modern period in Europe. In doing so, she successfully followed in the footsteps of personalities such as Maria Prophetissa, Cleopatra the Alchemist, Hypatia, and Isabella Cortese. After her failure to produce gold in Arnstadt, Düsseldorf, and Coburg, she moved back to the small town of Arnstadt in Thuringia, not far from her hometown of Weimar, and lived there the remaining fifteen years of her life, still being active as a chymist but no longer for remarkable clients. The parish registers of the town church of Arnstadt state that she, widow of a secretary, died there as a poor woman in February 1725 at the age of sixty-seven. She was buried on 25 February 1725.

Wallich's legacy is the thermochromism of various cobalt compounds that she discovered. Shortly after her death, this temperature-dependent effect began to be used for secret inks. At room temperature, the writing was practically illegible; if the document was warmed, it could be read easily. Later, in the nineteenth century, this effect began to be used as a humidity indicator, for example for desiccants but also for "weathermen", small figures that were placed in the home that changed color depending on the humidity. In the twentieth century, it was finally discovered that this so-called ligand exchange thermochromism is not limited to cobalt, but is also found in other transition metals, such as iron or nickel. A modern application is the use of this effect for adaptive sun protection glazing in architecture. At lower temperatures, such dynamic glazing has a high light transmittance. However, when it heats up in strong sunlight, it turns dark and thus lets less light through. This reduces the overheating of rooms and prevents glare in summer.

The ligand exchange thermochromism, discovered by Dorothea Juliana Wallich, is based on the mechanism that transition metal ions such as Ni^{2+} or Co^{2+} can be complexed by certain ligands. Such ligands are molecules or ions with free electron pairs, so-called Lewis bases. Depending on the ligand strength, the corresponding transition metal complexes exhibit different colors. These can be very pale, but also extremely dark colors. If different ligands are available, the temperature of the system and the concentration of corresponding substances determines which ligand is preferred for complex formation. Wallich discovered the thermochromism of the ligand exchange of the ligands water and chloride with the transition metal ion Co^{2+}. At lower temperatures, cobalt is complexed by water molecules. This complex shows a pale pink color. At higher temperatures, the water ligands are exchanged by chloride ions. The resulting complex is deep blue. With other transition metal ions and other ligands, all the colors of the world can be produced.

EDITION HISTORY

According to the prefaces, Wallich wrote her three books in the years 1704 and 1705 in her home town, Weimar. The books were published in 1705 and 1706 in Leipzig in Saxony by the publishing house of Johann Heinich's widow. This widow, Magdalena Heinich, had continued the publishing business of her husband Johann Heinich after his death in 1693 until 1719. A second edition of Dorothea Juliana Wallich's books was published in 1722 in Frankfurt am Main and Leipzig by Georg Christoph Wintzer. Wintzer had taken over the publishing house of Johann Heinich's widow in 1719. After 1724 the business was continued by the Heinsius family. In their publishing house, the third edition of Dorothea Juliana Wallich's three books appeared in 1763, thirty-eight years after the death of the author.

The fourth edition presented here contains not only the German text with carefully modernized spelling and grammar, but also the first translation of her three books into English. This is the first known complete translation of her works. In doing so, we are also honoring the three-hundredth anniversary of the death of this important Early Modern female alchemist. We hope that this edition will provide an impetus and also a basis for researchers in the history of alchemy, the natural sciences, and the history of religion and western esotericism to focus more intensively on this interesting and fascinating, but until now almost completely neglected personality and her writings.

The Mineral Gluten

DAS MINERALISCHE GLUTEN

Doppelter Schlangen-Stab

MERCURIUS PHILOSOPHORUM

Langer und kurzer Weg

ZUR

UNIVERSAL-TINCTUR

Deutlich und klärlich entdeckt und
angewiesen durch

D. I. W.

VON WEIMAR AUS THÜRINGEN

LEIPZIG
In Verlegung Joh. Heinichens Witwe
1705

THE MINERAL GLUTEN

Double Serpent Staff

MERCURIUS PHILOSOPHORUM

Long and Short Way

TO THE

UNIVERSAL TINCTURE

*Explicitly and Clearly Uncovered
and Instructed by*

D. I. W.

FROM WEIMAR IN THURINGIA

LEIPZIG
Published by Joh. Heinichen's Widow
1705

*Bist stumm, liebes Büchlein, gegen
alle Unwürdigen; schreie aber laut
gegen alle, die es würdig; Ja rufe
mit Trompeten aus:
Wer mich versteht, wird der Pluto
und Æsculapius sein.*

Be silent, dear little book,
to all who are unworthy;
but to all who are worthy cry aloud
indeed call out with clarions:
Whoever understands me
will be Pluto and Æsclepius.[1]

VORREDE AN DEN LESER

NACH STANDESGEBÜHR GEEHRTER LESER, DASS ICH DIESES geschrieben, ist aus Mitleiden geschehen, so ich habe gegen meinen Nächsten. Ich weiß, was ich für Mühe gehabt, wegen der dunklen Schriften der guten chymischen Bücher, ist auch in der ganzen Welt kein guter Freund anzutreffen, der einen unterrichtet. Der größte Schaden und Verhindernis kommt vollends daher, dass sich viele Irrwische finden, die von dieser Sache schreiben, selber aber nichts davon wissen und einen durch ihre Narration irre machen; Auch, wenn man sich schon etwas zurechtgefunden, wiederum auf Irrwege führen. Etliche haben so hochgelehrt und dunkel von der Sache geschrieben, dass sie kein Mensch verstehen soll, weil sie sich selber nicht verstanden, auch nicht gewusst, wovon sie schreiben; Die nichts können, als nur von gemeinen ☉ und ☽ schwatzen oder von ♂, ♀, ♃, ♄, ☿, zu welchen Körpern sie so viele *Menstrua* erdenken, als Tage im Jahre. Bald soll es sein ⊕,⊖,○,⊕, *Tartarus*, Weinstein, Salz, ♠, ♂, bald weißer Kieselstein, *Minera ♂ Solaris*, Blutstein, roter *Bolus*, *Terra Sigillata*, ♈︎, Maien-Tau, Essig, Kinder-Urin, Menschen-Speichel, &c. Wenn sie dann erschnappt haben, dass es ein metallisches *Menstruum* sein soll, so zerren sie den ☿ und ♂ herzu, da muss der Weinstein und *Spiritus Vini* herhalten; Und wer will alle diese unnötigen Dinge erzählen? Ich warne den Leser vor jetzt genannten, auch noch vor vielen, die mir jetzt nicht einfallen. Weil ich nun fleißigen Arbeitern gerne will dahin helfen, da sie längst gern hin gewesen wären, so habe dieses geschrie-

PREFACE TO THE READER

DEAR READER, HONOURED ACCORDING TO YOUR STATUS, I have written this out of compassion for my neighbour, since I know what difficulties the obscure writings in the good chymical books can present their readers, just as I am aware that it is impossible to find a good friend anywhere in the world who could enlighten you as to their meaning. The greatest harm and hindrance comes entirely from the fact that there are many flighty and whimsical persons who write about this matter, but know nothing about it themselves, and so confuse their readers. Even if one has already found one's bearings, one is lead back onto the wrong pathways. Some wrote about the matter in such dense and erudite prose, that no one could understand it, because the authors, too, did not understand or know about what they wrote. Such authors know nothing but how to prattle on about common ☉ and ☽ or about ♂, ♀, ♃, ♄, ☿, to which bodies they devise as many *menstrua* as there are days in a year. First it shall be ⌽,⊖,○,⊕, *tartarus*, cream of tartar, salt, ♄, ♂, soon to be white pebble stone, *minera ♂ solaris*, blood-stone, red *bolus, terra sigillata*, ℞, May dew, vinegar, children's urine, human saliva etc. When they realize that it has to be a metallic *menstruum*, they pull out ☿ and ☌,[2] and tartar and *spiritus vini* must serve there—and who wants to speak of all of these unnecessary things? I warn the reader of what was mentioned above and of much more that I cannot think of at the moment. Because I now want to help diligent workers reach the point where they wanted to go long ago, I wrote this

ben, in welchem niemand ein Schlaraffenland vorstellen will. Habe mich in solchem der Kürze beflissen, so viel es möglich gewesen, kein Wort umsonst noch vergebens gesetzt, sondern den schnurgeraden Weg zum hesperischen Garten und zum Paradies gewiesen, wo die güldenen Äpfel und die heilsamen Kräuter wachsen, zum Aufenthalt und Gesundheit des menschlichen Lebens, die alles Leid und Trauern vertreiben, habe dies nicht mir geschrieben, weil ich keines Buches bedarf, sondern meinen fleißigen Mit-Arbeitern, so viel Licht mitteilen wollen, als mir GOTTES Güte und Gnade gegeben. Von klugen Leuten, welche von diesen Sachen verstand haben, wird diese Schrift wohl unangetastet bleiben, wenn aber solches von Unerfahrenen gesehen wird, ist es keiner Antwort oder des Papiers wert. Gehab dich wohl, du lieber Mit-Arbeiter. Erkennt GOTT dein Herz und Gemüt, dass du Ihn und deinen Nächsten recht liebst, die Welt verachtest und nur nach dem Himmlischen trachtest, so wird dich GOTT auch hier erleuchten und dich würdig achten dies Licht zu sehen, dadurch du Armen und Kranken viel Nutzen schaffen und dem Geiz, welcher die Wurzel allen Übels, entrinnen kannst. Ihr aber, meine vortrefflichen Lehrmeister und *Adepti*, zürnt nicht mit mir, dass ich durch eure Unterweisung mich in den Irrgarten gefunden, durch das rechte Leitseil aber wieder heraus gekommen. Ich bin so vermessen nicht, dass mich euch an die Seiten stellen will, sondern krieche euch nur hernach; Noch einmal sage ich, zürnt nicht mit mir, dass ich so deutlich geschrieben, oder ich klage Euch an, dass ihr so dunkel geschrieben. Ob ihr dadurch wohl oder übel getan, weiß ich nicht. Ihr habt, durch Hinterhaltung, vielen Bösen vorkommen wollen und dadurch ist viel Gutes nachgeblieben und viel Böses befördert worden. Wie viele unzählige arme Menschen haben zu verbotenen Mitteln greifen müssen und sich dadurch mit Leib und Seele in die Hölle gestürzt? Wie viele unzählige arme kranke Menschen haben über ihren Schmerzen 1000 Ach und Weh schreien und in ihren Schmerzen und Elend untergehen müssen, weil kein Arzt war, der ihnen helfen konnte? Ihr habt es gro-

book, but no one should imagine this will be a land of milk and honey. As far as possible I have been diligent in keeping my prose concise, not adding any pointless words, but taking the direct path to the garden of the Hesperides and to the paradise where golden apples and healing herbs grow, for extension and health of human life, so that suffering and grief are dispelled. I did not write this for me, because I do not need a book, but to transmit as much light for my diligent co-workers as given to me by GOD's kindness and grace. This writing will probably remain untouched by clever people who have an understanding about these things, but if this book is seen by the inexperienced, it is not worth answering or even the paper it is written on. Farewell you dear co-worker. If GOD recognizes in your heart and mind that you really love him and your neighbor, despise the world and only strive for the heavenly, then GOD will also enlighten you here and make you worthy to see this light, whereby you will create much benefit for the poor and the sick and escape from avarice, which is the root of all evil. But you, my excellent teachers and *adepti*, should not be angry with me, that I found myself in the mad garden through your instruction, but came out again through the right guide rope. I am not so presumptuous that I want to stand by your side, but just creep after you. Once again I say, don't be angry with me that I wrote so clearly, or I accuse you of writing so obscurely. I don't know whether you did it for better or for worse. You tried to prevent a lot of evil by holding back, and in doing so much good has been prevented and much evil has been promoted. How many innumerable poor people have had to resort to forbidden means and thereby plunged their body and soul into hell? How many innumerable poor sick people have had to scream a thousand cries of agony over their pain and drown in their suffering and misery because there was no doctor who could help them? You wanted to withhold

ßen Herren hinterhalten wollen, damit nicht unnötige Kriege geführt und viel unschuldiges Blut vergossen würde, aber deswegen bleibt der Krieg doch nicht nach und wäre besser, große Herren hätten Geld genug, so dürften sich die Untertanen nicht so rupfen lassen und welcher große Herr hat doch wohl Zeit und Geduld, dieses Werk zu arbeiten? Er muss es durch seine Diener tun lassen: Wer aber von seinen Dienern soviel Verstand hat, dieses Werk zu arbeiten, der hat auch so viel Verstand, wenn er einen untugendhaften gottlosen Herren hat, dass er ihm das Werk nicht geben wird, sondern vielmehr sagen: Es geht nicht an. Kurz, ich habe meiner Feder kein Verbot geben können, es ist die letzte Zeit, da es soll offenbar werden. Dennoch wird GOTT seine Hand darüber halten. Wem der Schatz beschert ist, wird ihn finden, ein anderer, dem er nicht beschert, wird darüber weg tappen. Ihr wertesten *Adepti*, mit Recht kann ich Euch nicht beschuldigen, dass ihr zu dunkel geschrieben, jedoch habt ihr viele Dinge dazwischen gesetzt; Was zuvor weiß geschienen, mit schwarzer Farbe wieder vertuscht; Endlich habt ihr doch durch kluge *Remonstrationes* mich und andere auf den rechten Weg gebracht. Euch sei Dank für eure Information, dass ihr eure Schriften der Nachwelt gönnen und hinter euch lassen wollen. Gehabt Euch wohl!

GEGEBEN, WEIMAR DEN 1. NOVEMBER 1704

such knowledge from powerful noblemen, so that unnecessary wars would not be waged and much innocent blood would not be shed, but that doesn't prevent war and it would be better if powerful noblemen had enough money so that they would not need to extort their subjects like that—and in any case what powerful nobleman has time and patience enough to complete this work? He must have his servants do it, but whoever of his servants has the understanding to complete this work also has so much understanding, if he has a wicked godless master, that he will not show him any results, but rather say: It doesn't work. In short, I have not been able to restrain my pen, and it is finally time when it should be revealed. Yet GOD will hold his hand over it. Whoever the treasure is intended for will find it, someone else whom it is not intended for will grope over it. Dearest *adepti*, I cannot rightly accuse you of writing too darkly, but you have put a lot of things in between. What previously seemed to be white, you covered up again with black paint. At last you have brought me and others on the right path through clever remonstrations. Thank you for your information and that you indulged and left behind your writings for posterity. Fare you well!

PRESENTED IN WEIMAR ON 1 NOVEMBER 1704

DER INHALT DIESES BUCHS

IST IN 5 KAPITELN VERFASST NACH ART DER ARBEIT, welche in den 5 Zirkeln und in den 5 Tabellen abgebildet, welche die 5 Kapitel auslegen und die Arbeit zeigen, den alten langen Weg und den neuen kurzen Weg: Geheimnisse, so bei dieser Materia anzutreffen und in den 5 Kapiteln verfasst.

I. *Æs Hermetis*, *Subjectum* der Natur, erstes *Confusum Chaos*, *Materia Remota*, *Materia Lapidis Nostri*, ist der ♄ wie die *Magnesia* entgrobt, der Natur gemäß von seinen *Fecibus* gereinigt, in *Primam Materiam* gebracht und in sein erstes Wasser, der *Vitriol* in seinen eigenen vorigen Wassern solviert?
II. Wie aus diesen das viskosische ponderosische Wasser oder der einfache ☿ zu machen?
III. Wie aus dieser Jungfer-Milch der ☿ *Duplicatus* wird, erste Composition zur *Terra Foliata*?
IV. Wie dieser flüchtige Vogel durch *Ignis Fortissimus* zu binden, dass er zum weißen Stein wird, zur weißen Tinctur, zum Stein der ersten Ordnung?
V. Wie dieser gelb und in die höchste Röte gebracht wird? Wie dieser zu multiplicieren (welches das *Subjectum* der Kunst) und diese alle aus dem einzigen entspringen und nichts bedarf als der Solution und der Coagulation, welchen nichts fremdes zugesetzt wird?

Unsere *Materia* ist nur ein Ding und hat doch die Zahl 1, 2, 3, 4. Als: 1. Ein Ding *re & numero*. 2. Mann und Weib. 3. Die drei *Principia* ☉, ♃, ☿. 4. Die vier *Elementa*.

THE CONTENT OF THIS BOOK

IS WRITTEN IN FIVE CHAPTERS ACCORDING TO
the type of work being discussed, which are presented in the
five circles and in the five tables that explain the five
chapters and show the work in both the old long
way and the new short way: Secrets as found
in this materia and written
in the five chapters.

I. *Æs Hermetis, subjectum* of nature, first *confusum chaos, materia remota, materia lapidis nostri*, is the ♄ ground finely like the *magnesia*, purified from its feces according to nature, brought in *primam materiam* and into its first water, and the *vitriol* dissolved in its own previous waters?
II. How to make from these the viscous ponderous water or simple ☿?
III. How this virgin-milk becomes ☿ *duplicatus*, first composition towards *terra foliata*?
IV. How to bind this fleeting bird by *ignis fortissimus*, so that it becomes a white stone, a white tincture, a stone of the first order?
V. How it will be made yellow and then the highest red? How to multiply it (which is the *subjectum* of the art) and how they all arose from the only one, and nothing is allowed but solution and coagulation, to which nothing strange is added?

Our *materia* is only one thing, and yet has the numbers 1, 2, 3, 4. As one: one thing with respect to its nature and its number. As two: man and woman. As three: the three *principia* ☉, ♀, ☿. As four: the four *elementa*.[3]

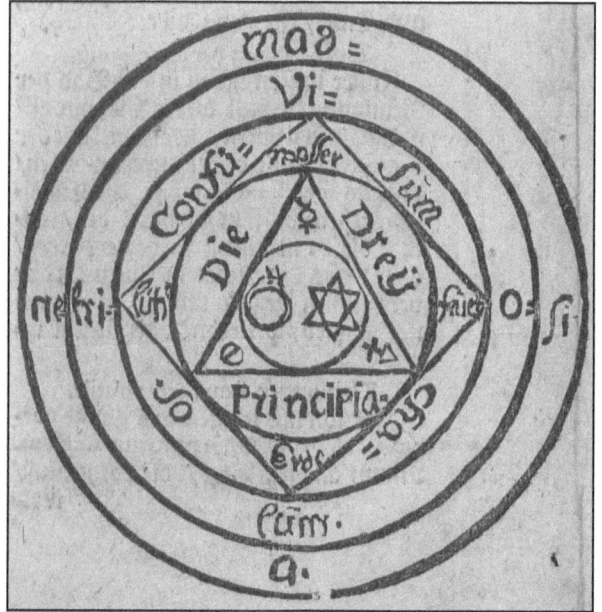

Figura Cabbalistica

Figura Cabbalistica, das ganze *Universal*, A und O, Anfang und Ende, Vor-Arbeit und Nach-Arbeit. Erste, andere und dritte Reinigung, Scheidung *purum ab impuro*, darinnen alle *Menstrua* entdeckt und das Paradies gezeigt, da sich der *Nilus* in 4 Ströme teilt: Einer führt ☉ und das ☉ desselben Landes ist köstlich; das andere, so um das ganze Mohren-Land fließt, das unsern Stein schwärzt. Wenn du diesen Stein nicht erst in Kopfe gemacht, wirst du ihn mit Händen wohl ungemacht lassen.

THE MINERAL GLUTEN

Figura Cabbalistica

Figura Cabbalistica,[4] the complete *universal*, A and O, beginning and end, preliminary work and after-work. First, second, and third purification, separation *purum ab impuro*, in which all *menstrua* are found, and paradise is shown, and there the *Nilus* is separated into four streams, one of which contains ☉ and the ☉ of this country is delicious; the other [stream] flows around the land of the Moors and blackens our stone. If you do not make this stone first in your head, you will not be able to make it with your hands.

Beschreibung des 1ten Zirkels, von seinem Namen

Das erste Chaos von der Natur, ☿ *Solut.*, *Æs Hermetis, Magnesia, Electrum Minerale Immaturum*, ☿ *Philosophorum*, Zeuge-Vater aller Götter, darinnen *Prima Materia* aller Metalle, der ☿ *Coagulatus*, das große *Alcale, Metallum Primum*, ☿ *Crudus Philosophorum*, der Felsen, der Öl, Wein, Blut und Milch gibt, ☉ Baum und seine Wurzel, Anfang und Ende, ein Ding *Re & Numero*.

Beschreibung der Reinigung

Führe unsern Alten in sein Bad der Reinigung, so hast du ein Geheimnis, nimm unsern Alten, der so viel mit der ♀ gespielt und daher unrein geworden, der das *Metallum Primum*, den korallenroten Saft in sich hat, der, welcher die ☉ in seinem Leibe verborgen oder unser ☉, welches in seinem Kerker verschlossen liegt und ♌ *Rubeus* genannt wird, solvier ihn in der Natur ihrem Essig.

Teilung in fixum und volatile

Ihr sollt mitnichten das ganze Corpus solvieren, sondern nur seine *Animam*. Nimm all sein Blut, die rote Nuss, worinnen unser ♂ und ♀, ☉ und ☽, ja alle 7 Metalle stecken. Nimm von diesem Mann, wenn er schläft, sein Weib, seine Rübe, destilliere das flüchtige Weib, das giftige ☿al Wasser von ihm, so hast du im Grunde den roten Ton, *Terram Adamicam*, so nah am Paradies, roten ♌, unser Zinnober-Erz.

Namen dieses Wassers oder Spiritus, auch des fixen Teils

Dieses Wasser heißt die empfangene Gottes-Gabe, Berg-Essig, Solvier-Wasser, giftige Schlange oder Drache, die Juno, roher unzeitiger *Spiritus* ☿, Schlange, so im grünen Grase liegt, Maien-Tau, destillierter Essig. Der fixe Teil, als der Mann, heißt unser ☉, das Gold GOTTES, unser ☉

THE MINERAL GLUTEN

Description of the first circle, from its name

The first chaos from nature, ☿ *solut.*, *æs Hermetis*, *magnesia*,[5] *electrum minerale immaturum*, ☿ *philosophorum*, begetter of all gods, in which is *prima materia* of all metals, ☿ *coagulatus*, the great *alcale*, *metallum primum*, ☿ *crudus philosophorum*, the rock, which gives oil, wine, blood, and milk, ☉ tree and its root, beginning and end, one thing *re & numero*.

Description of purification

Take our old man in its bath of purification, so you have a secret, take our old man which has played so much with ♀, and therefore has become impure, it which has *metallum primum*, the coral-red juice inside itself, it, which has the ☉ hidden in its body, or our ☉, which lies locked in its dungeon, and is called ♌ *rubeus*,[6] dissolve it in nature's vinegar.

Separation in fixum and volatile

You should by no means dissolve the whole corpus, only its *animam*. Take all its blood, the red nut, in which our ♂ and ♀, ☉ and ☽, yes all seven metals stick. Take from this man, when it sleeps, its wife, its turnip, distill the fugitive woman, the poisonous ☿al water from it, so you basically have the red clay, *terram Adamicam*, so close to paradise, red ♌, our cinnabar ore.

Names of this water or spiritus, also of the fixed part

This water is called the received gift of GOD, mountain vinegar, dissolving water, poisonous snake or dragon, the Juno, raw untimely *spiritus* ☿, serpent, which lies in green grass, May dew, distilled vinegar. The fixed part, this is the man, is our ☉, the gold of GOD, our ☉ and ☽

und ☽ Baum, der die roten und weißen ☉ und ☽ Blumen trägt, dieses muss in seiner eigenen Feuchtigkeit, seinem eigenen Wasser aufgelöst werden, damit weißes und rotes zum Vorschein kommt.

> Wenn du von GOTT dem HERRN dazu bist ausersehn,
> So kann der güldene Zweig mitnichten dir entstehn:
> Wo nicht, so hilft kein' Weisheit noch Verstand,
> Auch wird des Eisens Schärf vergeblich angewandt.
> Du weißt nicht was es kost, das güldne Fell zu ziehen
> Von dem bekannten Haupt, was du vor ein Bemühen
> Und Arbeit ohne Maß dir aufgebürdet hast,
> Bei diesem rohen Werk, bei dieser schweren Last.

Beschreibung des anderen Zirkels

Mache aus Mann und Weib einen ○, setze unsere *Materiam* flüchtig und fix zusammen, welches Mann und Weib bedeutet, die giftige Schlange, so unsern geheimen ⊕ solviert, so aus dem ♄ gekommen und der ♄ ist, welcher einen Stein für seinen Sohn den ♃ gefressen und wieder ausgespien, ist auch der Adler-Stein, der noch einen in sich hat, nämlich den Stein der Weisen. *Visitando Interiora Terræ Rectificando Invenies Occultum Lapidem Veram Medicinam*, unser *Vitriolum Generans*, welches den roten und weißen Geist in sich hat, den roten Mann und weißes Weib, der durch sein eigenes Wasser solviert, weil es heißt: Es solviert sich selber, schwärzt, weißt und rötet sich selber, &c. Wird nun unser *Confusum Chaos*, in welchem die *Elementa* verborgen stecken, und die vielen *Menstrua* hervor kommen.

Es werden aber diese Menstrua nicht gesehen, bis es dem Künstler gefällt, welcher sie muss zum Vorschein bringen, das ist, in die *Elementa* scheiden, woraus 2 Säulen gemacht werden, worauf das ganze Gebäude des Werks beruht und hierinnen die Vor-Arbeiten des Herculis begriffen, seine 12 *Labores*, es wird alles überstiegen, wenn das Chaos wohl eröffnet und ein Regenbogen in der Vorlage etliche Stunden zu sehen sein.

tree, which bears the red and white ☉ and ☽ flowers, this must be dissolved in its own moisture, its own water, so that white and red appear.

> If you are chosen by GOD the LORD
> This is how the golden branch can arise in your record:
> If not, no wisdom or understanding can help you to gain
> The sharpness of iron is also used in vain.⁷
> You don't know, how much it costs to pull the golden fur
> From the well-known head, what you did strive before
> And you have burdened yourself with work without measure
> With this rough work, with this heavy treasure.⁸

Description of the second circle

Make a ☉⁹ out of man and woman, put our volatile and fixed *materiam* together, which means man and woman, the poisonous serpent, which dissolves our secret ⊕,¹⁰ which comes from ♄, and which is ♄, which has eaten and spat out a stone for its son ♃, which is also the eagle-stone, which still has one in it, namely the stone of the philosophers. *Visitando Interiora Terræ Rectificando Invenies Occultum Lapidem Veram Medicinam*,¹¹ our *vitriolum generans*, which has the red and white spirit in itself, the red man and white woman, which is dissolved by its own water, because it is said: it dissolves itself, blackening, whitening, and reddening itself, etc. It now becomes our *confusum chaos*, in which the *elementa* are hidden, and from which many *menstrua* emerge.

But these menstrua are not seen until the artist likes it, who has to bring them to light, that is, divide into the *elementa*, from which two pillars are made, on which the whole building of the work rests, which includes the preliminary works of Hercules, his twelve *labores*, everything will be exceeded, when the chaos is well opened and a rainbow can be seen in the receiver for several hours.

Rebis ein einzig Ding nur ist,
doch aus 2 Dingen zugerüst,
Sonn und Mond solviert man ab,
das man ihren ersten Samen hat.

Nun hat dieser Spiritus schon eine andere Art als vorhin, weil sein fixer Teil darinnen gekocht, so hat diese metallische Schlange an sich genommen die ölige und fette Art seines Körpers, ist nun männlicher und weiblicher Natur, viskosisch, fett, ölig und leimiger Gummi.

Beschreibung des dritten Zirkels

Mache aus Mann und Weib ein □, scheide es in die 4 *Elementa*.

Elixir wird in gemein genannt,
das andere Stück im Werk bekannt,
in dem die schönen Corpora rein,
zu Wasser verkehret sein.

Elixir ist ein Tingier-Wasser, *Venenum Tingens*, unser schneeweißer unverbrennlicher ♁ und unser blutroter fixer ♁, der weiße Geist auf weiß, der rote Geist auf rot, Hitze von der Sonne, Speichel vom Mond, Sonnen ♁.
Aus einem Felsen hat GOTT klares Wasser gegeben und eine Menge Öl aus einem harten Stein. Wenn die *Elementa* geschieden, so sieht die Luft gelb aus, ist fett wie Öl, der Weisen ihr *Spiritus Vini*, weil er alle ♁ auflöst und mit über den Helm führt, heißt auch *Spiritus* ☿ und kann ohne diesen kein *Aurum potabile* gemacht werden, er tingiert alle Leiber in gelb, das ist, in ☉.
Das Wasser heißt die Jungfer-Milch, jungfräulicher ☿, Fasan, Schwan, *Menstruum Mundi*, in *Sphæra* und Kreis ☽, die *Lunaria*, Paradieswasser, Wasser aus einer Schlange gemacht, das viskosische pontische Wasser, Wasser aus dem *Nilo*, so um das ganze Mohren-Land

> *Rebis* only one thing is,
> but out of two things it consists,
> Sun and Moon are to be dissolved,
> so that their first semen is evolved.[12]

Now this spirit is of a different type than before, because its fixed part was boiled in it, so this metallic serpent has taken to itself the oily and fatty nature of its body, is now male and female in nature, viscous, fatty, oily, and glutinous rubber.

Description of the third circle

Make a □[13] out of man and woman, divide it into the four *elementa*.

> Elixir is commonly called
> the other piece known in the work,
> in which the beautiful *corpora* pure,
> are turned into water for sure.[14]

Elixir is a dyeing water, *venenum tingens*, our snow-white incombustible ♁, and our blood-red fixed ♁, the white spirit on white, the red spirit on red, heat from the sun, saliva from the moon, sun ♁.

GOD has given clear water from a rock and a lot of oil from a hard stone. When the *elementa* are separated, the air looks yellow, is as fat as oil, *spiritus vini* of the wise, because it dissolves all ♁ and carries them over the helmet, is also called *spiritus* ☿ and without this no *aurum potabile* can be made, it tinges all bodies in yellow, that is, in ☉.

The water is called the virgin milk, virgin ☿, pheasant, swan, *menstruum mundi*, in *sphæra* and circle ☽, the *Lunaria*, paradise water, water made from a serpent, viscous pontic water, water from the *Nilo*, which flows around the whole of the moors land, because it blackens the

fließt, weil es den Stein schwärzt und sobald es auf die Erde fällt zum Stein wird und sich dabei coaguliert, Mond-Speichel und einfacher ☿, Götter-Trank, ist sehr wohlriechend, die Beja genannt.

Das Feuer heißt ☿ Solaris, Hitze der Sonne, ♌ rubeus, brennender *Rubicundus*, roter Mann, rosenfarbenes gesegnetes Blut, roter Wein; Die Erde heißt die jungfräuliche Erde, so nie die Sonne beschienen, auch nie den Beischlaf erlitten, der erste Leib mit dem sich der *Universal*-Geist bekleidet, das den ☿ coaguliert, die Diana, das teure schneeweiße Salz, der Gabricus und grüner *Duenech* und Eidechse, das martialische solarische Salz, davon der Jungfern ☉ herkommt, Königin *Alma*.

Beschreibung des vierten Zirkels

Mache aus Mann und Weib ein △, aus den 4 Elementen mache die 3 Anfänge des Steins, ⊖, ♄, ☿, nimm unser ponderosisches Wasser, bringe es zu seiner geliebten Erde, so wird dieser ☿ von seinem fixen ☿ coaguliert, welches Mann und Weib, ist *Primus Concubitus*, woraus unser Hermaphrodit entsteht, die erste Composition, welche nicht wieder zu scheiden. Was du nun coagulierst, musst du wieder solvieren, auch flüchtig machen. Wenn es seine Zeichen gibt, dass es anhebt zu leuchten und zu funkeln, so sublimier es in die *Terram Foliatam*, in den doppelten ☿ *exuberatum*. Wenn dann ☽ die Erde 7-mal umlaufen, der Leib in der Erde verfault, muss er verklärt wieder auferstehen.

Wenn unser doppelter Leib in seine 7 *Systemata* gekommen, sich geläutert und in die geblätterte Erde sublimiert, welches der Weisen ihr ✶ ist, so mache diesen wieder fix, verbrenne diesen königlichen Leib durch *Ignis Fortissimus*. Unser Feuer verbrennt die Hände nicht, es ist der Drache, das wunder-seltsame Höllen-Bad der Natur, lass das güldene Büchlein fröhlich in die Fontinam fallen, die Adler den ♌ zerreißen und der ♌ die Adler töten, ist hernach der Wald der Nymphen ♀, darinnen die weißen Tauben der Diana fliegen, bis sie alle lebendig tot dahin gefallen und in die weiße Tinctur verwandelt.

stone and as soon as it falls on the earth it turns into a stone and coagulates itself, moon saliva and simple ☿, gods' drink, is very fragrant, which is called Beya.

The fire is called ☿ Solaris, heat of the sun, ♌ rubeus, burning rubicundus, red man, rose-colored, blessed blood, red wine; the earth is called the virgin earth, on which the sun has never shone, which never suffered coitus, the first body with which the *universal* spirit clothes itself, which coagulates the ☿, the Diana, the precious snow-white salt, the Gabricus, and green *Duenech* and lizard, the martial solar salt from which the virgin ☉ comes from, queen *Alma*.

Description of the fourth circle

Make a △[15] out of man and woman, from the four elements make the three beginnings of the stone, ⊖, ♁, ☿, take our ponderosian water, bring it to its beloved earth, this is coagulated by its fixed ☿, which is man and woman, is *primus concubitus*, from which our hermaphrodite arises, the first composition which cannot be separated again. What you are now coagulating, you must dissolve again, even make it volatile. If there are signs that it begins to shine and sparkle, then sublimate it in the *terram foliatam*, in the double ☿ *exuberatum*. If ☽ revolved seven times around the earth, the body rotten in the earth, it must rise again transfigured.

When our double body has come into its seven *systemata*, has purified itself and has been sublimated into the leafed earth,[16] which is ✶ of the wise, fix it again, burn this royal body with *ignis fortissimus*. Our fire does not burn the hands, it is the dragon, the wonder-strange hell-bath of nature, let the golden little book happily fall into the *fontinam*, the eagles tear the ♌ apart, and the ♌ kill the eagles, is afterwards the forest of the nymphs ♀, in which the white pigeons of Diana fly, until they all fall alive dead, and turn into the white tincture.

Wenn nun das gläserne Meer seinen Lauf vollendet, der ☿ bei der ♀ geschlafen, so gibt es den Smaragd von seltsamer Wirkung, darin der König, wie in den steinernen Palast, sicher gehen und seine Wohnung beständig haben kann. Wenn das Erdreich in seinem eigenen Wasser aufgelöst, zum trockenen Wasser geworden, zur flüchtigen Erde, so lass es im Meer seiner Ertränkung gehen, trockne durch die Luft das Wasser aus, vollbringe die 3 mal 7 fache magische Zahlen, 7 Circulationes mit dem Wasser, 7 mit der Luft, 7 mit dem Feuer, wenn es von der Luft lebt, so ist es der Chamäleon, so alle Farben an sich nimmt; Wenn er aber von der Luft allzusehr imprimiert wird, sieht das ☉, so davon kommt, zu gelb aus und ist weich-flüssig.

Beschreibung des fünften Zirkels

Mache aus Mann und Weib wieder einen ○.
Wenn du das Haupt fügst zu den Schwanz
hast du die Tinctur ganz.

Wenn er nun von der Luft sehr gelb geworden, so gib ihm das Wasser der Sonne, bis er sehr rot wird und alle Elementa ausgestritten, das *Agens* und *Patiens* sich verglichen und keine *Turba* mehr zu sehen, da denn die Welt, so von der Sintflut geblieben, nun durchs Feuer gerichtet wird, so wird der neue Himmel und die neue Erde erscheinen, die ewig bleibend ist und keiner Veränderung unterworfen. Unser Karfunkel, so im finstern leuchtet, der Salamander, so im Feuer lebt, unverbrennliches Öl, ewiges Licht.

Bericht von den 3 ☿ und den 3 ☉, auch 3 ♃, so in dieser Arbeit vorkommen aus diesem einzigen Ding und 1 vom ☿ welchen wir den gemeinen nennen und allen Artisten zur Hand ist, die andern 2 werden philosophische genannt, welche die unsern sind.

Der erste ☿ in seiner ersten Reinigung erscheint uns in einem wässri-

When the sea of glass[17] has completed its course, the ☿ slept with ♀, there is the emerald of strange effect, in which the king can be sure of staying in the stone palace and have its dwelling there constantly. When the earth is dissolved in its own water and has become dry water, the volatile earth, let it go in the sea of its drowning, dry the water through the air, do three times seven times the magic numbers, seven circulations with the water, seven with the air, seven with the fire, if it lives from the air, then it is the chamaeleon which takes all colors into itself; But if it is overly impressed by the air, the ☉ that comes off looks like it is too yellow, and it is soft and liquid.

Description of the fifth circle

Make again one ○[18] out of man and woman.
If you join head to tail,
you have the tincture without fail.[19]

If it has now become very yellow from the air, give it the water of the sun until it becomes very red and all elements have ended their quarrels, the *agens* and *patiens*[20] have settled and the *turba* cannot be seen anymore, there then the world, which has remained after the great deluge, is now judged by fire, then the new heaven and the new earth will appear, which are eternal and are not subject to any change. Our carbuncle that shines in the dark, the salamander who lives in the fire, incombustible oil, eternal light.

Report of the three ☿ and the three ⊖ also of the three ♁, which appear in this work which starts from only one thing, and one of ☿ which we call common, which is at all artists' fingertips, the other two are called philosophical, which are ours.

The first ☿ in its first purification appears to us in a watery body, as it

gen Leib, wie er von der Natur gemacht; In seinem anderen Wesen oder Reinigung ist er dick, viskosisch und zur Jungfer-Milch geworden, sein ♂ meist davon geschieden und ist seine innere Reinigung, die vorhergehende war nur seine äußerliche Reinigung, welches nun unser ☿; In seinem dritten Wesen erscheint er als der doppelte Schlangen-Stab, die sublimierte *Terra Foliata* und ☿ der Weisen, der Hermaphrodit, welcher seinen *fixum* ☿ bei sich hat und nun königliche Kronen und Zepter trägt, ☿ *exuberatus*, das trockene Wasser, das güldene Büchlein, so in die *Fontinam* fallen muss.

Die 3 ♃, so in der Arbeit zu sehen vorkommen.

Die erste ♃ ist das *Oleum* ♄, so aus dem ersten Wasser von der giftigen Schlange muss geschieden werden, ist das ♃ Öl, so den ☿ fix macht; Der andere ♃ ist der rote ♃ darinnen das philosophische ☉ steckt, die rote Nuss, welche durchscheinend muss gemacht werden: Der dritte ♃ ist der ☿ *Solaris*, das Feuer des Steins, der rote ♌, rosenfarbenes gesegnetes Blut, brennender *Rubicundus*, Hitze von der Sonne.

Die 3 Salze, so in der Arbeit vorkommen.

Das erste ist unser 🜔, so den roten und weißen Geist in sich hat. Wenn diese 2 davon geschieden, bleibt zurück das andere Salz, *Hyle*, das jungfräuliche Salz, die jungfräuliche Erde, so nie die Sonne beschienen, das spermatische Salz; Das dritte ist *Sal metallorum*, fix und flüchtig zusammen verbunden, welches unser Salz-Stein, das gläserne Meer.

Hier sind nun gezeigt alle Arbeiten und die 2 Compositiones: In der ersten Composition wird der doppelte ☿ gemacht und wachsen hier die philosophischen Metalle und wenn er den ☉ Grad auf der 7ten Stufe erstiegen, weil sich dieser ☿ selber in ☉ kocht, muss er wieder flüchtig und in die *Terram Foliatam* sublimiert werden, dass dieses weiße ☉ flüchtig wird zu unserem weißen ♃. Die andere Composition

is made by nature; In its other essence or purification it has become thick, viscous, and virgin milk, its ☍ mostly separated from it and is its internal purification, the previous one was only its external purification, which is now our ☿; In its third being it appears as the double serpent-staff, the sublimated *terra foliata*, and ☿ of the wise, the hermaphrodite, which has its *fixum* ☿ within itself and now carries royal crowns and scepter, ☿ *exuberatus*, the dry water, the little golden book-that must fall into the *fontinam*.

The three ♄ which appear in this work.

The first ♄ is the *oleum* ♄, which has to be separated from the first water by the poisonous serpent, is the ♄ oil, which fixes the ☿; The other ♄ is the red ♄ in which the philosophical ☉ is, the red nut, which has to be made translucent: the third ♄ is the ☿ *Solaris*, the fire of stone, the red ♌, rose-colored, blessed blood, burning *rubicundus*, heat from the sun.

The three salts which appear in the work.

The first [salt] is our 🜔, which has the red and white spirit in it. When these two are separated, what remains is the second salt, *hyle*, the virgin salt, the virgin earth, on which the sun has never shone, the spermatic salt; The third [salt] is *sal metallorum*, fixed and volatile connected together, which is our salt-stone, the sea of glass.

All works and the two compositions are shown here: In the first composition, the double ☿ is made, and here the philosophical metals grow and if it climbs to the ☉ degree on the seventh level, because this ☿ cooks itself into ☉, it must again be volatilized and sublimated into the *terram foliatam*, so that this white ☉ becomes volatile to our white ♄. The second composition is to make the stone, since this little golden book

ist den Stein zu machen, da dieses güldene Büchlein in die *Fontinam* fallen muss und wird hernach das gläserne Meer, das mit Feuer gemengt werden kann, die weiße Tinctur, ♌ *viridis*, welche endlich zur Gelbe und Röte gebracht wird. Zu diesem Werke nimmt man weder ☉ noch Silber, unser ☉ ist lebendig ☉, das gemeine ☉ ist tot, unser ☉ solviert das gemeine ☉ und ist nie corporalisches ☉ gewesen, unsere ☽, die weiße *Lunaria*, Mond-Speichel, der weiße und nicht brennende ♃, der feuchte Mond, gehört auch kein gemeiner ☿ dazu, sondern unser reiner und glänzender, das silberne Fischlein *Æscheneis*, dass im Meer schwimmt mit silbernen Schuppen und rote Floßfedern hat, seinen roten ♃ bei sich trägt und viel edler ist als der gemeine; Man nimmt auch nicht ♂ noch ♀, sondern die ♀, die so viel mit dem ♄ gespielt, *Metallum Primum*, die rote ♀, welche rot in der Solution, die gemeine ♀ ist grün und wenn der korallenrote Saft davon geschieden, ihren Bruder ♂, den grünen flüssigen *Duenech* zurück lässt, welcher den *Spiritum Generantem* bei sich hat; Man nimmt auch kein ♄, ♃ noch ☿ dazu, sondern unsern ♄, der den ☉ Geist in sich verschlungen hält und dieses alles steckt in unserem ♄.

Folgen die 5 Tabellen, da immer eine aus der anderen geht.
Als: Aus der 1ten entspringt die andere; Aus dieser die 3te;
Aus der 3ten die 4te und aus dieser die 5te.

must fall into the *fontinam*, and afterwards it becomes the sea of glass, which can be mingled with fire, the white tincture, ♌ *viridis*, which is finally brought to yellow and red. For this work one takes neither ☉ nor silver, our ☉ is alive ☉, the common ☉ is dead, our ☉ dissolves the common ☉, and has never been corporal ☉, our ☽, the white *Lunaria*, moon saliva, the white and not burning ♁, the wet moon, also no ordinary ☿ belongs to it, but our more pure and shiny [☿], the little silver fish *echeneis*,[21] which swims in the sea with silver scales and red fins, carries its red ♁ with it and is much nobler than the common one. Do not take either ♂ or ♀, but the ♀ which played so much with the ♄, *metallum primum*, the red ♀, which is red in the solution, the common ♀ is green and when the coral-red juice is separated from it, its brother ♂ leaves behind the green liquid *Duenech*, who has the *spiritum generantem* with it. Do not add ♄, ♃, nor ☿, but our ♄ which keeps the ☉ spirit engulfed in itself and all of this is in our ♄.

Follow the five tables, as one always goes out from the other.
That is to say: From the first the second arises; from this the third;
from the third the fourth; and from this the fifth.

I. TAB.

Opus ♄

Mache aus Mann und Weib einen ☉, wenn aus unserem Adam, welcher mit seiner Eva schwanger, sein flüchtiger Teil genommen wird, so reinige den roten *Laton*, welches unser verborgener 🜨, diesen resolviere in seinem giftigen Solvier-Wasser, so wird es der *Rebis*.

II. TAB.

Mache aus Mann und Weib einen □, scheide unseren *Rebis*, unseren 🜨 in die 4 *Elementa*, unser *Confusum Chaos*, den roten Wein und Götter-Trank, so hast du das Paradies-Wasser, Tingier-Wasser, das Elixir.

III. TAB.

Mache aus Mann und Weib ein △, aus dem einfachen Adler der Jungfer-Milch, mache durch das teure schneeweiße Salz, *Hyle*, den doppelten ☿, sublimier es in die *Terram Foliatam*, dies ist der Hermaphrodit.

IV. TAB.

Mache aus Mann und Weib wieder einen ☉, bringe Mann und Weib zusammen. Wenn du das weiße Weib gebracht zum roten Mann, so nehmen sie alsbald einander freundlich an, darauf empfängt das edle weiße Weib, die zuvor waren 2, sind geworden nur ein Leib. Stein der 1ten Ordnung.

V. TAB.

Sonnen-Sohn, wenn du das Haupt fügst zu dem Schwanz, so hast du die Tinctur ganz.
Illuminier es. Multiplicier es.

FIRST TABLE

Opus ♄

Make a ☉ out of man and woman, if from our Adam, who is pregnant with his Eve, his volatile part is taken, then purify the red loam, which is our hidden ⊕, dissolve this in its poisonous water of solvation, then it becomes the *Rebis*.

SECOND TABLE

Make a □ out of man and woman, separate our *Rebis*, our ⊕ into the four *elementa*, our *confusum chaos*, the red wine and divine drink, so you have the paradise water, tinging water, the elixir.

THIRD TABLE

Make a △ out of man and woman, from the ordinary eagle of virgin milk, make the costly snow-white salt, *hyle*, double ☿, sublimate it in the *terram foliatam*, this is the hermaphrodite.

FOURTH TABLE

Make again a ☉ from man and woman, bring man and woman together. When you bring the white woman to the red man, they immediately accept one another in a friendly manner, and then the noble white woman receives [him], and those who were previously two have become only one body. Stone of the first order.

FIFTH TABLE

Son of the sun, if you add the head to the tail,
then you have the tincture without fail.
Illuminate it. Multiply it.

Zu besserer Erklärung der 5 Zirkel und 5 Tabellen, als eine Repetierung der Vor-Arbeit und Nach-Arbeit, sind die 12 Arbeiten des Herculis ausgelegt.

1. Die erste weist uns die *Materiam crudam & remortam*, den ♄ nämlich, dass es sei von Juno, das ist, die metallische Natur herab gesandte 2 Schlangen, flüchtig und fix, so in einem Subjecto begriffen und giftig sind, welche wir in ihrer Rohigkeit nehmen müssen, wie sie uns die Natur gibt, ohne einigen fremden Zusatz, unser *Rebis*.

2. Die andere zeigt uns den ⊕ *materiam propinquam*, unser *Confusum Chaos*, wenn diese in der Fäulung steht und Sonne und Mond noch in ihren dunklen Schatten verborgen liegen, diese sind es, sie sind aber nicht zu sehen, bis es dem Künstler gefällt, welcher sie muss zum Vorschein bringen, das ist, in die *Elementa* scheiden, so ist es unser Elixir, unsere Tingier-Wässer, welche schwarz, weiß, gelb und rot färben, wovon hernach 2 Säulen aufgerichtet werden, auf welchen das ganze Gebäude des Werks beruht.

3. Diese zeigt uns den doppelten ☿, wie der einfache, als das metallische Wasser, soll mit seinem eigenen Leibe, der Erde, vereinigt werden und das Wasser zur Erde werden, das hernach durch öfteres imbibieren noch mehr Schlangen erwachsen, nämlich aus diesem Leibe und besteht aus Wasser und Erde, die Kröte, so auf der Erde geht und der Adler, so ihr zugesellt, die Begrabung des Leibes in der Erde.

4. Zeigt noch mehr von dieser Composition, welche nun *Prima Materia* aller Metalle genannt wird, auch *Prima Materia* des Steins, *Materia Proxima*, weil diese Erde, wenn sie solviert, grün wird, wird sie genannt animalisch, vegetabilisch und mineralisch, aber durch öfteres imbibieren wird die Schwärze neben anderen vermischten Farben davon gewaschen, bis sie zuletzt in die Weiße verkehrt, sehr volatilisch wird, sehr wohlriechend. Hier steht die versiegelte Mutter im Leibe ihres Kindes auf, das ist nun der Weisen ihr ✱, ihre *Terra Foliata*, sie tingiert aber noch nicht, sondern ist der Acker, darein das ☉ soll gesät werden und also muss sie als der Leib mit ihrer Seelen bekleidet und wieder vereinigt werden, dass diese Erde lebendig wird.

For a better explanation of the five circles and five tables, as a repetition of the preliminary work and after-work, the twelve labours of Hercules are interpreted.

1. The first [work] shows us the *materiam crudam & remortam*, namely the ♄, that it was from Juno, that is, the metallic nature sent down two serpents, volatile and fixed, so included in a *subjecto* and poisonous, which we must take in their rawness, as nature gives us, without any extraneous addition, our *Rebis*.

2. The second [work] shows us the ⊕ *materiam propinquam*, our *confusum chaos*, when it is putrefying and the sun and moon are still hidden in their dark shadows, these they are, but they cannot be seen until the artist likes it, who has to bring them to light, that means, separate in the *elementa*, so that it becomes our elixir, our tinging waters, which does color black, white, yellow and red, from which after that two pillars are erected, on which the complete building of the work is founded.

3. This [work] shows us the double ☿, as the ordinary [☿], the metallic water, is to be united with its own body, the earth, and the water becomes earth, that afterwards by frequent imbibing still more serpents arise, namely from this body and it consists of water and earth, the toad, as it walks on the earth, and the eagle, when it joins it, the burial of the body in the earth.

4. Shows even more of this composition, which is now called *prima materia* of all metals, also *prima materia* of the stone, *materia proxima*, because this earth, when it is dissolved, turns green, it is called animal, vegetable, and mineral, but through frequent imbibing, the blackness and other mixed colors are washed away until it finally turns white, very volatile, very fragrant. Here the sealed mother stands up in the body of its child, that is now the ✶ of the wise, their *terra foliata*, but it is not yet tinged, but is the field in which the ☉ is to be sown and so it must be clothed as the body with its soul and be reunited so that this earth may come to life.

5. Zeigt an, wie die zuvor behaltene Seele seinen Geist wiedergeben wird, als wodurch sie noch mehr erhöht und noch mehr glorifiziert, auch durch öfteres Eintränken stärker wird, von ihrem unvollkommenen Stand zur Vollkommenheit gelangt, zur Fixität.

6. Wenn man vom weißen Stein zum roten Stein schreiten will, muss durch öfteres Eintränken des 3ten *Menstrui*, die Juno, als Göttin des Reichtums, dahin gebracht werden, dass aus ihrer Weiße das Rote zum Vorschein kommt, das heißt, wenn nun des erwachsenen Knabens seine Wangen mit Blut gefärbt werden.

7. Wenn der Stein mit der solarischen Jungfer-Milch eingetränkt worden, werden sich viele vermischte Farben sehen lassen, als blau, violenblau, blitzblau, gelb, dunkelrot &c., welche noch der Unvollkommenheit Zeichen sind und bedeuten die Vögel *Harpyias*, die unvollkommenen philosophischen Metalle, wodurch endlich der Stein, durch mehrere Eintränkung, welches ist die natürliche Transmutation, im Glase höher steigt und von ♂, welcher mit einer Dotter-Gelbe geblüht, in ☉ kommt. Hier heißt es mit was für Banden soll ich doch den *Protheum* binden, der sich aus einer Gestalt in die andere verwandelt? Jetzt ist er Wasser, bald Feuer und der Drache schlupft ganz aus seiner alten Haut und wird endlich zur Medizin, so kann man dem giftigen Basilisken durch sein giftiges Herz sehen.

8. Zeigt, dass nunmehr diese Erde oder der noch nicht vollkommene Stein oft aus sich selber schmilzt und wird er sich bald coagulieren, bald wieder fließen, bis er endlich durchs philosophische Feuer überwunden zur gänzlichen Trockne und Reife kommt.

9. Zeigt die andere Composition, nämlich den Stein der ersten Ordnung, wenn er fertig und den Drachen in sich verschlungen hält, weil er nicht stirbt, er werde denn mit seinem Bruder oder Schwester umgebracht, welche sind Sonne und Mond, alsdann muss man die Äpfel aus dem hesperischen Garten holen, welche dieser Drache bewahrt. Es wird aber 1 Teil über 1000 Teile nicht tingieren und also noch von kleiner Kraft sein, bis er höher und weiter zur Multiplication gebracht wird.

5. Indicates how the previously retained soul will give its spirit back, as by which it will be even more exalted and even more glorified, also stronger through more frequent impregnations, from its imperfect state to perfection, to fixity.

6. If one wants to step from the white stone to the red stone, Juno, as the goddess of wealth, must be brought to the point by frequent impregnations with the third *menstrui* that the red comes out of its whiteness, that is, when the adult boy's cheeks are colored with blood.

7. When the stone has been soaked with the solar virgin milk, many mixed colors will appear, such as blue, violet blue, lightning blue, yellow, dark red, etc., which are still signs of imperfection and mean the birds *Harpyias*, the imperfect philosophical metals, whereby finally the stone, through several impregnations, which is the natural transmutation, rises higher in the glass and from ♂, which bloomed with the yellow of a yolk, comes into ☉. Here it says, with what ties I am supposed to bind the *protheum* that transforms from one shape into the other? Now it is water, soon fire, and the dragon slips completely out of its old skin and finally becomes medicine, so can the poisonous basilisk be seen through its poisonous heart.

8. Shows that now this earth or the not yet perfect stone often melts out of itself and it will soon coagulate, soon flow again, until it is finally overcome by the philosophical fire and comes to complete dryness and maturity.

9. Shows the other composition, namely the stone of the first order, when it is finished and has the dragon entwined, because it [the dragon] does not die, it will be killed with its brother or sister, which are the sun and the moon, then one must fetch the apples from the Hesperidian garden, which this dragon keeps. But it will not tinge one part over a thousand parts, and therefore still be of little power until it is brought higher and further to the multiplication.

10. Zeigt an, wenn er durch die rote Tinctur so weit gebracht wird, dass er 3 Gestalten angenommen, wodurch er auf 3 Arten hat müssen überwunden werden, als durch Wasser, Luft und Feuer, er sodann seinen Haufen Vieh, das ist Reichtum, davon gebracht.

11. Die Elfte repetiert in einer *Summa*, wie der philosophische, das ist der ☿ der Weisen, durch den ganzen *Zodiacum*, und durch die 7 Planeten durchlaufen müssen, welches nicht sowohl die Abwechselung der Farben und die Veränderung der Arbeit, sondern des mehrenteils seiner Erhöhung, wodurch es von der untersten Staffel zum königlichen Thron steigt und sich als ein Herr des ganzen mineralischen Reichs sehen lässt, dass ein Planet den anderen spiritualischer Weise verzehrt und nur die stärksten geblieben, als ☉ und ☽, wobei man die Gradus des philosophischen Feuers wohl zu observieren hat, dass nicht eins für das andere genommen und das ganze Werk verderbt werde.

12. Zeigt die Multiplication an, wenn dieser Starke immer stärker wird, so oft er die Erde berührt, wodurch er immer mehr und mehr zunimmt im Tingieren und muss der Künstler wissen, wie er den Schatz zu rechter Zeit erheben soll, weil die *Materia* zuletzt so flüssig wird, dass sie die Gefäße durchdringen möchte, also muss es aus der Arbeit zu Ruhe und Schlaf gebracht werden, kann aber durch den Arbeiter, wenn es ihm beliebt, wieder erweckt und vorige Arbeit aufs neue wiederholt werden, wodurch es an seiner Kraft unendlich zunimmt.

10. Indicates when it is brought so far by the red tincture that it takes on three forms, whereby it has to be overcome in three ways, by water, air and fire, then its herd of cattle, that is the wealth, is brought from it.

11. The eleventh [work] repeats in a *summa* how the philosophical [☿], that is the ☿ of the wise, must go through the whole *zodiacum*, and through the seven planets, which is not so much an alternation of colors and the change of the work, but the greater part of its elevation, whereby it rises from the lowest echelon to the royal throne and can be seen as a lord of the entire mineral kingdom, that one planet consumes the other in a spiritual way and only the strongest have remained, such as ☉ and ☽, whereby one has to observe the degrees of the philosophical fire so that one does not put one before the other and the whole work is spoiled.

12. Shows the multiplication when this strong one gets stronger and stronger, as often as it touches the earth, whereby it increases more and more in tinging, and the artist has to know how to raise the treasure at the right time, because in the end the *materia* becomes so fluid that it wants to penetrate the vessels, so it must be brought from work to rest and sleep, but can be awakened again by the worker, if he likes, and previous work can be repeated anew, whereby its own strength increases infinitely.

DAS 1. KAPITEL

Zeigt die erste Reinigung ♄,
Mannes und Weibes, wie solche
wieder in ☉ zu bringen

Opus ♄ quod Lapidis Pater nunquam concubuerit. Reinige unsere *Materiam Remotam* von seinen *Fecibus* durch der Natur Essig, nimm von diesem Mann, welcher mit seinem Weibe, als Adam, schwanger geht, sein Weib, seine Evam, so hast du unsere empfangene Gottesgabe, unser Solvier-Wasser, unser *Aquafort*, unseren unreifen unzeitigen *Spiritum* ☿, flüchtige Schlange; Zurück bleibt der rote *Laton*, ♌ *rubeus* oder unser ☉, unser verborgener *Vitriol*, unser Zinnober-Erz; Ziehe es zum Schlich, scheide das Reine von dem Unreinen, so hast du das rosenfarbene Blut, lass es zu Kristallen schießen. Diesen *Vitriol* oder unser ☉ solviere in seinem eigenen *Spiritu*, der empfangenen GOTTESgabe, seinem Weibe, wenn du zuvor das giftige Öl des ♄ von ihm geschieden hast, so ist Mann und Weib in ☉, heißt nun der *Rebis*.

THE FIRST CHAPTER

Shows the first purification of ♄,
of man and woman, how they can be
brought back to a ☉

Opus ♄ *quod Lapidis Pater nunquam concubuerit.*²² Purify our *materiam remotam* from its feces through nature's vinegar, take from this man, Adam, which is pregnant with its wife, its Eve, its wife, so you have received our gift of GOD, our water of dissolution, our *aquafort*, our immature untimely *spiritum* ☿, volatile serpent. What remains is the red loam, ♌ *rubeus*, or our ☉, our hidden *vitriol*, our cinnabar ore. Prepare sands out of it, separating the pure from the impure so that you have the rose-colored blood, and then let it shoot up into crystals. Dissolve this *vitriol* or our ☉ in its own *spiritu*, the received gift of GOD, its wife. If you have previously separated the poisonous oil of the ♄ from it, then man and woman are in a ☉ and they are now called the *Rebis*.

DAS 2. KAPITEL

*Zeigt den 🜨, das Confusum Chaos,
wie solches in die Elementa zu scheiden,
welches unsere Elixire sind*

Visitando Interiora Terrae Rectificando Invenies Occultum Lapidem Veram Medicinam. Mache aus Mann und Weib einen □, scheide es in die 4 *Elementa*, unser ♄, der einen Stein, den 🜨, gefressen, lass ihn solchen wieder ausspeien, den *Rebis*, welcher unseren ♃ bedeutet, unseren 🜨, den scheide in Wasser, Luft, Feuer und Erde, so hast du aus unserem roten Wein und Götter-Trank unsere Elixire, tingirende Öle und *Unguenta*, womit du die Götter auf ihrer Hochzeit, wenn sie beiliegen, speisen und tränken musst, welches uns der rohe mineralische Geist hat zuwege gebracht, welcher, als das Weib, hat solviert ihren Mann, seinen öligen und fetten Samen in sich gezogen und davon geistlich schwanger geworden, sie ist nun unsere Diana, unser jungfräulicher ☿, die Jungfer-Milch, welcher ☿ beiliegen muss, mit seinem fixen Teil, der jungfräulichen Erde, woraus unser *Sal Metallorum* wird, aus unserem geschiedenen *Confusum Chaos*, ist nun *Materia Propinqua*.

THE SECOND CHAPTER

*Shows the ☉, the confusum chaos,
how it has to be separated into the elements,
which are our elixirs*

Visitando Interiora Terrae Rectificando Invenies Occultum Lapidem Veram Medicinam. Make one □ out of man and woman, divide it into the four *elementa*, our ♄ which had eaten a stone, the ☉, let it spit it out again, the *Rebis*, which means our ♃, our ☉, separate it into water, air, fire and earth, so from our red wine and drink of the gods you have our elixirs, tinging oils, and *unguenta*, with which you have to feed and water the gods at their wedding, during their intercourse, which the raw mineral spirit has brought to us, which, as the woman, dissolving its husband, and has drawn its oily and fatty semen into itself and has become spiritually pregnant from it. It is now our Diana, our virgin ☿, the virgin milk with which ☿ must have intercourse with its fixed part, the virgin earth, from which our *sal metallorum* emerges, from our divided *confusum chaos*, and it is now *materia propinqua*.

DAS 3. KAPITEL

Zeigt die erste Composition, den doppelten ☿ zu machen, unsere geblätterte Erde, die Kröte, so auf der Erde geht und der Adler, so ihr zugesellt ist, das Meisterstück

Mache aus Mann und Weib einen Trigon, △, die Anfänge des Steins, ☉, ♁, ☿, die erste Composition, *Prima Materia* aller Metalle, *Primus Concubitus*. Nimm unsere jungfräuliche Erde, so nie die Sonne beschienen, auch das Anschauen der Männer nicht hat vertragen können, sondern sich in so viele Kleider versteckt, das erste *Hyle*, unser teures schneeweißes Salz, so durch sein eigen *Phlegma* gereinigt worden, bringe dazu den jungfräulichen ☿, tränke solches nach rechtem Gewicht mit der Jungfer-Milch, so wird das Wasser dabei gerinnen und sich coagulieren, welches unser grüner *Duenech* verursacht, wird auch *Vas Viride* ♄ genannt, worin dieser Wundervogel nisten und hecken soll, tränke es wieder ein mit unserem Gluten *Aquilæ*, nach und nach, bis es die dikke weiße Buttermilch gibt, die metallische Gur, welche sich als Butter schmieren lässt, heißt nun amalgamiert. Bei den ersten Eintränkungen hast du unseren sauberen Jungfern-Salpeter zu sehen, auch wird bald nach des ♃ Regiment der ☿ zu sehen sein, aber kaum 3 Stunden glänzend, weiß, viel weißer als Milch. Die erste Eintränkung gab ♄, die andere ♃, die dritte ☿, die vierte ☽, bis hierher hast du coaguliert, nun musst du solvieren, sobald als dieses Salz aufgelöst, hast du die Grüne zu sehen. Die Solution wird grün, welches der ♀ Regiment, bald wird sich das Glas mit einem güldenen Stück überziehen, welches der Gold-Spiegel genannt wird, weil diese Erde solarischer, martialischer und venerischer, auch saturnalischer Natur ist, obschon die Tincturen davon geschieden, bleibt doch ihr Salz zurück, ob es gleich auswendig

THE THIRD CHAPTER

Shows the first composition to prepare the double ☿, our foliated earth, the toad that walks the earth, and the eagle that accompanies it, the masterpiece

Make a trigon △ out of man and woman, the beginnings of the stone, ☉, ♃, ☿, the first composition, *prima materia* of all metals, *primus concubitus*. Take our virgin earth, on which the sun has never shown, it could also not stand the looks of the men, therefore it did hide in so many clothes, first *hyle*, our precious snow-white salt, which has been purified by its own *phlegma*, add to this virgin ☿, water it according to the right weight with the virgin milk, so the water will curdle and coagulate, which our green *Duenech* causes, it is also called *vas viride* ♄, in which this miracle bird should nest and hedge, soak it again with our gluten *aquilæ*, little by little, until there is the thick white buttermilk, the metallic gur,[23] which can be smeared like butter, which is now called amalgamated. During the first impregnations you have to see our clean virgin saltpeter, and soon after the ♃ regiment the ☿ will be seen, but barely three hours, shiny white, much whiter than milk. The first impregnation gave ♄, the second ♃, the third ☿, the fourth ☽, if you have coagulated up to this point, you have to dissolve, as soon as this salt is dissolved, you will see the greenness. The solution turns green, which is due to the regiment of ♀, soon the glass will be covered with a golden piece, which is called the gold-mirror, because this earth is solar, martial, venusian, and also saturnine in nature, although the tinctures remain separate from it, but their salt remains behind, although it is white on the outside, it is decorated inside with all colors,

weiß, ist es doch inwendig mit allen Farben geziert, auch blutrot. Gemeiniglich sieht man in der 4. Eintränkung die Schwärze, je eher du das Salz oder einen Stein auflöst, je eher die Schwärze kommt. Wenn es sich wieder will coagulieren, wird es als gelber Sand zu Boden fallen, bis es alles dick wird und fängt an Blasen zu bekommen; Wenn solche aufreißen, fährt ein schwarzer rußiger Rauch heraus, fällt zurück auf die *Materia*, sieht aus, als wenn schwarzer Pfeffer darauf gestreut wäre, welches die subtilen *Atomi* der Metalle sind, endlich wird es dicker und wie geschmolzenes schwarzes Pech. Die bunten Farben bekommt man nicht zu sehen, als nach der Schwärze, wenn die *Materia* anfängt trocken zu werden; Wenn die feuchten Dünste täten, würde man die bunten Farben eher zu sehen bekommen, welche wie subtiles Geflitter im Glase herum steuben, weil ihn der Wind gänzlich im Bauche tragen muss. Wenn die Materia beginnt trocken und aschegrau zu werden, so sieht man die bunten Farben nicht mehr, sondern weißere Rauche, doch sind die schwarzen Dünste, je rußiger sie sind, schon voll roter und gelber Dünste, muss also der Berg Ätna sehr rauchen. Wenn nun die weißen Wolken kommen, schlägt es zurück auf die Erde und macht einen subtilen Reif, höre nicht auf zu weißen, weil die ☽ nicht zu weiß werden kann, auch hier nichts verderbt wird, dies heißt die Metalle im Glase wachsend machen, den ♄ in ♃ und diesen in die *Lunam* verwandeln, die *Lunam* in ♀. Wenn sich die Grüne sehen lässt, kommt der ♀ Regiment, die ☽, wenn sie in den Bergwerken sich in ihr erstes Wasser resolviert, wird es ein absteigendes Erz und ♀ daraus, hier auch, nun siehst du, dass die Natur Meister ist, nicht du, sondern hilfst nur der Natur, hast den jungfräulichen ☿ und die metallische Erde, die den *Spiritum Generantem* bei sich hat, diese Erde feuchtest du zu rechter Zeit an mit ihrem Wasser, wo sie mehr trocken wird und mehr Feuer bekommt, als ihr gehört, bist du in großer Gefahr. Zeigt dir auch der Stein an der Farbe, als wenn ihm zu heiß geschehe durch seine Röte, wie wilder Mohn und die Leibfarbe eines Menschen, so ist der Stein in Todesnot, tränke diese Erde alsobald mit dem Lebens-Wasser, mit ihrer Jungfer-Milch, erlabe sie in ihrem Durst, dass sie wieder zu sich selber

including blood red. Generally, one can see the blackness in the fourth impregnation, and the sooner one dissolves the salt or a stone, the sooner the blackness comes. When it starts to coagulate again, it will fall to the ground as yellow sand until it all thickens and bubbles start to form. When these [bubbles] open up, a black sooty smoke comes out, falls back on the *materia* and looks as if black pepper had sprinkled on it, which are the subtle *atomi* of the metals, finally it becomes thicker and like melted black pitch. The bright colors are not seen as after the blackness, when the *materia* begins to dry out. When the damp vapors were done, one would get to see the variegated colors sooner, which move around like subtle glitter in the glass, because the wind has to carry it completely in its belly. When the *materia* begins to become dry and ash-gray, the bright colors can no longer be seen, but whiter smokes, but the black fumes, the sootier they are, are already full of red and yellow fumes, so Mount Etna must smoke a lot. When the white clouds come, it falls forcibly back on the earth and makes a subtle rime, do not stop whitening because the ☽ cannot become too white, nothing is spoiled here either, this means making the metals in the glass grow, transmuting ♄ into ♃ and this into *lunam*, *lunam* in ♀. When the greenness comes to be seen, the regiment of ♀ begins, ☽ when it dissolves itself in its first water in the mines, it becomes a descending ore and ♀ arises from it, and here, too, you now see that nature is master, not you. You can only help nature, and you do so by taking the virgin ☿ and the metallic earth, which has the *spiritum generantem* within itself. You then moisten this earth with its water at the right time. If it becomes too dry and gets too much fire than it should, you are in great danger. If the stone also shows you by turning a color that is red like wild poppy and the bodily color of a human that it has become too hot, then the stone is in distress. Impregnate this earth immediately with the water of life, with its virgin milk. In its thirst allow it to come to itself again, because its spirit

kommt, weil schon ihr Geist von ihr fliehen wollte. Wenn nun der ♀ ihr Regiment zu Ende und es wieder durch den Resolutions-Knall erweckt worden, so will dieser ☿, welcher schon geschickt genug, sich selber in ☉ kochen, er findet aber vor diesmal noch so viele Hindernisse:

1. Ist zu viel Erde da.
2. Ist noch zu viel ♃ da.
3. Des ☿ zu wenig. Wird also verhindert, dass er den ☉ Grad nicht erreichen kann und wird ♂ daraus, diesen kocht der ☿ hernach in ☉.

Nun weist du wohl, dass dem ☉ kein Element mehr Schaden tut, so wird ihm diese giftige metallische Schlange auch nichts anhaben, sondern ihn noch einen Grad höher führen, als die Natur kann. Wenn nun die 7 Gestalten der Natur vorbei, solches in seine 7 *Systemata* gekommen, der unreine Leib sich 7-mal im Jordan gewaschen, der Leib auch in der Erde verfault, muss er verklärt wieder aufstehen. Wenn er reif, so gibt er seine Zeichen und reißt das Samen-Band ab, begibt sich der Schimmer in Glimmer, hebt an zu funkeln und zu leuchten, wird ganz spiritualisch, er will fort, alle 7 Gestalten der Natur sind in ihrem größten Hunger, weil ihnen allen das Maul aufgetan und begehren gesättigt zu werden. Hier hast du die mächtigste Witterung zu sehen und will dieser ☿ fort, darüber sich sonst ein kluger Autor verwundert, dass dieser Stein so zuvor fix war, nun wiederum flüchtig wird, weiß keine andere Ursache zu geben, als dass sich dieses also nach GOTTES Willen zutrüge: Aber ich habe *Rationes* genug; Dieser ☿, welcher sich vorher in ☉ gekocht und dieses ☉ abermals in ☿ aufgelöst und in größerem Gewicht dazu kommt, so macht es diesen ☉ Leib flüchtig, dass er mit ihm in die Höhe steigt, zu unserm ☿ *sublimato*, zur *Terra Foliata*, wird, höchst-glänzend, in welcher lauter *Uniones* hervor leuchten. Hingegen, wenn du es, wenn es fix geworden, nicht wieder mit der Jungfer-Milch eintränken wolltest, würde es sich in Ewigkeit nicht in die Höhe begeben. Nun sollte ich dir die Zeit der Eintränkung melden. Jedwede Eintränkung bedeutet einen Monat, welches unser *Menstru-*

was already trying to flee from it. If now the regiment of ♀ has come to an end and it has been awakened again by the clamor of the resolutions, so this ☿ which is already skillful enough, wants to cook itself into ☉, but this time, it still finds so many obstacles:

1. There is too much earth.
2. There is still too much ♃ there.
3. Not enough ☿. So it is prevented from being able to reach the ☉ degree and it only becomes ♂ out of it, which is then cooked to ☉ by ☿.

Now you know well that no element can do harm to ☉ anymore, so this poisonous metallic serpent will not be able to harm it either, but will lead it one degree higher than nature can. When the seven forms of nature have passed by, and it has come into its seven *systemata*, the unclean body washed seven times in the Jordan, the body which also decays in the earth, it must rise again transfigured. When it is ripe, it gives its signs and tears off its seed-band; and the shimmer turns to glimmer, beginning to sparkle and shine. It becomes very spiritual, it wants to go, all seven forms of nature are in their greatest hunger because they all open their mouths and desire to be satisfied. Here you have the most powerful weather to see and this ☿ wants to go away, usually a wise author would be amazed about this fact, that this stone which was fixed before, is now volatile again, so he knows no other cause than that this should happen according to GOD's will: But I have enough reasons. This ☿, which was previously cooked into ☉ and this ☉ which is again dissolved in ☿ and added in larger weights, so it makes this ☉ body volatile with the result that it rises with it upwards, to our ☿ *sublimato* to *terra foliata*, and it becomes extremely shiny, in which many *uniones* shine out. On the other hand, if it has become fixed, then if you do not want to impregnate it again with virgin milk, it would never in eternity rise up. Now I should tell you the duration of the impregnation. Every impregnation means a month, which is our

um ☽ ist, welches so vielmals den Erdboden umlaufen muss, nämlich 7-mal, welches die 7 Monate zum Weißen sind. Es sind aber philosophische Monate und wird dich der Zeichen-Deuter schon unterrichten, *id est*, aus dem Werk erscheinende Zeichen werden dir schon sagen wenn der Monat aus ist. Das Gewicht will ich dir auch sagen: Nimm die Fundamental-Regel der *Chymiæ* in acht: Wenn du figieren willst, so muss das fixe Teil 4 mal schwerer sein als das flüchtige, willst du aber flüchtig machen, so müssen 4 Teile des flüchtigen ein Teil fixes in die Höhe heben. Viele Künstler dürfen nicht wägen, wenn sie ein scharfes Augenmaß haben.

menstruum ☽, which has to go around the earth so many times, namely seven times, which are the seven months of the wise men. But they are philosophical months and the soothsayer will teach you, i.e., signs appearing from the work will tell you when the month is over. I also want to tell you the weight: pay attention to the fundamental rule of *chymiæ*: If you want to make it fixed, the fixed part must be four times heavier than the volatile one, but if you want to make it volatile, four parts of the volatile must raise one fixed part in the air. Many artists may not need to weigh if they have a sharp eye.

DAS 4. KAPITEL

Zeigt die andere Composition, den Stein auf weiß und rot zu machen, den alten langen Weg, wie auch den neuen kurzen Weg

Mache aus Mann und Weib wieder einen ☉. Sät aus euer ☉ in die geblätterte Erde, das wohl gedüngte Land sollst du erst tapfer wenden, dann wird beim West-Wind sich die Fäulung bald vollenden. Du hast nun das *Sal Metallorum*, welches mit leichter Mühe wird *Lapis Philosophorum*, in einem Gefäß, in einem Ofen, ist nun die Nach-Arbeit und das Kinder-Spiel. Du hast dich aber auf den flüchtigen ☿ nicht zu verlassen, sondern auf den calcinierten. Verbrennt unsere Leiber durch das allerstärkste Feuer, welches unser seltsames Höllen-Bad ist und muss der König allein darein gehen und alle seine Diener, so noch auf ihn warten, zurück lassen. Nimm unseren gekrönten König, unsere 7 Adler, lass ihn mit dem ♌ streiten, unser gülden Büchlein in seine *Fontinam* fallen, welches du jetzt zum letzten Mal siehst, so wird die *Turba* zu sehen sein, der ♌ und der Adler werden einander zerreißen, mit Grimm anfallen, dein schönes Büchlein, welches weit subtiler als Blattgold, ist nicht mehr da, lass sie in ihrer Wut und Blut liegen, bis sie alle tot dahin gefallen, ihr Grab schwarz, faul und stinkend geworden, welches die andere und letzte Schwärze ist. Vorher aber wird in diesem Meer das Fischlein *Æscheneis* mit silbernen Schuppen und roten Floßfedern schwimmen, auch das Meer grün werden, wenn der ☿ bei ♀ schläft, wird es aus den Wolken ☉ regnen und ist der Wald der Nymphen ♀, darinnen die Tauben der Diana fliegen. Es wird dieses Meer gefrieren bei heißem Wetter, als ein glatter Spiegel, wenn die Eisschollen schwer werden, brechen sie, fallen auf den Bo-

THE FOURTH CHAPTER

*Shows the other composition to make the stone, white and red,
the old long way, as well as the new short way*

Make a ○ out of man and woman. Sow your ☉ in the leafed earth, you should first turn the well-fertilized land bravely, then the putrefaction will soon be complete with the west wind. You now have the *sal metallorum*, which becomes *lapis philosophorum* with little effort, in a vessel, in an oven, is now the after-work and the children's game. But you shall not rely on the volatile ☿, but on the calcinated one. Burn our bodies in the very strongest fire, which is our strange hell bath into which the king has to go in alone, leaving behind all its servants which are still waiting for it. Take our crowned king, our seven eagles and let them fight with the ♌. Our golden booklet falls into its *fontinam*, which you will see now for the last time, so the *turba* will be seen, the ♌ and the eagle will tear each other apart and attack ferociously. Your beautiful little book, which is far more subtle than gold leaf, is no longer there. Let them lie in their anger and blood until they have all fallen dead. Their grave has become black, putrefying and smelly, which is the second and the last blackness. Before that, however, the little fish *echeneis* with silver scales and red fins will swim in this sea, and the sea will also turn green, when ☿ sleeps with ♀, it will rain ☉ from the clouds und and that is the forest of the nymphs ♀ in which the doves of Diana fly. This sea will freeze in hot weather as a smooth mirror, when the ice floes become heavy, they break, fall to the ground, soon others will come again, until finally the blackness breaks in, the sorrowful dark night, the soul of the king separated from the body and after the dampness the body

den, bald kommen wieder andere, bis endlich die Schwärze einbricht, die traurige finstere Nacht, die Seele des Königs vom Leibe geschieden und nach der Feuchte der Leib wieder anfängt trocken zu werden, so kommt der bunte Pfauenschwanz, hernach die weißen Strömlein der ☽, wenn sie noch trockener werden, so ist kein Tau noch Regen mehr, sondern trockene Dünste, Reif und Schnee, weil die ☽ kalt in der Hitze alles congeliert und gefriert. Nun musst du dich vor dem Glasmachen hüten, wenn die *Materia* in der Höhe zu Glase geworden, ist sie nicht wohl wieder herunter zu bringen, weil dieser Wasser- oder Salz-Stein der Weisen jetzt etwas hart ist zur Schmelzung, Ursache, weil seine Öle noch nicht mit ihn vereinbart, als wodurch er erweicht, güssig und flüssig wird, jetzt aber ist es nur Salz und ☿, unser weißer ♄, wenn aber ein wenig Reif sich oben anhängt, sieht es wie Duft im Keller. Wenn nun solcher schwer wird, bricht es, fällt herunter, nimmt seine Kameraden neben sich mit herunter, welche auf der Erde wieder schmelzen und andere Dünste geben, welches Bäume und Sträucher vorstellen wird, ist überaus lustig zu sehen. Bald wirst du denken, du siehst Brunnen-Wasser im Glase vor lauter Klarheit und ist doch ein helles Glas, bis es zuletzt das gläserne Meer geworden. Zuvor aber, ehe es ganz hart wird, wird es sich in unterschiedene Farben verändern, deren jede wie ein schönes Edelgestein glänzen wird. Wenn nun der steinerne Palast erbaut, ☉ und ☽ im Drachen-Haupt verfinstert, die Ungewitter hinweg, das Gnadenzeichen der Regenbogen erschienen und die Weiße, der volle Mond, so tränke die Luft auch 7-mal ein, drehe das Rad zum anderen mal herum, mit dem Feuer zum 3ten mal, bis du alles, was aus den einzigen gekommen, wieder in eins gebracht. Wenn du das Haupt fügest zu den Schwanz, so hast du die Tinctur ganz. Dies ist nun der Alten ihr langer Weg, da sie die 3-mal 7-fachen magischen Zahlen erfüllt. Wenn du aber den kurzen Weg gehen willst, so nimm nach des Theophrasti Geheimnis, vom roten ♌ das rosenfarbene Blut und von dem Adler das weiße Gluten, nimm unsern höchstglänzenden ☿, die 7 Adler, solviere dieses metallische Gluten in seiner Luft oder Öl, welches diesen Schnee alsobald zerschmelzt, so ist es das

starts to dry out again, so comes the colored peacock's tail, afterwards the white little stream of the ☽ when they get even drier, there is no more dew or rain anymore, but dry steam, frost, and snow, because the ☽ cold congeals and freezes everything in the heat. Now you have to be careful not to make glass, for if the *materia* has turned to glass in height it cannot be brought down again because this water or salt stone of the wise is now a little bit to hard to melt. The reason is that its oils do not fit together so that it becomes soft, pourable and liquid, but now it is just salt and ☿, our white ♃, but when a little frost clings to the top, it looks like a scent in the cellar. When it becomes heavy, it breaks, falls down, it also takes down its comrades next to it, which then melt on the earth again and give off other vapors. These will resemble trees and bushes, and it is extremely odd to see. Soon you will think that you see well water in the glass because of the sheer clarity and yet it is a light glass until it finally becomes the sea of glass. But before that, before it gets really hard it will change into different colors, each of which will shine like a beautiful jewel. When the stone palace is built, and when ☉ and ☽ are darkened in the dragon's head and the storms have gone, and the sign of grace, the rainbow, appears, as well as the whiteness of the full moon, then impregnate in the air seven times, turn the wheel the other time around, with the fire for the third time, until you have brought everything that came out of the only one back into one. If you join head to tail, you have the tincture without fail. This is now the long way of the ancients, since it fulfills three times the sevenfold magic numbers. But if you want to go the short way, then, according to Theophrasti's[24] secret, take from the red ♌ the rose-colored blood and from the eagle the white gluten, take our extremely shiny ☿, the seven eagles, dissolve this metallic gluten in its air or oil, which immediately

philosophische ♅, des Königs Wasser-Bad, das metallische Wasser, *Aqua Benedicta*, *Aqua Permanens*, in solchem koche das gesegnete Blut, unser Feuer im Wasser, so ist die Tinctur in einem Monat fertig; Ist aber nicht von solcher Kraft als jene, habe auch nicht so viel Wunderwerke der Kunst und der Natur zu sehen, wie bei dem langen Wege.

melts this snow, it is the philosophical ▽, the king's water bath, the metallic water, *aqua benedicta, aqua permanens*, in such [water] boil the blessed blood, our fire in the water, so the tincture is ready in one month. But it is not of such strength as that, nor have I seen so many miracles of art and nature as with the long way.

DAS 5. KAPITEL

Ist eine weitere Erklärung aller Kapitel, worin die Multiplication weitläufig ausgeführt und die Ursachen seiner Erhöhung gewiesen worden, auch allerhand Tincturen, wie auch der kurze Weg nochmals ganz deutlich gezeigt

Wenn du die *Elementa* geschieden, so hast du den Schlüssel zum hesperischen Garten und kannst die güldenen Äpfel brechen von welchem Baum du willst. Du hast das *Universal Menstruum* zum ganzen metallischen Reiche und ist das metallische und vegetabilische Wasser, darinnen die Metalle, als in ihrer Mutter Leib wieder gehen und neu geboren werden, ist auch nicht wieder von ihm zu scheiden, weil es ein metallisches Wasser ist und wegen seiner Reinigkeit sich mit den Metallen *radicaliter* vereinigt, du kannst sie darinnen als Eis in warmen Wasser zerschmelzen, alle metallische und mineralische ♃ in diesem Öl, welches die Luft, auch *Spiritus* ☿ genannt wird und ölig ist, auflösen und mit über den Helm führen, welches Trinkgold genannt wird, auch nimmermehr in ein *Corpus* zu reducieren ist: Kannst solche Tinctur gebrauchen nach allen deinen Gefallen, ist auch kein *Particular* in der Welt erklecklich, es gehe denn durch diesen Brunnen des *Universals*. Du hast das Wasser, kannst Fleisch und Fische kochen, wie du willst. Kannst du der Alten langen Weg nicht treffen, so befleißige dich das Wasser-Bad des Königs zu machen, erlerne unseren ☿ *exuberatum* in die *Terram Foliatam* zu sublimieren, diesen ✶ wirf in sein Öl, nach rechtem Gewicht, welches diesen Schnee auflöst, so ist es des Königs Wasser-Bad, so solchen Leib alsbald zerschmelzt. Wenn das ☉ solviert, tue sein eigenes Ferment dazu, koche es bis es fix wird, bei des Adlers Gluten. Kannst du diesen auch nicht treffen, so will ich dir den 3ten lernen: Nimm unsere Jungfer-Milch, solviere darinnen das jungfräuli-

THE FIFTH CHAPTER

Is another explanation of all chapters, in which the multiplication is extensively described, and the causes of its increase have been shown, also all sorts of tinctures, as well as the short way once again clearly shown

When you have separated the *elementa*, you have the key to the Hesperidian garden, and you can break the golden apples from whatever tree you want. You have the *universal menstruum* for the whole metallic realm and that is the metallic and vegetable water, in which the metals go again as into their mother's womb, and are reborn, it cannot be separated from it again, because it is a metallic water and because of its purity it combines with the metals *radicaliter*, you can melt them in it as ice in warm water, you can dissolve all metallic and mineral ♄ in this oil, which is called the air, also *spiritus* ☿, and is oily, and carries it over the helm, which is called potable gold, also can never again be reduced to a *corpus*. You can use such tincture as you please, there is also no *particular* in the world, which has not gone through this fountain of the *universal*. You have the water, you can cook meat and fish as you want. If you cannot find the long way of the ancients, make the king's water-bath, learn to sublimate our ☿ *exuberatum* into the *terram foliatam*, throw this ✶ into its oil according to the right weight, which dissolves this snow, so it is the king's waterbath which soon melts this body. When the ☉ dissolves, add its own ferment, boil it until it becomes fixed with the eagle's gluten. If you cannot find this one either, I want to teach you the third: take our virgin milk, dissolve the virgin salt in

che Salz, diesen nassen ☿ præcipitiere mit güldischem ♁, so wird es ein wunderschöner Præcipitat, solchen solvier in seinem Öle, tu das Feuer dazu, koche es bis es fix wird, du musst aber in den *Astris* eine Wissenschaft haben, dass du erkennst aller Wasser Natur. Der ♄ gibt dir so viel *Menstrua*, als der ☉, du hast aus demselben ein klares Wasser, auch etlichermaßen ein gelbes Öl, das rote aber ist ein großes Corrosiv, welches keinen Nutzen in der Kunst hat, als nur den ☿ fix zu machen, es ist der Alte, welcher mit seiner Sense dem ☿ die Flügel abhaut, ist auch der jungen Kinder Blut, darinnen sich ☉ und ☽ zu baden pflegen. Es ist nicht nötig dieses zu scheiden, als nur das einzige Corrosiv, der Cherub mit dem feurigen Schwert zu bewahren den Weg zu dem Baum des Lebens. Durch dieses Feuer-Gericht muss der ☿ hindurch und ist das Hermetische Siegel. Gehe geschwind und schließ zu. In dem vitriolischen *Spiritu* hast du abermals so viele *Menstrua*, welche viel herrlicher als vorhin, aus welchen die erste Composition, nämlich des doppelten ☿ zu machen, wie auch die andere Composition des Steins, hernach solches in seinem eigenen tingierenden Elixiren und Unguenten erhöhen, färben und zum Karfunkel-Stein, so im finstern leuchtet, machen. Zum weißen Stein brauchst du keine roten, sondern nur die weißen *Elementa*, unsern grünen ♌, welcher den Streit mit dem gestirnten Adler halten muss. Die weißen, nassen und trocknen Geister sind fast lauter arsenikalische Geister mit etwas reiferen ☿ialischen Geistern vermischt und durch die vielen Einträngungen erweicht, welche so viele *Sublimationes* bedeuten. Nun weist du wohl, dass in den Bergwerken aus den arsenikalischen Geistern ☉ und ☽ wachsen, aus dem gemeinen ♁ ♂ und ♀, weil aber diese weißen Geister von des Goldes Natur allzuweit entlegen und du wolltest gleich nach Scheidung der Elementen solche nasse und trockne Geister zusammen schütten und das sollte die Composition des Steins werden, wird es unmöglich sein und hast nur unser *Confusum Chaos*, wie vorher. Wenn du nun solches bis auf den Jüngsten Tag kochst, wird es stets zu scheiden sein und sich nimmermehr vereinigen. Du musst in dieser Schöpfung der großen Schöpfung nachdenken, ist nicht das Feuer von

it, this wet ☿ precipitate with gold-bearing ♃ so it becomes a wonderful precipitate, dissolve this in its oil, add the fire to this, cook it until it becomes fixed, but you must have a science in the *astris* so that you can recognize the nature of all waters. The ♄ gives you as much *menstrua* as the 🜨, you have from them a clear water, also to some extent a yellow oil, but the red one is a great corrosive, which has no use in art than only to fix ☿, it is the old man, which cuts off the wings of ☿ with its scythe, it is also the blood of the young children, in which ☉ and ☽ tend to bathe. It is not necessary to separate this, only the corrosive one, the cherub with the fiery sword, to preserve the way to the tree of life. The ☿ must pass through this fiery judgment and is the Hermetic seal. Go quickly and lock it. In the vitriolic *spiritu* you have again so many *menstrua*, which are much more splendid than before, from which the first composition, namely the double ☿, has to be made, as well as the second composition of the stone; afterwards elevate this in its own tinging elixirs and unguents, and color it into the carbuncle stone, so glowing in the dark. For the white stone you do not need red, just white *elementa*, our green ♌, which has to quarrel with the starry eagle. The white, wet, and dry spirits are almost nothing but arsenical spirits mixed with somewhat more mature ☿al spirits, and softened by the many impregnations which mean so many *sublimationes*. Now you know well, that in the mines ☉ and ☽ grow from the arsenical spirits, from the common ♃ ♂ and ♀, but because these white spirits are too far removed from the nature of gold and you wanted to pour together such wet and dry spirits immediately after the separation of the elements, which should be the composition of the stone, it will be impossible and you only have our *confusum chaos*, as before. If you now cook such a thing until the Judgment Day, it will always be possible to separate it, but never again to combine it. You must contemplate: in this creation of the great creation, is not the fire separated from the earth

der Erde geschieden und steht zu höchst über uns? Die Wasser auch um uns, in der Luft leben wir, bis GOTT alles der verklärten Erde wiedergeben wird, wenn sie vorher durchs Feuer wohl calciniert worden. Jetzt sind wir in dem *Confusum Chaos*, da bald Hitze, bald Frost, bis in der Calcination die *Turba* gar kommt, solcher Sachen denken nach, scheide die *Elementa* so rein du kannst, wird dir doch solche unmöglich ganz rein zu scheiden sein, die Wasser aber sind von dem ☉ allzuweit entfernt, so musst du nun dieses metallische Wasser kochen, bis es in ☉ kommt, welchen ☉ Grad er erst auf der 7ten Stufe ersteigen muss, dies ist eben die Wachsung der Metalle im Glase, durch diese trocknen und nassen Geister, welche zuvor durch viele Arbeit mussten dahin gebracht werden, dass sie in ihre ponderosische Feuchtigkeit gingen, welche nun geschickt, Metalle zu gebären, die viel höher, reiner und vollkommener als die gemeinen, weil hier kein verbrennlicher ♄ zuschlagen kann, auch keine unreine Erde. Wenn nun der König in seiner Exaltation den königlichen Thron erstiegen, so salbe das Blatt mit Gift, lege ihm über sein weißes Hemd das gelbe güldene Stück an, welches nur Königen zu tragen gebührt, hernach gib ihm den roten Purpur-Mantel um, so ist es der König der Herrlichkeit, so über den ganzen Erdkreis herrscht, von Orient bis Occident. Es sind bei diesem Werke 4 Geheimnisse, das erste *Arcanum* ist der ☿, unsere Jungfer-Milch. Das 2te *Arcanum* ist der doppelte ☿, *Prima Materia* aller Metalle. Das 3te ist unser Stein der ersten Ordnung auf weiß. Das 4te *Arcanum* ist die Tinctur auf rot. Diese 4 sind mehr engelhaft als menschlich, werden von GOTT gegeben, darinnen so viele Tugenden stecken, dass es nicht auszusprechen ist, wie die Tinctur alle Körper der Metalle durchgeht und sie in ☉ verwandelt, so tingiert sie die Krankheit in Gesundheit. 3 Dinge sind, die das Werk verrichten, ♌ *Viridis, Aqua fœtida, Fumus Albus*. Wenn du ihn nun fertig hast, so gib ihm seine eigenen *Fermenta*, dadurch er sich in viel 1000 multipliciert. Wenn du ihn einmal gemacht hast, so ist es nicht nötig, ihn noch einmal zu machen, sondern kannst bei diesem Licht viele 1000 Lichter anzünden und ihn unendlich vermehren und ist dieser Brunnen unerschöpflich, wenn

and now high above us? The waters are around us too, in the air we live until GOD will give everything back to the transfigured earth, if it has been well calcinated beforehand by fire. Now we are in the *confusum chaos*, there is now heat, now frost, until the *turba* comes in the calcination; think about such things, separate the *elementa* as pure as you can, it will be impossible for you to separate them completely purely. But the waters are too far away from the ☉, so you now have to boil this metallic water until it comes into ☉, which ☉ degree it must first climb on the seventh level, this is the growth of the metals in the glass, through these dry and wet spirits, which previously had to be brought into their ponderosic dampness through a lot of work, which are now adept at giving birth to metals that are much higher, purer, and more perfect than the common ones, because here no combustible ♄ can strike, not even any impure earth. When the king ascended the royal throne in its exaltation, anointed the leaf with poison, put the yellow golden piece over its white tunic, which only kings should wear, then put the red purple cloak on it, so it is the king of glory which rules over the whole world, from orient to occident. There are four secrets in this work. The first *arcanum* is the ☿ of our virgin milk. The second *arcanum* is the double ☿, *prima materia* of all metals. The third is our stone of the first order in white. The fourth *arcanum* is the tincture in red. These four are more angelic than human, are given by GOD, in which there are so many virtues, that it cannot be expressed, how the tincture goes through all the bodies of metals and transforms them into ☉, so it tinges the disease into health. Three things are doing the work: ♌ *viridis, aqua fœtida,*[25] *fumus albus.*[26] When you have finished it, give it its own *fermenta*, by which it is multiplied into many thousands. If you have made it once, it is not necessary to make it again, because you can light many thousand lights with this light and increase it infinitely and this fountain is inexhaustible, if you only boil it again in its fiery ☿, the

du ihn nur in seinem feurigen ☿ wieder kochst, den weißen zur weißen Tinctur, den roten zur roten Tinctur, doch muss er allezeit, wenn er ausgestritten und lass auf Erden ruht, mit gemeinem Golde versetzt werden, damit dieser feurige ☿ ein neues *Patiens* bekommt, dass er als ein *Agens* darin wirken kann und sich dabei erhitzen, wodurch ein neuer *Modus* wird, sobald er ihn berührt, wird er glühend heiß, wird auch nicht aufhören zu arbeiten, bis er ihn wieder erhöht, ohne dies hat er keine Bewegung. Bist du nun ein Verständiger von der Sache, so wirst du selber sehen, wann du aufhören sollst mit Kochen und Eintränken, wenn es genug gesättigt. Aus dem Werk erscheinende Zeichen werden dich schon lehren, dass du zu rechter Zeit es zur Ruhe bringst, auch wieder erweckst, dass ein neuer *Modus* darein gebracht wird und nicht eher seine Arbeit aus ist, seine völlige Sättigung erscheint denn. Wenn du aber unser ☉ gar nicht erkennst, so lerne nur unsere *Menstrua* machen, das ponderosische, viskosische Wasser, unseren schlackenreichen ☿ und arbeite in dem gemeinen ☉, doch kann solches die Stelle eines Ehemanns nicht vertreten, als wie unsere Erde oder Salz, *Hyle* genannt, ist auch ein sehr langwieriger Weg von weniger Kraft, da du in diesem 100 Teile hast, musst du dich des Kohlen-Feuers bedienen, in dem natürlichen Werk aber, unseren feurigen ☿, wiewohl du den zu machen auch das Herdfeuer dazu brauchen musst, in den *Herculis*-Arbeiten, die *Materiam* geschickt zu machen, da hat es Mühe, da hat es Not, auch musst du wissen, dass aus allen 7 philosophischen Metallen der Stein zu machen ist, koche nur unsern einfachen ☿ bei seiner Erde, bis er ein *Corpus* aus den 7 bekommt, seine schöne weiße Zeltlein, solche brauche anstatt der 7 Adler zum Fundament des Steins. Nimmst du nur einen Adler, so kannst du nur ☽ und ♄ in ☉ transmutieren, doch wird der ganze Körper des ♄ nicht in ☉ verwandelt, bekommst nur eine *Materia*, darinnen du viel ☉ siehst und muss erst auf dem Test abgetrieben werden. Nimmst du aber 4 bis 5 Adler, so kannst du den ♃, ♄, ☿ und ♀ in ☉ verwandeln. Nimmst du aber die 7 Adler, den ☿, so 7-mal sublimiert, so kannst du den ♂ und das ganze metallische Reich in ☉ verwandeln, ja die unzeitigen Mine-

white to the white tincture, the red to the red tincture, but it must always be replenished with common gold when it has fought, and let it rest on earth so that this fiery ☿ receives a new *patiens*, so that it can act as an *agens* in it, thereby heating itself, which creates a new *modus* and as soon as it touches it, it becomes red-hot and will not stop working until it is elevated again; without this it has no movement. If you are now someone who understands the matter, you will see for yourself when you should stop cooking and impregnating, when it is saturated enough. Signs appearing from the work will teach you that you will bring it to rest at the right time, and also awaken it again, so that a new *modus* will be introduced and that its work will not be over until its complete saturation appears. But if you do not recognize our ☉ at all, learn to prepare our *menstrua*, the ponderosic, viscous water, our slag-rich ☿, and work with common ☉, but such things cannot take the place of a husband, as like our earth, or salt, called *hyle*, is also a very tedious way of less strength, since you have a hundred parts in this, you must use the coal fire, but in the natural work, [you use] our fiery ☿, although you also need the stove-fire to prepare it, in the works of *Herculis*, to make the *materiam* skillful, there is trouble, there is need, you also have to know that the stone is to be made from all seven philosophical metals, just cook our common ☿ on its earth, until it becomes one *corpus* out from the seven [metals], its beautiful little white *trochisci*, which you should use instead of the seven eagles for the foundation of the stone. If you only take one eagle, you can only transmute ☽ and ♄ into ☉, but if the whole body of the ♄ is not transformed into ☉, you only get one *materia* in which you can see much ☉ and which first must be refined in the cupel.[27] But if you take four to five eagles, you can change the ♃, ♄, ☿ and ♀ into ☉. But if you take the seven eagles, the ☿ which has been sublimated seven times, you can transform ♂ and the whole metallic kingdom into ☉, and also the untimely

ralien. Also warte lieber, bis du unsern höchstglänzenden ☿ hast, welches der beste zu diesem Werk ist, auch die Reife des ☉ hat, ja noch viele Grade darüber gestiegen und zur weißen Tinctur, zum weißen ☉ geworden, so ist an diesem königlichen Leibe kein *Spolium* mehr, sondern seine von ihm geschiedenen Gifte sterben bei ihm und werden zur Tinctur, und er wird durch sie als seine Knechte und Untertanen herrlich gekleidet, wenn er in seinem Blut und Schweiß badet. Nimmst du aber einen oder 2 Adler, illuminierst sie mit eben diesen Farben, wirst du doch noch lange nicht eine solche hohe Tinctur haben. Zum Gleichnis, färbe den ☿, ♄ oder ♃ mit eben diesen güldischen ♀, so tun sie nicht mehr, als dass diese Körper ☉ werden, nach langer Mühe und Zeit, hergegen färbe mit eben diesen ♀ ☉, so hast du eine Tinctur, die andere metallische Körper in ☉ transmutiert. Hier kannst du den Unterschied sehen, auch seine Erhöhung, wenn dieser königliche Leib zur Tinctur geworden und du löst abermals seine Glieder auf, zerschmelzt seinen Leib in seinem eigenen metallischen Wasser, darinnen es als in seiner Mutter Leib neu geboren wird, so ist ihm seine Haut entzwei geschlagen, Beine und Adern alles zerschmolzen, so wird dieses edle *Menstruum* gleichfalls diesen geistlichen Leib töten und nicht eher daran aufhören zu arbeiten, bis es ihm einen viel herrlicheren Leib wieder gegeben, besser als vorher und zum anderen und neuen Leben wieder auferweckt und dieses *in infinitum*. Wodurch du klärlich seine Erhöhung siehst, die *Menstrua* aber sind sehr unterschiedlicher Art und Kraft, das Wasser ist das allerschwächste, darum löst es auch so langsam seinen eigenen Leib auf, weil es von keiner Schärfe, sondern von Geschmack wie Regenwasser und darum solviert es auch das gemeine ☉ nicht. Wenn es aber seinen Leib solviert und mit diesem nun eins geworden, so ist es der ✶, so zu des Königs Wasserbad genommen wird, wenn es in unser Gradier-Öl geworfen, welches diesen Schnee alsobald zerschmelzt, so ist es das *Aqua Permanens, Aqua Benedicta*, der Schlüssel der Kunst, so die *Metalla philosophice* solviert, das vegetabilische *Menstruum*, dass die Metalle wachsend macht und sie darinnen, als in ihrer Mutter Leib neu geboren werden und sehr hoch erhöht. Unser

THE MINERAL GLUTEN

minerals. So you are better to wait until you have our most brilliant ☿, which is the best for this work and also has the maturity of ☉; indeed it has risen many degrees higher and has become a white tincture, a white ☉, for there is no *spolium* on this royal body anymore, but rather, the poisons separated from it die with it and become tincture, and it is splendidly clothed by them as its servants and subjects when it bathes in its blood and sweat. But if you take one or two eagles and illuminate them with these same colors, you will not have such a high tincture for a long time. As a parable, if you color ☿, ♄, or ♃ with this gold-bearing ♀, these bodies do nothing more than become ☉ after much time and effort; in contrast, if you color ☉ with the same ♀☉, you will have a tincture that transmutes other metallic bodies into ☉. Here you can see the difference, also its exaltation, when this royal body has become tincture and you dissolve its limbs again, melt its body in its own metallic water, in which it is reborn as in its mother's body, its skin is beaten in two, its legs and veins are all melted; this noble *menstruum* will likewise kill this spiritual body and will not stop working on it until it has been given a much more glorious body again, better than before, and resurrected to another new life, and this *ad infinitum*. Through which you can clearly see its increase, but the *menstrua* are of a very different nature and strength, the water is the very weakest, that is why it slowly dissolves its own body because it does not have a sharpness, but a taste as rain water and therefore it does not dissolve common ☉. But if it dissolves its body, and has now become one with it, it is the ✶ that is used for the king's water bath, when it is thrown into our gradating oil, which immediately melts this snow, it is the *aqua permanens, aqua benedicta*, the key of art that dissolves the *metalla philosophice*, the vegetabilic *menstruum* that makes the metals grow and in which they are reborn as in their mother's womb and are exalted very

anderes *Menstruum* ist unser Gradier-🜖, Luft oder Öl, unser *Spiritus* ☿ unser ♃ Öl, tingiert alle Leiber in gelb. Nun haben wir kein *Menstruum* mehr, als unser Feuer, Sonnen ♃, unser gesegnetes Blut, welches als brennende Kohlen leuchtet, hat auch im Munde die Hitze als Feuer. Du musst aber wissen, in welchem Wasser du deinen Stein auflösen sollst und in welchem du deinen weißen oder roten Stein kochen musst und wenn du den kurzen Weg gehen willst, wie du Feuer im Wasser kochen musst. Ich habe es dir vielmals ohne einige Verdeckung in diesem Buche gesagt, weiß dich nicht besser zu unterrichten, habe dir das Recept geschrieben, welches du wohl zurichten musst, bist du ein guter Koch, so kochst du eine herrliche Mahlzeit für die Götter und Menschen. Ich will dir auch noch was zum besten von ♄ und ☉ schreiben, damit du noch mehr unterrichtet wirst und sehen kannst, ob du zum Anfang auf rechtem Wege wandelst. Wenn du in deiner Arbeit das nicht findest, was ich es hier beschreibe, so höre auf, es ist vergeblich, nimm unser ♂, nicht das schwarze gemeine, sondern unser *Electrum*, welches von außen die Farben der Taubenhälse hat, inwendig, wenn man es zerschlägt, hat es viel gelbe Risslein und ist mit einem weiß-grauen Kittelchen bedeckt, solches mache zum allerkleinsten Pulver, ehe es eines Feuers teilhaftig geworden, soliviere es in der Natur ihrem Essig, der ♄ will mit starken Getränken trunken gemacht sein, bis er seinen Geist übergibt, nimm alle seine Röte von ihm, welche rosenfarben aussieht, den Essig lass abrauchen in gelinder Wärme, bis es eine Honig-Dicke bekommt, unten hat es ein Sediment, als *Sacharum* ♄ aber nicht weiß, sondern gefärbt, als wie kandierter Honig, darunter auch rote Kristallchen, als Spinetten, etliche grün, als Smaragde, etliches wie Amethysten, etliches weiß in langen Spießchen, als Salpeter angeschossen, daraus du schon seine vielen metallischen Arten sehen kannst, so er bei sich führt und diese 7 Blumen auf einem Stil stehen, aber jede *a parte*, keine sich noch mit der anderen vereinigt hat, dieses destilliere *gradatim*, es gibt feurige *Spiritus* mit vielem Brausen und Sturm-Winden, wenn es halb übergestiegen, siedet es als Pech in die Höhe, wegen seiner Fettigkeit, läuft über und stößt die Gefäße

high. Our other *menstruum* is our gradating ▽, air or oil, our *spiritus* ☿, our ♃ oil, [which] tinges all bodies in yellow. Now we have no other *menstruum* than our fire, ♃ of suns, our blessed blood, which shines like burning coals and also has the heat of fire in its mouth. But you have to know in which water you should dissolve your stone and in which you have to boil your white or red stone and if you want to go the short way, how you have to cook fire in water. I have told you that many times in this book without any concealment; I do not know how to teach you better; I wrote down the recipe that you must prepare; if you are a good cook, you will make a wonderful meal for gods and humans. I also want to write you something about the best of ♄ and ☉, so that you will learn even more and be able to see whether you are walking on the right path right from the beginning. As you proceed with the work, if you do not find what I am describing here then stop, for you are working to no avail. Take our ☿—not the black common [one], but our *electrum*, which has the colors of pigeon necks on the outside—but inside, if you crush it, it has many small yellow cracks and is covered with a white and little gray tunic. Make this into the smallest powder before it is steeped in fire and dissolve it in the vinegar of nature. ♄ wants to be made drunken with strong drinks until it surrenders its spirit; take all its redness from it, which looks rose-colored, let the vinegar smoke off in mild warmth until it becomes thick as honey, and at the bottom has a sediment, like *sacharum* ♄, not white, but colored like candied honey, including small red crystals like spinets, some green, like emeralds, some like amethysts, some white with lengthy spikes, shot like saltpeter, from which you can already see the many metallic species that it contains in itself. These seven flowers stand on one stem, but each apart, for none of them has united with another. Distill this *gradatim*. It gives a fiery *spiritus* with much storm and bluster, when it is half risen, it boils upwards as pitch, because of its fatness, it overflows and breaks the vessels in two, so be careful. But when its ☿al

entzwei, sei vorsichtig. Wenn aber sein ♄al Wasser anfängt zu gehen und die Tropfen auf das Glas fallen, so schlägt es im Augenblick ein Loch Talers groß in die Vorlage und du musst den dicken schneeweißen Geist fahren lassen, welcher mit solcher Gewalt und Ungestüm, wie die dicksten Wolken, schneeweiß zur Retorte heraus und zur Feuermauer hinaus fahren, als wie mir begegnet, da ich 3 andere Vorlagen anstieß und in alle Löcher schlug, weil aber keine von solcher Größe mehr hatte, musste ihn fahren lassen und zusehen, wie die Feuer-Esse, so weit sie war, von weißen Wolken angefüllt wurde, mein ganzes Werk war verloren, musste es wieder von neuem anfangen und die Vorlage anlegen, dass die Tropfen nicht auf das Glas fielen, sondern in den übergegangenen Spiritum, welche mit solchem Brausen sich darinnen löschten, als wenn eine eiserne Musketen-Kugel geglüht und in Wasser gelöscht würde, so hat man keine Gefahr. Halte mit dem Feuer an, bis die Retorte glüht und gar keine Dünste mehr gibt, den *Spiritum* rectificiere im Sand, so lange als etwas übersteigen will und ein schweres rotes *Oleum* zurück bleibt, solches rectificiere in großem Feuer und verwahre es vor der Luft, sobald es die empfindet, verraucht es alles, die Erde, so zurück bleibt, tu zu dem roten Ton, so in der Retorten ist. Das Wasser heißt die empfangene GOTTESgabe:

> Das Bergwerk gibt einen roten Ton,
> wer ihn recht kennt, hat großen Lohn,
> wenn er wird gar wohl præpariert,
> ein weißer Geist dann aus ihm wird,
> dazu ein schönes blutrotes Gold,
> das gibt den Meister großen Sold,
> wenn er es kann zu dem Geiste fügen,
> damit der Geist bleibt bei ihm liegen,
> mehr sag ich nicht, es ist zuletzt,
> was dich in dieser Welt ergötzt,
> es gibt Nahrung und gesunde Tag,
> mit Wahrheit ich solches sagen mag,

water begins to go and the drops fall on the glass, it instantly makes a hole the size of a thaler in the receiver, and you have to release the thick snow-white spirit—which pushes out with such violence and impetuosity, like the thickest of snow-white clouds—out of the retort and to the fire wall. When I encountered this, I hit three other receivers and punched holes in all of them; and because I had no more of that size, I had to let it go and watch the chimney, as wide as it was, become filled with white clouds. My whole work was lost, and had to be started again. It was necessary to place the receiver such that the drops could not fall on the glass but into the distilled *spiritum*. The drops then extinguished themselves with the effervescence of a glowing iron musket ball quenched in water, until there was no longer any danger. Continue with the fire until the retort glows and there are no more vapors. Rectify the *spiritum* in the sand as long as something wants to go over and a heavy red *oleum* will remain. Rectify it in a great fire and keep it from the air, for as soon as it senses it [the air], it will completely evaporate. Add the earth that remains behind to the red clay which is in the retort. The water is called the received gift of GOD:

> The mine gives you a red clay,
> who knows it right, has great rewarded pay,
> if it is well prepared by us,
> it becomes a white *spiritus*,
> in addition a beautiful blood-red gold,
> which makes its master rich and bold,
> if he can add it to the *spiritus*,
> so that the spirit stays with us,
> that's all I say, it is ultimately
> what delights you in this world lately,
> it gives food and a healthy day,
> with truth such a thing I may say,

findest du das, so danke mir,
dass ich eröffnet dir diese Tür,
gib GOTT das seine, die Armen bedenk,
so legst du wohl an dieses Geschenk.

Hier habe ich dir unseren ersten ☿ gewiesen, welchen uns die Natur in einem wässrigen Leibe in seinem ersten Wesen gegeben, hast nun das Weib vom Mann geschieden, von unserem Adam, so mit seiner Even schwanger gegangen. Nun will ich dir den ☿ zeigen, der von uns bereitet wird, der fette und schlackenreiche. Nimm unsere rote Erde, nahe am Paradies, weil aus dieser das Paradies-Wasser gemacht wird, unser Zinnober-Erz, ziehe es zum Schlich, koch es etliche Stunden in destillierten Regenwasser, so hast du eine schöne rosenfarbene Solution, viel schöner als vorhin, solche lass zum *Vitriol* anschießen, den resolvier in unserer giftigen Schlange, seinem Weibe, welche ihres Mannes Samen begierig an sich zieht und sich geistlich davon schwängert, solches destilliere der Kunst nach, so hast du das Rote und Weiße in unserem Chaos, dieses scheide in die *Elementa*, und mache aus der Jungfer-Milch unserem weißen ♄, welchen du hernach mit den gelben und roten färben kannst. Die Jungfer-Milch ist nun unser ☿, der viskosische in einem öligen Leib, welcher von uns bereitet wird. Der 3te ist der doppelte, die *Terra Foliata*. Was ich dir zu gute offenbart, hat kein Mensch in der Welt getan, alle haben von Scheidung der Elemente angefangen, daher sich so viele 1000 Versucher gestoßen und mit dem ersten rohen *Confusum Chaos*, von der Natur wollen zu Werke gehen, welches ganz vergebens, weil es nur ein Wasser-Glanz, ob es schon einen arsenikalischen Geist, ein gelbes Öl und einen ☿ bei sich hat, so ist es doch noch nicht die Grund-Feuchtigkeit, so im Feuer bleibt, sondern muss erstlich seinen eigenen Leib solvieren und in diesem *Decocto* an sich ziehen die ölige und schleimige Art seines Körpers, welches kein corrosivischer Geist tun konnte, als nur dieses Wasser, weil es schon etwas leimig und glutinosisch ist. Große Herren werden es nicht machen, weil sie keine Zeit noch Geduld dazu haben, die Armen auch

if you find that, thank me true
for opening this door for you,
give GOD his part, consider the poor,
then you will probably invest this gift well sure.²⁸

Here I have shown you our first ☿, which nature gave us in a watery body in its first essence, you have now separated the woman from the man, from our Adam, who became pregnant with his Eve. Now I want to show you the ☿ that is prepared by us, which is fat and slag-rich. Take our red earth, close to paradise, because from this the paradise water is made, our cinnabar ore. Prepare sands out of it, boil it for several hours in distilled rain water so that you have a beautiful rose-colored solution, much more beautiful than before. Let it shoot up to *vitriol*, resolve this [*vitriol*] in our poisonous serpent, its wife, which eagerly draws its husband's seeds into itself and spiritually impregnates itself with it. Distill this according to the art, so you have the red and white in our chaos. Divide this into the *elementa* and make our white ♃ out of the virgin milk, which you can then color with the yellow and red. The virgin milk is now our ☿, the viscous one in an oily body, which is prepared by us. The third [*mercurius*] is the double [*mercurius*], the *terra foliata*. What I have revealed to you to your benefit, no one in the world has done. Everyone began from the separation of the elements, hence so many thousand of seekers stumbled, for they want to go to work with the first raw *confusum chaos* of nature, which is completely in vain, because it is just a splendour of water. Whether it already has an arsenical spirit, a yellow oil and a ☿ with it, it is not yet the basic moisture that remains in the fire, but must first of all dissolve its own body and in this *decocto* draw in the oily and slimy nature of its body, which no corrosive spirit could do but this water, because it is already somewhat loamy and glutinous. Great lords will not do it because they have no time or patience, neither will the poor. Although

nicht, obschon das ganze Werk kaum 3 oder 4 Taler kostet, so haben die Armen doch nicht Lebens-Mittel, dass sie so lange bei den Gefäßen sitzen können. Auch erfordert dies Werk einen ganzen Menschen, der kein Amt und Hantierung hat und von allen Menschen abgesondert sein muss, die Ankunft eines guten Freundes kann einen um das ganze Werk bringen. Ich weiß, was mir 5-mal bei Verfertigung dieses Werks begegnet und wäre höchst not, dass ihrer 2 vollkommene Wissenschaft des Werks hätten und einander secundierten. Auch hüte dich, dass du von seinem Gift nicht Schaden leidest, zumal wenn es in seiner Exaltation und flüchtig ist, es ist ein sehr trockenes und hitziges Gift, wenn die Gefäße brechen und du dabei bleibst, bist du des Todes. Es ist ein *Venenum Tingens*, und wenn es auch schon fix und zur heilsamen Medizin geworden, würde es doch durch seine überschwengliche Hitze in großer Quantität gebraucht, dir das *Humidum Radicale* austrocknen und dich töten; Wiewohl der doppelte ☿ das *Humidum Radicale* stärkt, auch die alten Bäume fruchtbar macht, die grauen Haare auswirft und die ganze Natur renoviert, so hat er zu dieser Zeit sein Feuer noch nicht bei sich, wodurch er so hitzig und zur Röte gebrannt wird. Wenn aber alle 4 *Elementa* in der Erde ruhen und in rote Erde verwandelt, ist es nichts denn lauter Feuer. Wirfst du es in einen Brunnen oder Teich, so wird das ganze Wasser sieden, als wenn ungelöschter Kalk sich im Wasser löschte. In Reinigung und Rectificierung der Elemente musst du dich vor allen Dingen in acht nehmen, dass sie wohl von ihrem unnützen *Phlegmate* gereinigt, worin du sein Salz auflösen und reinigen kannst. Wenn dieses *Phlegma* dabei bliebe, würde es das ganze Werk verderben oder doch sehr verhindern oder doch verzögern. Das Wasser darfst du kaum einmal rectificieren, sonst würde es sich coagulieren, weil es keine Geister hat und die Luft rein davon geschieden, auch wenn trockene Geister aufsteigen, als weißes Salz oder Sublimat, muss es in die milchfarbenen Tropfen getan werden, das weiße gibt das Gluten *Aquilæ*, oder mineralische Gluten, wenn es mit seinem weißen jungfräulichen Salz, mit der reinen Königin *Alma* vereinigt und darinnen aufgelöst worden, gibt es die metallische Gur

the entire work hardly costs three or four thalers, the poor do not have the means to pay for food such that they can sit beside the vessels for so long. This work also requires a complete person who has no office or business and who has to be separated from all people, for the arrival of a good friend can deprive you of the whole work. I know what happened to me five times while producing this work and it would be great if two [persons] had a perfect knowledge of the work and that they second each other. Also be careful that you do not suffer from its poison, especially when it is in its exaltation and fleeting, for then it is a very dry and hot poison, such that if the vessels break and you are near them, you are dead. It is a *venenum tingens*,[29] and even if it had already become a fixed and salutary medicine, it would nevertheless dry out your *humidum radicale* and kill you due to its exuberant heat, if used in large quantities. Although the double ☿ strengthens the *humidum radicale*, it also makes the old trees fertile, throws out gray hair and renovates the whole of nature. At this time it did not have its fire within itself, by which it becomes so hot and is burned to redness; but when all four *elementa* rest in the earth and are transformed into red earth, it is nothing but pure fire. If you throw it into a well or pond, all the water will boil, as if quicklime was slaked in the water. In purifying and rectifying the elements, you must above all take care that they are cleansed of their useless *phlegmate*, in which you can dissolve and purify its salt. If this *phlegma* remained [in the work], it would spoil the whole work, or at least prevent it or delay it. You are barely even allowed to rectify the water once, lest it coagulate, because it has no spirits and the air is purely separated from it. Even if dry spirits rise as white salt or sublimate, it must be put into the milk-colored drops; the white therein gives gluten *aquilæ*, or mineral gluten; when it has been combined with its white virgin salt, with the pure queen *Alma*, and dissolved therein, there is the metallic gur or thick buttermilk, which

oder dicke Buttermilch, lässt sich schmieren wie Butter und ist die Grund-Feuchtigkeit, so im Feuer bleibt, heißt unseren ☿ amalgamiert, dieses gibt die rechte Schminke oder *Oleum Talci*, und ist der rechte Perlen-Grund, woraus die teuersten Perlen können gemacht werden. Man nimmt nur Perlen, stampft die an einen güldenen Draht ins Gefäß, wenn dieser Leib getränkt worden, so werden diese doppelten Geister in die Höhe steigen und auf der Perle niedersitzen, auch sich darauf coagulieren, solche kann man so groß machen als man will, wird sich alles anlegen als Zwiebelschalen, wie die natürlich-gewachsenen Perlen und ist kein Unterschied, nur dass diese größer, weil sie in den Perl-Muscheln so groß nicht wachsen können. Du musst auch die *Auctores* vorher alle wohl verstehen, bis du sie in allen übereinstimmig findest. Wenn du diesen Stein nicht erst im Kopfe machen kannst, wirst du ihn mit den Händen wohl ungemacht lassen. Wenn sie so unterschiedlich schreiben von so vielen *Menstruis*, ist es wahr, in seinem ersten Wesen seiner Reinigung erscheint es in einem wässrigen Leibe, ist unser himmlisches Gnaden-Wasser, die Gabe GOTTES, unsere giftige Schlange, Solvier-Essig, des Goldes sein eigenes Wasser. In seinem anderen Wesen oder Reinigung ist es ölig, leimig und viskosisch. Wenn der rote *Laton*, unser ☉, darinnen aufgelöst worden, schon zeitiger als vorhin, unser einfacher ☿ in seinem 3ten Wesen, wenn er seines eigenen fixen *Corpus* oder Salz aufgelöst und mit ihm eins geworden und anhebt zu leuchten und zu funkeln, Sonne und Mond zusammen getreten, ist es das *Sal Metallorum*, unser doppelter ☿ und trockenes Wasser. Wenn sie von so vielen ☿ reden, ist es auch wahr. Erst siehst du ihn als ein Wasser aus einem Brunnen geholt, welches unsere *Materia Remota*; in seinem anderen, wenn *purum ab impuro* geschieden, siehst du den fetten und schlackenreichen, unsere *Materiam Propinquam*; In seinem 3ten ist es unser erstandener Sieges-König, *Materia Proxima* des Steins und eben so viele Arten ☉ und ☽ hast du auch in dieser *Materia*, in jeder Reinigung zu finden, unsere rote Erde, welches unser ⊕ Salz ist, welches unser ☉ in sich hält, in der anderen Reinigung bleibt zurück das spermatische Salz, *Hyle*. Zum 3. hast du das *Sal Metallorum*,

can be smeared like butter and is the basic moisture that remains in the fire; this is called our amalgamated ☿, which gives the right cosmetic or *oleum talci*, and is the proper pearl base from which the most expensive pearls can be made. You simply take pearls, push them on a golden wire into the vessel, and when the body has been soaked, these double spirits will rise up and settle on the pearl and coagulate on it. You can make them as large as you like; they all build up like the layers of an onion, as with naturally grown pearls, and there is no difference, only that these are bigger because they cannot grow that big in the oyster shells. You must also properly understand all of the older *auctores* until you find them to be in agreement in everything. If you cannot make this stone first in your head, you will not be able to make it with your hands. If they write so differently about so many *menstruis*, it is true: in the first essence of its purification it appears in a watery body; it is our heavenly water of grace, the gift of GOD, our poisonous serpent, solvent vinegar, gold's own water. In its second essence or purification it is oily, glutinous, and viscous. When the red loam, our ☉ has been dissolved in it, earlier than before, our simple ☿ in its third essence, when it has dissolved its own fixed *corpus* or salt and has become one with it and begins to shine and sparkle, sun and moon come together, it is the *sal metallorum*, our double ☿ and dry water. If they talk about so many ☿, it is also true. First you see it as water drawn from a well, which is our *materia remota*. In its second [essence], when *purum* is separated *ab impuro*, you see the fat and slag-rich one, our *materiam propinquam*. In its third [essence] it is our risen king of victory, *materia proxima* of the stone, and you also have just as many types of ☍ and ♄ in this *materia*, to find our red earth in every purification, which is our ⊕ salt, which holds our ☉ in itself, in the other purification the spermatic salt remains back, *hyle*. Thirdly, you have the *sal metallorum, sal alembrot* of the wise. From this you have to make two

Sal Alembrot der Weisen. Von diesen musst du 2 Teile machen, einen zum Fundament des Steins und den anderen zur Vermehrung. Mache nun die 2 *Compositiones* zum Steine mit unserm ☿ vom ☿, lass das güldene Büchlein in seine *Fontinam* fallen, wenn du ein Blättlein kostest, ist es süßer als Honig und sehr fett anzugreifen, wie geblätterter Talk, riecht überaus schön. Dieses Büchlein siehst du jetzt zum letzten Mal. Die *Fontina*, worin dieser königliche Leib ertrunken, wird schwarz, ☉ und ☽ im Drachen-Haupt verfinstert, welches die andere und letzte Schwärze ist, bis das Gnaden-Zeichen, der Regenbogen kommt und danach der weiße Mondschein hervorbricht, wenn die weißen Dünste in die Höhe steigen und sich setzen, sind sie alsobald fix, weil sie diese metallischen Leiber in ihrer höchsten Reinigkeit bei sich haben und glutinosisch sind, haben sich durch die lange Zeit von ihren wässrigen Teilen oder Feuchtigkeiten los gemacht; Dies ist eben die lange Zeit, da man Geduld haben muss, solche unctuosische Kalche zu machen, einen weißen, einen gelben und einen roten, der hernachmals als ein flüssiges Salz oder Glas im Feuer bleibt, auch einen Ingress in die Metalle hat, ihr *Centrum* durchdringen und solche verbessern kann, welches kein metallisches Wesen und kein Salz tun kann. Siehst du also, dass diese Narren sind, die aus Maien-Tau, Essig, ⊕, ○, ⊖, ◐, ✶, ♀ und dergleichen wollen den Stein machen. Gesetzt, dass diese Geister leicht können durch ihre *Salia* coaguliert und zum flüssigen Salz gebracht werden, ist denn dieses unser metallisches Salz, welches sich *radicaliter* mit den Metallen vereinigt? Zerstören kann es solche wohl; Und überdies alles haben diese keine tingierenden Geister bei sich, weder unseren roten, noch weißen ♄, auch unser unverbrennliches Öl nicht, womit sollten sie sich denn färben, den Schnee gelb und hernach blutrot machen? Davon 1 Gran einen Becher Wein als Blut färben kann und süß wird, dies lassen sie wohl bleiben, es muss es nur unser *Sal Metallorum* tun, wenn es sich von seinem *Veneno Tingente* satt getrunken. Ich warne dich, dass du zu deinem Werk keine unrechte *Materiam* nimmst, weder Pulver noch Wasser, alles kommt aus dem einzigen und ist das herrliche Land, so alles Guten voll. Unsere *Materia* ist

parts, one for the foundation of the stone and the other for multiplication. Now make the two compositions for the stone with our ☿ from ☿, let the little golden book fall into its *fontinam*, if you taste a small leaf, it is sweeter than honey and very greasy to touch, like flaked talc, and smells extremely nice. This is the last time you see this little book. The *fontina*, in which this royal body is drowned, becomes black, ☉ and ☽ in the darkened dragon's head, which is the second and last blackness, until the rainbow, the sign of grace, comes and then the white moonlight breaks out; and when the white fumes rise up and settle down, they are immediately fixed because they have these metallic bodies with them in their highest purity and are glutinous, and have gotten rid of their moisture or watery parts during this long time. This simply takes a long time and one must have patience to make such unctuous limes—one white, one yellow, and one red—which remains hereafter in the fire as a liquid salt or glass, and also has an ingress into the metals, can penetrate their *centrum* and improve them, which only a metallic being and no salt can do. So you see that they are fools who want to make the stone out of May dew, vinegar, ⊕, ○, ⊖, ⌽, ✶, ♃ and the like. Assuming that these spirits can easily be coagulated by their *salia* and brought to the liquid salt, is this our metallic salt, which combines *radicaliter* with the metals? It can destroy them indeed; and what is more, they have no tinging spirits with them, neither our red nor white ♄ not even our incombustible oil. Therefore what should they color themselves with to turn the snow yellow and then blood-red? Of this, one grain can color a cup of wine as blood and it becomes sweet; they do not do this, our *sal metallorum* only has to do it, when it has drunk itself full with its *veneno tingente*. I warn you not to use any incorrect *materiam* for your work, neither powder nor water, everything comes from the only one and is the glorious land, so is full of everything good. Our *materia* is not ☉ but dissoles the ☉, not silver, but our heavenly

kein ☉, sondern solviert das ☉, kein Silber, sondern unsere himmlische *Lunaria* in *Sphæra* und Kreis *Luna*, der weiße ♃ auf weiß, der rote ♃ auf rot, so aus unserem 🜨 kommen, der den *Spiritum Generantem* bei sich hat, unseren roten Mann und weißes Weib, unseren schneeweißen unverbrennlichen ♃ und unseren blutroten fixen ♃, so alle in unserem *Electro* stecken. Es ist nicht der gemeine ♄ oder dessen *Minera*, kein Bleiglanz, weder ♃, ♂ noch ♀, auch nicht ☿, wenn sie diesen nennen ist es *figurate* gemeint, *ænigmaticè* ist es wahr und kannst was in Gleichnis lernen von diesem roten *Oleo*, so den ☿ zum fixen *Præcipitat* macht, du wirst aber das Fischlein *Æscheneis* mit silbernen Schuppen hier nicht finden, das unverbrennliche Öl auch nicht, das rosenfarbene oder gesegnete Blut auch nicht, sondern einen irdischen verbrennlichen unreinen ♃, welcher höchst unrein. Wer aber einen andern reinigen will, muss selber rein sein, welches in dem ganzen mineralischen Reiche nicht zu finden, als in unserer *Magnesia*. Reinige unser *Subjectum* aufs höchste, zerstöre solches, nimm seinen Geist davon, den Leib, die rote Asche, reinige, tu seinen reinen Geist dazu, bringe daraus den roten und weißen Geist, den weißen sublimier 7-mal mit seiner reinen weißen Erde, mache ihn zum schönen Sublimat, in welchem *Uniones* hervorleuchten, diesen verbrenne mit dem größten Feuer, dass das fliegende Erdreich in das rote Meer fallen und stürzen muss, hernach trockne durch die Luft das Wasser aus, dass gelbe Erde wird die Erde, so die Sintflut erlitten, muss auch durchs Feuer gerichtet werden, so wird der neue Himmel und neue Erde da sein, das alte vergangen und alles neu geworden, welches kein Element mehr zerstören kann. Du musst auch den ♌ aus den Klauen erkennen lernen, der dir Wasser zutragen muss, dem beide Hände gerecht sind, einer kommt von Mitternacht, vom kalten Sternenbär, wird auch der grüne ♌ genannt, in dessen Bauchs innersten Eingeweiden die roten Gestirne ihren Sitz haben, welche du deinem schönen *Apollini* bringen musst. Nun weißt du wohl, dass die grüne Farbe der ♀ zugetan, der wütende Apollo die 7 Adler bedeutet, den Sternenadler, welche mit dem ♌ ihren Kampf halten müssen, der ☿ die ♀ beschlafen, dann regnet es ☉ aus den Wolken,

Lunaria in *sphæra*, and the circle *Luna*, the white ♀ on white, the red ♀ on red, so come from our ⊕, which has the *spiritum generantem* with itself, our red man, and white woman, our snow-white incombustible ♀ and our blood-red fixed ♀, which all contained in our *electro*. It is not the common ♄ or its *minera*, no lead lustre, neither ♃, ♂ nor ♀ also not ☿, when they name this [☿], it is meant *figurate*, *ænigmaticè* it is true and you can learn something in parable from this red *oleo*, like that which made ☿ to a fixed precipitate, but you will not find the little fish *echeneis* with its silver scales here, nor the incombustible oil, nor the rose-colored or blessed blood, but an earthly, burnable, impure ♀ which is highly impure. But whatever wants to purify something else must itself be pure, which is not to be found in the whole mineral kingdom but only in our *magnesia*. Purify our *subjectum* to the highest, destroy it, take away its spirit, the body, the red ashes, purify, add its pure spirit, bring the red and white spirit out of it, sublimate the white [spirit] seven times with its pure white earth, make it into a beautiful sublimate, in which *uniones* shine out, this one burn with the greatest fire so that the flying earth must fall and tumble into the red sea. Afterwards dry out the water with the air so that the yellow earth becomes the earth which has suffered the great deluge, and which must also be judged by the fire, so there will be a new heaven and a new earth, so that the past is gone and everything becomes new, something which no element can destroy anymore. You also have to learn to recognize the ♌ by its claws, which has to bring you water, to whom both hands are the same; one comes from midnight, from the cold, starry bear,[30] and which is also called the green ♌, in whose belly's innermost bowels the red stars have their seat, which you must bring to your beautiful *Apollini*. Now you know well, that the green color is fond of the ♀, the angry Apollo means the seven eagles, the starry eagle, which have to fight their battle with the ♌, ☿ slept with ♀, because it rains ☉ from the clouds, from the

aus den Dünsten, so von dem Stein aufsteigen. Sät euer ☉ in die geblätterte Erde, gibt den Smaragd von seltsamer Wirkung, den glasförmigen *Azoth* des Lulli und das gläserne Meer, das mit Feuer gemengt werden kann, der steinerne Palast, darin der König gehen und seine Wohnung beständig haben kann. Unser *Sal Metallorum* ist unser *Lapis Philosophorum*, *Aurum Perspectibulum* und der Stein der ersten Ordnung, unser *Marcipan* und Himmels-Brot, wiewohl die geblätterte Erde vorher einen süßen Geschmack, nun aber den *Lapidem Herbalem* bedeutet, welcher Geschmack als Feuer oder geschmolzenes Fett, sich aber hernach in eine große Süßigkeit endigt. Der andere ♌ kommt vom Orient, von Aufgang der Sonne, ist der rote ♌, das Feuer des Steins, brennender *Rubicundus*, das Feuer der Sonne. Wenn du aber unseren Adler nicht erst mit dem ♌ aus dem Norden streiten lässt, wird die Sonne keinen Ingress in den Mond haben, sondern obenauf schwimmen, denn diese fette Erde erst wohl zubereitet werden muss, das wohlgedüngte Land sollst du erst tapfer wenden, dann wird beim West-Wind sich die Fäulung bald vollenden. Nun siehst du, dass unser *Electrum* alle Farben in sich hat und von GOTT mit einem grauen geringen Kittelchen bedeckt, damit die großen und stolzen in der Welt seine Macht und Kraft nicht sehen sollen, auch kein *Subjectum* in der Welt unseren roten und weißen ☿ in sich hat, als nur dieses, weil es ganz rein und mit keinem verbrennlichen ☿ adulteriert. Ja ☉ und ☽ sind selbst damit befleckt und unrein gegen dieses zu achten und ist dieser Stein in elender Gestalt und hat doch die ganze Natur in sich, hat auswendig ein armseliges graues Kittelchen an, das alle Versucher sein äußerliches Ansehen erschreckt, auch, wenn sie sich als Verwegene darüber gemacht und gleich Anfangs seine innerliche Schönheiten und die vielen *Menstrua*, tingierende nasse und trockene Geister nicht zu sehen bekommen, lassen sie es als verzweifelt liegen, da sie erst den rechten Anfang machen sollten. Wer sollte denken, dass dieser Stein nach der Kuh geworfen, mehr wert wäre, als die Kuh? Wer sollte denken, dass in diesem Stein rot und weiß, ja alle Farben steckten? Wer sollte denken, dass in diesem roten und weißen das *Aureum Vellus* ver-

mist, which rise from the stones. Sow your ☉ in the leafed earth. This gives an emerald of strange effect, the glass-shaped *azoth*[31] of Lulli,[32] and the sea of glass mingled with fire,[33] the stone palace in which the king can walk and have its dwelling permanently. Our *sal metallorum* is our *lapis philosophorum, aurum perspectibulum* and the stone of the first order, our *marcipan* and heavenly bread, although the leafed earth previously had a sweet taste, but now it means the *lapidem herbalem*, which tastes like fire or melted fat, but afterwards finishes like a great confectionary. The other ♌ comes from the orient, from the rising of the sun, is the red ♌, the fire of the stone, burning *rubicundus*, the fire of the sun. But if you do not let our eagle first fight with the ♌ from the north, the sun will not have ingress into the moon, but will float on top. Because this fat earth must first be prepared, you should first bravely turn the well-fertilized land, then the decay will soon be completed in the westerly wind. Now you see that our *electrum* has all the colors in itself and is covered by GOD with a little gray tunic, so that the great and proud in the world should not see its power and strength, also that no *subjectum* in the world but this alone has our red and white ♃ in itself, because it is completely pure and not adulterated with any combustible ♃. Indeed, even ☉ and ☽ are tainted and impure compared to it; and although this stone is in a miserable shape, it has the whole of nature within itself. On its outside, however, it has a squalid gray coat which terrifies all tempters with its outward appearance, even if they began boldly, unless they saw its inner beauties right from the start: the many *menstrua*, tinging wet and dry spirits. They let it lie in despair when they should simply make the proper beginning. Who would think that this stone, thrown at the cow, would be worth more than the cow?[34] Who would think that this stone contains red and white, indeed, all colors? Who would think that the *aureum vellus* is hidden in this red

borgen? Ja Diamanten, Smaragden und Rubine, die teure schneeweiße Perle? Ja gar der Karfunkelstein steckte und verborgen wäre? Wie ich dich lehre, hat kein Mensch getan, es ist kein Wort vergeblich oder umsonst geschrieben und GOTT will, dass seine Wunder, die er der Natur, seinen Geschöpfen verliehen, sollen offenbar werden, damit wir mit unseren dunklen Augen einen Blick ins Paradies tun, solches sehen, wie es vor dem Fluch gewesen, nun aber nach dem Fluch seine erste und reine Gestalt verloren, welchen Fluch und Tod wir wieder davon scheiden müssen, dass sein reiner und paradiesischer Leib offenbar wird, unseren großen Schöpfer, seinen schönen Thron und Herrlichkeit nur in etwas als durch einen Spiegel erblicken. Die Propheten, so solches gesehen, haben in der Schrift auch etwas davon geschrieben, aber des Teufels Neid wird solches den Menschen nicht gönnen und wird sich ein goldgeiziger boshafter Mensch finden, der aus Neid alles alleine haben will, dies Buch wegkaufen, verhehlen und verstecken, wie schon anderen guten Autoren mehr ergangen oder solches verändern, was deutlich geschrieben, heraus lassen, andere Verwirrungen darein schmieren und das ganze Werk verderben. Da GOTT der Sünden wegen den Erdboden verfluchte und der arme gefallene Mensch des himmlischen Bildes abstarb, verschaffte GOTT gleich Mittel sein Geschöpf zu erhalten, verhieß den Seelen des Menschen den Erlöser, des Weibes Samen und gab den Menschen den Verstand die Medizin zu bereiten, dadurch die Alten viele 100 Jahre gelebt und bei den Erz-Vätern sehr in Schwang gegangen und zu großem Reichtum gekommen. Es ist das Feuer, dadurch Moses das güldene Kalb verbrannt, welches er den Kindern Israel zu trinken gegeben, dadurch hat Moses in der Wüsten die Stifts-Hütte, Salomo den Tempel erbaut, auch solche Schätze in die königlichen Gräber beigelegt, welche wohl bis auf den Jüngsten Tag liegen werden. Da sie die Chaldäer holen wollten und nur die 2 ersten vorderen Kammern eröffnet, ist Feuer heraus gefahren, welches die Chaldäer verbrannt, da doch solcher Kammern noch 7 uneröffnet geblieben. Du musst auch wissen, dass du Edelgesteine damit machen kannst, auch das Glas, dass man

and this white? Indeed, diamonds, emeralds, and rubies, the costly snow-white pearl? Indeed, that the carbuncle stone would be contained hidden inside? As I teach you, nobody has done before, not a word is written in vain or to no avail. And GOD wants his miracles—which he bestowed on nature, on his creatures—to be revealed so that we can look into paradise with our dark eyes to see how such [a stone] was before the malediction, but after the malediction has now lost its first, pure form. It is this curse and death that we must once again separate from it so that its pure and paradisiacal body is revealed; to glimpse [thereby], somewhat as through a mirror, the beautiful throne and glory of our great creator. The prophets, who saw such things, also wrote something about it in the scriptures. But the devil's envy will not grant people such a thing and a spiteful person will be found, greedy for gold, who out of envy wants everything to themselves. This person will buy up [all copies of] this book to disguise and conceal it, as has happened to other good authors; or alter something that was clearly written, leave it out, or smear other confusions into the book and spoil the whole work. Since GOD cursed the earth because of sin and the poor fallen human made after his heavenly image had died, GOD immediately provided means to preserve his creature, promising the redeemer to the souls of humans, of woman, seeds, and gave human beings the ability to prepare the medicine which allowed the ancients to live for many hundreds of years and which became very popular among the arch-fathers, who came to great wealth. It is the fire by which Moses burned the golden calf, and which he gave the children of Israel to drink, and through which he built the tabernacle in the desert. Likewise, Solomon used it to build the temple, thereafter placing such treasures in the royal tombs, where they will likely remain until Judgment Day. When the Chaldeans wanted to fetch them and only opened the two first front chambers, fire sprung forth and burned the Chaldeans, so that seven such chambers remained unopened. You

es biegen kann und noch weit mehr Wunder damit ausrichten, als ich hier beschrieben. Dies ist das allergeringste. Wenn ich lebe, so erwarte noch ein Buch von mir, darinnen will ich dir Sachen schreiben, so weder du noch ich gedacht hätten, absonderlich von seinen Tugenden in der Medizin. Wir machen 3 unctuosische Kalke, einen weißen, einen gelben und einen roten. Den weißen bekommst du nach der schwarzen und finsteren Nacht zu sehen, wenn es beginnt Tag, das ist Licht zu werden, im vollen Mondschein und sich der Schimmer in den Glimmer begibt, von seinem Samen-Bande abreißt und anhebt zu funkeln und zu leuchten, auch alle 7 Gestalten der Natur in ihrem größten Hunger zum weißen Kalk reverberiert, begehren gesättigt zu werden, zur weißen Tinctur. Den gelben bekommst du zu sehen bei der schönen Morgenröte, wenn der Himmel erst gelb wird, bis die Sonne gegen Mittag ganz rot und blutig scheint, so hast du den roten. Auf einmal kannst du den roten und gelben Kalk nicht machen, Ursache, der Mond ist zu voll, die weiße Lilie hat die Oberhand und ist zu viel des Saffrans, aber noch zu wenig dabei, so geht der weiße Mondschein vor. Gleichnis: Wenn ein wenig Safran in Milch getan wird, sieht man ihn nicht, bei mehrerer Einwerfung desselben aber wird die Milch erst gelb, endlich sehr gelb und zuletzt rot; Also mit unserem Gewürz. Ist nun genug geredet von dem Gewächse der 3 Principien. Als 1. von der äußeren Welt. 2. von der inneren und Lichtwelt. 3. Von der engelhaften und himmlischen Welt. Mir beliebt der lange Weg weit mehr als der kurze, weil ich in solchen die Wunder der Natur zu sehen bekomme und mir die ganze Natur darinnen entdeckt ist, darinnen die Schöpfung zu sehen, der Fall Adams, seine Wiederbringung durch unseren Erlöser, unser Tod, Sterben und Auferstehung unsers verklärten Leibes, auch die Wiedervereinigung unser Seelen mit dem verklärten Leibe, welche vor der Faulung nicht ins Kadaver zu bringen und wie solcher verklärte Leib und Seele erst GOTT anschauen kann, da es der unreine sündige Körper nicht tun konnte, bis er seinen paradiesischen Leib wieder bekam, den er vor dem Fall hatte. Im kurzen Wege sieht man dieses nicht so wohl. Bitte GOTT um Erleuchtung, ich habe mehr

must also know that you can use it to make precious stones, also the glass that can be bent, and you can do even more miracles with it than I have described here. This is the very least. If I live, expect another book from me, in which I will write things for you that neither you nor I would have thought, especially about its medical virtues. We make three unctuous limes, one white, one yellow, and a red one. You get to see the white one after the black and dark night when day begins, which is when it becomes light, in the full moonlight, and the shimmer turns into a glimmer, departs from its seed band and begins to sparkle and shine. Also all seven shapes of nature in their greatest hunger, reverberated to white lime, desire to be saturated in the white tincture. You get to see the yellow one with the beautiful morning dawn, when the sky first turns yellow, until the sun shines completely red and bloody around noon, then you have the red one. You cannot make the red and yellow lime all at once, because the moon is too full, the white lily has the upper hand and there is too much of the saffron, but still too little, so the white moonlight proceeds first. Parable: if a little bit of saffron is given into milk, one does not see it, but if it is thrown in several times, the milk first turns yellow, eventually very yellow, and finally red; so too with our seasoning. Enough has now been said about the growth of the three principles. First, about the outer world; second about the inner and luminous world; third about the angelic and heavenly world. I like the long way far more than the short one, because in it I see the wonders of nature and all of nature is included in it. In this the creation can be seen, the fall of Adam, his return through our Savior, our death, dying, and the resurrection of our transfigured body, as well as the reunification of our souls with the transfigured body, which cannot be brought into the carcass before decay, and how such a transfigured body and soul can now look at GOD, which the impure sinful body could not do until it regained its heavenly body, which it had before the fall. In the short way you do not see this so well. Ask GOD for enlightenment, I have written more than I can answer to the world,

geschrieben, als ich bei der Welt verantworten kann, doch dünkt mir, es ist dem Willen GOTTES nicht zuwider, denn will ich, so viel mir möglich, gern erfüllen und dich in den letzten Zeiten unterrichten. Findet dies Werk einen bösen verruchten Menschen, so macht es daraus einen Frommen; findet es aber einen Frommen, der wird gar ein Heiliger; Auch alle Augenblicke, wenn ihm diese Erleuchtung einfällt, wird er nicht unterlassen GOTT auf den Knien zu danken, weil ein solcher Mensch stets lebt, als wenn er im Paradies wäre, die Welt, ihre Hoheit und Pracht für nichts achtet, sich stets nach dem Himmel sehnt und auf seine eigene Verklärung denkt, das traurige Schreckensbild des zeitlichen Todes für nichts achtet, sondern nur als die Tür und Eingang nach dem himmlischen und ewigen Leben betrachtet, da wir durch das finstere Todes-Tal ins himmlische Licht in den ewigen Freuden-Saal gelangen, da keine *Turba* mehr ist, weder Leid noch Geschrei, sondern ewige Ruhe und Stille ohne ENDE.

but it seems to me that it is not against the will of GOD, which I want to fulfill as much as I can and teach you in the last times. If this work finds a wicked person, it makes them pious; but if it finds someone pious, it makes them a saint. Every time this enlightenment happens to them, they will not fail to thank GOD on their knees, because such a person always lives as if they were in paradise, holding no regard for the world and its majesty and splendor, but instead always longing after that heaven and thinking about their own transfiguration, paying no attention to the sorrowful, terrifying image of temporal death, but considering it merely as the door and entrance to heavenly and eternal life. For we pass through the valley of the shadow of death into heavenly light, into the eternal hall of joy, where there is no more *turba*, neither suffering nor wailing, but eternal peace and stillness without END.

APPENDIX
Über das Büchlein

Ob es wohl nicht nötig noch etwas zu setzen, so hat mich doch zweierlei Ursache dazu bewogen:

1. Das nicht allen Menschen gegeben, Gleichnisse zu verstehen, wiewohl ich nichts im Gleichnis, sondern offenbar geschrieben, dass es jedermann verstehen kann; Weil aber dieser himmlische Stein von GOTT gegeben wird, so ist es fatal und sollen ihn nicht alle Menschen haben. Wer sich nun dieses hohen und großen Werks untersteht und fehlt, der gerät wieder in das alte Labyrinth, in die *Particularia*, einen Weg darinnen viel Tonnen, ja Millionen Goldes verschwendet worden und unter 100 kaum einer ein kleines Stücklein Brot damit gewonnen, weil ein Blinder den anderen geführt und viele Fuder solcher verlogener Prozesse und Bücher geschrieben von Gelehrten und Ungelehrten, hat also wieder ein anderer aus diesem Lügenkram neue Sachen hervorgesucht, was er vermeint das beste zu sein, solches zusammengesetzt und wieder eine neue Scharteke drucken lassen, da dann die Leute häufig zugefallen, es vor lauter *Oracula* gehalten und tapfer in den Beutel gegriffen, weil es alles ☉ und ☽ sein müssen, da sie doch ☉ und ☽ in geringen Dingen haben, welches nicht teuer und besser ist, als das gemeine ☉ und ☽, dadurch mancher an den Bettelstab geraten.

2. Weilen nun gesehen, dass der Geldmangel immer größer, sowohl bei Hohen und den Gelehrten, als bei den Handwerksleuten und Tagelöhnern; Also habe aus Mitleiden und Erbarmen gegen meinen Nächsten diese Zugabe anhängen wollen, als höchst-nötige Aphorismos, sowohl *universaliter* als *particulariter*, darinnen ich deutlich, treulich und redlich geschrieben, nicht aus anderen Büchern, weil solche in keinem Buche zu finden, sondern weil ich alle selbst mit eigenen Händen gearbeitet. Das ich aber alle und jede Handgriffe bei allen und jeden Arbeiten mit Weitläufigkeit beschreiben könne, ist unmöglich, es muss es ein jedweder selber angreifen. Wer aber nicht darinnen gleich fortkommen kann, der muss es lernen, ich habe es auch nicht flugs

APPENDIX
About the Little Book

Although it is not necessary to add anything else, two reasons have moved me to do so:

1. That it is not given to all people to understand parables, although I have written nothing in parables but rather clearly, so that everyone can understand it. But because this heavenly stone is given by GOD, it is fatal, and not all people should have it. Anyone who tries and fails at this great and exalted work ends up again in the old labyrinth, in the *particularia*, a path in which tons, even millions of [measures of] gold have been wasted and hardly a single person out of a hundred has gained a crumb of bread with it because here we have the blind leading the blind. Many such deceitful processes and books were written by both the learned and the unlearned. One author would search for novelties in this fabric of lies, adopting what he thought best and throwing all this together to let a new tattered volume to be printed. When this happened people would often fall for it, thinking such stuff nothing short of oracular, and confidently reach for their wallets because it must all be ☉ and ☽. Yet they have ☉ and ☽ in small and lowly things, which are not expensive and better than the common ☉ and ☽, whose expenditure might make beggars out of some of these people.

2. Because I have now seen that lack of money is growing ever greater, both among the high and learned, as among the craftsmen and daily wage earners. Out of compassion and mercy towards my neighbor I wanted to add this appendix, being highly necessary aphorisms, *universal* as well as *particular*, in which I have written clearly, faithfully, and honestly. I have not copied from other books because this cannot be found in any book, since I have worked everything with my own hands. It is, however, impossible for me to describe every detail in each and every work, and so everyone must experiment for themselves. But if you cannot get on with it straight away, you have to learn it. I could not do it quickly either and that is why nobody should berate me for

gekonnt und darf mich deshalb niemand lästern, dass ich ihn von den Irrwegen auf den rechten Weg geführt. Wer auch besseren Unterricht von mir begehrt, wenn es ihm in der Arbeit nicht gleich geraten will, der lasse es mich durch Briefe wissen, will ihn gerne unterrichten und mache nur nicht andere Sudeleien. Ich bin auch 20 Jahre durch diese Wildnis gewandert, Zeit, Mühe und Geld vergeblich angewendet und habe viel Nächte nicht geschlafen, bis ich durch die vielen Irrtümer mit Schaden klüger geworden. Fleißig gebetet, fleißig studiert, fleißig gearbeitet, bringt schon den Nutzen zu rechter Zeit. Ich weiß, dass ihrer sehr viele durch Lesen und Arbeiten meiner Schriften glückselig werden und, wenn ich lange verfault, mich erst rühmen und loben werden. Welcher sich nun hieraus bessert, dem gönne ich es von Herzen gern und bitte ihn, solches zu GOTTES Ehre und dem mitleidenden Nächsten zum besten anzuwenden. Es kann es ein jeder im kleinen versuchen, ehe er was im großen anfängt und erst an solchem kleinen den *Modum Tractandi* lernen, so behält er sein Geld im Beutel, mit 8 oder 10 Talern sind sie alle im kleinen probiert. Ich wünsche dem Liebhaber zu solchen Dingen von GOTT Glück, Heil und Segen. Wenn er die Arbeit wohl trifft und findet, was ich gefunden, so wird er vor Freuden verwundert stehen, dass ich solche geheimen Dinge in die Welt hinein geschrieben; Aber es ist die Zeit da, da solche Geheimnisse sollen offenbar werden, die armen Menschen brauchen es auch in der Welt. Lieber Leser, lebe wohl, ich halte mich dir weiter durch Schriften zu dienen verpflichtet.

D.I.W.

leading them from the wrong paths to the right path. Anyone who wants better lessons from me, if they do not soon find their own answers while doing the work, should let me know by sending me a letter, for I am happy to instruct them so that they do not continue to make a mess of it all. I have also wandered this wilderness for twenty years, using time, effort, and money in vain; and I did not sleep many nights until I became wiser through many costly mistakes. Praying diligently, studying diligently, and working diligently will bring benefit at the right time. I know that very many of them will be blessed by reading and working through my writings, and, when I have long since passed away, I shall only then receive their praises and obtain a great reputation. I will gladly grant [this knowledge] to whomever my writings may benefit, and I ask them to use this book for the honor of GOD and to benefit his compassionate neighbor as best he may. Everyone can try it on a small scale, before he starts something on a large scale, first learning the *modum tractandi* only on such small experiments, so that they keep their money in their purse. All experiments may be performed on a small scale for the price of eight or ten thalers. I wish the lovers of such things happiness, salvation, and blessings from GOD. If one does well at work and finds what I have found, one will be amazed with joy that I have written such secret things to the world. But the time has come when such secrets are to be revealed, the poor people also need it in the world. Dear reader, farewell, I continue to be obliged to serve you through my scriptures.

D.I.W.

NOTWENDIGE CHYMISCHE LEHRSÄTZE

*Oder Grundregeln vom Universal und
Particular und zwar erstlich
vom Universal*

1. Das *Universal* oder das große natürliche Werk, der Stein der Weisen, geht aus einem einzigen Dinge, darinnen ein metallisches ☉, ♁, ☿ ist, wenn diese 3 der Natur gemäß aufgeschlossen, gereinigt und wieder zusammengesetzt, wird es das große Werk genannt. Wer aber diese 3 nicht wieder zusammensetzen kann, der sucht nur diese *Menstrua* zu ☉ und ☽ zu gebrauchen, welches *Menstruum* des ☉ *Prima Materia*, auch das *Menstruum Universale* genannt wird, welches fett, ölig und leimig, das einzige unter der Sonne ist, so die Metalle *radicaliter* auflöst, welches auch nicht wieder davon zu scheiden ist und sie darinnen als in ihrer Mutter Leibe neu geboren und zu einem tingierenden Stein werden. Aber es ist solches nicht das große Werk, sondern nur eine Neben-Tinctur, die in der Projection nicht so viel tingiert, als das natürliche Werk, welches ohne Zusatz des ☉ und ☽ gemacht wird. Es dient aber zu dem großen Werk keines von den 7 Metallen, weder ☉, ☽, ♂, ♀, ♄, ♃, ☿, weil alle diese schon mit ihrem ☿ und ♁ vereinigt und gleichsam beigelegen haben, auch von der Natur ihre Grund-Feuchtigkeit schon vertrocknet und reif ist; Ja der ☿ selbst ist nicht mehr frei, hat sich mit seinem eigenen ♁ in einen Leib zur Trockenheit begeben und können nicht wieder geschieden werden, auch nicht der ☿ aus den Metallen, weil alle diese nicht der fette und schlackenreiche ☿ sind, welche das ☉ *radicaliter* solvieren und sich nimmer wieder davon scheidet, die erstere aber solches veredelt und zur Tinctur macht und solches so wohl in *Quantitate* als *Qualitate* vermehrt.

2. Dienen ☉ und ☽ nicht dazu, weil sie beide keine Tinctur haben und selber arm sind, nur so viel ♁, dass sie sich färben, kann man daher keine Tinctur von ihnen nehmen, andere zu färben.

NECESSARY CHYMICAL PROPOSITIONS

*Or basic rules of the universal and the
particular, and first of all from
the universal*

1. The *universal*, or that great natural work, the philosophers' stone, comes from a single thing in which there is a metallic ☉, ♄, and ☿, if these three are decomposed, purified, and reunited according to nature, it is called the great work. But whoever cannot reunite these three is only looking to use these *menstrua* for ☉ and ☽. This *menstruum* of ☉ is *prima materia*, also called the *menstruum universale*, which is fat, oily, and gluey, and is the only one under the sun which dissolves the metals *radicaliter*, and which also cannot be separated from it again and they are reborn in it as in their mother's womb and become a tingeing stone. But this is not the great work, but only a secondary tincture, which in the projection does not tinge as much as the natural work, which is done without the addition of ☉ and ☽. However, none of the seven metals is used for the great work, neither ☉, ☽, ♂, ♀, ♄, ♃, or ☿, because all of these have already combined with their ☿ and ♄ and, as it were, married; and their basic moisture is already dried up and ripe by nature. Yes, the ☿ itself is no longer free, has gone into a dry body with its own ♄ and cannot be separated again, not even the ☿ of metals, because all these are not the fat and slag-rich ☿ which dissolves ☉ *radicaliter* and never parts from it again, but the former refines such things and makes them tincture, and so certainly increases such in *quantitate* as in *qualitate*.

2. ☉ and ☽ do not serve this purpose, because neither have tincture and they are both poor themselves, having only enough ♄ to color themselves; therefore one cannot take any tincture from them to color others.

3. Kommen ☉ und ☽ nur als ein Ferment dazu, wenn die Tinctur fertig ist; 1 Teil der Tinctur, 4 Teile ☉ oder ☽, so macht die Tinctur solches zu Glas, wodurch sie einen Halt bekommt, damit man sie auf den ☿ tragen kann.

4. Die Tinctur wird deswegen auf den ☿ getragen, dass dieselbe einen Steig oder Leiter hat, auf welcher sie in die anderen Metalle eingeht, wozu der ☿ unter den 7 Metallen das geschickteste ist, weil er die Metalle öffnet, sie durchdringt und die Tinctur darein führt, auch solche den Metallen mitteilt und ohne den ☿ ist es unmöglich, die Transmutation zu verrichten, weil die Tinctur zu subtil, auch in kleiner Quantität sich in den Metallen nicht so ausbreiten kann, weil sie nicht so eröffnet, als der ☿, welcher die Tinctur in einem Augenblick in sich nimmt, davon erstarrt und gleichsam gefriert.

5. Dienen Blei und Zinn auch nicht dazu, weil kein tingierender 🜍 bei ihnen ist und diese beiden auch schon mit ihrem 🜍 vereinigt, haben auch einen bösen ☿. Der gemeine ☿ taugt auch nicht, ob er sich schon mit ☉ und ☽ amalgamiert, so vereinigt er sich doch nicht radicaliter mit ihm und ist wieder davon zu scheiden, lässt ☉ und ☽ unverändert zurück. Ergo, so ist er nicht das wahre *Menstruum* des ☉, kann solches nicht verbessern, taugt selber nichts.

6. ♂ und ♀ haben wohl viel tingierenden 🜍, aber zu wenig ☿, böse, grob, hart, flüssig und der 🜍 hat zu viel verbrennlichen 🜍 bei sich, der viel irdische grobe Unreinigkeit bei sich hat.

7. Haben ☉ und ☽ wohl reinen ☿, sie nehmen aber ihren eigenen 🜍 nicht an, sind auch nicht der fette ☿ und sind zum *Universal* nichts nütze, sowohl als der gemeine ☿.

8. Auch nicht ihre *Vitriola*.

9. Auch nicht der *Vitriol* von ♀ oder ♂.

10. Weder zyprischer noch ungarischer *Vitriol*.

11. Auch nicht *Minera Martis Solaris*.

12. Auch nicht der Blut-Stein.

13. Keine Blei-Minera oder Mennige, sie habe Namen wie sie wolle.

14. Kein rotgülden Erz.

3. Only add ☉ and ☽ as a ferment when the tincture is ready. One part of the tincture, four parts of ☉ or ☽, this is how the tincture turns it into glass, which gives it a hold so that it can be carried onto ☿.

4. The reason why the tincture is carried on the ☿ is that it has a step or ladder on which it enters the other metals, for which ☿ is the most skillful of the seven metals, because it opens the metals, penetrates them, and leads the tincture into them; it also communicates such to the metals, and without ☿ it is impossible to carry out the transmutation because the tincture is too subtle, even in small quantities, cannot spread inside the metals, because it does not open them like ☿ [does], which instantly takes the tincture into itself, and then solidifies and freezes, so to speak.

5. Lead and tin do not serve this purpose either, because there is no tingeing ♃ with them and these two have already combined with their ♃, they also have a bad ☿. The common ☿ is also not good. Although it can be amalgamated with ☉ and ☽, it does not unite *radicaliter* with them and can be separated from them again, leaving ☉ and ☽ unchanged. Ergo, it is not the true *menstruum* of ☉, cannot improve it, and is itself unsuitable.

6. ♂ and ♀ have a lot of tingeing ♃ but not enough ☿, bad, coarse, hard, liquid, and the ♃ has too much combustible ♃ within itself, which has a lot of gross, earthly impurities within itself.

7. ☉ and ☽ probably have pure ☿, but they do not accept their own ♃, [they] are also not the fat ☿ and [they] are of no use for the *universal*, just like the common ☿.

8. Nor does their *vitriola*.

9. Nor does the *vitriol* of ♀ or ♂.

10. Neither Cypriot nor Hungarian *vitriol*.

11. Nor does *minera martis solaris*.

12. Nor does the blood stone.

13. Not lead *minera* or red lead, they may have names as they please.

14. Not red-gold ore.

15. Kein Zinnober, auch nicht der *Cinnabaris nativa*.
16. Kein Grünspan, *Atrament* oder *Bolus*.
17. Kein ♃, auch nicht das rote Rauschgelb.
18. Kein Auripigment.
19. Kein Kobalt.
20. Kein Salz, weder ✶, ○, ⊖, ⊙, ⊕, Meersalz, Weinstein oder das Weinstein-Salz, *in summa*, es habe Namen wie es wolle, denn die *Salia* sind nicht metallische *Menstrua*, welche die Metalle *radicaliter* solvieren, sind auch wieder von den Metallen zu scheiden. Das wahre *Menstruum* muss metallisch sein und ist nicht wieder davon zu scheiden, jedoch ist ein Salz, davon die Metalle in der Erde wachsen, welches unter das Element *Aquæ* prædestiniert, mit dem selbigen kann die rechte Materia in sein Bad der Reinigung geführt werden und ist der Natur Essig, welches der Materien ♃ und ☿ nicht verbrennt, sondern lebendig macht, darinnen die *Materia*, der Natur gemäß, entgrobt und von den groben *Fecibus* gereinigt wird, welches kein ⛢ noch ⛢ tut, sondern rein und unrein zugleich wegfrisst. Diese wilden Geister lassen sich auch mit den Metallen nicht coagulieren, sind auch nicht die Wasser der Gesundheit, sondern des Menschen Gift und Tod, der Natur Essig aber ist schon eine Arznei für sich selbst.
21. Auch nicht das schwarze ♁, wenn es auch gleich ungarisch wäre und noch so langstreifig, weil es viel von unreinem verbrennlichen ♃ hat, auch schon sich mit seinem ☿ vereinigt und beide höchst unrein und räuberisch sind.
22. Das *Antimonium* ist zweierlei, das schwarze ist das Männchen, das weiße ist das Weibchen, rein von verbrennlichen stinkenden Schwefel, hat auch den reinsten und schönsten ☿.
23. Unter allen metallischen und mineralischen Körpern ist nur ein einziges, dass zum *Universal* dient.
24. Solches muss den Metallen nahe verwandt sein.
25. Es muss einen metallischen ☿, ♃ und ⊖ haben.
26. Und eine sehr hohe Tinctur zum weißen und zum roten.

15. Not vermilion, not even *cinnabaris nativa*.
16. Not verdigris, *atrament*, or *bolus*.
17. Not ♃, not even the red orpiment.
18. Not auripigment.
19. Not cobalt.
20. Neither salt, nor ✶, ☉, ☽, ☉, ♁, sea salt, tartar, or tartar salt; in sum, its names do not matter, because the *salia* are non-metallic *menstrua* which dissolve the metals *radicaliter*; [they can] also be separated from the metals again. The true *menstruum* must be metallic and cannot be separated from them again, but is a salt, from which the metals grow in the earth, which is predestined under the element *aquæ*, with which the right *materia* can be led into its bath of purification and is that nature's vinegar, which does not burn the material's ♃ and ☿, but brings them to life, in which the *materia*, in accordance with nature, is de-coarsened and purified from the coarse *fecibus*, which no ♅ or ♆ does, which instead eat away pure and impure alike. These wild spirits cannot be coagulated with metals either, nor are they the waters of health, but rather of human poison and death, but nature's vinegar is a medicine in itself.

21. Not the black ♂, even if it were Hungarian and no matter if long streaky, because it has a lot of impure combustible ♃, also combines with its ☿, and both are extremely impure and predatory.

22. *Antimonium* is two different things, the black is the male, the white is the female, pure from combustible, stinking sulphur, and it also has the purest and most beautiful ☿.

23. Among all metallic and mineral bodies there is only one that serves for the *universal*.

24. Such thing must be closely related to metals.

25. It must have a metallic ☿, ♃, and ☽.

26. And a very high tincture for the white and the red.

27. Einen schneeweißen 🜍, der erstlich flüchtig doch unverbrennlich ist.

28. Und einen blutroten, fixen 🜍, welcher wie brennende Kohlen leuchtet.

29. Welche auch beide wohlriechend.

30. Und alle beide erstlich in wässriger Gestalt erscheinen, bis sie endlich mit ihrem eigenen fixen Leibe vermischt in ihre viskosische ponderosische Feuchtigkeit gehen, ölig, fett, schmierig und glutinosisch werden, welches die Grund-Feuchtigkeit, so im Feuer bleibt und aller Metallen Mutter ist, solche neu gebärt.

31. Welchen nichts fremdes zugesetzt wird, weder im Anfang, Mittel noch Ende, bis die Tinctur fertig ist.

32. Weil es alles selbst bei sich hat, was es bedarf, sich vollkommen zu machen.

33. Weil es der Gold-Baum und seine Wurzel ist, so die silbernen und güldenen Früchte trägt, welche der *Saturnus* befeuchten und waschen muss.

34. Der ♄, ♃, ♂, ☉, ♀, ☿, ☽ sind alle in diesem *Subjecto*.

35. Welches das *Electrum Minerale Immaturum* genannt wird.

36. Auch *Magnesia*.

37. Auch der ♄ und Zeuge-Vater aller Götter, weil die güldene *Genealogia* der Götter von ihm herstammt.

38. Es ist der ♄, so einen Stein vor seinen Sohn, dem ♃, gefressen und wieder ausgespieen, wodurch die Arbeit angezeigt wird.

39. Es ist auch der Adler-Stein, der noch einen anderen in sich trägt, nämlich den Stein der Weisen.

40. Dieser ♄ frisst alle seine Kinder und gebiert sie aufs Neue in einer besseren Gestalt als zuvor.

41. Es ist auch in diesem Werk oder Schöpfung dieser neuen Welt der erste Adam, welcher mit seinem Weibe, seiner Even, schwanger geht.

42. Ist *Venenum Tingens*.

27. A snow-white ♆ that is initially volatile but incombustible.

28. And a blood-red, fixed ♆, which shines like burning coals.

29. Which are also both fragrant.

30. And both of them initially appear in a watery form until, mixed with their own fixed bodies, they finally go into their viscous, ponderosic moisture, become oily, fatty, greasy, and glutinous, which is the basic moisture which remains in the fire and is the mother of all metals and gives birth to new ones.

31. To which nothing foreign is added, either in the beginning, middle, or end, until the tincture is finished.

32. Because it has everything within itself which is necessary to make itself perfect.

33. Because it is the gold tree and its root, which bear silver and golden fruits, which *Saturnus* must moisten and wash.

34. The ♄, ♃, ♂, ☉, ♀, ☿, ☽ are all in this *subjecto*.

35. Which is called the *electrum minerale immaturum*.

36. Also *magnesia*.

37. Also the ♄, the begetting father of all gods, because the golden *genealogia* of the gods comes from it.

38. It is the ♄ which has eaten a stone for its son, ♃, and spat it out again, which indicates the work.

39. It is also the eagle stone that has another within it, namely the philosophers' stone.

40. This ♄ eats all its children and gives birth again to them in a better shape than before.

41. There is also in this work, or the creation of this new world, the first Adam, who is pregnant with his wife, his Eve.

42. Is *venenum tingens*.

43. Weder ☉, ☽, ♀ noch ♂ hat Gift oder ist giftig, auch der ☿ nicht. In großer Verstopfung kann er pfundweise getrunken werden; Wie können nun diese giftigen flüchtigen Schlange, Drache oder Basiliske sein, derer Kopf den Schwanz frisst, sich selbst tötet, auch sich selbst lebendig macht?

44. Der ☿, er sei in Zinnober oder Erz, er sei güldisch oder silberreich, er sei noch so oft sublimiert oder animiert, ist er doch nicht der ☿ *Philosophorum*, ob er schon mit güldischem ♃ kann animiert werden, dass er die ☽ gelb färbt, so taugt er doch zum Werk nichts. Warum?

45. Wenn er soll mit den Metallen coaguliert werden, so solviert er 1. sie nicht, dass er sich radicaliter mit ihnen vereinigt. Zum 2. wird er bei ihnen zu einem bösen hart-flüssigen Præcipitat, er werde *per se*, oder durch ♃, Gradier-Wasser, *Spiritus Vitrioli*, ♅ oder dergleichen dazu gebracht, so bleibt er doch ☿ wie vorher, kommt er aber ins große Feuer oder wird durch Eisen oder *Sal Tartari* reviviciert, kommt er in seinem ersten Kleide wieder gewandelt und ist der ☿ wie zuvor.

46. Hat der ☿ seinen eigenen ♃ nicht bei sich, dass er sich selber in ☉ kochen kann, wie der ☿ *Philosophorum*.

47. In dem gemeinen ☿ ist nicht was die Weisen suchen, er ist nicht der Hermaphrodit, er hat auch nicht die 4 *Elementa* bei sich, auch nicht die 3 *Principia*, ist auch nicht die *Lunaria*, wie der philosophische, er hat weder Salz noch Schmalz zur Tinctur, sondern ist selber ein armer Teufel und Götterbote, der nichts zu nütze als die Tinctur, wenn solche fertig, in die Metalle zu tragen.

48. Das gemeine ☉ ist auch nicht das philosophische ☉, denn dieses solviert das gemeine ☉ und färbt es sehr hoch rot, das gemeine Gold hat nur so viel ♃, als es selbst bedarf, dass es ☉ ist, auch ist das gemeine ☉ tot, das philosophische aber lebendig und viel edler.

49. Das gemeine ☉ kann nicht wieder in die *Elementa* geschieden werden, weil es einmal zum Ende und Reife des ☉ gekommen und seine *Spiritus* in der Erde und in dem Feuer verloren, auch kann sein ☿ nicht die geblätterte Erde sein, welche unctuosisch und glutinosisch ist, er mag zubereitet werden wie er will.

43. Neither ☉, ☽, ♀ nor ♂ has poison or is poisonous, not even the ☿. In cases of severe constipation it can be drunk by the pound, so how can this be the poisonous fugitive serpent, dragon, or basilisk, whose head eats its tail, kills itself, and also brings itself to life?

44. The ☿, it is in vermilion or in ore, it is gold-bearing or rich in silver, no matter how often it is sublimated or animated, it is not the ☿ *philosophorum*, whether it can already be animated with gold-bearing ♃, so that it colors ☽ in yellow,³⁵ so it is no good for the work. Why?

45. If it is to be coagulated with metals, it (1) does not dissolve them, because it unites itself radicaliter with them. (2) it becomes an evil hard-liquid precipitate with it, it is brought to it *per se*, or by ♃, gradating water, *spiritus vitrioli*, ▽ or the like, so it remains ☿ as before, but if it comes into the great fire or if revived by iron or *sal tartari*, it comes again in its first dress and is the ☿ as before.

46.³⁶ The ☿ does not have its own ♃ with itself that it can cook itself into ☉, like the ☿ *philosophorum*.

47. In common ☿ is not what the wise are looking for, it is not the hermaphrodite, it does not have the four *elementa* with itself, nor the three *principia*, nor is it the *Lunaria*, like the philosophical one, it has neither salt nor lard for the tincture, but is itself a poor devil and messenger of gods, which is of no use but to carry the tincture into the metals.

48. The common ☉ is also not the philosophical ☉, because this dissolves the common ☉ and colors it very intensely red, the common gold only has as much ♃ as it needs for itself, the common ☉ is also dead, but the philosophical [☉] lively and much more noble.

49. The common ☉ cannot be separated again into the *elementa*, once it has come to the end and maturity of the ☉ and its spirits are lost in the earth and in the fire, and its ☿ cannot be the leafed earth, which is unctuous and is glutinous; it may be prepared as it will.

50. Ingleichen die ☽ ist nicht das große Werk zum weißen, weil sie jetzt gedachte Mängel alle so wohl als das ☉ hat und daher untüchtig zum *Universal*.

51. ♂ und ♀ sind ganz davon ausgemerzt, wegen ihrer harten, strengen, groben und unreinen Leiber, auch sind sie schon Metalle und reif, welche gleichfalls nicht wieder zurückzubringen und der güssige Wachs, flüssige Stein aus ihnen nicht zu machen.

52. Auch nicht ihre *Vitriola* oder selbst gewachsener *Vitriol*, weil solche nichts anders in sich haben, als ♀ und ♂ Art, kann doch der schleimige ☿ und die *Terra Foliata* nicht aus ihnen gebracht werden, weder die weiße noch die rote Tinctur, auch hat ihr Salz nichts metallisches in sich, es werde noch so oft calciniert und solviert. Wenn zuletzt alle Röte oder Erde davon geschieden, ist es nichts mehr als ein klein wenig Alaun-Salz.

53. Das Salz aus dem philosophischen *Vitriol*, wenn das weiße Weib und der rote Mann davon geschieden, bleibt es schneeweiß zurück, glänzend und fett, viel weißer als weißer Talk und ist ein rechtes metallisches Salz, welches sich durch kein Wasser solvieren lässt, der kleine grüne ♌, weil es inwendig grün, auch *Hyle*, der erste Leib mit dem sich der *Universal*-Geist bekleidet, sein eigenes Wasser, das teure jungfräuliche Salz, dass nie die Sonne beschienen und von klugen Arbeitern seiner Kleider beraubt, dass es nackig da stehen muss, wird auch *Sal Martis* genannt, auch ☉ Salz, des ☿ fixer Teil, *Gabricus* oder Mann, welcher sein Weib, seine Schwester coaguliert und wird aus ihnen beiden der doppelte ☿ und durch 7 *Sublimationes* in die *Terram Foliatam* verwandelt, welches der weiße ♄ ist und ✶ der Weisen, davon des Königs Wasserbad gemacht wird, wenn er in seinem Öl solviert worden, so ist es das *Aqua Benedicta, Aqua Permanens*.

54. Dieser einfache ☿ ist schon sehr unterschieden von dem gemeinen, geschweige der doppelte, weil er die Metalle viel anders solviert und nicht wieder davon zu scheiden ist, als wenn man Wasser unter Wasser gießt, so bleiben sie beisammen, er nimmt sie mit in die Höhe oder wird fix und eins mit ihnen, danach einer seine Arbeit anstellt,

50. At the same time, the ☽ is not the great work for the white, because it now has all imagined defects as well as the ☉ and is therefore unfit for the *universal*.

51. ♂ and ♀ are completely eradicated from it, because of their hard, strict, coarse, and impure bodies; they are also already metals and ripe, which likewise cannot be brought back again, and the pourable wax, or liquid stone, cannot be made out of them.

52. Not their *vitriola* or self-grown *vitriol*, because these have nothing else in them than the ♀ and ♂ kind, and cannot bring the slimy ☿ and *terra foliate* out of them; neither the white nor red tincture and their salt have anything metallic in them, no matter how often it is calcined and solvated. When all the redness or earth has finally separated from it, it is nothing more than a tiny bit of alum salt.

53. The salt from the philosophical *vitriol*, when the white woman and the red man are separated from it, it remains snow-white, shiny, and fat, much whiter than white talc, and is quite a metallic salt, which cannot be dissolved by any water, the little green ♌, because it is green inside, also *hyle*, the first body with which the *universal* spirit clothes itself, its own water, the expensive virgin salt, on which the sun never shone and is robbed of its clothes by clever workers, so that it has to stand there naked, is also called *sal martis*, also ☉ salt, the ☿ fixed part, *Gabricus* or man who coagulates his wife, his sister, and is transformed from both of them into the double ☿ and through seven *sublimationes* into the *terram foliatam*, which is the white ♃ and ✶ of the wise, from which the king's water bath is made when he has been dissolved in its oil, so [that] it is the *aqua benedicta*, *aqua permanens*.

54. This simple ☿ is very different from the common one, let alone the double [☿], because it dissolves the metals very differently, and cannot be separated from them again, as when one pours water under water, so they remain together. It takes them up with itself or becomes fixed and one with them. If someone sets up their work accordingly,

ist wunderbar, von der Natur eingewickelt, kommt den Artisten erstlich zu Gesicht als ein Brunnen-Wasser, nimmt hernach von seinem eigenen Körper so viel zu sich, dass er dick, schwer, leimig und fett wird, zur Jungfer-Milch und endlich, wenn er als das rechte Wasser, der Speichel des Mondes, mit seiner Erde, dem teuren weißen Salz, eingetränkt, das ist, 7-mal getötet und wieder lebendig geworden, welches 7 *Sublimationes* bedeutet, ist es der ☿ *Exuberatus*, welcher königliche Kronen und Zepter trägt, der doppelte ☿, die *Lunaria, Terra Foliata* und weißer ♃, riecht überaus schön und ist zum weißen und hungrigen Kalk gebrannt, das Silber, so durchs Feuer 7-mal geläutert und nun einen Stein gebiert, hier ist es in seine 7 *Systemata* gekommen und die 7 Grade der Natur vorbei, kann nun höher nicht. Nun muss er von dem Künstler Hilfe haben, er ist in seinem größten Hunger, wenn er nicht bekommt, was ihm gehört, so wird er sich selber fressen und aus dem *Folio* ein *Spolium* werden. Es ist das Fischlein *Æscheneis* mit silbernen Schuppen, das bisher im philosophischen Meer geschwommen ist, sieht wie geblätterter Talk aus, aber wunderschön glänzend, in welchem lauter *Uniones* hervorleuchten.

55. Der *Sulphur* aus dem philosophischen *Vitriol*, ist der rote Mann, der rote ♃ auch recht, das philosophische ☉, so das gemeine ☉ solviert und nie corporalisches ☉ gewesen, auch viel besser als das ☉ ist, so der ♄ in seinem Kerker verschlossen gehalten, der rote ♌, rosenfarbenes gesegnetes Blut, das Gold GOTTES, Hitze der Sonne, brennender *Rubicundus*, ☿ *Solaris*. Seine Gestalt ist brennend, rot als eine glühende Kohle. Wenn es in einem Glase ist und die Sonne davor scheint, so wird man denken, das Gemach ist lauter feurige Spiegel.

56. Der *Spiritus Vini* ist nichts nütze zum philosophischen Werk, sondern ihr *Spiritus Vini* ist der *Spiritus* ♄, welcher ihr *Spiritus* ☿, Luft und Öl genannt wird, womit sein eigener angeborener ♃ extrahiert wird, weil es heißt: Es solviert sich selber, coaguliert sich selber, schwärzt, weißt und rötet sich selber, tötet sich selber und macht sich auch selber lebendig, dem nichts fremdes zugesetzt wird und dieses

it is marvelous, enveloped by nature, appears first to the artist as fountain water, then takes in so much of its own body that it becomes thick, heavy, glutinous, and fat, becomes the virgin milk and finally, when it is the real water, the saliva of the moon, soaked in its earth, the costly white salt, that is, killed seven times and come to life again, which means seven *sublimationes*, it is the ☿ *exuberatus*, which wears royal crowns and scepter, the double ☿, the *Lunaria*, *terra foliata* and white ♃, smells extremely beautiful and is burned to white and hungry lime, the silver, so refined through fire seven times and now gives birth to a stone, here it came into its seven *systemata* and the seven degrees of nature are over, and now it cannot get higher. Now it must get help from the artist. It is in its greatest hunger, and if it does not get what it wants, it will devour itself and turn the *folio* into a *spolium*. It is the little fish *echeneis* with silver scales that has swum in the philosophical sea so far, looks like flaked talc, but beautifully shiny, in which many *uniones* shine out.

55. The sulphur from the philosophical *vitriol* is the red man, the red ♃ is also right, the philosophical ☉ which dissolves the common ☉, and has never been a corporal ☉, and is also much better than the ☉, so the ♄ kept locked in its dungeon, the red ♌, rose-colored blessed blood, the gold of GOD, heat of the sun, burning *rubicundus*, ☿ *solaris*. Its body is burning, red like a glowing coal. If it is in a glass and the sun shines on it, one will think that the room is full of fiery mirrors.

56. The *spiritus vini* is of no use for the philosophical work, but their *spiritus vini* is the *spiritus* ♄, which is called their *spiritus* ☿, air and oil, with which its own innate ♃ is extracted, because it is said: it dissolves itself, coagulates itself, blackens, whitens, and reddens itself, kills itself and also makes itself alive, to which nothing foreign is added and all

alles kommt aus dem philosophischen ♄, der ist der Schnitzer dieses Wunder-Gebäudes.

57. Nur muss der Alte in sein Bad der Reinigung geführt werden, weil er sehr unrein. Er hat wohl die Sonne in seinem Leibe verborgen, auch den Mond, sind aber mit dicken Nebeln bedeckt, dass man sie nicht sehen, ingleichen unsern ♂ und ♀, hat außen eine andere Farbe, ist mit einem weißen ☿ angeflogen, dass man seine inwendigen Schönheiten und die vielen Farben nicht sehen kann, deswegen heißt es auch *Electrum Minerale Immaturum*.

58. Hierzu braucht man kein Scheidewasser, kein *Aquam Regis* und *Spiritum Salis*, weil diese keiner metallischen Natur, die Metalle auch nicht davon wachsen, sich auch diese mit den Metallen nicht vereinigen, sondern wieder davon scheiden, auch solche die Metalle nicht verbessern können.

59. An dem gemeinen ♄ kann man die Arbeit des philosophischen ♄ lernen, wenn er extrahiert. Man darf zum gemeinen ♄ nur destillierten Wein-Essig nehmen, zum philosophischen ist er zu schwach, dass er sollte seinen ☿, 🜍, ⊖ extrahieren.

60. Wenn der ♄ extrahiert, und der Essig evaporiert bis in Honig-Gestalt, so destilliert man es, zuletzt gibt man starkes Feuer, dass die Retorte glüht, die *Spiritus* rectificiert man, so bekommt man einen *Spiritum*, welcher als Branntwein brennt und ein *Menstruum* ist, die metallischen 🜍 zu extrahieren, man bekommt auch ein weißliches *Phlegma*, welches von ☿ des ♄ participiert, zuletzt kommt ein rotes, gelbes Öl, dass den ☿ figiert und zum fixen *Præcipitat* macht, in dem *Capite Mortuo*, von rechten ♄ bleibt zurück der rote Ton, *Terra Adamica*.

61. Den roten Ton koche etliche Stunden in Regenwasser oft gerührt, solch Wasser lässt man abrauchen, so bleibt der rote *Vitriol*, diesen solviert man in dem *Spiritu* ☿, so von dem ♄ gemacht worden, destilliert es, alsdann steigt es alles herüber als Regenbogen-Farben; Ist das Salz noch nicht weiß, so muss man seine Tinctur vollends extrahieren und überführen, bis es alles übergegangen und die Erde ohne Tinctur zurück bleibt. Diese wird wieder in *Spiritu* ☿ aufgelöst, alsdann solchen

this comes from the philosophical ♄, which is the carver of this miraculous building.

57. Only the old man has to be led into his cleaning bath, because he is very impure. He has hidden the sun in his body, also the moon, but they are covered with thick fog so that you cannot see them, and like our ♂ and ♀, it has a different color on the outside, being covered with a white ☿ so that you cannot see its internal beauties and diverse colors. This is why it is called *electrum minerale immaturum*.

58. For this one does not need a separating water,[37] no *aquam regis* and *spiritum salis*, because these do not have a metallic nature, and the metals also do not grow from them. They do not unite with the metals either, but separate from them again, and they also cannot improve the metals.

59. The common ♄ can be used to learn the work of the philosophical ♄, if it is extracted. You can only use distilled wine vinegar with the common ♄, it is too weak for the philosophical one to extract its ☿, ♃, and ⊖.

60. When the ♄ has been extracted and the vinegar has evaporated until it is honey-shaped, it has to be distilled, finally a strong fire is put on so that the retort glows, the *spiritus* is to be rectified so one gets a *spiritum*, which burns like brandy, [which] is a *menstruum* to extract the metallic ♃; one also gets a whitish *phlegma*, which participates from the ☿ of ♄; at last comes a red, yellow oil, which fixes the ☿, and makes it a fixed precipitate, in the *capite mortuo*, from the right ♄, what remains is the red clay, *terra Adamica*.

61. Boil the red clay for several hours in rain water, often stirred, such water is left to evaporate, so the red *vitriol* remains back, this one dissolve in the *spiritu* ☿, which has been made from ♄, distill it, then everything rises over in rainbow colors. If the salt is not yet white, its tincture has to be extracted completely and transferred until it has all passed over and the earth remains without tincture. This is dissolved

davon geschieden, bleibt die Erde schneeweiß zurück.

62. Das *Confusum Chaos* muss in die *Elementa* geschieden werden, in Erde, Luft, Wasser und Feuer.

63. Die erste *Composition* wird gemacht zum doppelten ☿, Wasser und Erde. Wenn das Wasser alles zu Erde geworden, wird es sublimiert, der Sublimat geteilt, 1 Teil zum Stein, solches fix zu machen, den anderen Teil zur Vermehrung. Die 2te *Composition* ist, wenn der Stein gemacht wird, da dieser ☿ mit dem *oleo* ♄ figiert wird, ins gläserne Meer verwandelt, der steinerne Palast zubereitet; Der andere Teil wird in der Luft als seinem Öl resolviert, welches das *Menstruum* ist, den weißen Stein in Quantität als Qualität zu vermehren, mit dem Feuer aber das noch allein übrig, macht man ihn rot.

64. Diese Dinge sind kurz und einfältig beschrieben, brauchen aber großen Verstand, lange Arbeit und geschickte Hände, auch ist notwendig zu wissen, dass das giftige *Oleum* ♄ keinen Nutzen hat, als nur seinen eigenen ☿ zu figieren und in ein grünliches Glas zu verwandeln, weil dieser ♄ mit der ♀ gespielt und solch *Oleum* von ♄ Art ist, welches die Grüne zeigt.

65. Große Unkosten braucht man nicht, mit 3 oder 4 Talern kann das ganze Werk gearbeitet werden, ohne die Gefäße, welche in der Vor-Arbeit seiner Reinigung und Scheidung gebraucht werden, in der Nach-Arbeit aber, in der *Composition*, braucht man nur ein Gefäß. Viele solvierende *Menstrua* braucht man nicht, weil man keine Metalle zerstören darf, sondern diese *Materia* hat ihr eigen Wasser und Geist bei sich, dadurch es sich selbst vollkommen macht und ist das herrliche Land so allen gutes voll. Alle diese vielen *Menstrua*, nasse und trockene tingierende Geister kommen von diesem einzigen, dem nichts fremdes zugesetzt wird, es ist ein Ding, *Re & Numero*.

66. Warum aber aus diesem einzigen Dinge so viele nasse und trockene Geister in der Arbeit hervor kommen, ist die Antwort, dass sie alle vorher darinnen gesteckt und immer eins durch das andere hervorgebracht wird und nie miteinander in ihren Wurzeln vereinigt gewesen, lassen sich auch alle wieder voneinander scheiden und besteht

again in *spiritu* ☿, when such is separated from it, the earth remains snow-white.

62. The *confusum chaos* must be separated into the *elementa*, earth, air, water, and fire.

63. The first composition is made to the double ☿, water and earth. When the water has turned completely to earth, it is to be sublimated, the sublimate divided, one part for the stone, to make it fix, the other part for the multiplication. The second composition is when the stone is made, as ☿ is fixed with the *oleo* ♄, transformed into the sea of glass, the stone palace is prepared. The other part is resolved in the air as its oil, which is the *menstruum*, to increase the white stone in quantity as in quality; but with the fire that alone is left, one makes it red.

64. These things are described briefly and simply, but need great understanding, long work and skillful hands, it is also necessary to know that the poisonous *oleum* ♄ has no use than just fixing its own ☿ and turning it into a greenish glass, because this ♄ has played with ♀ and is such *oleum* of ♄ property is, which shows the greenness.

65. You do not need a great deal of money, and with three or four thalers the whole work can be done without the vessels that are used in the preliminary work of its purification and separation; but in the after-work, in the composition, one only needs a single vessel. One does not need numerous dissolving *menstrua*, because one is not allowed to destroy metals, but this *materia* has its own water and spirit with it, through which it makes itself perfect and is the wonderful land full of all good things. All these many *menstrua*, wet and dry tinging spirits come from this only one [*materia*], to which nothing foreign is added, it is one thing, *re & numero*.

66. But as to why so many wet and dry spirits come out of this single thing in the work, the answer is that they were all contained within it beforehand; one is always brought forth through the other, and they have never been united in their roots; they can also be separated from each other again and the art consists in uniting those [things] that

eben die Kunst darinnen, solche zu vereinigen, welches eben die größte Mühe braucht, werden alle durch ihren rohen unreifen *Spiritum* ☿ extrahiert, wer sie aber der Natur gemäß kann voneinander scheiden und die widerwärtigen *Elementa*, *Agens* und *Patiens*, vereinigen, der hat das *Centrum* in der Scheibe getroffen.

67. Dass aber aus diesem einzigen Dinge dieses hohe und große Werk zu machen, ist nicht zu wundern, weil GOTT es vor allen anderen in dem ganzen mineralischen Reiche mit lauter Vortrefflichkeiten begabt und von außen mit einem grauen Kittlein bedeckt, dass die Unwürdigen seine innerlichen Schönheiten nicht sehen sollen. Sein ♃ übertrifft alle anderen, ja des Goldes selbst wie die helle Sonne die kleinen Sterne, über welche er alle seinen Glanz weit hervor streckt, weil er ganz rein, von allen gemeinen irdischen verbrennlichen ♃, so leuchtet und funkelt er als der rechte Karfunkel-Stein, das *Urim* und *Thumim*, welches aus diesen glänzenden Erz gemacht worden, *Blachmal* genannt. Zum andern, wegen seines allerreinsten Jungfer ☿, der sich nie mit der groben und unreinen Erden vermengt, wenn er aber mit seiner reinen metallischen Erde in einem Leibe gewachsen, ist es das leuchtende ☿al Wasser, höchst-glänzend, sehr unterschieden von dem gemeinen, der sich selbst in ☉ kocht, weil er von dem *Primo Ente* der himmlischen Sonne oder des Goldes in sich hat. Wenn er nun durch die 7 Grade der Natur, durch seine eigene Exaltation, den königlichen Thron erstiegen und zum güldenen Büchlein geworden, so ist er so zart wie ein Gestiebe, durchsichtig als der zarteste Flor, weil es ein geistlicher Leib ist, dieser nimmt sein eigenes färbendes Gift gern an, wegen seines Hungers und Trockenheit, welches ihm auch im Augenblick durchgeht als Öl das Löschpapier und ihn gradiert, hierinnen geschehen die natürlichen *Transmutationes*, in diesen subtilen und offenen ☿. Wenn er nun gelb genug, so wird ihm sein eigenes Feuer zugesetzt, dann wird er brennend rot, der blutige Stein.

68. Die Scheidung *purum ab impuro* ist dreimal in diesem Werk. I. Da die *Materia Cruda*, unser Alter in sein Bad der Reinigung geführt wird, da unsere Rosenfarbe zum ersten Mal gesehen wird, des Alten

simply need the greatest effort. All are extracted by their raw immature *spiritum* ☿, but whoever can separate them according to nature and unite the unwilling *elementa*—*agens* and *patiens*—will hit the center of the target.

67. But it is not surprising that we can make this high and great work from this single thing because GOD endowed it, above all others in the whole mineral kingdom, with sheer excellence, and covered it from the outside with a gray tunic so that the unworthy should not see its inner beauties. Its ♀ surpasses all others, indeed even of gold itself, just as the bright sun outshines the tiny stars over which all its splendour stretches. This is because it is completely pure from all common, earthly, combustible ♃; it shines and sparkles like the fair carbuncle stone, the *Urim* and *Thumim*, which were made of this brilliant ore called *blachmal*.[38] On the other hand, because of its very purest virgin ☿, which never mixes with the coarse and impure earth, but when it is grown in one body with its pure metallic earth, it is the luminous ☿al water, extremely shiny, very different from the common one, which cooks itself into ☉ because it has the *primo ente* of the heavenly sun or gold in itself. If it has now ascended the royal throne through the seven degrees of nature, through its own exaltation, and has become a golden book; it is then as delicate as a dust, transparent as the most delicate gauze, because it is a spiritual body. This one willingly accepts its own coloring poison because of its hunger and dryness, which also passes through it instantly like oil through blotting paper, bringing it to a higher degree. This is where the natural *transmutationes* happen, in these subtle and open ☿. When it is yellow enough, its own fire is added to it, then it becomes burning red, the bloody stone.

68. The separation *purum ab impuro* is threefold in this work. First, where the *materia cruda*, our old man, is led into its bath of purification, where our rose color is seen for the first time: the old man's

sein Blut. Die andere Scheidung, da aus dem trockenen roten ♌ das rosenfarbene Blut extrahiert wird, schon reiner als vorhin. Die 3te Scheidung, wenn die *Elementa* geschieden werden, so ist es das kleine himmlische Feuer, die himmlische Sonne und der Mond, hell und klar, auch durchsichtig, nicht die gemeine ☉ oder ☽, denn aus den beiden ☉ oder ☽, so wir in Händen corporalisch haben, wird der Stein nicht gemacht, sind auch diese nicht das *Agens* und *Patiens*, weder der Mann noch das Weib, wie sich die Unwissenden träumen lassen.

VON PARTICULARIEN

69. Die *Particularia* anbelangt, ist keines erklecklich, sie gehen denn durch diesen Brunnen des *Universals*, dass sie durch ihr *Universal Menstruum* aufgeschlossen, aufs neue wieder in ihrer Mutter Leib gehen und darinnen wieder geboren werden, dazu denn ☉ und ☽ die geschicktesten. Es sind aber nur kleine Zweiglein, welche bei weitem nicht so hoch tingieren als das große Werk.

70. Wer nun nicht so glücklich oder zu dumm von Verstande, das große Werk zu finden, hat aber die *Elementa* zu scheiden gelernt, der kann alle güldischen ♃ dem *spiritui* ☿ zufügen, sie darinnen, wenn solche erst vorhero wohl aufgeschlossen, besser als in dem *Spiritu Vini* auflösen, über den Helm treiben und potabel machen, hernach den nassen ☿, wenn er sein Salz solviert, damit præcipitieren und mit seinem eigenen *Oleo* fix machen, welches das herrlichste *Particular*.

71. Die metallischen *Species*, so tingierende ♃ haben, sind roter Talk, Granate, ☉ Kies, güldische Marcasiten, gelber *Vitriol*-Kies, ☍, *Minera ♂ Solaris Hassiaca*, so außen rund und inwendig spießig ist, Blut-Stein, Eisen und Kupfer, auch die gemeinen *Vitriola*, wenn sie aus guten Kiesen gesotten.

72. Diese alle erfordern absonderliche Arbeiten, da immer in einem der ♃ fester verbunden als in anderen. Wer nun die ♃ haben will, muss die ☿ los machen von den ♃ Banden oder ein *Menstruum* wissen, solche

blood. The second separation then occurs, when the rose-colored blood is extracted from the dry red ♌, already purer than before. The third separation is when the *elementa* are separated, and it is the little heavenly fire, the heavenly sun and the moon, bright, clear, and transparent, not the common ☉ or ☽, because the stone is not made from both ☉ or ☽ such as we have them corporally in our hands; and these are also not the *agens* and *patiens*, neither the man nor the woman, as the ignorant dream.

OF PARTICULARS

69. As far as the *particularia* are concerned, none are considerable, for they pass through the fountain of the *universal* so that they are decomposed by their *universal menstruum*, go back into their mother's womb, and are born there again. For this the ☉ and ☽ are the most suitable. But they are only small sprigs, which by far do not tinge as fully as in the great work.

70. Anyone who has not had the fortune to find the great work, or who is too dim of understanding but has learned to separate the *elementa*, can add all the gold-bearing ♃ to the *spiritui* ☿ to dissolve them after they have been broken up, better than in *spiritu vini*, carry them over the helm, make them potable, then precipitate the wet ☿ when it dissolves its salt, and make it fix with its own *oleo*, which is the most wonderful *particular*.

71. The metallic *species* that have tinging ♃ are red talc, garnets, ☉ pyrites, gold-bearing marcasites, yellow *vitriol* pyrites, ♄, *minera* ♂ *solaris Hassiaca*,[39] which is round on the outside and spiked on the inside, bloodstone, iron and copper, and also the common *vitriola* if boiled from good pyrites.

72. All of these require exceptional work, because in one of them the ♃ is always more firmly bonded than in others. If you want to have the ♃, you have to free the ☿ from the ♃ bonds or know a *menstruum* to ex-

davon zu extrahieren, welches der nasse Weg ist. In trockenen Wege aber ist keines besser als der Arsenik, welcher alle metallischen ☿ in sich nimmt, die ihm leicht wieder abzunehmen sind.

73. Der ☿ ist in keinem so fest verbunden als Granaten, welche erst müssen zum kleinsten Pulver gemacht werden, mit ☿ calciniert und denn mit ☌ geschmelzt, so sucht der flüchtige güldische *Sulphur* den anderen fixen und vereinigen sich, alsdann kann man sie durch ♅ extrahieren oder durch den trockenen Weg, wie einen beliebt.

74. Der Blut-Stein ist auch sehr schwer zu solvieren und muss ein starkes ♅ sein, dass was von ihm solviert, wenn er aber vorher mit 3 oder 4 Teilen *Sulphur* abgebrannt, löst er sich besser auf.

75. In den Marcasiten aber und ☉ Kiesen braucht man der Calcination nicht, weil solche mit einem flüchtigen güldischen *Sulphur* angeflogen und viel dadurch verloren geht, muss man sie derhalben nur durch ♅ extrahieren, wenn sie viel ♀ bei sich haben, sieht die erste Solution grün, die anderen aber gelb und rot aus.

76. In dem roten Talk ist er auch sehr fest verbunden, weil ihn kein ♅ wegen seiner Fettigkeit solvieren kann, so muss man mit ihm verfahren, wie mit den Granaten N. 73 oder auf folgende Weise: Man stößt ihn, wenn er heiß gemacht worden, in einem warmen Mörsel, sonst stößt es sich übel, nimmt so schwer als er wiegt Kochsalz und 2-mal so schwer Salpeter, vermischt es, drückt einen Schmelztiegel voll, setzt es ins Feuer zum wenigsten 16 Stunden, dass ihn die *Salia* wohl calcinieren, dann solviert man die *Salia* davon, gießt ♅ darauf und solviert ihn, doch haben die *Salia* ein gut Teil von seinem *Sulphur* in sich gefasst, welches man sehen kann, wenn man das Wasser lässt zur Trockne abrauchen und ist der obige besser mit dem ☌, weil bei diesem nichts verloren geht.

77. Die ♅ zu diesen Dingen werden also gemacht: 4 Pfund des stärksten Scheidewassers, es muss gelb aussehen, wenn es recht stark ist, dies gießt man auf 1 Pfund Kochsalz, zieht es wieder herüber, dass die Retorte zuletzt glüht, die gemeinen ♅, da man nur Salz oder ✶ in ♒ wirft, dienen hierzu nicht.

tract such from it, which is the wet way. In the dry way, however, there is nothing better than arsenic, which takes in itself all the metallic ♃, which can easily be removed from it again.

73. The ♃ is in none [of them] so firmly bonded as in the garnets, which must first be made into the finest powder, calcined with ♃, and then melted with ☉; then the volatile golden sulphur looks for the other [sulphur] and they unite. They can then be extracted by ♋︎ or by the dry way, as one prefers.

74. The bloodstone is also very difficult to dissolve and it has to be a strong ♋︎ that dissolves something of it, but if it has been burned off beforehand with three or four parts of sulphur, it dissolves better.

75. In the marcasites and ☉ pyrites,[40] however, you do not need the calcination, because these are entered with a volatile gold-bearing sulphur and much is lost as a result. Therefore you only have to extract them through ♋︎; and if they have a lot of ♀ with them, the first solution looks green, but the others are yellow and red.

76. It [the ♃] is also very tightly bound in the red talc, because no ♋︎ can dissolve it due to its fatness, so you have to deal with it as with the garnets [of] No. 73, or in the following way: you pound it when it has been made hot, in a warm mortar, otherwise it will be difficult. Take table salt as heavily as it weighs and twice as heavy of saltpeter, mix it, press it in a melting pot, put it on the fire for at least sixteen hours so that the *salia* will calcine it, then dissolve the *salia* from it, pour ♋︎ on it and dissolve it, but the *salia* have captured a good part of its sulphur, which can be seen when the water is left to smoke off and the above is better with ☉, because nothing is lost in this case.

77. The ♋︎ for these things are made: four pounds of the strongest separating water. It must look yellow if it is quite strong. This is poured onto one pound of table salt and distilled over again so that the retort finally glows. The ordinary ♋︎, which is prepared by simply throwing salt or ✶ into ♒︎, does not serve this purpose.

78. ♂ und ♀ sind leicht aufzulösen: Man besprengt nur Eisen-Feil mit Tau oder Regenwasser, auch mit ♅, vermischt es wohl, tut solches oft, so wird es zum Safran, diesen digeriert man in destilliertem Wein-Essig, so nimmt er das subtilste zu sich, den Essig lässt man abrauchen, so hat man den *Sulphur* aus Eisen.

79. Den *Sulphur* von ♀ zu scheiden, ist der leichteste Weg: Man macht daraus ein *Æs Ustum* per se, nimmt 3 Teile ☉, 1 Teil Arsenik, lässt solche eine Stunde zusammen sieden, dass es nicht überläuft, so wird das ☉ fix bei dem ♂, das gieße man aus, so ist es ein blauer Salz-Stein, diesen 1 Teil, des *Æsis Usti* etwas weniger, solches lässt man 12 bis 16 Stunden im Feuer stehen, wenn es recht gemacht und erkaltet, so steht unten ein grünlicher *Regulus*, oben der rote *Sulphur* in dem Salz und ♂, welche man durch zugießen Wassers davonlaugen kann, so hat man den roten *Sulphur* aus dem ♀.

80. Den *Sulphur* im ⊕ macht also: Der ⊕ wird zur Röte calciniert, vermischt ihn alsdann mit ✶ und sublimiert ihn.

81. Den trockenen Weg alle diese *Sulphur* zu haben ist also: Man nimmt welche man will, vermischt sie mit ☿, so viel als die metallischen *Species*, nimmt 4-mal so viel gelben *Sulphur* dazu und auch 4-mal so schwer ☉, vermischt sie, drückt einen Schmelztiegel davon voll, setzt es 16 Stunden in Kohlen, fest zugedeckt, das der *Sulphur* nicht helle brennen kann, so solviert er mit seiner Säure sein *Acidum*, das ☉, welches am ersten dem *Spiritum* vom *Sulphur* in sich zieht und zum trockenen *Alcahest* wird, die metallischen *Sulphure* gehen in Rauch weg, der fixe Teil aber, so güldisch ist, wird im Salz behalten und muss deshalb 4-mal mehr Salz und *Sulphur* zugesetzt werden, dass die metallischen Leiber ganz destruiert werden, alsdann wird es gepulvert, mit gleich schwer ✶ vermischt und sublimiert, vermutet man noch etwas vom fixen *Sulphur* in *Capite Mortuo*, kocht man es in destilliertem Wein-Essig, lässt ihn evaporieren, solches kann man wieder sublimieren.

82. Es sind sehr viel Wege, die metallischen *Sulphure* zu haben, habe aber nur die Kürzesten und Besten beschrieben, weil viele *Superflua* in anderen getan werden.

78. ♂ and ♀ are easy to dissolve: You just sprinkle iron filings with dew or rain water and mix it well with ℞. If you do this often, it turns into saffron. This is digested in distilled wine vinegar, so it takes the most subtle unto itself. Let the vinegar smoke off so that you have the sulphur from iron.

79. Separating the sulphur from ♀ is the easiest way: you make an *æs ustum*[41] *per se*, taking three parts ☉ and one part arsenic. Let it simmer for an hour so that it does not overflow, so the ☉ becomes fixed with the ♂. If you pour it out, it is a blue salt stone, of this one part, of the *æsis usti*[42] a little less, this is left in the fire for twelve to sixteen hours. If done right and cooled down, there is a greenish *regulus* below, above the red sulphur in the salt and ♂ which can be leached from it by pouring water over it, then one has the red sulphur from the ♀.

80. The sulphur in the ♁ is thus made: the ♁ is calcinated to redness, then mixed with ✶ and sublimated.

81. The dry way to have all these sulphurs is: you take which one you want, mix it with ☿ as much as the metallic *species*, add four times as much yellow sulphur, and also mix it as four times as it is heavy with ⊖, press in a melting pot to make it full, set it in the coals for sixteen hours, tightly covered, so that the sulphur cannot burn brightly, so with its acid, its *acidum*, it dissolves the ⊖, which first draws the *spiritum* from the sulphur into itself and becomes the dry *alcahest*. The metallic sulphurs go away in the smoke, but the fixed part, which is gold-bearing, remains in the salt and therefore four times more salt and sulphur must be added, so that the metallic bodies are completely destroyed; then it is powdered, mixed with the same weight of ✶ and sublimated. If you suspect something of fixed sulphur still to be in the *capite mortuo*, boil it in distilled wine vinegar and let it evaporate. You can sublimate it again.

82. There are many ways to have the metallic sulphurs, but I have only described the shortest and best, because a lot of *superflua* is done in others.

83. Diese extrahierten güldischen *Sulphure* geben nun *Particularia*, da man bei diesen das *Universal* nicht findet, solche güldische *Sulphure* geben nun ☉, wenn sie der Künstler in die ☽ zu bringen weiß, dazu viele Wege sind. 1. Kann man den gemeinen ☿ damit animieren, dass er solches ☉ in sich nimmt, hernach mit der ☽ amalgamiert, davon die ☽ gelb wird, weil aber diese Gelbe nur auswendig an die ☽ angeflogen, setzt es noch keine großen Stücke, sondern der ☿ muss noch weiter animiert und mit der ☽ amalgiert werden, solche Arbeit so oft wiederholt, bis die ☽ reich genug, so gibt es dann Ausbeute; Aber dass dieser animierte ☿ soll ☉ und ☽ in eine Tinctur verwandeln, solches ist lächerlich, wenn sich gleich etwas durch die lange Zeit von den ☿ bei ☉ oder ☽ præcipitiert, ist es doch keine Tinctur, sondern nur ein hartes trockenes Pulver, welches etwas an Silber oder Gold gibt, mehr kann man ihn nicht anmuten.

84. Die ☽ kann aus vielerlei Art zu Gold gebracht werden, durch ▽, welche voll *Sulphur Solis* sind, durch Gradier-Öle, durch *Cementa*, da die güldischen *Sulphure* wohl aufgeschlossen, dass sie spiritualisch sind, so durchgehen sie die ☽ und bleiben daran als einem fixen *Corpore* hängen.

85. Die gemeinen *Cementationes* sind nicht, dass man ⊕ ⊕, roten *Bolus*, Blut-Stein, ♂ Feil, ♁, ⊙, ✷ und dergleichen nimmt und cementiert die ☽ damit, als wenn diese groben Körper könnten in die ☽ gehen und solche verbessern, welches nicht sein kann, wenn aber diese Körper aufgeschlossen, ihre tingierende *Anima* von ihnen genommen, solche der ☽ einverleibt, dann kann es was tun, ein Leib kann einen andern Leib nicht durchdringen, ein Geist kann es tun.

86. Das gemeine Gradier-Öl vom corrosivischen Sublimat und ♁ tut nichts oder wenig bei der ☽, weil es nur der ☿ des ♁ ist, mit etwas flüchtigem *Sulphur*, sein tingierender fixer *Sulphur* bleibt zurück. Wenn man aber solchen zurückgebliebenen *Sulphur* solviert, in diese gelbe Solution des *Butyrum* ♁ gießt, solches gelinde herüber zieht, bis gelbe Tropfen kommen, welche man allein fangen kann, die weißen *Spiritus* gießt man in einen solvierten Eisen-Safran oder sonst

83. These extracted gold-bearing sulphurs now give *particularia*, since one does not find the *universal* in them. Such gold-bearing sulphurs now give ☉ if the artist knows how to bring them into the ☽, and there are many ways to do it. First, you can animate the common ☿ so that it takes such ☉ in itself, then amalgamate [it] with the ☽. In this amalgamation the ☽ becomes yellow, but because this yellowness is only superficially deep on the surface of the ☽, it is not much of an accomplishment yet, and the ☿ still has to be animated further and amalgamated with the ☽. Such work has to be repeated until the ☽ is rich enough, and this will give good yield. But it is ridiculous that this animated ☿ should transform ☉ and ☽ into a tincture, if something is precipitated after a long time from the ☿ at ☉ or ☽, it is not a tincture, but just a hard dry powder, which gives something of silver or gold, and nothing more can be seen from it.

84. The ☽ can be brought to gold by many ways, through ▽, which are full of *sulphur solis*, through gradating oils, through *cementa*, in which the gold-bearing sulphur is well unlocked, so that they are spiritual, so they go through the ☽ and stick to it as if they are hanging on a fixed *corpore*.

85. The common *cementationes* are nothing, in that you take ⊕,⊕, red *bolus*, blood stone, ♂ filings, 🜍, ☉, ✱, and the like, and you cement the ☽ with them, as if these coarse bodies could go into the ☽ and improve it, which cannot be; but if these bodies are decomposed, their tinging *anima* is taken from them and this [*anima*] is incorporated into the ☽. Then it can accomplish something, for although a body cannot penetrate another body, a spirit can do so.

86. The common gradating oil from the corrosive sublimate and 🜍 does little or nothing with the ☽, because it is only the ☿ of 🜍, with some volatile sulphur, and its tinging fixed sulphur remains behind. But if one dissolves such sulphur residue and then pours *butyrum* 🜍 into this yellow solution, gently distills it until yellow drops appear, which one can catch on their own. The white *spiritus* is poured into an

solvierte güldische *Crocus*, zieht es abermals über, die gefärbten Tropfen tut man zum vorigen, dies ist ein rechtes Gradier-Öl. Nun macht man ein Sublimat vom Gradier-*Aquafort*, welche hochgelbe aussehen und lauter *Sulphur Solis* sein, tut solchen Sublimat in solches Gradier-♑︎, zieht es gelinde herüber, dies so oft wiederholt, bis der ☿ *Sublimatus* als ein gelbes Öl zurück bleibt, welches in der Kälte ein hochgelbes Salz ist, worüber man ⊕ Öl gießt, bis der Sublimat fix, welchen man in dem Gradier-Öl auf solvieren muss, bis es zusammen ein blutroter Stein und fix geworden, welchen man mit Gold schmelzen und auf die ☽ tragen muss, dieser Sublimat wird schon mehr geben, als man einem gemeinen *Particular* zumuten kann, man kann auch die ☽ in diesem Gradier-Öl zu ☉ machen, je länger sie darinnen liegt, je besser ist es.

87. Die Gradier-Wässer dürfen die ☽ nicht solvieren, sobald sie in ♑︎ solviert und in ein Gradier-Wasser oder Öl geschüttet, fällt sie zu Boden und nimmt so viel von güldischen *Sulphur* zu sich, als sie halten kann, solche lässt man 20 ja 30 bis 40 Tage darinnen, so wird sie immer besser, doch kann solche des Tages einmal aufgerührt oder umgeschwenkt werden, auch kann man das Gradier-Wasser abgießen, die ☽ gelinde trocknen, etwas wieder vom Gradier-Wasser zugießen, bis es wieder gelinde ertrocknet, dies wiederholt so viel als es nötig, doch alles verschlossen, so wird sie schwarz, weiß, gelb, und rot, ganz in ☉ verkehrt, wenn das Gradier-Wasser gut gewesen.

88. Die ☽ ist im trockenen Weg durch *Vitra* dahin zu bringen, dass sie güldisch wird, auf diese Weise macht man eine Extraction von vorgenannten metallischen *Speciebus* mit ♑︎, weil noch das Salz dabei ist, welches mit Essig kann solviert werden und das ♑︎ darein gießen, weil einige Tinctur zu sehen, dies lässt man abrauchen, muss aber, wenn es will dick werden, stets mit einem gläsernen Stab gerührt werden, dieses Pulvers 1 Teil vermischt man mit 2 Teilen *Vitro* ♄, so von 2 Teilen Glätte und 1 Teil Kieselstein gemacht worden, dies lässt man etliche Stunden zusammenfließen, schüttet es aus, alsdann lässt man die ☽ fließen, wirft auf 1 Lot ☽ 1 Lot Glas, lässt es etliche Stunden fließen,

iron saffron solution, or otherwise dissolved gold-bearing *crocus*, and then one distills it over again, and the colored drops are added to the previous one. This is quite a gradating oil. Now one makes a sublimate of gradating *aquafort*, which looks bright yellow, and is nothing but *sulphur solis*. Add such a sublimate to such a gradating ▽, pull it over gently. Repeat this until the ☿ *sublimatus* remains as a yellow oil, which when cooled is a brightly yellow salt, over which ⊕ oil is poured until the sublimate becomes fixed. You are then to dissolve this in the gradating oil until it has become a blood-red stone and fix, which you have to melt with gold and put on the ☽. This sublimate will give more than one can expect from a common *particular*, and one can also make the ☽ into ☉ in this gradating oil. The longer it lies in it, the better it is.

87. The gradating waters must not dissolve the ☽, as soon as it dissolves in ▽ and is poured into a gradating water or oil, it falls to the ground and takes as much of the gold-bearing sulphur as it can hold. Let it steep for twenty or even up to thirty or forty days, while it gets better and better. It can be stirred up or swirled around once a day, and you can also pour off the gradating water, dry the ☽ gently, pour some of the gradating water back on it, until it dries up again. Repeat this as much as it is necessary, but everything closed,[43] so it becomes black, white, yellow and red, completely transmuted to ☉, if the gradating water has been good.

88. The ☽ is to be brought in the dry way through *vitra* so that it becomes gold-bearing, in this way one makes an extraction of the aforementioned metallic *speciebus* with ▽, because the salt is still there and can be dissolved with vinegar; and the ▽ is poured into it, because you can see some tincture. You can smoke it off, but if it starts to become thick, you must always stir it with a glass rod. One part of this powder is mixed with two parts of *vitro* ♄, which has been made from two parts of litharge[44] and one part of pebble stone. You let this flow together for several hours before pouring it out, and then let the ☽ flow, throwing on one lot of ☽ and one lot of glass. Let it flow for several hours,

alsdann wirft man 1 halbes Lot Eisen-Feil zur Præcipitation darauf, gibt stärker Feuer als vorher noch eine Stunde, lässt es hernach erkalten, den Silber-König schlägt man ab, schmelzt ihn abermals mit dem Glase und præcipitiert es als vorher, dies kann man so oft tun als man will, so wird das Silber immer reicher von ☉, auch vermehrt sich das ☽, welchen Zuwachs man schon auf der Capelle erfahren kann, das Silber tut man in Scheidewasser, setzt man aber etwas Gold zu, so bekommt man noch mehr.

89. Der allerleichteste Weg mit den wenigsten Unkosten die ☽ zu cementieren ist also: *Crocus* als vorher beschrieben N. 79, Eisen-Safran, calcinierte Granate durch *Sulphur*, ☉ Kiese, ☉ Marcasit, roter Talk, gelber *Vitriol* Kies, ☍, ☿ *Solut.*, jedes *ana*, 4-mal so schwer ⊖ und auch 4-mal so viel ♄, diese werden fest zusammen gedrückt und 16 Stunden im Feuer gehalten, dass es stets glüht, fleißig zugedeckt, das der Sulphur nicht brennen kann, hernach solviert man es in Wasser, lässt es wohl aufkochen, erkalten und gießt es vom Sediment, das zurück gebliebene solviert man in *Aqua Regis*, weil noch eine Tinctur darinnen, gießt solches in die solvierte Salz-Sulze und lässt es zum schönen Farben-Salz abrauchen, bis zur Trockne, darunter vermengt man ✶ und cementiert die ☽ damit, je öfter je besser. Wenn aber das Feuer zu stark, fließt es alles zusammen, so nimmt man den Sublimat der in der Höhe ist, vermischt ihn wieder mit dem untersten, bis es alles zusammen fix bleibt, alsdann lässt man Blei in einem Scherben fließen, wirft diese *Materia* einzeln ins Blei, bis es alles darein gegangen. Wenn es anfängt zu verschlacken, wirft man Eisen-Feil dünngezettelt oben darauf und lässt es verschlacken, den König treibt man ab, die ☽ nur mit Kochsalz oft cementiert, macht sie zur ☽ *Fixa* oder weißen ☉, hier aber werden dem Salz noch die tingierenden *Sulphure* zugesetzt, welche sie auch zugleich färben, welches weit mehr tut, als das Salz alleine.

90. Das beste und leichteste Gradier-Öl, dies jetzige mit Salz und ♄ corrodierte metallische *Mixtum* wird mit ✶ sublimiert, zuletzt mit starkem Feuer, diesen schönen Sublimat vermischt man mit 6-mal so schwer calcinierten *Vitriol*, und etwas Kohlen-Staub, dass es nicht

then throw a half lot of iron filings on it for precipitation, and give it more fire than before for an hour. Let it cool down afterwards. One then knocks off the silver king and melts it again with the glass. It is then precipitated as before. You can do this as often as you want so that the silver becomes richer and richer in ☉; and in addition the amount of ☽ also increases, an increase one can already perceive on the cupel. The silver is put in the separating water, but if one adds a little gold, one gets even more.

89. The easiest and least expensive way to cement the ☽ is therefore: *Crocus* (as previously described in No. 79 [above]), iron saffron, calcined garnets through sulphur, ☉ pyrites, ☉ marcasite, red talc, yellow *vitriol* pyrites, ♁, ☿ *solut.*, each *ana*, four times as heavy ⊖ and also four times as much ♀. These are pressed tightly together and kept in the fire for sixteen hours, so that it glows constantly while also being diligently covered so that the sulphur cannot burn. Thereafter you dissolve it in water, which you then bring to a boil. You then cool it and pour it off from the sediment. Dissolve the remnant in *aqua regis* (because there is still a tincture in it) and pour this into the dissolved salt brine. Let it smoke off to the beautiful colored salt until it is dry. Mix ✳ under it and cement the ☽ with it, the more frequently the better. But if the fire is too strong and it all flows together, then you take the sublimate that is on top and mix it again with the lowest until it all remains fixed together. Then you let lead flow in a shard and throw this *materia* one by one into the lead until it has all gone in. When it starts to slag, you throw a paper-thin layer of iron filings on top of it and let it slag, and you drive off the king. If the ☽ is cemented many times only with table salt, it turns into ☽ *fixa* or white ☉, but here the tinging sulphur is added to the salt. This tinging sulphur also colors it at the same time, doing far more than the salt alone.

90. The best and easiest gradating oil, this current metallic *mixtum* corroded with salt and ♀ is sublimated with ✳, and then finally with a strong fire. This beautiful sublimate is mixed with calcinated *vitriol* which is six times as heavy, as well as some coal dust so that it

schmelzt, destilliert es als einen *Spiritum* ☉, das hinterbliebene, davon der Sublimat gekommen, wird in destilliertem Essig aufgesotten, weil einige Farbe darinnen, solches raucht man gelinde ab, bis es zur Honig-Dicke wird, so schüttet man das Gradier-Öl als die flüchtige ♁ dazu, zieht es als bald durch den Helm herüber, den übergegangenen *Spiritum* zieht man halb wieder ab, so geht nur eine weiße Farbe, diese schüttet man auf das *Caput Mortuum*, so solviert sich das reinste Teil, solches lässt man eine Nacht digerieren, scheidet es vom Sediment und destilliert es abermals zuletzt mit starkem Feuer über den Helm, von welchen *Spiritu* man abermals das *Phlegma* abzieht, das Salz darinnen solviert und in das Gradier-Öl tut, die wird darinnen nicht allein die ☽, sondern auch den ♄ verbessern, auch kann man dieses Gradier-Öl zu ☉ gebrauchen, solches darinnen solvieren, die Feuchtigkeit gelinde abrauchen lassen, weil das Wasser weiß geht, hernach wieder mit solchem Gradier-Öl solviert, und gelinde abrauchen lassen, die Arbeit wiederholt bis das ☉ genug mit den güldischen ♁ gesättigt und gelbe Tropfen davon steigen wollen, alsdann wird es verschlossen in der *Phiolen*, so wird es durch die Farben gehen und zum Stein oder flüssigen Salz werden, welcher auf ☽ getragen, solche in ☉ verwandelt, doch muss es in ziemlicher Menge darauf geworfen werden, weil es nicht die *Universal-Tinctur* sondern eine *Particular* ist.

91. Eine Tinctur so auch nur ein *Particular*, ist roter Talk, calcinierte Granaten durch *Sulphur*, ☉ Kies, der sehr gelb, ☉ Markasiten, gelber ♁ Kies, *Mercur Solut.*, ☿, dieses zum allerkleinsten Pulver mit 4-mal so viel *Sulphur* corrodiert, fest verdeckt, dass der *Sulphur* nicht helle brennt, bis es wohl calciniert, solches solviert man in ♅ als hierinnen N. 77 beschrieben, man gießt so lange ♅ darauf, weil einige Tinctur zu sehen, das gefärbte ♅ zieht man gelinde herüber, bis es sich anfängt zu färben und in gelben Tropfen anfängt überzusteigen, solche fängt man, allein es muss über den Helm übersteigen, riecht als der schönste Safran, zuletzt bis der Kolben glüht das Weiß übergegangen, ♅ gießt man wieder auf das *Caput Mortuum*, kocht es darinnen, muss aber vorher wohl kleingerieben sein, so solviert es das reineste Teil, welches

does not melt. Distill it like a *spiritum* ⊕, the residuum from which the sublimate came is then boiled up in distilled vinegar. Because there is some color in it, this is gently smoked off until it becomes thick as honey; and then the gradating oil is poured on it as well as the volatile ♀. Distill it with the alembic, and soon half of the *spiritum* that has passed over is distilled off again, so only a white color passes over. This is poured onto the *caput mortuum*, so the purest part is dissolved. One then lets it digest for one night before separating it from the sediment and finally distilling it again through the alembic with strong fire. From this *spiritu* the *phlegma* is again removed, the salt dissolved in it and added into the gradating oil. It will not only improve the ☽, but also the ♄, and you can also use this gradating oil for ☉, by dissolving some in it. Let the moisture fume off gently, because the water goes white, then again dissolve it with such gradating oil, and let it smoke off gently. The work is repeated until the ☉ is saturated enough with the gold-bearing ♀ and yellow drops start to rise from it. Then it is sealed in the *phiol*, where it will go through the colors and become a stone or liquid salt, which if thrown on ☽ transforms such into ☉; but it has to be thrown on it in considerable quantities, because it is not the *universal* tincture but a *particular* one.

91. A tincture that is also only a *particular* one is red talc, calcined garnets through sulphur, ☉ pyrite, which is very yellow, ☉ marcasites, yellow ⊕ pyrite, mercury *solut.*, ☿, this to the smallest powder with four times as much sulphur corroded, firmly covered so that the sulphur does not burn brightly until it is well calcined. This is then dissolved in ♈ as is described in No. 77 [above]. ♈ is poured on it for so long as some tincture can be seen. You distill the colored ♈ gently over until it starts to color and begins to rise up in yellow drops which you will catch, though this is to be done with the alembic. It smells like the most beautiful saffron until the flask eventually glows white. You then pour the ♈ that has passed over into the [receiving] flask onto the *caput mortuum* again. Boil it in the *caput mortuum*, but first it has to be ground into small bits so that its purest part is dissolved, which you

man auf die vorige Weise destilliert, die roten Tropfen tut man zusammen, hernach solviert man beide *Capita Mortua* in neuem 🜄 bis keine Farbe mehr zu sehen in 🜄, darein wirft man Sublimat, zieht das 🜄 gelinde ab und sublimiert den ☿, so nimmt er den güldischen *Sulphur* mit in die Höhe, ehe er sich sublimiert, kommen auch noch gefärbte Tropfen, welche man zum vorigen tut, zuletzt gießt man *Spiritum* 🜨, welcher sehr stark sein muss, über das *Caput Mortuum*, davon der ☿ sublimiert, das Salz zu extrahieren, gießt die Solution, welche durch die Digestion ihre Erde fallen gelassen, auf den Sublimat lässt ihn gelinde abrauchen bis zur Trockne, darauf gießt man von den gelben wohlriechenden Tropfen, lässt es aber gelinde abrauchen, so behält er die Tinctur bei sich, darauf gießt man noch mehr solcher tingierenden Tropfen, lässt das weiße gelinde davon rauchen, bis der Sublimat gesättigt und die *Spiritus* gelb davon gehen wollen, so schließt man es zu, hält es im Feuer bis es fix und zum rechten Glas geworden, welches mit ☉ geschmelzt werden muss und auf die ☽ getragen, solviert man aber ☉ in diesen gefärbten Tropfen und übergießt diesen Sublimat damit, so figiert er sich eher, darf ihn auch hernach nicht mit ☉ fermentieren, sondern trägt es auf ☿ und diesen auf die ☽, so gibt es ☉ genug.

92. Noch eine andere Art: Man solviert den ☿ in 🜅, zieht es wieder herüber, gießt ein anderen 🜅 darauf, zieht es wieder herüber bis der ☿ rot wird, darüber gießt man obige tingierende Tropfen, welche er in sich schluckt, dies kann man etliche Mal tun, bis der Præcipitat von der schönsten Farbe und fix, solchen vermischt man mit Wachs, wirf ihn an kleinen Stückchen im Fluss auf die ☽ oder cementiere die ☽ damit.

93. Das leichteste und beste Gradier-Wasser wird also gemacht mit geringen Kosten: ♂ Feil, Grünspan, gelber 🜨 Kies, *Minera Martis Solaris* in runden Stücken, *Sulphur*, Rauschgelb, Auripigment, diese 2 letzten werden mit Fleiß dazu genommen, weil sie viel Arsenik haben, welche die *Sulphur* rauben und mit in die Höhe nehmen, doch muss des gemeinen *Sulphur* das meiste sein, dies tut man wohl gepulvert in eine Retorte, destilliert es aufs stärkste, so steigt etwas Wasser über,

then distill in the previous way. The red drops are put together, and then you dissolve both *capita mortua* in new ♅ until no more color can be seen in ♅. One then throws sublimate into it, gently distilling ♅ off and sublimating the ☿ so that it takes the golden sulphur upwards with it. Colored drops also come up before it sublimates, and these you should add to the previous one, until finally you pour *spiritum* ⊕ (which must be very strong) over the *caput mortuum*, from which the ☿ sublimates to extract the salt. Pour the solution (which has dropped its earth through digestion) onto the sublimate, and then let it smoke off gently until it dries. On top of this one then pours the yellow, fragrant drops, but let it smoke off gently, so it keeps the tincture. Pour more of these tingeing drops onto it and let the white smoke off from it gently until the sublimate is saturated and the *spiritus* want to go from it yellow, such that you then close it and keep it in the fire until it is fixed and has become a proper glass. This glass then has to be melted with ☉ and must be carried on the ☽, but if you dissolve ☉ in these colored drops and pour it over this sublimate, it more readily becomes fixed; but one must not ferment it with ☉ afterwards, but rather give it onto ☿ and this onto the ☽ so that it gives enough ☉.

92. Another way: dissolve the ☿ in ᛝ, distill it over again, pour another ᛝ on it, distilll it over again until the ☿ turns red. Then pour the above tingling drops over it, so that it absorbs them. You can do this several times until the precipitate is of the most beautiful color and is fixed. You then mix it with wax and throw it in small pieces in the flux on the ☽, or else cement the ☽ with it.

93. The easiest and best gradating water is made with low costs: ♂ filings, verdigris, yellow ⊕ pyrite, *minera martis solaris* in round pieces, sulphur, orpiment, and auripigment.[45] These last two are frequently used because they have a lot of arsenic which robs the sulphur and evaporates it with itself, though in this case it has to be mostly common sulphur. This is then reduced to a fine powder and placed in a retort, where it is distilled to the greatest possible extent, so that some

das übrige ist ein Sublimat von Sulphur und ☌ aber als ein rotes Glas, wenn es Feuer genug hat, das *Caput Mortuum* öffnet man noch besser, sublimiert es mit gleich schwer ✶, diesen schönen Sublimat und das vorige Glas vermischt man, wenn es zum kleinsten gerieben mit 4-mal so schwer ⊖ und 4-mal so schwer calcinierten ⊕, destilliert solches in einer Vorlage, darinnen das vorige Wasser vorgeschlagen wurde, doch muss es mit etwas Kieselsteinen vermischt werden, sonst fließt es sehr zusammen und gibt keine Spiritus. Dies ist der besten Gradier-Wässer eines, auf die ☽ und auf den ♄, man kann auch den ☿ darinnen solvieren und zum Præcipitat machen solchen mit der ☽ cementieren oder im Fluss darauf werfen.

94. N.B. Alle Tincturen, Gradier-Öle oder Gradier-Wässer, welche nur aus dem flüchtigen *Sulphur* und ☿ialischen Teilen gemacht werden, tun nichts oder wenig, wenn man aber mit den flüchtigen *Sulphur* die fixen in die Höhe führen und [in] die Tincturen, Gradier-Wasser und Gradier-Öle bringt, dann tut es mehr, als man der Kunst zugemutet, hält auch der fixe *Sulphur* den flüchtigen und wird durch ihn figiert und beständig gemacht.

95. Diese Dinge, ob sie wohl alle ganz deutlich beschrieben, so sind sie doch für keinen Anfänger, sondern erfordern einen geschickten Laboranten, welcher Nachsinnen haben muss, weil das Feuer Meister in diesen Dingen, zu viel und zu wenig bringt Schaden, auch muss Wissenschaft sein, die *Feces* von den übergestiegenen Tincturen, Ölen und Wässern zu scheiden, Hefen von den Salien, damit nur die reinen Teile genommen werden, sonst bringt es Schaden und verhindert die Transmutation, auch bei den *Cementationibus*, da oft die ☽ schmelzt, muss solche wieder laminiert oder wenn sie corrodiert und gefressen worden, muss man Verstand haben, sie durch Sicherung aus dem Cement zu bringen, dass sie nicht verloren geht, wiewohl diese *Cementa* nicht so grob als die Gemeinen, welche die ☽ rauben und irreducibel machen, diese aber, so hier beschrieben, sind alle solviert, und nur der reinste Teil, die ☽ aber auf einem anderen Weg aus dem Cement zu haben, kann man das ganze Cement-Pulver reducieren auf diese

water rises above. The rest is a sublimate, made of sulphur and ♂, but as a red glass if there was enough fire, the *caput mortuum* is opened even better, it is sublimated with equal weight of ✱. This beautiful sublimate and the previous glass is mixed, after it was scrubbed down to the smallest extent possible, with four times as heavy ⊖ and four times as heavy calcinated ⊕. Distill this into a receiver into which the previous water was poured in, but it must be mixed with some pebbles, since otherwise it will flow very much together and there will be no spirit. This is the best gradating water, one on the ☽ and on the ♄. One can also dissolve the ☿ in it and make it a precipitate, cement it with the ☽ or throw influx onto it.

94. N.B.[46] All tinctures, gradating oils or gradating waters, which are made only from the volatile sulphur and ☿ial parts do little or nothing, but if the volatile sulphur is used to bring the fixed one up into the air and into the tinctures, gradating waters and gradating oils, because it does more than one would have expected of the art, the fixed sulphur also retains the volatile, being made fixed and made permanent by it.

95. Although these things are all clearly described, they are not for a beginner, but require a skilled laboratory worker capable of careful consideration and caution; for fire is the master in these things, and too much or too little of it causes damage. Such a worker must also know how to separate the feces from the distilled tinctures, oils and waters, and the yeasts from the *salia*, so that only the pure parts are taken, as otherwise it will cause damage and prevent transmutation. This is the case even with the *cementationibus*, since the ☽ often melts, and then has to be laminated again; or if it has been corroded and eaten away, one has to know that one should take it out of the cement to secure it, so that it is not lost, although these *cementa* are not as coarse as the common ones which rob the ☽ and make it irreducible, but which are described here are all dissolved, and only the purest part, but to have the ☽ out of the cement in a different way, the whole cement powder can be reduced in this way: quicklime, finely ground, ⊖, potash, shiny

Weise: Ungelöschter Kalk, klein gestoßen, ⊖, Pottasche, glänzender Ofen-Ruß, jedes *ana*, dieses wird in Urin aufgesotten, dass es sich alles solviert, solches seigt man rein ab, lässt es zum Salz abrauchen, nimmt gleich so schwer als das Cement mit den Silber, vermischt es mit etwas Kohlen-Staub, besserer *Præcipitation* halber und schmelzt es mit geschwindem Feuer, wenn es als Wasser fließt, wirft man noch eine glühende Kohle darauf, so geschieht eine *Ebullition*, alsdann setzt es sich und fließt als Öl, so lässt man es erkalten, den König treibt man auf der Capelle ab.

96. Ein kluger Artist wird sich wundern, dass ich den Zinnober nicht unter die *Species* gesetzt, weil solcher fast von allen Bücherschreibern mit dazu genommen wird, ich lasse es zu unter solche grobe Arbeit und Mixturen wie sie machen, ich aber schließe die Körper auf, mache den flüchtigen und fixen *Sulphur* los von ihren Banden, das sie spiritualisch werden, damit sie in die Wasser gehen und ihre *Tincturen* darein geben können. Wenn ich nun den Zinnober darunter nehme, so ist es mir nicht dienlich, er ist nichts als *Sulphur* und ☿ Schwefel habe schon genug dabei, der ☿ dabei bringt mir Schaden, weil er die güldischen *Sulphure* lieber in sich fasst als den Gemeinen, sich damit præcipitiert und unter den groben *Fecibus* bleibt und zu Schanden wird, wie können denn die güldischen *Sulphure*, welche er zu sich genommen in die Gradier-Wässer-Öle oder Tincturen gehen und spiritualisch werden? Ich sage, der Zinnober verderbt es und macht vergebene Unkosten, weil er teuer ist.

97. Doch auf den Zinnober auch eine nützliche Arbeit zu zeigen, muss man den Zinnober selber machen, man lässt 2 Teile *Sulphur* zergehen, dass er nicht brennt, schüttet einen Teil ☿ da rein, rührt solches über dem Feuer, bis es sich mit dem *Sulphur* vereinigt, dicke und zähe wird, als dickes Pech und kein ☿ mehr zu sehen ist, diesen sublimiert man zum Zinnober, reibt ihn klein, übergießt ihn mit *Oleo* ⊕, digeriert etliche Tage, das *Oleum* ⊕ muss stark sein, wenn es trocken und das *Phlegma* weg gerauchet, darauf gießt man das rote *Oleum* ☉, welches

oven soot, each *ana*.⁴⁷ This is boiled away in urine so that everything is dissolved, and it is then drained off pure, left to smoke off until it becomes salt and it takes from it as much mass of silver as the cement does, mix it with some coal dust, for better precipitation, and melt it with quick fire, when it flows as water, one throws another glowing coal on it, so an *ebullition* happens, then it sits down and flows as an oil, so one leaves it to cool down, the king is driven off on the cupel.

96. A clever artist will be surprised that I do not include cinnabar among the *species*, because almost all authors do so. I leave it under such rough work and mixtures as they do, but I decompose the bodies, so that the volatile and fixed sulphurs shed their bonds and become spiritual, so that they can go into the waters and produce their tinctures in them. It would not serve my purposes to include cinnabar among them now, since it is nothing but sulphur and ☿ sulphur; and I have already got enough sulphur with me, while the ☿ is doing me harm because it prefers to have the gold-bearing sulphur in itself rather than the common one, with which it is precipitated and remains under the coarse *fecibus* and is put to harm, how can the gold-bearing sulphurs, which it consumed, go into the gradating waters, oils or tinctures so that they become spiritual? I say the cinnabar spoils it and makes a wasted expense because it is expensive.

97. But to do something useful with cinnabar one has to make the cinnabar oneself. You melt two parts of sulphur in such wise that it does not burn, and then you pour one part of ☿ on it, stirring it over the fire until it unites with the sulphur, becoming thick and tough like thick pitch, until no more ☿ can be seen. This is sublimated to cinnabar and pounded into small pieces. Pour *oleo* ⊕ over it and digest it for several days. The *oleum* ⊕ must be strong. When it is dry and the *phlegma* smoked away, pour over it the red *oleum* ☉ which was strengthened

durch einen solvierten Eisen-Safran gestärkt, in N. 86 beschrieben, so solviert es den Zinnober, er muss zerrieben werden in der Wärme, sonst wieder nass und schmutzig, hernach hält man ihn im Feuer, bis er zum fixen Stein geflossen, welchen man hernach mit ☉ und ☽ versetzen kann, er wird guten Teils mehr ☉ geben, als er schwer ist.

98. Den allerleuchtesten Weg, welcher auch der kürzeste ist eine Tinctur zu haben, dadurch der ☿ in ☉ nur durch Kochen in einer Stunde kann gebracht werden, ist also: Man nimmt ☿, ♂, ☉ Kies, roten Talk, jedes *ana*, 4-mal so schwer ⊖ und 4-mal so schwer Rauschgelb, füllt es in eine Retorte und destilliert es, so geht etwas an Tropfen, aber viel Sublimat, das *Caput Mortuum* öffnet man noch besser mit gleich schwer ✶, sublimiert solches, diese beiden *Sublimata* wiegt man, nimmt 8-mal so schwer, trocken calcinierten ⊕ mit Kohlen-Staub vermischt dazu, destilliert solches als ein *Oleum Vitrioli*, so steigt nur der reinste Teil von diesen Geistern über, das Wasser oder Öl, so vorher übergegangen, wird vorgeschlagen, das erste *Caput Mortuum*, davon der Salmiak sublimiert, kocht man in destilliertem Wein-Essig, weil es eine Farbe geben will, den lässt man gelinde abrauchen, so bleibt ein schönes Salz zurück, auf solches gießt man das Gradier-Öl, zieht es herüber, gießt es wieder darüber, bis es alles bei dem Salz bleibt und nur ein *Phlegma* ohne Geschmack übergeht, das Salz aber ist flüssig und güssig, von schönen Farben, nun nimmt man ein halbes Lot starkes ⋎, wirft darein eine *Quente* ⊖, welches man mit etwas *Spiritu* ⊕ anfeuchtet, tut es in eine gläserne Retorte, wirft 1 *Quente* ☉ dazu, lässt es solvieren, wenn es solviert, tut man von diesem güssigen Salz 2 Lot darein, zieht das ⋎ herüber, welches man wieder darauf gießen muss und gelinde wieder überziehen, bis man sieht, dass es weiß ohne Tinctur übersteigt, das Salz lässt man 3 Stunden fließen, danach solviert man es in Wasser, kocht ☿ 2 oder 3 Stunden darinnen, das Wasser gießt man ab, schmelzt ☽ oder ☉ und wirft den ☿ nach und nach darauf, man kocht so lange ☿ darinnen, bis keiner mehr darinnen hart werden will, wenn das süße Wasser einkocht, muss man wieder anderes nachgießen und es dünne erhalten, damit es nicht scharf wird und den ☿

by a dissolved iron saffron (as described in No. 86) so it dissolves the cinnabar. It has to be grounded in the warmth, otherwise it becomes wet and dirty again, afterwards it is kept in the fire until it has flowed to the fixed stone, which you can then mix with ☉ and ☽, it will give a good deal more ☉ than it is heavy.

98. The most illustrious way to have a tincture, which is also the shortest one, in which the ☿ can only be brought into ☉ by boiling for an hour is this: one takes ☿, ♂, ☉ pyrite, red talc, each *ana*, four times the weight of ⊖ and four times the weight of orpiment; fill a retort and distill it; some passes with drops, but there is much sublimate. The *caput mortuum* is opened more easily with the same weight of ✻, sublimate it, these two *sublimata* are weighed, take eight times the weight of dry calcinated ⊕ mixed with coal dust, distill it as an *oleum vitrioli*; only the purest part of these spirits rises. The water or oil which has been distilled beforehand is put in the receiver. The first *caput mortuum*, from which the salmiac has sublimated, is boiled in distilled wine vinegar, because it wants to give a color. Lets it smoke off gently so a nice salt remains, pour the gradating oil over it, distill it over, and pour it over it again until everything remains with the salt and only a *phlegma* without taste passes over, but the salt is liquid and pourable with beautiful colors. Now take a half lot of strong ▽, throw a *quente* ⊖ into it, which is moistened with a little *spiritu* ⊕, put it in a glass retort, add one *quente* ☉, let it dissolve. When it is dissolved, put two lots of this pourable salt in it, distill over the ▽, which has to be poured back on it, and gently distill over again, until one sees that it passes over white without tincture. The salt is allowed to flow for three hours, then dissolve it in water, boil ☿ in it for two or three hours, pour the water off, melt ☽ or ☉, and then gradually throw ☿ on it. Boil ☿ in it until it no longer wants to harden. When the sweet water boils down, other [water] must be poured on it in again to keep it thin so that it does not become strong and dissolve

solviert, die Salze dienen alle wieder aufs neue.

99. N.B. N.B. Diese Dinge zu arbeiten muss man die Nase nicht drüber halten, denn es sind färbende Gifte und spiritualisch, auch muss es unter einer Feuer-Mauer gearbeitet werden. Wo diese Gefäße in einer Stuben brechen, wäre man des Todes und wenn man sie schon ganz behält und feste in der Destillation verlutiert, dringen sie doch durch und machen Haupt und Herz-Weh, laedieren alle Nerven, machen *Convulsiones*, die Zunge im Munde kalt, auch alle Zähne an der Wurzel los, als wären sie mit einem Stein losgeschlagen, machen auch Lenden-Lahm und Lenden-Fieber, bösen Magen und stetiges Herz-Weh, welche Zufälle ich mit herrlichen Medicamenten kaum in 14 Tagen können los werden, da ich erstes mal so begierig zugesehen, nicht vermeint, dass es so giftig wäre, hernach aber bedacht, wie es erstlich durch *Sulphur*, ♂ und ☉ zerrissen, durch den ✶ noch in kleinere *Atomos* gebracht und sodann durch den *Spiritum* ⊕ in die Potabilität, welche Salz-Geister schon geschickt genug sie durch den Atem ins Gehirn, Herz, Lunge und Leber zu führen, sich im ganzen Leibe auszuteilen, die Gesundheit zu zerstören und des Lebens-Licht auszublasen. Habe es dem Nach-Arbeiter zur Warnung nicht verhalten wollen.

100. Es sind noch viele *Particular*-Wege da durch der ☿, ☽, ♄ kann zu ☉ gebracht werden, nämlich: wenn man güldische *Species* nimmt, solche im trockenen Wege aufschließt, in ☿, ♄ oder ☽ bringt. Es hat aber der Leser genug an diesen, doch kann ein trockener Weg durch ☿ vorgenommen werden, dadurch ☉ Kiese, arme ☉ Erze, ☉ Marcasiten können dahin gebracht werden, dass sie ihr flüchtiges ☉ corporalisch geben, auch durch figierende Flüsse, welches bei Bergwerken, ein sehr großes austragen kann, absonderlich durch *Cementationes* und Zuschläge, dadurch eins das andere verbessert und sich daran præcipitiert, welches sonsten alles in Rauch weg geht, ist auch nicht kostbar und wohl zu haben mit geringen Kosten.

N.B. Bei allen Cementationen mit Salz und *Sulphur* muss gelinde Feuer sein, dass es nicht schmilzt und nur die *Species* corrodiert, sonsten reduciert es die metallischen Leiber.

the ☿. All the salts can be used again.

99. N.B. N.B. You should not hold your nose over it if you work these things, because they are coloring poisons and spiritual, and you have to work under a fire wall. If these vessels break in a room you would be dead, and if you keep them unbroken and well lutated in the distillation, they [i.e., the fumes] penetrate nevertheless and cause headache and heartache, damage all nerves and causing convulsions. The tongue in the mouth becomes cold, also all teeth become loose at the roots, as if they had been knocked loose with a stone. It also causes lumbar lameness and lumbar fever, bad stomachaches and constant heartache. These symptoms were so strong that I could hardly get rid of them in fourteen days with the help of wonderful medicaments. When I watched so eagerly the first time, not assuming that it would be so poisonous, but afterwards considered, that it was first torn apart by sulphur, ♂ and ⊖, brought into smaller *atomos* by ✳ and then by the *spiritum* ⊕ into potability, these salt spirits are already strong enough to pass through the breath into the brain, heart, lungs and liver, thence to be distributed throughout the body to destroy health and to extinguish the light of life. I did not want to withhold this from the after-worker as a warning.

100. There are still many *particular* ways through which ☿, ☽, ♄ can be brought to ☉, namely: if one takes gold-bearing *species*, decompose them in the dry way, bring them into ☿, ♄ or ☽. But the reader has had enough of these, and a dry way can be made through ☿, thereby ☉ pyrites, poor ☉ ores, ☉ marcasites can be brought to the point that they give their volatile ☉ corporally, also through figuring fluxes, which can give very large [results] in mines, particularly through *cementationes* and additions, thereby improving one and another thing and precipitate it, which would otherwise go up in smoke, is also not valuable and can be had at low cost.

N.B. In all cementations with salt and sulphur, the fire must be mild so that it does not melt and only corrode the *species*, otherwise it will reduce the metallic bodies.

☿ *Solut.*, dies Signum muss man kennen lernen, es ist kein ☿ auch nicht der Sublimat, sondern ein mineralisches Wesen, welches das meiste bei der Sache tut.

Das ganze *Universal* ist in diesen Reime vorgestellt, die Vorarbeit und Nacharbeit:

* *
*

Such nur das einzige Ein, den Ursprung aller Dinge,
   Damit es Anfangs dir, nicht auch wie mir misslinge.

Die Nacharbeit ist schlecht, die aber vorgeht, schwer,
   Ein kluger hat zu tun, dass er durchschwimmt das Meer,

   Hast du dann dieses Ding, so scheide davon ab,
     Nur das unreine Teil, das reine wirf ins Grab,

   Nachdem es seine Zeit im Grabe hat gelegen,
     So führ es aus der Gruft, verwahre diesen Segen,

   In ihm liegt der Schatz, auf ihm das Sternenfeld,
     Schau, dass er wiederkehrt zu unsrer kleinen Welt.

Umsonst wär sonst die Müh; Nun fange an zu scheiden,
Den Geist von Seel und Leib, dabei magst du wohl meiden,

   Das große Kohlen-Feuer, bis Seele, Geist und Leib
     Nach Kunst geschieden sind, so gib dem Mann sein Weib.

   Er wird sie halten fest, sie herzlich Ihn umfangen,
     Ein jedes stillet hier, sein brünstiges Verlangen,

THE MINERAL GLUTEN

☿ *solut.*: you have to get to know this sign, it is not a ☿ also not the sublimate, but a mineral being that does most in the thing.

The whole *universal* is presented in these rhymes, the preliminary work and the after-work:

* *
*

SEARCH FOR THE ONLY ONE, the origin of all things,
So that you do not fail, like me, when it begins.

The after-work is bad, the first work not to see,
a clever one has a lot to do, to cross the mighty sea

If you finally have this thing, then separate it brave,
only the impure part, the pure you throw into the grave

After it lay in the grave for its time,
You put in the cave, to keep it fine.

The treasure is in it, above it the stars shine all,
But make sure to come back to our world so small.

to no avail would be all your effort, so now start to separate
the spirit from soul and body, but you have to obviate

the large coal fire, until soul, spirit and body separate
according to the art, and give to the male its female mate

He will hold her tight, she will embrace him to
So they will still their ardent desire though

Und werden wiederum mit schwarzem Tuch bedeckt,
Worunter ganz gewiss die Weiß und Röte steckt.

Hier fahre säuberlich und eile ja mit Weile,
Damit die weiße Braut den Perlenschmuck erteile.

Denn wo du ihn verscherzt, so ist der Schatz verloren,
Den sich der König selbst, zum Schatz hat aus erkoren.

Hast du du die Perlen-Kron, so wird der König kommen,
Mit Purpur angetan und, wie ich hab vernommen,

vergelten deine Müh: Wohl dir du hast es gut,
Bewahre diesen Schatz, so lieb dir Leib und Blut.

Schau so wird der von GOTT in dieser Welt gesegnet,
Der ihm mit reiner Lieb, zu aller Zeit begegnet,

Die reine GOTTES-Furcht erlanget solchen Lohn,
Die GOTTES-Furcht allein, krönt einen Tugend-Sohn.

ENDE

And they are again covered with a black cloth,
under which are hidden whiteness and redness both.

Here now continue cleanly, fast but no to hastily,
So that the white bride gives you, the pearl jewelry

But if you do forfeit it, the treasure will be lost,
the treasure for which the king wanted to be host.

But if you have acquired the pearl crown, the king will come
dressed in purple and he will give you some

recompense for your effort, you're doing well and have it good
but save this treasure as you save your body and blood.

Look, you will be blessed by GOD so fine,
if you loved GOD all the time,

Only the fear of GOD brings such reward,
only the fear of GOD crowns a son of virtue, this is the art.[48]

END

The Philosophical Pearl Tree

DER
Philosophische Perl-Baum

DAS GEWÄCHS DER DREI
PRINCIPIEN

zu

DEUTLICHER ERKLÄRUNG DES
STEINS DER WEISEN

Wie er mit seinen Wurzeln in der äußeren und finstern Welt,
mit seiner Blüte aber in der paradiesischen und Licht-
Welt und mit seiner reifen Frucht in der
engelhaften und himmlischen Welt
steht und wächst.

BESCHRIEBEN DURCH

D. I. W.

VON WEIMAR AUS THÜRINGEN

LEIPZIG
Verlegt von Johann Heinrichs Witwe
1705

THE
Philosophical Pearl Tree

THE GROWTH OF THE THREE
PRINCIPLES

for

CLEAR EXPLICATION OF THE
PHILOSOPHER'S STONE

How it is Standing and Growing with its Roots in the Outer and

Dark World, but with its Blossom in the Paradisical and

Luminous World, and with its Ripe Fruits

in the Angelic and Heavenly

World

DESCRIBED BY

D.I.W.

FROM WEIMAR IN THURINGIA

LEIPZIG
Published by Johann Heinich's Widow
1705

VORREDE

ICH HABE IN MEINEM VORIGEN TRAKTÄTLEIN, WELCHES genannt wird *Das Mineralische Gluten, Doppelter Schlangen-Stab, Mercurius Philosophorum, Langer und kurzer Weg zur Universal-Tinctur*, versprochen, noch ein Traktätlein herauszugeben von der Tinctur Kraft und Wirkung, absonderlich von seinem Nutzen in der Medizin. Nachdem nun fast am Ende sotanen Büchleins mich dieser Formalien gebraucht: Ist also genug geredet von dem Gewächs der 3 Principien, als 1. von der äußeren Welt, 2. von der inneren und Lichtwelt und 3. von der engelhaften und himmlischen Welt; Solches aber nach diesen 3 Welten nicht ausgeführt, auch wegen der Multiplication mich nicht weitläufig erklärt und daher sich der Leser verwundern und wissen möchte, was doch dadurch gemeint? So habe solches versprochene Traktätlein noch zur Zeit zurück stellen und erstlich wegen der Multiplication dieses Gewächses der 3 Principien nach allen 3 Welten mich besser erklären wollen, welche 3 Welten ich *philosophicé* verstehe in unserem Werke, weil es sowohl als die große Welt 3 Reiche besitzt. Als 1. das mineralische, weil es seinen Anfang und Ursprung aus diesem hat. 2. Das vegetabilische, weil es wächst, denn es wachsen alle 7 Metalle aus ihm und aus solchen die leuchtenden Planeten als helle funkelnde Sterne, die allerlautersten Edelgesteine und Tincturen. 3. Das animalische, weil es lauter lebendige Geister hat, als Erd-Geister, Wasser-Geister, Luft-Geister, Feuer-Geister. Es möchte mir aber jemand sagen, warum ich denn Himmel und Paradies eine Welt nenne? So dient zur

PREFACE

IN MY PREVIOUS TREATISE, WHICH IS CALLED *THE MINERAL Gluten, Double Serpent-Staff, Mercurius Philosophorum, Long and Short Way to the Universal Tincture*, I promised to publish another treatise on the tincture's power and effect, especially in medicine. Now that I am almost at the end of this little book, I have used these formalities: thus enough has been said about the growth[49] of the three principles[50] from: (1) the outer world, (2) the inner and luminous world, and (3) the angelic and heavenly world;[51] but I have not explained these three worlds, nor have I explained the multiplication in detail, and therefore the reader may be amazed and might wish to know what I meant by this. Therefore I have put this promised treatise on hold for the time being. I have done so because, first of all, I want to explain the multiplication of this growth of the three principles with respect to all three worlds, which three worlds I understand philosophically in our work, which also has three realms just like the macrocosm. The first world is the mineral [kingdom], because it has its beginning and origin from this. The second is the vegetabilic [kingdom], because it experiences growth, and since all seven metals grow out of it and out of these metals the shining planets as bright, sparkling stars, and the most immaculate jewels and tinctures. The third is the animal [kingdom], because it has many living spirits, such as earth spirits, water spirits, air spirits, and fire spirits. But if someone would like to ask me why I call heaven and paradise a single world I would answer that between our heavenly and

Antwort: Dass zwischen unserer himmlischen und paradiesischen Welt und dem göttlichen Himmel und Paradies ein großer Unterschied und unseres nur ein Spiegel, darinnen wir jenes erblicken, weilen wohl wissend, dass GOTT nicht einen Thron der heiligen Engel, sondern viele geschaffen. Dass auch der Himmel keine Welt ist, wie er jetzt steht mit der unzählbaren Menge der Sterne, da jeder seinen sonderlichen Himmel hat, darinnen er schwebt (als wie der Dotter im Ei, welches den Dotter trägt) und zwar mit einer sotanen unermesslichen Größe, dass unsere Welt kaum so groß dagegen, als eine Kugel, die man in der Hand halten kann. Ob nun solche großen Körper auch bewohnt mit Engeln oder anderen Geistern (weil ja kein leerer Ort in der ganzen Natur) oder ob sie etwa mit anderen Geschöpfen besetzt? Dies wissen wir nicht, es ist GOTT allein bekannt. Von unterschiedlichen Himmeln aber wissen wir aus der Heiligen Schrift: Als 1. Buch der Könige, 8 Kap. V. 27 steht: Siehe der Himmel und aller Himmel Himmel mögen dich nicht versorgen. Hiob am 15 Kap. V. 15: Die Himmel sind nicht rein vor Ihm. 19. Psalm V. 1: Die Himmel erzählen die Ehre GOTTES. 89. Psalm V. 5: Die Himmel werden, HERR, deine Wunder preisen. Hiob am 41. Kap. V. 11 : Es ist mein, was unter allen Himmeln ist. 102. Psalm V. 25 : Die Himmel sind deiner Hände Werk. 136. Psalm V. 5: Der die Himmel ordentlich gemacht hat. 148. Psalm V. 1: Lobet ihr Himmel den HERRN, V. 4: Lobet Ihn ihr Himmel allenthalben. 5. Buch Moses am 2. Kap. V. 25: Alle Völker unter allen Himmeln sollen erschrecken, wenn sie von dir hören, 2. Cor. 12. Kap. V. 2: Ward Paulus entzückt bis in den dritten Himmel. Und an mehr Orten der Heiligen Schrift. Von unserem Himmel aber, unter welchem wir leben so weit unsere Augen reichen, haben wir aus Heiliger Schrift diese Wissenschaft aus dem 1. Kapitel des ersten Buches Moses, dass er aus dem besten, lautersten und lichten Teil der Wasser geschaffen und die Feste oder Himmel genannt worden und aus dem reinsten Teil Licht und Feuer die Sterne. Es hat uns zwar Moses kein Licht vorher gegeben, sondern nur von Schöpfung dieser Welt geschrieben, worauf wir leben. Es war aber die Erde, worauf wir jetzt gehen und die Höhe des Him-

paradisiacal world and the divine heaven and paradise there is a great difference, and ours is only a mirror in which we see the other's reflection, while knowing that GOD created not one throne of holy angels, but many. I would also answer that the sky is also not a single world, as it stands now with the innumerable multitude of stars, since each [star] has its own sky, in which it floats (like the yolk in the egg, which carries the yolk) and with such an immeasurable size that our world [by comparison] is barely as big as a ball that you can hold in your hand. As to whether such large bodies are also inhabited by angels or other spirits (because there is no empty place in all of nature) or whether they are occupied by other creatures, I say that we do not know this, GOD alone knows. But we know of different heavens from the Holy Scriptures, as for example in the First Book of Kings 8:27,[52] where it is written "The heaven and heaven of heavens cannot contain you". Or Job 15:15: "The heavens are not clean in his sight". Then again Psalm 19:1: "The heavens declare the glory of GOD". Or again Psalm 89:5:[53] "And the heavens shall praise your wonders, O LORD". One reads in Job 41:11:[54] "Whatsoever is under the whole heaven is mine". And in Psalm 102:25:[55] "The heavens are the work of your hands". Or again in Psalm 136:5: "To him that by wisdom made the heaven". We read in Psalm 148:1: "Praise ye the LORD from the heavens" and in 148:4: "Praise him, ye heavens of heavens". The Fifth Book of Moses 2:25 says: "I put fear of you upon the nations that are under the whole heaven, who shall hear report of you". And finally in 2 Corinthians 12:2: "Paulus was swept up to the third heaven". One could produce more citations from Holy Scripture. But about our heaven, under which we live as far as our eyes can see, we know from the first chapter of the First Book of Moses that it was created from the best, clearest, and brightest part of the waters, and that it was then called the firmament or heaven; and from the purest part of light and fire the stars [were created]. It is true that Moses did not tell us about what was before, but only wrote of the creation of this world on which we live, the earth on which we are now walking; and he wrote of the height of heav-

mels so weit unsere Augen den Himmel und die Sterne erreichen, der Thron, Sitz und Himmel des Lucifer, worin keine Sterne, weder Sonne noch Mond, sondern dieses Licht war, mit klarem lauteren Wasser der lautersten Erden vermischt, alles ein lauterer schöner Thron und Himmel, hell leuchtend und glänzend, worin Lucifer erschaffen, dass er diesen Himmel oder Thron ewig beherrschen sollte. Dieser war nun der schönste Fürsten- und Thron-Engel aus dem Licht des Sohnes GOTTES erschaffen und damit verklärt, als er aber dies Licht in sich erblickt, verwunderte er sich über seine Schönheit, fing an hoffärtig zu werden, wollte noch stolzer sein und sich selber noch schöner erschaffen und noch mit größerem Licht verklären, dass er dem Sohn GOTTES möchte gleich sein oder gar über ihm, zündete daher sein selbst eigenes Licht an in allen seinen Quell-Geistern, da doch seine Geburt und Schöpfung von GOTT war. Wenn er hätte sollen mehr verklärt und verherrlicht werden, hätte es aus GOTT geschehen müssen, denn GOTT allein ist Schöpfer. Alsobald verlosch das göttliche Licht in ihnen und seine ganze Legion Engel mit ihrem Himmel gerieten ins Brennen, denn der Himmel brannte sowohl in seinen 7 Quell-Geistern als auch die Engel selbst, in diesen Anzündungen schied sich alles und reducierte sich der ganze Himmel in das *Confusum Chaos*. Weil aber GOTTES Allmacht und Wille musste bestehen, so schied GOTT doch aus diesem *Confuso Chao* erstlich das lichte Wasser von dem dunklen, er schied das Licht von der Finsternis, er schied das geseelte Wasser von der Erde aus dem Mittel des Wassers, woraus er die Feste, den Himmel machte und schied das Wasser unter der Feste von dem Wasser über der Feste, welches der obere Himmel, darinnen die Sterne schweben, als ein Dotter im Ei. Aus dem angezündeten Licht und Feuer schaffte GOTT die Sterne und aus dem groben Teil die Erde, darauf wir gehen. Dieses nun und der ganze Himmel, so weit wir sehen, darinnen die Sterne laufen, ist das ganze Reich des Lucifer, darinnen er so übel gehaushaltet, dass es nun alles in der quälenden Angst herumlaufen muss, im Bösen und Guten, in Verderbungen und Anfeindungen, in Schiedligkeiten und im drehenden Angst-Rade der Natur, in wel-

en as far as our eyes reach up to the stars, and of the throne, seat and heaven of Lucifer, in which there were no stars, neither sun nor moon, but this light was mixed together with clear, pure water of the purest earths, all a pure beautiful throne and heaven, brightly shining and glittering, in which Lucifer was created to rule this heaven or throne forever. He now was the most beautiful prince and throne angel created out of the light of the son of GOD and thus transfigured. Yet when he saw this light in himself, he was amazed at his own beauty and began to be arrogant. He wanted to be even more proud, and to make himself even more beautiful and transfigured with an even greater light, wishing to be equal to the son of GOD, or even to stand above him. Accordingly he lit his own light in all his source-spirits[56], although his birth and creation was from GOD. If he were really to have been more transfigured and glorified, however, this should have come from GOD, for GOD alone is the creator. Soon the divine light in them went out and his whole legion of angels with their heaven caught on fire, because heaven burned with its seven source-spirits as well as with the angels themselves; and in these ignitions everything was separated and the whole sky was reduced to the *confusum chaos*. But because GOD's omnipotence and must prevail, GOD first separated from the *confusum chaos* the clear water from the dark, and He separated the light from the darkness, and He separated the soulful water from the earth from the middle of the water, from which He made the firmament, the heaven. He then separated the water under the firmament from the water above the firmament, which is the sky above, in which the stars float like a yolk in an egg. From the kindled light and fire GOD created the stars and from the coarse part He made the earth upon which we walk. This earth now, and the whole sky, as far as we can see the stars running in it, is the whole kingdom of Lucifer, in which he kept house so badly that the world has been plunged into tormenting fear, in a clash evil and good, in corruption and hostilities, in divisiveness[57] and the rotating fear-wheels of nature in which they are driven about. Their burning pain and agonizing fear-birth will en-

cher sie also herumgetrieben werden. In welcher Entzündung und quälenden Angst-Geburt es bleiben wird bis ans Ende der Welt, da sich alle Creaturen ängstigen und sehnen nach ihrer Befreiung und gerne los sein wollten von der Eitelkeit, der sie unterworfen sind, von den Banden der Schiedligkeit in Böse und Gut, in welche sie durch Anzündung des Lucifer geraten und gerne wieder mit allen Kräften in einen Geist im vorigen lauteren klaren Himmel und Splendor vor GOTT erscheinen und himmlische beständige Früchte gebären wollten, für die jetzigen vergänglichen und tödlichen, so wird sich auch GOTT der Schöpfer Himmels und der Erden erbarmen und sich der Natur und Creatur Ängstigen und Seufzen lassen zu Herzen gehen, wenn die Zahl der gefallenen Engel erfüllt und die Gottlosen die *Turbam Magnam* auch entzündet durch ihre bösen Taten, so wird GOTT die 7 Schalen des Zorns GOTTES, welches die letzten Plagen sind, über die Erde und Elemente schütten, sie in ihrem eigenen Feuer verbrennen und der reinen verklärten Erde die himmlischen Wasser wiedergeben, sie mit ihren eigenen Sternen, Lichtern und Feuer wieder vereinigen und zum vorigen Himmel und klaren Thron verwandeln, aus einer Gestalt in die andere, aus dem reinen gläsernen Meer in das gläserne Meer mit Feuer gemengt, da kein ängstliches Umlaufen mehr der Himmel noch Sterne, sondern die neue Stadt, das neue Jerusalem, weder Sonne noch Mond bedarf, denn die Sterne sind nur Kräfte der ängstlichen Ausgeburt. Die reine Wiedergeburt aber der Himmel bedarf weder Sonne noch Mond, läuft nicht mehr, sondern ist ein selbstständiges Wesen, der lauterste schöne Thron und Himmel, lauter Licht und Glanz, welchen die Frommen und Auserwählten mit *Christo*, ihrem Thron-Fürsten, ewig besitzen, welches der ewige Sabbat und Ruhe-Tag ist. Sollte GOTT das unmöglich sein, der Erde ihren *Splendorem* und Schein wieder zu geben? Johannes sagt: Ich sahe einen neuen Himmel und eine neue Erde, denn der erste Himmel und die erste Erde verging und der auf dem Stuhl saß sprach: Siehe, ich mache alles neu. Auch sagt Petrus: Wir warten eines neuen Himmels und einer neuen Erden, in welcher Gerechtigkeit wohnet. 2. Epist. Petri am 3.

dure until the end of the world, since all creatures both fear and yet also desire their liberation, and wishing to become free from the vanity to which they are subject; and they yearn to be liberated from the bonds of the difference between evil and good, into which they fell when Lucifer was struck down by God's lightning. They want to appear again in all their original spiritual power and glory in the former clear sky and splendor before GOD, when they will constantly bear heavenly fruits instead of the impermanent and deadly [fruits] they presently produce. GOD, the creator of heaven and earth, will have mercy, and let go to heart natures and creatures fearing and sighing, when the number of fallen angels is fulfilled and when the godless have set on fire the *turbam magnam* through their evil deeds, so GOD will pour out the seven bowls of the wrath of GOD, which are the last plagues, above the earth and the elements. He will burn them in their own fires and give back the heavenly waters to the pure transfigured earth, reuniting them with their own stars, lights and fire, transforming them into the previous sky and clear throne, from one figure to the other, from the pure sea of glass into the sea of glass mingled with fire, so that there will no longer be the fearful circling of the sky or the stars, but the new city, the new Jerusalem, which will need neither sun nor moon, for the stars are only forces of fearful spawning. The pure rebirth of heaven does not need sun or moon, no longer runs, but is an independent being, the purest beautiful throne and heaven, all light and splendor, which the pious and chosen with Christ, their princethrone, have forever which is the eternal Sabbath and rest day. Should it be impossible for GOD to give the earth its *splendorem* and appearance again? John said: "And I saw a new heaven and a new earth: for the first heaven and the first earth were passed away.[58] And he that sat upon the throne said, Behold, I make all things new".[59] Peter also says: "We look for new heavens and a new earth, wherein dwelleth

Kap. V. 10. Es wird des HERRN Tag kommen wie ein Dieb in der Nacht, in welchem die Himmel zergehen werden mit großem Krachen. Und im 102. Psalm, sie werden vergehen, aber du bleibst, sie werden veralten wie ein Kleid, wenn du sie verwandeln wirst. Zur selben Zeit wird Fürst Lucifer seinen Thron einnehmen, mit allen Verdammten, den er sich selber erbaut im finsteren Höllen-Reich, da sie in der *Terra Damnata* ewig ihre Wohnung haben werden. Es möchte mir jemand übel vorhalten, dass in diesem Büchlein dieses Geheimnis oft mit dem göttlichen, himmlischen und engelhaften verglichen, *item*, das geistliche mit dem weltlichen vermischt? So dient zur Antwort, dass der Sohn GOTTES selbst die himmlische Lehre im Gleichnis geredet. Ja das Himmelreich dem Senfkorn, Sämann, Haus-Vater, Weingärtner und dergleichen vergleicht, so habe dieses auch als ein Gleichnis und Spiegel gebraucht, darinnen ich dann besser GOTT, Himmel, Engel, Menschen, Natur und Creatur gezeigt: Das aber geistliche Auslegung dabei geführt, ist geschehen, dass heutigen Tages fast wenig Menschen sich um die geistlichen, sondern um die leiblichen Güter bekümmern. Weil es aber bei dieser *Materia* lauter schöne, reine Geister gibt, welche mir immer schöne und herrliche Gelegenheit an die Hand geben, der himmlischen und geistlichen Güter dabei zu gedenken, damit ich nicht den Schweinen gleich, so die Eicheln unter den Baume fressen und nicht über sich sehen, wer sie ihnen gibt, habe daher bei allen Arbeiten, wo es Materie und Gelegenheit gegeben, den Menschen mit seinem Herzen und Sinn erst in GOTT einzuleiten gesucht und ihm gewiesen, wie herrlich er erschaffen, wie tief er gefallen und wie er nach Absterben seines tierischen Fleisches und Blutes, nach Tötung des alten Adams, den neuen Menschen wiederfinden und sich noch hier mit GOTT vereinigen könne, dass er hier im Geist *Christi* lebt und ein rechtes Kind GOTTES ist, so will uns der Heilige Geist, der Geist der Weisheit und des Verstandes, des Rats und der Stärke in seine Schule nehmen und uns alles lehren, welches die rechte hohe Schule ist, weit über alle Weisheit der Ägypter, Chaldäer und Perser. Solches sehen wir an Moses, Joseph, David, Salomo, Daniel, dieselben haben

righteousness."⁶⁰ 2. Epist. Petri 3:10: "But the day of the LORD will come as a thief in the night, in which the heavens shall pass away with great noise". And in Psalm 102: "They shall perish, but you shall endure: all of them shall wax old like a garment, as a vesture shall you change them".⁶¹ At the same time Prince Lucifer will take his throne which he built for himself in the dark kingdom of hell. He will be accompanied by all the damned, since they will have their dwelling in the *terra damnata* forever. Maybe someone would resent me that in this little book this secret is often compared with the divine, heavenly and angelic, *item*,⁶² the spiritual mixed with the worldly? So the answer is that since the son of GOD himself told the heavenly doctrine in parable when He compared the kingdom of heaven to a grain of mustard seed, and spoke of the sower, householder, vine-growers and the like, so I, too, used this as a parable and mirror, in which I better show GOD, heaven, angels, people, nature and creature. The fact that spiritual interpretation was cited has happened because today only very few people care about spiritual as opposed to material goods. Yet because within this *materia* there are nothing but beautiful, pure spirits, which always give me beautiful and glorious opportunities to commemorate the heavenly and spiritual goods, and in order that I do not act like pigs which eat the acorns under a tree without a glance upwards to see who gave them this food, in all work where there is matter and opportunity, I first tried to introduce people's hearts and minds to GOD, showing them how wonderfully they were created, and how deeply they did fall and how after the death of their animal flesh and blood and after the killing of old Adam, they can find the new human again and still be able to unite with GOD here. I tried to show them that they live here in the spirit of Christ and as a true child of GOD, the holy spirit, the spirit of wisdom and understanding, advice and strength wants to take us into his school and teach us everything, which is the real high school, far beyond all the wisdom of the Egyptians, Chaldeans, and Persians. We see such things in Moses, Joseph, David, Solomon, Daniel, for they all had supernatural wisdom. Because those who live

alle übernatürliche Weisheit gehabt. Denn die im neuen Menschen leben, erleuchtet GOTT mit seinem Heiligen Geist und ist die neue Geburt aus GOTT geboren, derer Himmel und Inclination GOTT selber ist und die heiligen Engel sind ihre Sterne, die haben mit dem natürlichen Sternen-Himmel nichts zu tun, sie herrschen über denselben und ihre Werke haben einen höheren Ursprung aus GOTT. Die heiligen Apostel und Propheten waren auch solche, mit dem Heiligen Geist erfüllt, mit Licht und Kraft aus der Höhe angetan, welches weit mehr ist, als der Geist der großen Welt, der Einfluss der Sterne. Sie waren in der Schule des Heiligen Geistes gelehrt, sollten die ewige himmlische Weisheit verkündigen, welche die Welt-Weisen und Welt-Gelehrten nicht verstanden, denn sie lebten nicht in diesem Himmel und Licht, sondern hatten ihr Leben und Influenz von dem äußeren Sternen-Himmel, der sie in das irdische unter die Welt-Gelehrten warf und waren nur vom natürlichen Licht erleuchtet, wussten aber vom göttlichen Licht nichts, sondern nur was der äußere Himmel bewirkt, die Gestirne und Planeten, da doch der innere Himmel ganz andere *Præsagia* hat, die den äußerlichen Menschen unbekannt, welche nur die verstehen, die in der neuen Geburt leben. Und warum sollte ich stille schweigen und das göttliche Licht verdecken oder verstecken? Da ich doch zum Lob GOTTES erschaffen, welches Licht er so reichlich in alle seine Geschöpfe eingesenkt und allen Himmeln, Engeln, Menschen, Natur und Creatur mitgeteilt, dass wir ihn als sein göttliches Licht in und außer uns suchen und finden sollen. Er will uns auch gern unseren finsteren Verstand erleuchten und uns den Heiligen Geist geben, wenn wir ihn darum bitten. Auch wenn jemand will die Wissenschaft des natürlichen Werks erlernen, so trachte er am ersten nach dem Reiche GOTTES und nach seiner Gerechtigkeit, so wird ihm das andere alles zufallen. Er bete und arbeite, habe den beständigen Vorsatz, solches nicht für sich zu gebrauchen, sondern seinem notleidenden Nächsten damit zu helfen. Es darf aber keine Heuchelei sein, denn GOTT prüft Herz und Nieren. Ich wünsche dem Leser, dass er sich selbst hierinnen suchen und finden möge, als den neuen Menschen

in the new human being are illuminated by GOD with his holy spirit and the new birth is born from GOD, whose heaven and inclination GOD is himself and the holy angels are their stars, they have nothing to do with the natural starry sky, they rule over it and their works have a higher origin from GOD. The holy apostles and prophets were also filled with the holy spirit, endowed with light and power from above, which is far more than the spirit of the great world, the influence of the stars. They were taught in the school of the holy spirit to proclaim the eternal heavenly wisdom, which the worldly sages and worldly scholars did not understand because they did not live in this heaven and light; but had their life and influence from the outer starry sky which threw them into the mundane among the worldly scholars who were only enlightened by natural light, knowing nothing of divine light. They know only about the outer heaven, its stars and planets, but the inner heaven has completely different *præsagia*, unknown to outward people, and only understood by those who live in the new birth. And why should I be silent and cover or hide the Divine Light? Since I was created for the praise of GOD, what light He so abundantly sunk into all his creatures and communicated to all heavens, angels, people, nature and creatures, was deposited there so that we should seek and find him as his divine light both within and outside of us. He also wants to illuminate our dark minds and to grant us the holy spirit when we ask him. Even if someone wants to learn the science of the natural work, if he strives first for the kingdom of GOD and for his righteousness, everything else will fall to him. Pray and work, have the constant resolve not to use this for yourself, but to help your needy neighbor with it. But such help must not be hypocrisy, because GOD tests hearts and kidneys. I wish the reader to search and find themselves here as the new human and to discover the kingdom of GOD within themselves

und in solchem das Reich GOTTES. Im alten Adam, der irdischen Erde, ist es nicht. Auch wünsche, dass er in der neuen Erde, die Perle, den Schatz im Acker finde, damit er mit seinen Neben-Christen die Ausbeute teilen möge. GOTT gebe einem jedweden was ihm hier und dort selig und nützlich ist! Ich verbleibe, jedem nach Standesgebühr christlich zu dienen, geneigt und willig.

GEGEBEN WEIMAR DEN 2. MAI 1705

as such. It is not to be found in the old Adam, the worldy earth. I also wish that [the reader] will find the pearl, the treasure in the field, in the new earth, and that they might share the yield with their fellow Christians. GOD grant everyone what is soothing and useful for them here and there! I remain inclined and willing to serve as a Christian to everyone according to their status.

PRESENTED IN WEIMAR ON 2 MAY, 1705

INHALT DES ERSTEN TRAKTATS

Erster Traktat vom *Universal* nach dem Gewächs der 3 Principien, wie es sich mit allen 3 Welten, der äußeren und finsteren Welt, der Licht- und paradiesischen und der engelhaften und himmlischen Welt vergleicht und in diesem universaliter seine ganze Schöpfung nach allen 3 Welten in einer *Summa* berührt wird, von seiner groben, unreinen und finsteren Art beschrieben, was in der äußeren Welt für Gestalt, Geruch, Geschmack, auch alle 7 Quell-Geister in ihrer herben, bitteren, sauren und grimmigen Qualität darinnen enthalten, von welcher Unart es muss geschieden werden und in das Licht versetzt, in das leuchtende gesunde Lebens-Wasser, da alle 7 Quell-Geister rein, weiß, süß, hell, leuchtend, auch zuletzt alle in einem ausgehen, begreiflich und zur vollkommenen Natur werden, welches die Bauung des Paradieses genannt wird. Auch wie dieser paradiesische Leib in die engelhafte und himmlische Welt versetzt, da diese 7 Gestalten der Natur in der engelhaften Welt durch die reinen Feuer-Geister, welches die 7 Quell-Geister in der engelhaften Welt sind, diese, als die paradiesischen Leiber in hell leuchtende Sterne und Edelgesteine verwandeln, dass unsere 7 Metalle die planetischen Gestirne werden, so auf den hohen Achsen fahren und in lauter himmlische Geister verwandelt werden, durch die Luft und Himmel der engelhaften Welt, welches das dreifache magische Rad genannt wird und die dreimal siebenfachen magischen Zahlen erfüllt, wenn das Rad dreimal herumgedreht wird, durch seine 7 Speichen, die 7 Quell-Geister in allen 3 Welten, durch welche 3 Sphären oder Himmel dieser *Universal*-Stein durchlaufen muss, ehe er verewigt und verherrlicht wird.

CONTENT OF THE FIRST TREATISE

This first treatise is on the *universal* according to the growth of the three principles, as it compares with all three worlds: the outer and dark world, the luminous and paradisiacal world, and the angelic and celestial world; and in these universally, its whole creation, according to all three worlds, is touched in *summa*, described in its coarse, impure, and dark kind, what form, smell, taste are contained in the outer word, also all seven source-spirits in their tart, bitter, sour, and grim quality. From that bad habit it has to be separated and moved into the light, into the shining, healthy water of life, where all seven source-spirits are pure, white, sweet, bright, shining, and ultimately all proceed in one, comprehensibly, and become of perfect nature, which is called the building of paradise. Also how this paradisiacal body moves into the angelic and heavenly world, because these seven figures of nature in the angelic world, the heavenly bodies in light, are transformed into shining stars and jewels by the pure fire spirits, which are the seven source-spirits in the angelic world, so that our seven metals become the planetary celestial bodies that travel on the high axes, and are transformed into nothing but heavenly spirits by the air and heaven of the angelic world, which is called the threefold magical wheel by which the three-times and seven-fold magical numbers are fulfilled, when the wheel is turned three times, through its seven spokes, the seven source-spirits in all three worlds, through which three spheres or heavens this *universal* stone must pass before it is immortalized and glorified.

DER ERSTE TRAKTAT

Sein Land liegt in dem Segen des Herrn, da sind edle Früchte vom Himmel, vom Tau und von der Tiefe, die unten liegt. Das sind edle Früchte von der Sonne und edle reife Früchte des Mondes, von den hohen Bergen gegen Morgen und von den Hügeln für und für, von den edlen reifen Früchten von der Erde und was drinnen ist. Dies ist der dreifache Segen in den Büchern Moses. Erstlich segnete Isaac seinen Sohn Jacob damit, auch segnete Jacob den Joseph und gab ihm diesen Segen, auch segnete Moses vor seinem Tode abermals den Stamm Joseph damit. Will nun jemand diesen Segen auch ererben, der bitte GOTT, dass er ihn wolle erkennen lassen, was dieses für ein Himmel und Erde, Sonne und Mond und dessen edle Früchte sind? Er darf nicht bei den Heiden in die Schule gehen, ob sie es wohl vollkommen gewusst, auch in ihren Büchern wahr und recht beschrieben, sondern er nehme die Bibel für sich, das Neue und Alte Testament, darinnen Unterricht genug. Die Arbeit in Schaffung dieser kleinen Welt ist uns vorgeschrieben im 1. Buch Moses am 1. Kap. in der großen Schöpfung der Welt, welcher wir auch nachkommen müssen, eben diesen Prozess halten und diesen Weg gehen, denn das Wort *Fiat* ist noch heutigen Tages in *esse*, welches Fürst Lucifer wohl verstund und sich zu einem GOTT, dem Sohn GOTTES gleich, wollte erschaffen, weil er sein Licht in ihm erblickte, so wollte er sich noch mehr damit verklären, welches aber nicht sein konnte, es musste nur ein GOTT bleiben und keine Neben-Götter haben, so nahm GOTT den Schein,

THE FIRST TREATISE

"Blessed of the Lord be his land, for the precious things of heaven, with the dew, and from the deep lying beneath. And with the choice yield of the sun, and with the choice produce of the months. And with the best things of the ancient mountains, and with the choice things of the everlasting hills".[63] This is the threefold blessing in the books of Moses. First, Isaac blessed his son Jacob with it, Jacob also blessed Joseph and gave him this blessing, and Moses also blessed the tribe of Joseph with it again before his death. Someone who now wants to inherit this blessing should ask GOD to show him what the heaven and earth, the sun and moon and their noble fruits are. He shall not learn from the pagans, even if they knew it perfectly, and described it truly and correctly in their books, but he should take the Bible for himself, the New and Old Testament, in which are lessons enough. The work in creating this little world is prescribed to us in the First Book of Moses in the first chapter wherein is described the great creation of the world. This we also have to follow, keeping to this process and going this way, because the word *Fiat* is still in *esse* today, which Prince Lucifer well understood and wanted to become a GOD, or like the son of GOD, because he saw his light in him and he wanted to transfigure himself even more with it, but this could not be, for the divine being just had to remain one GOD and not have any lesser gods, so GOD took the appearance of the light of the son, as his heart and light, from Lucifer, there he became a devil, his seven source-spirits, which he had in himself, as well as in

das Lichts des Sohnes, als sein Herz und Licht, von Lucifer, da war er ein Teufel, seine 7 Quell-Geister, die er in sich, wie auch in seinem ganzen Thron-Himmel in der Natur entzündet hatte, verbrannten ihn zu einer schwarzen dürren Kohle und höllischen Löschbrand, denn das süße Licht-Wasser des Sohnes GOTTES war ihm entzogen. Weil aber GOTTES heiliger Rat beschlossen, dass diese Region woraus Lucifer gestoßen und solche in grimmigen Quell-Geistern angezündet hatte, danach sollte erhalten und mit den neuen Geschöpfen der Menschen ewig erfüllt und bewohnt werden, so schuf GOTT aus diesem wüsten Klumpen so sich in dieser Entzündung reduziert hatte in solch hartes, grimmiges, finsteres und dunkles Wesen, die jetzige Welt, darauf wir leben, mit dem Himmel und 7 Planeten über uns. Die Erde war zuvor der schön leuchtende Thron des Lucifer, welche GOTT auch einmal wieder erneuern wird. Er sagt: Siehe, ich schaffe alles neu. Ich will einen neuen Himmel und eine neue Erde schaffen, in welcher Gerechtigkeit wohnt. Auch sagt Er zu Abraham: Hebe deine Augen auf und sieh gegen Morgen, gegen Mittag, gegen Abend, alles dies Land will ich deinem Samen ewig zu besitzen geben. Wir wissen aber, dass die Welt im Feuer vergehen soll, aber GOTT kann nicht lügen, was er zusagt, das hält er gewiss. Wie Lucifer die ganze Natur in allen 7 Quell-Geistern entzündet und brennend gemacht, in welchem es noch steht, in böse und gut, in lauter Anfeindung und Widerwärtigkeit, so hat doch GOTT bisher erhalten, dass die Natur nicht zu Grunde geht und wenn die Zahl der gefallenen Engel durch die Menschen erfüllt und *Christus* unser Thron-Fürst mit uns das Reich einnehmen und ewig besitzen soll, so müssen die Schiedligkeiten alle weg, böse und gut, alle 7 Quell-Geister, wieder in einen Geist der Natur inqualieren und wird GOTT diese Quell-Geister selber anzünden, dass alle diese Schiedligkeit in ihrem eigenen Geist und Feuer ausbrennen, das Böse so vom Teufel eingeführt, abgeschieden und ihm ewig zur Wohnung eingeräumt, da er mit allen Verdammten und Gottlosen seinen Kerker und Gefängnis hat, welches das Höllen-Reich genannt wird. Das Gute

his whole throne heaven in nature, burned him to a black dry coal and hellish extinguishing fire, because the sweet water of light of the son of GOD was withdrawn from him. But because GOD's holy counsel decided that this region from which Lucifer had been expelled and had set in fire with fierce source-spirits, should then be preserved and forever filled and inhabited with the new human creatures, so GOD created the present world in which we live, with the sky and seven planets above us from this desolate clump which had been reduced by inflammation to such a hard, grim, glowering, and dark thing. The earth was previously the beautifully shining throne of Lucifer, which GOD will also once again make new. He says: "See, I am creating everything anew. I want to create new heavens and a new earth in which righteousness dwells".[64] He also said to Abraham: "Lift up now your eyes, and look from the place where you are northward, and southward, and eastward, and westward: For all the land which you see, to you will I give it, and to your seed forever".[65] But we know that the world should perish in fire, and that GOD cannot lie and that he will surely keep his promises. Just as Lucifer ignited and burned the whole of nature with all seven source-spirits, in which it still stands, in evil and good, in sheer hostility and repugnance, so GOD has thus far achieved that nature does not perish and if the number of fallen angels is filled by humans, and Christ, our throne-prince, is to occupy the kingdom with us and possess it for ever, then the divisiveness must all go away, evil and good, all seven source-spirits, must be transformed[66] back into one spirit of nature and GOD will set fire to these source-spirits himself so that all this divisiveness shall burn out in their own spirits and fire, evil thus introduced from the devil, separated and given to him forever as a dwelling, where he has his dungeon and prison with all the damned and wicked, which is called the kingdom of hell. But GOD will renew the good, which is the kingdom of heaven that we will have with Christ

aber wird GOTT erneuern, welches das Himmelreich ist, dass wir mit *Christo* ewig besitzen werden. Möchte nun jemand sagen: Lucifer, da er schaffen wollte, ward ein Teufel und der Mensch will sich unterstehen GOTTES Affe zu werden, eine neue Welt zu schaffen, welche mit allen 3 Reichen sich vergleichen soll, mit allen accordieren, auch alle Kräfte der obersten und untersten an sich haben, dies ist ja ein törichtes Vornehmen. So dient zur Antwort: Dass dieses Vornehmen nicht aus der Ursache geschieht, dass der Mensch dadurch will GOTT gleich werden, sondern er nimmt dieses Werk für seine Liebe gegen GOTT, er sucht darinnen die Wunder GOTTES, seine Kraft und Allmacht, die er in die Natur, sein Geschöpf, gelegt, die er zum Lob GOTTES rühmen und preisen will, sich daraus als in einem Spiegel erkennen, die Creatur, auch die Natur, wie alles aus GOTT kommt und er sich in alles eingedrückt und eingesenkt und seine göttliche Kraft denselben verliehen, zumal in dieser *Materia*, welche er sonderlich dem gefallenen Menschen, den viele Krankheiten und Armut plagt, soll zugute und Nutzen dienen, solches dem heiligen Menschen gezeigt und gewiesen. Weil aber die heiligen Erzväter sahen, dass die meisten Menschen in des Teufels und der Hölle Reich, als in dem Zorn-Reich GOTTES lebten, so achteten sie solche des Geheimnisses nicht würdig, wollten es doch aber auf die nachkommenden gläubigen Kinder GOTTES pflanzen, dass diese Wissenschaft nicht untergehen sollte, daher beschrieben sie solches in hieroglyphischen Figuren und verdeckten Schriften, damit nur die Auserwählten dieses göttliche Studium üben, die anderen wilden tierischen Menschen sollten davon ausgeschlossen sein. Ist also keine Sünde noch wieder GOTT, sich darinnen zu üben, sich solches zu GOTTES Ehre und dem Nächsten zu Nutz zu gebrauchen. Wir wollen den Stoff und Zeug, da GOTT selbst Schöpfer gewesen, für uns nehmen, welche uns GOTT in unsere armen Hände gegeben, dass wir den Fluch und Tod davon scheiden müssen, welche es auch müssen über sich ergehen lassen, gleich der großen Welt, da die Erde Dorn und Disteln tragen muss, dies ist unser Anfang, unser *Confusum Chaos*, von der Natur oder vielmehr von der Natur Schöpfer.

forever. Now someone might say that when Lucifer wanted to create he became a devil and that now humans want to dare to become GOD's ape, by creating a new world, which is to compare with all three realms, to accord with all, also to have all powers of the highest and the lowest to itself, and that this is a foolish undertaking. The rebuttal to this objection is that this is not done for the reason that a person wants to become like GOD by doing this work, but because he considers this work as an act of love towards GOD, seeking in it the miracles of GOD, his strength and omnipotence, of which nature bears the imprint. This the operator wants to acclaim and glorify this nature for the praise of GOD, and to recognize himself in it as though looking into a mirror, to observe how all natural creation comes from GOD. For God impressed and imbued his Divine Power into all creation, and especially in this *materia*, which should serve in particular to benefit people in this fallen world who are plagued by many diseases and poverty, such things are shown and directed to holy people. But because the holy patriarchs saw that most of the people lived in the devil's kingdom and the kingdom of hell, that is the kingdom of the wrath of GOD, they did not respect such people or think them to be worthy of these secret, but wanted to implant it in the minds of future generations of believing children of GOD in order that this science should not perish. They therefore described such things in hieroglyphic figures and occult writings, so that only the chosen ones could practice this divine study, while the other wild and animalistic people would be excluded from it. So it is not a sin, nor is it against GOD, to practice using such things for GOD's glory and for the benefit of our neighbors. We want to take for ourselves the *materia* and stuff of which GOD himself was the creator, which GOD has given into our poor hands, so that we have to separate the malediction and death from it, which we also have to endure, like in the great world, as the earth has to bear thorns and thistles. This is our beginning, our *confusum chaos*, from nature or rather from nature's

Der Text sagt im 1. Buch Moses am 1. Kapitel:

> Im Anfang schuf GOTT Himmel und Erde und die Erde ward wüst und leer und es war finster auf der Tiefen und der Geist GOTTES schwebte auf dem Wasser und GOTT sprach: Es werde Licht und es ward Licht, da scheidet GOTT das Licht von Finsternis und nennt das Licht Tag und die Finsternis Nacht, da ward aus Abend und Morgen der erste Tag.

Dieses Tagwerk GOTTES ist auch unser erstes Tagwerk, unser Anfang, da die Erde noch mit dem Wasser vermischt, auch alle *Elementa* beisammen in dem wüsten Klumpen, der Geist aber dieser Welt auf dem Wasser schwebt und selbige zeitigt, das dunkle, trübe, unreine und hefenhaftige zu Boden schlägt. Aus diesem Werk ist nun zu ersehen, dass in der Schöpfung das Wort *Ruach*, nicht den Geist GOTTES als GOTT, den Heiligen Geist bedeutet, so auf dem Wasser geschwebt, sondern den Geist der Natur, als den Geist der großen Welt, nach allen 7 Quell-Geistern, welche in diesem *Confuso Chao* staken und aus den 7 Quell-Geistern GOTTES ihren Anfang genommen, welches der Welt ihr selbst eigener Separator und Scheider sein sollte, war auch von GOTT dazu erschaffen, hatte ihm auch das Vermögen dazu gegeben, dass es durch ihn getan werden könnte, weil dieser Welt-Geist nur in Fruchtbarkeit der Welt bestehen sollte. Und was sollte auch der Heilige Geist GOTTES auf diesen Wasser schweben? Er durfte es nicht erleuchten noch heiligen, weil keine göttliche Geburt daraus werden sollte, als wie in dem Leib der Jungfrau Maria, welche er überschattete, sondern es war nur die Ausgeburt dieser großen Welt, so durch Fürst Lucifer verderbt, welche er aus der Verderbnis neu schaffen wollte, aus dem wüsten Klumpen, welche auch 7 Tage, das ist 7000 Jahre stehen soll, in ihren 7 Quell-Geistern nach Zahl der 7 Planeten und wenn solche ihre großen 6 Tage in Mühe und Arbeit, ihr Tagwerk vollendet, so wird der siebente Tag ihr Sabbat sein und in dem siebenten Quell-

creator. The text of the First Book of Moses, in the first chapter, reads:

> In the beginning GOD created the heaven and the earth. And the earth was without form, and void; and darkness was upon the face of the deep. And the Spirit of GOD moved upon the face of the waters. And GOD said, Let there be light: and there was light. And GOD divided the light from the darkness and he called the light Day, and the darkness he called Night. And the evening and the morning were the first day.

This day of work of GOD is also our first day of work, our beginning, because the earth is still mixed with water, and all the *elementa* are together in the desolate lump; but the spirit of this world, which produced the same, floats above the water and hits the dark, cloudy, impure and yeast-like to the ground. From this work it can now be seen, that in Genesis the word *Ruach*[67] does not mean the spirit of GOD as GOD means the holy spirit, floating above the water, but means instead the spirit of nature, as in the spirit of the macrocosm, after all seven source-spirits, which were stuck in this *confuso chao* and which started from the seven source-spirits of GOD, which should be the world's own divider and separator, being also created by GOD for this purpose, had also given him the ability to do it through him, because this world-spirit should only exist in the fertility of the world. And why should the holy spirit of GOD also float above this water? He was not allowed to illuminate it nor to sanctify it, because it was not to result in a divine birth, as in the body of the Virgin Mary, which he overshadowed; but it was only the offspring of this great world, so corrupted by Prince Lucifer, which he wanted to create anew out of corruption, out of the desolate lump, which should also endure for seven days, that is seven-thousand years, in their seven source-spirits according to the number of the seven planets; and if such people complete their great six days of labor and work, then that becomes the seventh day of be-

Geist in göttlicher Beschaulichkeit stehen. Ach wer wird leben, wenn GOTT dieses tun wird? Es ist der Natur Ende, worauf das ewige Reich angeht. GOTT sprach:

> Es werde Licht, da scheidet GOTT das Licht von Finsternis und nennt das Licht Tag und die Finsternis Nacht.

Das *Confusum Chaos*, wenn es durch seinen Geist als dem rechten Separator gezeitigt, dass er kann geschieden werden, das Licht von diesem finsteren Klumpen, so wird der lichte Geist Tag genannt, der zurückbleibende hefenartige ist die Nacht, ist das erste Tagwerk und wird aus Abend und Morgen der erste Tag. Diesmal war noch weder Sonne noch Mond geschaffen, weder Tag noch Nacht, wird allhier der helle und lautere Geist Tag genannt, die finstere Erde die Nacht und ein Tagwerk absolviert. Ist dem Nacharbeiter sein erster Tag, aber noch keine Scheidung der Elemente, sondern nur eine Geschickt- und Zeitigmachung, dass solches kann in die Elemente geschieden werden, weil nur das Licht von der Finsternis geschieden, die anderen Elemente aber noch alle bei der Erde sind. Und GOTT sprach:

> Es werde eine Feste zwischen dem Wasser und die sei ein Unterschied zwischen den Wasser; da macht GOTT die Feste und scheidet das Wasser unter der Feste von dem Wasser über der Feste und nennt die Feste Himmel, da war aus Abend und Morgen der andere Tag.

Die Feste zwischen dem Wasser und der Unterschied zwischen dem Wasser geschieht durch eben den Geist, welcher Tag, das ist Licht, genannt worden, welches der rechte Scheider ist und die reinen Wasser von diesem wüsten Klumpen aufzieht, das himmlische Wasser davon scheidet, welches die Feste oder Himmel genannt und von den gröberen Wassern geschieden, da jenes das beseelte Wasser wird, denn das Wasser unter der Feste wurde geschieden von dem Wasser über

ing your Sabbath and standing in the seventh source-spirit in divine contemplation. O, who will live when GOD will do this? It is the end of nature, after that the eternal kingdom begins. GOD said:

> Let there be light, and GOD divided the light from the darkness and called the light day and the darkness night.[68]

The *confusum chaos*, when it is brought about by his spirit as the right separator, so that it can separate the light from this dark lump, at which point the bright spirit is called day and the yeast-like that remains back is the night, is the first day's work and evening and morning become the first day. At this time neither sun nor moon were created, neither day nor night, here the bright and pure spirit is called day, the dark earth is called night and a day's work is complete. This is the first day for the after-worker, but there is not yet a separation of the elements, but only a preparation and ripening so that such things can be separated into the elements, because only the light is separated from the darkness, but the other elements are all still on earth. And GOD said:

> Let there be a firmament in the midst of the waters, and let it divide the waters from the waters. And GOD made the firmament, and divided the waters which were under the firmament from the waters which were above the firmament, and GOD called the firmament Heaven. And the evening and the morning were the second day.[69]

The firmament between the water and the distinction between the water happens through the very spirit, which has been called day (that is to say light) which is the correct separator and draws up the pure waters from this desolate lump, separating the heavenly water from that which is called the firmament or heaven, separating it from the coarser waters, since that becomes the ensouled water, as the water

der Feste, welches in die Höhe geführt wurde und der Wasser-Schatz ist, welchen GOTT in seinem Kasten hatte, da er in der Sintflut die Fenster des Himmels auftat, die ersten bösen Menschen zu ersäufen in der Sintflut, denn da brachen alle Brunnen der großen Tiefe auf und die Fenster des Himmels. Welche Wasserscheidung, da die Wasser von dem Wasser geschieden, wir auch vornehmen müssen, ein Teil zum himmlischen beseelten Wasser aufführen, das andere bleibende Wasser bei der Erde, so ist aus Abend und Morgen der andere Tag, die andere Arbeit vollbracht.

> Und GOTT sprach: Es sammle sich das Wasser unter dem Himmel an einem besonderen Orte, dass man das Trockne sehe und nennt das Trockne Erde und die Sammlung der Wasser nennt er Meer; und GOTT sprach: Es lasse die Erde aufgehen Gras und Kraut, dass sich besamt und fruchtbare Bäume, das ein jegliches nach seiner Art Frucht trage und habe seinen eigenen Samen bei sich selbst auf Erden; da ward aus Abend und Morgen der dritte Tag.

Hier wird nun von völliger Scheidung der Elemente geredet, welches genannt wird eine Zerstreuung der Elemente. Weil nun die Erde aus dem Wasser bestanden, so wird es davon geschieden und auch wieder dadurch befeuchtet als durchs Meer-Wasser, welches sich in den Wolken in die Höhe zieht und durch den Regen wieder herunter kommt, dadurch die Erde fruchtbar wird, dass alles aus ihr wächst und jedes seinen Samen bei sich selbst hat, dadurch es sich vermehrt. Ob nun wohl die *Elementa* sowohl in der großen als in unserer kleinen Welt schienen rein geschieden zu sein, so hat dennoch die Erde noch alle *Elementa* in sich, das Wasser ist auch voller Luft, Feuer und schmutziger Früchte, die Luft hat auch in sich Wasser, Feuer und Erde, das Feuer hat auch in sich Erde, Wasser und Luft, muss auch also sein, denn die simplen *Elementa* wären ohne Kraft und ein totes Wesen, welches

under the firmament has been separated from the water above the firmament, which was raised up, and which is the treasure of water that GOD had in his case when he opened the windows of heaven to drown the first wicked people in the deluge, for all the wells of the great deep broke and the windows of heaven, too. This separation of the waters is something we also have to undertake, adding one part to the heavenly ensouled water while the other water remains with the earth, and thus evening and morning are the second day, the second work is done.

> And GOD said: let the waters under the heaven be gathered together unto one place, and let the dry land appear, and GOD called the dry land Earth, and the gathering together of the waters called the Seas. And GOD said: Let the earth bring forth grass, the herb yielding seed, and the fruit tree yielding fruit after his kind, whose seed is in itself, upon the earth, and the evening and the morning were the third day.[70]

Here we speak of a complete separation of the elements, which is called a dispersion of the elements. Because the earth now consisted of the water, it is separated from it and also moistened again by it, as through sea water, which rises in the clouds and comes down again through rain, whereby the earth becomes fertile, that everything grows out of it and each has its own seed, whereby it multiplies. Even though the *elementa* seemed to be purely separated in the macrocosm as well as in our microcosm, the earth still has all the *elementa* in itself, the water is also full of air, fire, and dirty fruits; the air also has in itself water, fire, and earth; fire also has earth, water, and air in itself. It has to be this way, because the simple *elementa* would be without power and a dead

in der Nacharbeit alles zu finden und zu sehen, wenn der Künstler die Elementa geschieden und die Erde durch Anfeuchtung ihres eigenen und bleibenden Wassers fruchtbar gemacht, so ist aus Abend und Morgen der dritte Tag, die dritte Arbeit vollbracht. Hier muss sich der Nacharbeiter vorsehen, dass er die erste Welt nicht flugs mit der Sintflut ersäufe, sondern dies bedenken, dass GOTT nicht flugs regnen ließ auf Erden, es musste solche erst wohl austrocknen und ging ein Strom aus zu wässern den Garten, das teilte sich daselbst aus in 4 Haupt-Wasser, diese hat der Künstler vor allem in acht zu nehmen,

> und GOTT sprach: Es werden Lichter an der Festen des Himmels und GOTT macht 2 große Lichter, ein großes Licht, dass den Tag regiert und ein kleines Licht, dass die Nacht regiert, dazu auch Sterne und scheidet das Licht von der Finsternis, da ward aus Abend und Morgen der vierte Tag.

Der Nacharbeiter muss nun aus dem reinen Licht-Wasser, den reinen Elementen, das Gestirne schaffen, unserer kleinen Welt ihre 7 Planeten, dass wir auch Sonne und Mond haben, welche stets unsere Erde bescheinen und wird noch eine Finsternis zu sehen sein, eine Nacht, welche wir davon scheiden müssen ehe diese Lichter in ihrer Ordnung im Glanz aufgehen und leuchten werden, die *Luna* muss durch ihre Kälte und Luft die Erde erfrischen, die Sonne solche trocknen, so ist dem Künstler der vierte Tag, das vierte Tagwerk, vollbracht. Und GOTT sprach:

> Es errege sich das Wasser mit webenden und lebenden Tieren, mit Gevögel, dass auf Erden unter der Feste des Himmels fleucht und schuf große Walfische und allerlei Tiere das vom Wasser erregt war, da ward aus Abend und Morgen der fünfte Tag.

being, which can be found and seen in the after-work, if the artist has separated the elements and made the earth fertile by moistening with its own and permanent water, so this is the third evening and morning, the third day, the third work is done. Here the after-worker must be careful that he does not drown the first world in a hurry with the deluge, and should now consider this, that GOD did not let rain quickly on earth. Rather, it first had to dry out and a stream went out to water the garden that divided itself from there into four principal waters. The artist has to be careful about these.

> And GOD said: Let there be lights in the firmament of the heaven. And GOD made two great lights, the greater light to rule the day, and the lesser light to rule the night: He made stars also and divided the light from the darkness, and the evening and the morning were the fourth day.[71]

The after-worker must now create the stars from the pure light-water, the pure elements, so that our microcosm has its seven planets, including sun and moon, which always shine on our earth and there will still be a darkness to be seen, a night which we must separate from it before these lights rise in their order, in their splendor, and shine. The *Luna* must refresh the earth with its cold and air, and the sun must dry it, so that the fourth day, the fourth day's work, is accomplished by the artist. And GOD said:

> Let the waters bring forth abundantly the moving creature that hath life, and fowl that may fly above the earth in the open firmament of heaven and GOD created great whales, and every living creature that moveth, which the waters brought forth abundantly, and the evening and the morning were the fifth day.[72]

Hier wird das fruchtbare Lebens-Wasser beschrieben, so in Geist, Seele und Leib besteht, darinnen sich alles multipliciert, in diesem gesegneten Wasser. Der Nacharbeiter schaffe nun aus dem Wasser das Fischlein *Æscheneis*, dass es zum Vorschein kommt mit silbernen Schuppen, wie auch leuchtenden und funkelnden Augen, damit das Meer damit erfüllt. Zuvor aber müssen die Wasser-Vögel in der Luft über diesem Meer fliegen, sonderlich muss der silberne weiße Schwan, der rechte Wasser-Vogel, auf diesem Meer schwimmen, so hat er das fünfte Tagwerk auch vollbracht.

> Und GOTT sprach: Die Erde bringe hervor lebendige Tiere und sprach: Lasset uns Menschen machen, ein Bild das uns gleich sei, die herrschen über die Fische im Meer, über die Vögel unter dem Himmel und über alles was sich regt auf Erden und sprach: Seid fruchtbar und mehret euch und füllet die Erde. Da ward aus Abend und Morgen der sechste Tag.

Der Nacharbeiter sieht nun, dass er mit der Quintessenz lebendige Geschöpfe zeugen soll, die ihr Geschlecht in viele 1000 multiplicieren, durch die hermaphroditische Natur, welche der erst-geschaffene Mensch Adam war, sein Weib, seine Matrix, in seinem Leibe hatte, dadurch er sein Geschlecht fortpflanzen sollte, welche große Könige und Herren der Welt werden, die großen Reichtum besitzen, so wird mit dem sechsten Tagwerk Arbeit aufhören und der siebente Tag sein Ruhe-Tag sein und wird der kluge Arbeiter sehen, wie es in der Multiplication dieser herrschenden Könige hergehen wird, dass großes Blutvergießen, Krieg und Streit erregt, einer den anderen vom Thron stößt, solchen einnimmt, sich zum Herrn und Regenten macht, bald das Unterste das Oberste und das Oberste das Unterste wird, aber vorher muss Adam und Eva im Paradies sein, in Lust und Freude, werden aber durch den Cherub in die äußere corrosivische Welt getrieben, da lauter Elend und Jammer, Kummer und Schmerzen, zeugen auch böse

Here the fertile water of life is described as consisting of spirit, soul, and body, in which everything multiplies in this blessed water. The after-worker now creates the little fish *echeneis* out of the water, so that it appears with silver scales, as well as shining and sparkling eyes, so that the sea is filled with it. Before that, however, the water birds have to fly in the air over this sea, and the silvery white swan, the proper water bird, has to swim on this sea in particular, so it has also completed the fifth day's work.

> And GOD said: Let the earth bring forth the living creature, and GOD said: Let us make humans in our image, after our likeness: and let them have dominion over the fish of the sea, and over the fowl of the air, and over every creeping thing that creepeth upon the earth, and GOD said: Be fruitful, and multiply, and replenish the earth. And the evening and the morning were the sixth day.[73]

The after-worker now sees that he is to beget living creatures with the quintessence, multiplying their lineage by many thousands by the hermaphrodite nature which was the first-created human Adam, who had his wife, his matrix, in his own body, whereby he should propagate a generation that would become great kings and lords of the world, and who would possess great wealth. Then work will cease with the sixth day's work and the seventh day will be the day of rest. The wise worker will then see how the multiplication of these reigning kings will proceed. There will be great bloodshed, war and strife, as one pushes the other off the throne and takes it, making himself lord and regent. Soon the lowest will be the highest and the highest will be the lowest, but before that Adam and Eve must be in paradise, in pleasure and joy until they are driven by the cherub into the outer corrosive world, with sheer misery and lamentation, sorrow and pain, there to beget wicked

Kinder, Kain den Bruder-Mörder, bis letztlich der aus dem sechsten Stamm oder Glied, Enoch im göttlichem Leben blieb: Also in unserem Werk die Kinder, welche von Adam und Eva gezeugt werden, sind nicht gleich große, herrschende Könige, sondern unsere 7 Metalle, welche alle nicht die Probe bestehen, das ist die Capelle, bis der sechste, die 6 weißen Felder erfüllt, durch die 6 Grade der Natur durchgegangen und den siebenten erreicht, den Sabbat und Ruhe-Tag, da man das Rad umwenden muss, das ist, aufhören mit kochen, in welcher Zeit die Sintflut wird kommen und der Regenbogen, das Gnaden-Zeichen, erscheinen. Wie aber GOTT die erst geschaffene Welt nicht gar im Wasser aufgelöst oder zergehen lassen, so muss es der Nacharbeiter wohl bedenken, dass er die Erde nicht zu dürre lasse austrocknen, welche nichts nütze wäre, auch durch Anfeuchten und Regen nicht gar auflöse oder zerschmelze, sondern noch fruchtbar bleibe, dass, wenn die Wasser der Sintflut eingetrocknet, das Gnaden-Zeichen, der Regenbogen, erscheinen könne, welche Arbeit noch besser und weitläufiger erklären will, wenn erstlich seine vielen Namen ausgelegt, wodurch es der Leser erkennen lernt, als unser *Confusum Chaos* von GOTT dem Schöpfer, welches uns die Natur gegeben, die *Materia Cruda*, welche mit so vielen Namen genannt wird, als:

1. *Azoth*, weil daraus der rechte Wurzel-Essig gemacht wird.
2. ♄, der Zeuge-Vater aller Götter, da die güldene *Genealogia* der Götter herstammt, welcher alle seine Kinder frisst.
3. *Electrum*, weil es alle 7 Metalle in sich hat und das rechte *Electrum artificiale* aus ihm zu machen, von außen aber mit einem weißen ☿ angeflogen ist, dass es eine andere Farbe hat, damit man seine innerlichen Schönheiten nicht sieht und es vor so vielen Nachstellern verdeckt bleibt.
4. *Magnesia*, weil es wegen seiner hitzigen und trockenen Natur alle *Acida* an sich zieht und was es an sich zieht, ist ihr Gefangener, absonderlich den *Spiritum Mundi*, das *Nitrum Aëreum*.
5. Unser Magnetstein, weil sein innerlicher centralischer *Polus* oder

children such as the fratricidal Cain, until finally in the sixth generation Enoch remained in divine life. So in our work the children who are begotten by Adam and Eve are not great ruling kings at first, but our seven metals, none of which pass the test, that is the cupel, until the sixth fills the six white fields, [having] passed through the six degrees of nature and reached the seventh, the Sabbath and day of rest, when one must turn the wheel, that is, stop cooking, in which time the deluge will come and the rainbow, the sign of grace, will appear. But just as GOD does not let the first created world dissolve or melt in water, so the after-worker must carefully ensure that he does not let the earth dry out too much, which would be of no use. He must ensure that even if it is moistened and rained on, it does not dissolve or melt at all, but still remains fruitful, so that when the waters of the deluge are dried up the sign of grace, the rainbow, can appear. I will explain this work even better and more extensively, if first its many names are interpreted, through which the reader learns to recognize it, as our *confusum chaos* of GOD the creator, which was given to us by nature, the *materia cruda*, which is called by so many names as:

1. *Azoth*, because the right root vinegar is made from it.
2. ♄, the procreating father of all the gods, from whom the golden genealogy of the gods came forth, who devoured all his children.
3. *Electrum*, because it has all seven metals in it and the right *electrum artificiale* can be made out of it, but has a white ☿ attached to it from the outside, so that it has a different color, so that its inner beauties cannot be seen and it remains hidden before for so many followers.
4. *Magnesia*, because due to its hot and dry nature it attracts all *acida*, and what it attracts is its prisoner, especially the *spiritum mundi*, the *nitrum aëreum*.
5. Our magnet stone, because its inner central *polus* or salt is our

Salz unsere kleine Welt ist und alle seine gereinigten Teile als ein Magnet an sich zieht und in Gestalt eines trockenen und verklärten Leibes in sich behält und zum festen Stein coaguliert, wenn die gereinigten Elementa wieder zusammengesetzt werden, welches hernach nimmermehr zu scheiden ist.

6. *Æs Hermetis*, das hermetische Erz, weil die philosophischen Metalle hier stecken, Hermes heißt auch die Schlange und ist es die metallische giftige Schlange.

7. *Prima Materia* des Steins, weil es der erste Stoff oder Zeug dazu ist.

8. Der ☿ *coagulatus*, weil der ☿ in diesen *Subjecto* gleichsam nur niedergesessen und mit seinem 🜍 noch nicht radicaliter vereinigt und in der Scheidung jedes *á parte* zu haben und die ganze Kunst hierinnen besteht, diese beiden zu vereinigen, welches die größte Mühe in der ganzen Arbeit und ist die Braut darum man tanzt, wie man den roten Knecht mit dem weißen Weibe vereinigen soll, wer dieses kann, der hat gewonnen.

9. Der Stein oder Felsen, der Öl, Wein, Blut und Milch gibt, die 4 *Elementa* und die vielen *Menstrua*, das Land, darinnen Milch und Honig fließen.

10. *Metallum Primum*, weil die rote ♀ überflüssig in diesen *Subjecto* ist und das erste Metall genannt wird.

11. Ein Stein, weil es ein hartes, trockenes und irdisches Wesen, hernach ist es unser 🜨 Stein, weiter der glasförmige *Azoth* oder Kristall-Stein des ☿. Endlich die weiße Tinctur, Wasser-Stein der Weisen, zuletzt wird es der Rubin und Karfunkelstein.

12. Gold-Baum und seine Wurzel, es hat kein ☉ noch ☽, führt aber viel flüchtigen ☽ Schwefel und von dem blut-roten Purpur-Schwefel des Goldes und hat den einzigen ☿ bei sich, der sich gern gradieren lässt von seinem eigenen Schwefel, dieser ☿ ist der Baum, der die Blüte der Sonne und des Mondes trägt, welche auch auf ihm reif werden und sich viel tausendfältig vermehren.

13. *Prima Materia* aller Metalle, erstlich weil alle 7 philosophischen Metalle daraus gezeugt werden. Wenn aber die *Elementa* geschie-

microcosm and attracts all its purified parts like a magnet and keeps it in the form of a dry and transfigured body and coagulates into a solid stone, when the purified elements are put together again, which cannot be separated afterwards.

6. *Æs Hermetis*, the Hermetic ore, because the philosophical metals are contained here; Hermes is also called the serpent and he is the poisonous, metallic serpent.
7. *Prima Materia* of the stone, because it is the first material or stuff of it.
8. The ☿ *coagulatus*, because the ☿ just sat down in this *subjecto* and has not yet united *radicaliter* with its 🜍, and to get each *á parte* in the separation, and the whole art consists in uniting these two, which is the greatest effort in the whole work and the bride around which all are dancing. How to unite the red servant with the white woman? Whoever can do this has succeeded.[74]
9. The stone or rock that gives oil, wine, blood and milk, the four *elementa*, and the many *menstrua*, the land in which milk and honey are flowing.
10. *Metallum primum* because the red ♀ is superfluous in this *subjecto* and is called the first metal.
11. A stone because it has a hard, dry, and earthly nature, henceforth it is our 🜨 stone, further the glassy *azoth* or crystal stone of the ☿. Finally the white tincture, the water-stone of the wise,[75] in the end the ruby and carbuncle stone.
12. Gold-tree and its root, it has no ☉ nor ☽, but carries much volatile ☽ sulphur and from the blood-red purple sulphur of gold and has with it the only ☿ that likes to be gradated by its own sulphur. This ☿ is the tree that bears the blossoms of the sun and the moon, which also ripens on it and multiply many thousandfold.
13. *Prima materia* of all metals, firstly because all seven philosophical metals are created from it. But if the *elementa* are separated and

den und die erste Composition vorgenommen wird, dann ist es *Prima Materia* aller Metalle, weil hier die Planeten müssen durch den ganzen *Zodiacum* durchlaufen, bis die Sonne in ihr eigenes Haus im ♌ kommt und der ♌ die Sonne gefressen, weil er seinen Anfang, Ursprung und Fortgang von dieser *Materia* hat.

14. Pyrites, Feuerstein, weil der Stein *Ignis* daraus zu machen, der feurige Karfunkel so im Finstern leuchtet, das feurige glänzende *Urim* und *Thumim*, das coagulierte und erhärtete Feuer.

15. *Terra Lemnia*, roter Ton, weil aus dieser roten Erde, der rote ♀, der rote Mann und das weiße Weib vorkommen, weil es aber noch viele andere solche Namen und diese viele Namen zur Sache nichts helfen, lässt man billig solche fahren und befleißigt sich nur seine Arbeit zu lernen, da die Reinigung vorher gehen muss, worauf die Scheidung und Zusammensetzung erfolgt.

Es ist nur ein GOTT im Himmel und nur eine *Materia* in der Welt, woraus dieses zu machen, welches hat:

1. Die erste *Materiam*,
2. Die zwei Schwefel, Mann und Weib, roten und weißen ♀,
3. Die 3 *Principia*,
4. Die 4 *Elementa*, diese alle gereinigt, geschieden und wieder zusammen gesucht, so ist die Arbeit geschehen, aber ohne die aufgesperrte Pforte der Elemente ist nichts zu tun.

Von der Erde, worauf wir gehen, muss unsere Erde, unser Laton, genommen werden, welche zweierlei ist: 1. Der rote Ton, Letten oder Erde, die andere ist das jungfräuliche Salz, *Hyle* genannt und kommt beides aus einem, mit dem roten *Laton* wird keine Composition gemacht, weil es wieder davon zu scheiden ist, auch geschieden werden muss, weil sich Feuer und Wasser nicht vereinigen lassen, mit dem reinen Salz oder der reinen Erde und Wasser geschieht die erste Composition. Die Luft und das gemeine Wasser sind es auch nicht, denn so

the first composition is made, then it is *prima materia* of all metals, because here the planets have to go through the whole *zodiacum* until the sun comes into its own house in the ♌ and the ♌ has eaten the sun because it has its beginning, origin, and progression from this *materia*.

14. Pyrites, flint stone, because the stone *ignis* is made of it, the fiery carbuncle, which glows in the dark, the fiery lustrous *Urim* and *Thumim*,[76] the coagulated and hardened fire.
15. *Terra lemnia*, red clay, because the red ♁, the red man and the white woman come out of this red earth, but because there are many other such names and these many names do not help the matter, you simply let them go and just make an effort to learn this work, as the purification must go first, after which the separation and composition follows.

There is only one GOD in heaven and only one *materia* in the world from which this is to be made, which has:

1. The first *materiam*,
2. The two sulphurs, male and female, red and white ♁,
3. The three *principia*,
4. The four *elementa*, all purified, separated and put together again, so the work is done; but without the unlocked gates of the elements, nothing can be done.

From the earth on which we walk, our earth must be taken, our loam, which is of two kinds: The first is the red clay, potter's clay, or earth; the second is the virgin salt called *Hyle* and both come from one, no composition is made with the red loam, because it is to be separated again, also must be separated, because fire and water do not combine, the first composition occurs with pure salt or pure earth and water. It is also not air and common water, because our metallic seed is not set

weit ist unser metallischer Same nicht zurückgesetzt, würde auch eines Menschen Leben zu kurz sein, aus solchen einen tingierenden Stein zu machen, der fix und unverbrennlich im Feuer wäre, weil unsere *Materia* muss unverbrennlich sein. Aus Metallen, durch Metalle und mit Metallen wird der Stein gemacht, wiewohl die *Materia* kein eigentliches Metall, doch die Blume der Metalle, auch kein eigentliches Mineral, doch den Mineralien am nächsten verwandt; vegetabilisch ist es, weil es wächst und mit einer lebendigen Seele begabt, auch das vegetabilische lebendig-machende Wasser daraus zubereitet wird, sich auch auf alle 3 Reiche applicieren lässt. Dem mineralischen Reiche ist es sehr nütze, weil es solche Leiber in ☉ und ☽ verwandelt, dem vegetabilischen noch mehr, dass alle alten Bäume dadurch fruchtbar werden und kann man viele tausend Wunder in diesem Reiche damit anrichten, weil durch dieses Feuer alle vegetabilischen Essentien in kurzer Zeit können damit geschieden werden. Zum animalischen Reiche ist es am herrlichsten. Was geht über die Gesundheit? Von innen und außen heilt dieser Balsam. Wenn GOTT nicht den Tod der Sünden wegen gesetzt, würden die Menschen dadurch unsterblich sein. Diese reinen *Elementa* nun scheinen wohl rein und geschieden, da es doch unmöglich ist, solche ganz rein zu scheiden, die Erde hat noch von sulphurischer Fettigkeit, der ☿ oder Wasser noch mehr, die Luft hat auch ☿ oder Wasser bei sich, das Feuer hat viel irdisches als Erde, welches in der Composition alles zu sehen. Diese *Materia* muss sehr wohl gereinigt werden, ihr grober irdischer Leib dient dazu nicht, nur mit den reinen Teilen wird das Werk gemacht, weil nicht das ganze *Corpus* genommen wird, sondern nur des Alten sein Blut und Seele, wenn er in sein Bad der Reinigung geführt, dass sein äußerlicher Aussatz abgewaschen, so ist es erst unser Adam, der die Welt vermehren soll, dieser ist mit seinem Weibe, seiner Eva, schwanger, welche von ihm geschieden wird, so bleibt Adam, unser roter Ton, *Laton* oder Letten zurück, aus welchem *Laton* das Paradies-Wasser gemacht wird, der Strom, so aus dem Paradies entspringt und sich in 4 Ströme teilt. Der erste, *Pison*, ist der Ganges in Indien, so viel Gold bei sich führt und

back that far, even a human life would be too short to make a tingeing stone from them, a tingeing stone that would be fixed and incombustible in fire, because our *materia* must be incombustible. From metals, through metals, and with metals, the stone is made, although the *materia* is not a metal proper, yet the flower of metals, not a mineral proper either, yet is closest to minerals. It is vegetabilic because it grows and is endowed with a living soul, the vegetabilic life-giving water is also prepared from it, and can also be applied to all three kingdoms. It is very useful for the mineral kingdom, because it transforms such bodies into ☉ and ☽, for the vegetabilic kingdom even more so, because all old trees become fertile and one can work many thousands of miracles in this kingdom with it, because through this fire all vegetabilic essences can be separated with it in a short time. To the animal kingdom it is most glorious. What is more important than health? This balm heals from the inside and outside. If GOD had not instituted death because of our sins, human beings would be immortal because of it. These pure *elementa* now seem to be pure and separated, although it is impossible to separate such completely pure, the earth still has a sulphurous fatness, and the ☿ or water even more. The air also has ☿ or water with it, and the fire has a lot terrestrial such as earth, which all can be seen in the composition. This *materia* must be very well purified, its coarse earthly body is not useful for this, and the work is only done with the pure parts, because the whole *corpus* is not taken, but only the blood and soul of the old man when it is led to his bath of purification, so that its external leprosy is washed off. Thus it is only our Adam who is to multiply the world, he is pregnant with his wife, his Eve, who will be separated from him, so Adam remains our red clay, loam, or potter's clay, from which loam the water of paradise is made. The river thus springs from paradise and divides into four rivers, the first of which, *Pison*, is the Ganges in India, which carries so much gold with it—and

das Gold desselben Landes köstlich ist und den Edelgestein Onyx und Bedellion. Der andere, *Gihon*, fließt um das Mohrenland, ist der *Nilus* in Ägypten, da die große Schlange, das Krokodil. Der 3te heißt *Hidekkel*, ist der Tigris in Assyrien. Der 4te ist der *Phrath* oder Euphrat in Syrien, diese beiden Ströme sind in einem Lande, sie gehören auch zusammen, der andere ist die giftige Schlange, *Venenum Tingens*, unser ♂alisches Öl, das erste unser güldener ♁, dadurch der Stein zu ☉ wird, dass durchsichtig ist, leuchtend als ein Edelgestein, ☉ in Glasgestalt. Wenn nun dies Weib vom Mann geschieden, die Eva von Adam genommen, so muss sie ihm wieder ehelich beigelegt werden, dass sie von ihm empfängt und schwanger wird. Dies ist unser anderes *Confusum Chaos*, durch Kunst gemacht, darinnen die 4 Elemente stecken und die vielen *Menstrua* hervorkommen, sie werden aber nicht wohl zu scheiden sein, wenn es nicht vorher wohl verfault und putrificiert, alsdann scheidet man sie, jetzt ist es die äußere finstere Welt, Fluch und Tod, Gift und Mord, wenn Eva von Adam genommen, so ist sie zwar geschaffen, sie kann aber nicht schwanger werden ohne Adam, die Erde, welche gereinigt werden muss, auch muss man wohl bedenken was die giftige Schlange sei, welche sich zur Even gesellt, sie zu verführen: Es ist das giftige *Oleum* ♄, welches am allermeisten der ♀ zugetan und bisher innerlich in ♄ gespielt und nun von ihm geschieden worden, diesen Schlangen-Stich werden hernach diese beiden Eheleute schon fühlen, wenn sie aus dem Paradies getrieben, aus der Lust und Freude in die äußere Mord- und corrosivische Welt, da sie beide den Tod leiden müssen, weil aber ihre Kinder der Welt und nicht göttlich lebten, bis Seth erst im 6. Glied in einem göttlichen Leben blieb, so werden diese Kinder auch nicht in der Feuer-Probe bestehen, als der 6te, welcher der ☿ *Philosophorum* wird, so Krone und Zepter trägt, auch unsere Sonne und Mond ist, doch muss er leiden und sterben durch der Schlangen Stich, durch welchen Tod er erst verklärt zum weißen Stein wird, welcher hernach in die himmlische und engelhafte Welt versetzt, dass er nimmermehr wieder sterben kann, sondern ewig lebend ist. Diese *Materia* ist von sehr widerwärtiger Natur,

the gold of that land is good, and there is also bdellium and the onyx stone. The name of the second river is *Gihon*, which flows around the land of Ethiopia, and this is the *Nilus* in Egypt, where the great serpent is, the crocodile. And the name of the third river is *Hiddekel*, which is the Tigris in Assyria. The fourth river is the *Phrath* or Euphrates in Syria, these two rivers are in one country, and they also belong together; the other is the poisonous serpent, *venenum tingens*, our ♂al oil, the first our golden ♃, by which the stone becomes ☉, which is transparent, shining like a jewel, ☉ in the form of glass. Now if this woman is separated from her man, if Eve is taken from Adam, then she must have marital intercourse with him again, so that she conceives from him and becomes pregnant. This is our other *confusum chaos*, made by art, in which the four elements are contained and from which the many *menstrua* emerge. But they will not be easy to separate. If they did not rot and putrify beforehand, and you separate them at that point, it is the outer dark world, malediction and death, poison and murder, when Eve was taken from Adam. She is indeed created, but she cannot become pregnant without Adam, the earth, which has to be purified. One must also carefully consider what the poisonous serpent was that approached Eve to seduce her: It is the poisonous *oleum* ♄, which is most attached to ♀ and has up to now played inwardly in ♄, and has now been separated from it. These two spouses will feel this serpent's sting afterwards, when they are driven out of paradise through lust and joy into the murderous and corrosive external world, where they must both suffer death. Yet because their children lived in the world and were not divine, until only Seth in the sixth generation remained in a divine life, these children will not live to pass the test of fire, as the sixth, which becomes the ☿ *philosophorum*, wearing a crown and scepter, is also our sun and moon. But it must suffer and die through the sting of the serpent, through which death it first becomes transfigured into a white stone which afterwards is transposed into the heavenly and angelic world where it can never die again, but is ever living. This *materia* is of a highly repulsive nature, for when the *elementa* are separated, all

wenn die *Elementa* geschieden, welche alle einander anfeinden und widerwärtig sind, nach Art der äußeren finsteren Welt, die Luft ist das ♃ Öl, gelb und inwendig blau, weil es ♃ ☽ in großer Menge in sich hat, tingiert alle ☿ialischen Leiber in gelb, das Wasser ist ein metallisches Wasser, glutinosisch und die Jungfrau, so von dem geistlichen Samen ihres Mannes schwanger worden, durch die Imagination, welche nur geschickt ist beizuliegen ihrem rechten Mann, dem reinen Adam, der paradiesischen Erde, dem centralischen Salz, *Hyle*, so aus dem roten *Laton* kommen und mit des roten ♄ Blut bedeckt gehalten und unser ☉ war, davon der rote und weiße Geist geschieden worden, dieses ist das solarisch-martialische und feurige Salz, welches seinen eigenen ☿ coaguliert und fixer als ☉ ist, gras-grün inwendig und sehr fett, kann durch viele Reinigung dahin gebracht werden, dass es viel gleißender als Perlmutt, ja mit solchen zarten Farben spielt, welche alle paradiesisch scheinen und wird alle Gefäße zertrümmern, die jungfräuliche Erde, die so tief muss gegraben werden, unsere kleine Erde und centralischer *Polus*, da unsere Planeten herum gewälzt werden. Die Künstler lassen sich begnügen, wenn sie solches trocken und schneeweiß zum Werk haben, welches nun alle seine gereinigten Teile wieder an sich zieht, welche ewig nicht wieder von ihm zu scheiden, weil rein und rein zusammenkommen, welche beide ein himmlisches Wesen zeugen, so nicht mehr irdisch, sondern paradiesisch und der Baum des Lebens ist, wegen der Menschen Sünde aber mit dem Fluch und Tod bedeckt, dass sein himmlisches *Ens* erschreckt und zurück in sein *Centrum* getreten, nun aber in seiner paradiesischen, spiritualischen Gestalt wieder hervorgekommen, weil es im Paradies nicht corporalisch gewesen und es Adam auch nur spiritualisch daraus getragen, wiewohl er eben von diesem roten Ton gemacht worden, denn die 7 Quell-Geister im Paradies ist eben das Licht-Wasser, von oberen, mittleren und unteren Wassern und im siebenten in Licht und Klarheit ausgangen, die Licht-Welt und begreifliche paradiesische Welt worden in ihrer Diaphanität, woraus Adam und Eva getrieben und der Cherub mit dem feurigen Schwert davor gelegt, zu bewahren den Weg zu dem Baum des Lebens.

of which are hostile and repulsive to one another, in the manner of the outer dark world. The air is the ♃ oil, yellow and blue within, and because it has ♃ ☽ in itself in large quantities, it tinges all ☿al bodies in yellow. The water is a metallic water, glutinous, and the virgin conceived by her husband's spiritual seed, by the imagination, which is only skillful to accompany her right husband, the pure Adam, from the paradisiacal earth, the central salt, *Hyle*, thus came from the red loam, and kept covered with the red ♌ blood and was our ☉, from which the red and white spirit has been separated. This is the solar-martial and fiery salt, which coagulates its own ☿, and is more fixed than ☉. It is grass-green inside and very rich, and by undergoing many purifications it can be brought to the point that it is much more dazzling than mother-of-pearl, indeed playing with such delicate colors, which all seem paradisiacal and will smash all vessels, the virgin earth which must be dug so deep, our little earth and central *polus* around which our planets are rolled. The artists are content when they have something dry and snow-white for their work, which now draws back all its purified parts, which can never be separated from it again, because pure and pure came together, both of which beget a heavenly being, which is no longer earthly but paradisiacal, and the tree of life, but covered with the malediction and death because of human sin, that its heavenly *ens* is frightened and has stepped back into its *centrum*; but it now comes out again in its paradisiacal, spiritual form because it was not corporal in paradise, and Adam only carried it spiritually from it although he was made of this red clay, because the seven source-spirits in paradise are the light water, from upper, middle, and lower waters and in the seventh went out in light and clarity. It has become the world of light in its diaphaneity, the comprehensible, paradisiacal world from which Adam and Eve were expelled and the cherub were put before to guard the way to the tree of life with the fiery sword. The water of paradise, our virgin, is the most beautiful creature,

Das Paradies-Wasser, unsere Jungfrau, ist das allerschönste Geschöpf, weiß, klar und durchsichtig, als ein Opal, schwer, wohlriechend, Milch der Vögel, Milch des ☿, ohne welche kein Mensch leben kann, welchen wir unsichtbarer als das Wasser in der groben elementischen Luft in uns ziehen, ist ohne Schärfe. Das Feuer ist sehr scharf, brennend und leuchtend als eine glühende Kohle, welches wir Menschen nach den animalischen Quell-Geistern auch in uns haben und in den Augen, als im Licht am meisten seinen Sitz hat, solches sieht man wohl, wenn man in die Augen geschlagen wird, wie es heraus springt und leuchtet, dieses Feuer hat nun die Kraft, das gläserne Meer in den Karfunkel zu verwandeln, wenn es mit diesem Feuer gemengt wird, welches das *Urim* und *Thumim*, Licht und Recht bei den Israeliten gewesen und aus einem funkelnden Erz gemacht worden, welches geleuchtet. Es leuchtet auch unser doppelter ☿, unser ☿ Wasser, als ein Licht, weiß, helle, der rote Karfunkel-Stein mit Blicken, als wenn ein Metall auf dem Test abgeht, auch hat dies Feuer schon einen leuchtenden Glanz, weil es noch in Gestalt des solarischen ☿ ist, in Gestalt des fliegenden roten Löwen oder Drachen erscheint, wenn man es in die Sonne setzt, so wird ein ganz gemach Leuchten als lauter feurige Spiegel. Alle diese widerwärtigen Dinge stecken nun in diesem einzigen *Confuso Chao*, welche, wenn sie alle höchst gereinigt wieder zu einem einzigen unsterblichen Leibe durch die Zusammenkunft werden müssen. Es hat den Fluch als die große Welt auch über sich ergehen lassen, jedoch hat GOTT sein himmlisches Wesen und Kraft darinnen gelassen, dem gebrechlichen Menschen zum besten, wie er auch Adam den göttlichen Odem als den Geist gelassen, ob er wohl des himmlischen Bildes abstarb, des Tages, da er das Gebot GOTTES übertreten, verhieß er ihm gleich dem Erlöser des Weibes Samen, durch welchen Geist er wieder im neuen Menschen erweckt ward, da er ihn im Glauben erfasst und bis Adam, wie vorher, den freien Willen, ob er in GOTT, oder der Welt und seinen eigenen Willen leben wollte. Wie nun das Paradies durch den Sohn GOTTES wieder geöffnet und Adam das verlorene himmlische Bild wieder erlangt, welches wir hier im neuen Menschen durch

white, clear, and transparent, like an opal, heavy, fragrant, milk of the birds, milk of the ☿, without which no human being can live, which we draw into us, invisible as water in the coarse elemental air, being without sharpness. The fire is very sharp, burning and gleaming like a glowing coal, which we humans also have in us after the animal source-spirits, being mostly located in the eyes as in the light. One can see it when one looks into one's own eyes, being struck as it springs forth and shines. This fire now has the power to turn the sea of glass into the carbuncle when mixed with these fires, which have been the *Urim* and *Thumim*, light and justice among the Israelites, made of a sparkling ore that shone. Our double ☿, our ☿ water, also shines as a white, bright light; the red carbuncle stone which looks like a metal coming off the cupel, this fire already has a radiant shine; because it is still in the form of the solar ☿, it appears in the form of a flying red lion or dragon, if placed in the sun the whole room will shine with nothing but fiery mirrors. All these repelling things are now contained in this single *confuso chao*, which, when they are all highly purified, must again become a single immortal body through their reunion. It has also endured the malediction just as the great world has, but GOD has left his heavenly being and power within it for the benefit of frail human beings, just as he also left the divine breath as the spirit to Adam, even though died of the heavenly image. For the day when he transgressed the commandment of GOD, GOD promised him the redeemer, the seed of the woman, by whose spirit he would be again awakened into a new human being, since he grasped him in faith, until Adam, according to his free will, decided whether he wanted to live in GOD or in the world and according to his own will. Now when paradise has been reopened by the son of GOD and Adam has regained the lost heavenly image, which we are renewing here in the new human being

ein heiliges Leben in GOTT wieder erneuern und den neuen Menschen im Geist *Christi* anziehen, welches hier nicht eher geschehen kann, er muss denn den alten Adam töten und kreuzigen, damit der neue Mensch zum Leben kommen kann, so wird der Geist *Christi* sich mit seinem reinen Geist noch hier vereinigen und mit ihm in einem Geist leben, denn der Geist GOTTES kommt nicht in eine boshafte Seele und wohnt nicht in einem Leibe den Sünden unterworfen: Also auch unser Stein muss die *Terram Damnatam* verlassen, damit er in die neue Erde zum paradiesischen reinen Licht-Leben gelangt. Wenn er in die reine Erde versetzt, so wird er anheben zu leuchten und zu funkeln vor Tugend und Klarheit, dass heißt, das Paradies bauen, doch muss dieser Leib, wenn er soll verklärt werden, wie der Mensch erst sterben, in die Erde begraben, darinnen verfaulen und wieder auferstehen und die Seele in dem verklärten Leibe viel herrlicher leuchten und ewig darinnen wohnen, weil der Leib auch ein Geist, das ist ein Engel, geworden und kann man den giftigen Basilisken durch sein giftiges Herz sehen, worin kein Gift mehr, sondern ein herrlicher Lebens-Balsam, weil alle 7 Quell-Geister lebendig in ihrer Klarheit und höchster Vollkommenheit sind, so kann dieser geistliche Leib alle Geister der Metalle durchdringen, solche neu gebären, auch unseren animalischen Geist durchdringen, als der Rauch im Haus und durch seinen Lebens-Geruch unsere Geister ins Leben erwecken, Tod und Krankheit von uns scheiden. Wenn wir nicht wegen der Sünden sterben müssten, würden wir ewig dadurch leben, aber GOTTES Macht-Spruch geht über alles im Himmel, auf Erden und unter der Erden. Also hat auch der Sohn GOTTES, weil er unser Fleisch an sich genommen, müssen leiden und sterben und unsere Schuld tragen, weil er aber eben mit unserem Leib wieder auferstanden und verklärt worden, so wird er uns am Jüngsten Tage auch von unserem Tode wieder auferwecken und unser Fleisch als sein Fleisch auch verklären und mit seinem Geist erleuchten, damit uns sein Blut, seine göttliche Tinctur tingieren kann, dass wir dem Leib des Erlösers gleich werden, dass wir ähnlich sind seinem verklärten Leibe, welches die Hochzeit des Lammes ist, da wir

through a holy life in GOD, and putting on the new human being in the spirit of Christ, which cannot happen here before he has killed and crucified the old Adam, so that the new human being can come to life, so the spirit of Christ will unite here with his pure spirit and live with him in one spirit, because the spirit of GOD does not come into a malicious soul and does not dwell in a body subject to sin. So, too, our stone must leave the *terram damnatam* so that it can reach the new earth for the paradisiacal pure light-life. When it is transferred to the pure earth, it will begin to shine and sparkle with virtue and clarity—that is to say it will build paradise, but if this body is to be transfigured, it must be buried in the earth, just as a human who dies rots in the earth before rising again, the soul shining much more gloriously in the transfigured body and dwelling in it forever, because the body has also become a spirit that is an angel which can see through the poisonous basilisk's venomous heart, in which there is no poison anymore, but a glorious balsam of life. Because all seven source-spirits are alive in their clarity and highest perfection, this spiritual body can penetrate all spirits of metals, can give birth to such metals, and can also penetrate our animal spirit as smoke penetrates a house, and by its living scent awaken our spirits to life, separating death and sickness from us. If we did not have to die because of our sins, we would live by it forever; but GOD's claim to power is supreme in heaven, on earth, and beneath the earth. So also the son of GOD, because he took our flesh, had to suffer and die and bear our guilt; but because he rose again with our body and was transfigured, he will also raise us from our death again on the last day and also glorify our flesh as his flesh, and enlighten with his spirit, that his blood, his divine tincture, may tinge us, and that we may become like the body of the redeemer, that we may be like his glorified body, which is the marriage of the lamb, since we are to

uns hier als sein Weib zubereiten sollen, durch einen Tugend-Wandel, hat uns auch hier Pfand und Siegel darauf gegeben, in seinem Wort, Taufe und Abendmahl, da wir durchs Wasser als durchs heilige Element von Sünden gewaschen, unser innerer neuer Mensch mit seinem Leib und Blut gespeist und getränkt wird und dadurch bis zum ewigen Leben erhalten, welches ist das Himmels-Brot so er geben wird für die Sünde der Welt und solch Abendmahl mit uns neu halten wird in seines Vaters Reich, wenn wir nach dem Tode in seinem Bilde erwacht sein werden und das Bild GOTTES wieder vollkommen haben, welches große Abendmahl die Hochzeit des Lammes ist, da diese Freude und Wonne, Herrlichkeit und Klarheit keines Menschen Zunge aussprechen kann, die Freude, die GOTT bereitet hat, denen die ihn lieben, welches kein Auge gesehen, kein Ohr gehört, ist auch in keines Menschen Herz gekommen: Eya wär ich da, Eya wären wir alle da!

Wieder auf unser Werk zu kommen, so müssen diese Kinder von Adam gezeugt, die 4 *Elementa* so von Adam und Eva kommen, einander heiraten, es muss in dieser Welt zugehen wie in der großen Welt, es muss Bruder und Schwester einander heiraten und Blut-Schande begehen, daher werden sie auch für ihre Missetat gestraft, sie müssen durchs Feuer-Gericht, der Cherub mit dem Schwert muss ihre Bosheit abschneiden, was nicht durch dieses Feuer geläutert wird, kann kein reiner Engel werden, noch in dem himmlischen reinen Feuer leben, worin die Engel und reinen Geister sind. Wenn nur der Fluch und Tod davon geschieden, wird sich der paradiesische Leib schon zeigen und anheben zu leuchten und zu funkeln und alsdann sich in das innere *Principium* versetzen lassen, wenn die neue Stadt davon gebaut, von Perlen und Jaspis, das Pflaster von durchsichtigem Golde, da die Stadt keiner Sonne mehr bedarf, weil sie selber leuchtet und glänzt. Welches eine Abbildung des himmlischen Jerusalems, da GOTT ihr Licht und Glanz ist. Es können zwar von diesen Kindern, von Adam und Eva gezeugt, Neben-Ehen gemacht werden, wer die Geheimnisse des großen Werks nicht finden kann, so muss er Tincturen von den gemeinen Metallen machen, als wenn man das gemeine Gold und *Luna*, mit un-

prepare ourselves here as his wife through a change of virtue. He has given us a pledge and seal on it here too, by his word expressed in the sacraments of baptism and communion, because our sins are washed away from us by the water as the holy element, our new inner human being is fed and drunk with his body and blood and thereby preserved until eternal life, which is the bread from heaven which he will give for the sin of the world, and such communion with us will take place anew in his father's kingdom, when we have awakened after death in his image and have the image of GOD perfect again, which great supper is the wedding of the lamb, since no human tongue can express this joy and delight, glory and clarity, the joy that GOD has prepared for those who love him, which no eye has seen, no ear has heard, and which has also not come into any human heart: O, that I were there; O, that we were there![77]

Coming back to our work, these children fathered by Adam, the four *elementa* that came from Adam and Eve must marry each other. Things in this world must be like in the macrocosm, brother and sister must marry each other and commit incest, thereby also receiving punishment for their misdeeds. They must go through the judgment of fire, and the cherub with the sword must cut off their wickedness. That which is not purified by this fire cannot become a pure angel, nor live in the pure heavenly fire, in which angels and pure spirits exist. But when malediction and death are separated from it, the paradisiacal body will already show itself and rise to shine and sparkle and then let itself be transferred into the inner *principium*, when the new city is built of it, of pearls and jasper, the pavement of transparent gold, since the city no longer needs the sun because it itself shines and glows. Which is a picture of the heavenly Jerusalem, where GOD is its light and splendor. It is true that side marriages can be made by these children begotten by Adam and Eve, whoever cannot find the secrets of the great work must make tinctures of the common metals, as if one were to mix common gold and *Luna* with our lunar and solar water,

serem lunarischen und solarischen Wasser vermischt, es sind aber dieses nicht die rechten Kinder, sondern von einem anderen Geschlecht gezeugt und Bastarde aus einem Konkubinat, daher bekommen sie nicht das rechte Erbe, sondern nur den Mägde-Teil, sie tun sich in der Projection nicht hervor wie das große Werk, sondern sind nur kleine Zweiglein von kleiner Kraft. Wenn aber unsere Sonne mit unserem Mond vermählt wird, ist es das rechte Ehe-Bett, welche beide viel Mühe haben, ehe sie vollkommen werden und sich rein baden, wenn sie aber ihre große Finsternis ausgestanden, sind sie hell leuchtend, auch braucht es viel Mühe, diese beiden in einen Leib zu bringen. Ohne diese Vereinigung ist alles vergebens, wäre es auch vorher noch so wohl gemacht, in der Multiplication aber muss die Sonne mit der Sonne vereinigt werden, welches abermals nicht die gemeine Sonne oder ☉ ist, sondern unsere Sonne, ist die ausgekochte figierte *Materia* des Steins selbst, so sich durch seinen eigenen ☿ in Gold gekocht, auch solches Gold durch sein eigenes Fleisch und Blut gesättigt, lauter feurige, hitzige und rote Tinctur ist, weit höher als das gemeine Gold, kein höheres, besseres, vollkommeneres Gold ist in der ganzen Natur zu finden. Dieses Gold oder Sonnen-Glanz wird mit dem Namen des gekrönten roten ♌ belegt und ist der rote blutige Stein der ersten Ordnung, welchen wir mit unserer anderen Sonne, dem ☿ *solare*, unserem roten fliegenden ♌ vermählen. Auch haben wir 2 Monde in der Multiplication zur weißen Tinctur, der erste Mond ist die ausgekochte trokkene figierte Substanz unseres Steins der ersten Ordnung, welches die weiße gekrönte Königin genannt wird, soll nun diese multipliciert werden, muss sie mit dem gestirnten Adler, mit dem unvergänglichen lebendigen Wasser, welches in Leib, Seele und Geist besteht, wieder resolviert werden, dass sie in ihrem blühenden Blut und Schweiß badet, welches beides *in infinitum* kann getan werden. Wenn unser Adam mit seiner Eva schwanger ist, so ist unsere Sonne und unser Mond in dunklen Schatten und dickem Vorhang, ist trübe, dunkel und finster, schwarz, das *Confusum Chaos*, weil es aber den ♄ bedeutet, der die Götter zeugen soll, so ist er von Natur böse, unfreundlich, gebrechlich,

but these are not the proper children, but are begotten of another lineage, bastards from a concubinage, therefore they do not receive the proper inheritance, but only the maiden's part; they do not distinguish themselves in the projection like in the great work, but are small sprigs possessing little power. But when our sun is married to our moon, it is the proper marriage bed. While both our sun and moon have much difficulty before they become perfect and bathe themselves clean, after they endure their great darkness, they will be brightly shining nevertheless; and it also takes a lot of effort to bring these two into one body. Without this union everything is in vain, no matter how well it was made beforehand, but in the multiplication the sun must be united with the sun, which again is not the common sun or ☉, but our sun is the boiled fixed *materia* of the stone itself, which has boiled itself into gold through its own ☿, having also saturated such gold through its own flesh and blood, which is nothing but fiery, hot, and red tincture, far higher than common gold. There is no higher, better, more perfect gold to be found in all of nature. This gold or solar luster is given the name of the crowned red ♌. It is the red bloody stone of the first order, which we wed with our other sun, the ☿ *solare*, our red flying ♌. Also we have two moons in the multiplication to the white tincture. The first moon is the boiled, dry, fixed substance of our stone of the first order, which is called the white crowned queen. Now if this is to be multiplied, it must be dissolved by the starry eagle, by the imperishable living water, which is body, soul, and spirit, so that they bathe in their blooming blood and sweat, both of which can be done *ad infinitum*. When our Adam is pregnant with his Eve, our sun and our moon are in dark shadows and behind thick curtains, being cloudy, dim, and gloomy, black, the *confusum chaos*. Yet because it means the ♄, which will beget the gods, it is evil by nature, unfriendly, frail, lame, and

lahm und hinkend, er ist auch arm, kann keine Gaben austeilen, ob er schon große Herren und Könige zeugen soll, so ist es doch jetzt armselige Bettelei, die Kleider sind dünn, nicht fest gewebt, darinnen sie von Frost viel Anstoß leiden müssen. Es ist weder ☉ noch *Luna* in diesen Kleidern zu sehen, aber nach seinem Tode verlässt er ein herrliches Erbe, dass seine Kinder und Kindes-Kinder Kleidung genug haben, er aber besitzt die grobe irdische Erde, da doch die paradiesische, ja die engelhafte Welt seine Kinder besitzen werden und er als der Oberste und ihrer aller Vater diese in seinem Leibe oder Zirkel hat, sie leben elend, ihre Speise ist grob, sauer, bitter, nagend und beißend und noch nicht viel heilsames Kraut bei ihnen zu finden, da doch ihre Kinder schon mehr vermögen und ihnen an nichts mangeln wird, das Paradies-Wasser aber ist nicht nur die unsaubere Eva, sondern die reine Maria, welche sich mit der groben Erde, dem Adam, vermischt, daher sie auch eine viel herrlichere Geburt zeugt, weil sie nur geistlich geschwängert wird, wenn die ☉ im Wasser bei der ☽ schläft, gebärt die *Luna* in Wolken ihr Kind, in den Dünsten wird der Stein geboren, wenn er in Gefäßen anfängt zu arbeiten, das Wasser hält sich zur Erde und macht solche rund, das ist laufend, die Erde hebt an zu rauchen, gibt feurige schwarze Wolken, in Wolken wird der Geist gezeugt, der erst auf dem Wasser schwebte im Anfang der kleinen Welt und als ein weißes Wölklein erscheint. Wenn diese Wolken durch Ungeschicklichkeit des Arbeiters verrauchen, ist der Stein verloren und bleibt eine dürre Erde, welche ohne dieses Wasser nicht kann zur viskosischen Feuchtigkeit gebracht werden, welches die Grundfeuchtigkeit wird, so im Feuer bleibt, das Wasser und die Erde sind Mann und Weib und doch sehr widerwärtiger Natur, die Erde, welche für den ♂, auch das ☉ genommen wird, ist sehr feuriger, trockener und hitziger Natur, das Wasser, welches für des ♄ Tochter, auch die *Luna* oder Juno genommen wird, ist kalt und feucht, sie hassen und fliehen einander, sind die widerwärtigen Eheleute, auch die beiden widerwärtigen Fechter ♄ und ♂, die beiden Drachen, flüchtig und fix, die Kröte und Adler. Sobald nun das Weib den Mann berührt, fängt sie an zu toben, brudelt und siedet

limping. He is also poor and cannot hand out any gifts, and despite the fact that he will certainly beget great lords and kings, he is for the present still a poor beggar whose clothes are thin and loosely woven, in which he must suffer much from frost. Neither ☉ nor *Luna* can be seen in these clothes, but after its death he leaves a glorious inheritance so that his children and his children's children have enough clothing. Yet he owns the rough terrestrial earth, whereas his children will possess the paradisiacal, indeed, the angelic world, and as the chief and father of all, he has them in his womb or compass; they live miserably, their food is coarse, sour, bitter, gnawing, and biting and few healing herbs can be found among them. But their children will have more wealth, and they will lack for nothing. The water of paradise, however, is not only the unclean Eve, but the pure Mary which mixes with the rough earth, Adam. Therefore it also begets a much more glorious birth, because it is only spiritually impregnated. When ☉ sleeps in the water with ☽, the *Luna* gives birth to its child in the cloud, the stone is born in the vapors. When it begins to work in vessels, the water stays close to the earth and makes it round (that is to say running), and the earth begins to smoke; and there are fiery black clouds. The spirit is conceived in the clouds which first floated above the water at the beginning of the microcosm and which appeared as white cloudlings. When these clouds smoke off through the clumsiness of the worker, the stone is lost, and what remains is a dry earth. Without this water it cannot be made into the viscous moisture (which then becomes the basic moisture), and so it remains in the fire. This water and the earth are man and woman and are yet of a highly repugnant nature; the earth which is taken for the ♂ and the ☉ is of a very fiery, dry, and hot nature, while the water which is taken for the ♄ daughter, also *Luna* or Juno, is cold and damp. They hate and flee from each other, being mutually alienated spouses, and also the two odious swordsmen: ♄ and ♂, the two dragons, volatile and fixed, the toad and the eagle. As soon as the woman touches the man, she begins to rage, simmer, and boil and the cloud

und erscheint die Wolke, die von den Dünsten in die Höhe steigt, aber GOTT hat nach Erschaffung der Erde nicht flugs regnen lassen auf Erden, sondern es ging ein Strom aus zu wässern den Garten, diesem muss man auch folgen, wenn der Regen zu zeitig kommt, so wird die Sintflut und die Erde verderbt, so muss nun dieser Fluss die Erde befeuchten, endlich durch öftere Befeuchtung gibt es grausame Sturmwinde, welches knallt, kracht, donnert und blitzt, wodurch die Erde erregt und die Seele vom Leib geschieden, da man durchs Weib getötet und mit ihr flüchtig wird, so fingen die Wolken an, Farben zu bekommen, schwarz, gelb und rot, bitter und entzündend, fährt man aber zu geschwind und ungeschickt, dass die Wolken zu dick und finster durcheinander laufen, als bei schwerem Hagel und Schlossen-Wetter, dass man hört wie es knallt und kracht im Glase, so hat man Gefahr, dass der Donnerschlag geschieht, wodurch der Geist im Brausen des Windes entgeht und ist das Werk unwiederbringlich verloren, diese 2 widerwärtigen nennt man 2 Drachen, einer hat Flügel, der andere keine, welches die Erde, der geflügelte ist das Wasser, der ☿, diese sind nun als der Adler und die Kröte mit einer Kette zusammen verbunden, der Adler flieht stets in die Höhe, die Kröte zieht ihn durch die Kette wieder zurück in die Erde. Nimm was du mit Füßen trittst, wirst dich unterstehen ohne Leitern zu steigen, so versichere ich dir, dass du auf deinen Kopf fallen wirst. Die Leiter ist die Hilfe vom Wasser mit so vielen Sprossen, die vielen Teile der Einträkungen, ist die Vermählung des Bruders mit der Schwester, welchen ein Becher Liebe zugetrunken wird, oft müssen sie angefeuchtet werden, bis sie im inneren Leib gewachsen, welches der Hermaphrodit männlicher und weiblicher Natur ist und sich mit allen verheiraten kann und das leuchtende ☿al Wasser. Es hat auch kein Ding in der ganzen Welt einen doppelten Schwefel bei sich als dieses und hat 2 Naturen, flüchtig und fix, ich sage nichts unrecht, es hat 4 ☿, erstlich hat es den einfachen jungfräulichen ☿, zum 2. hat es den fixen, die jungfräuliche Erde, 3tens hat es den *Spiritum* ☿, welches sein Öl ist, 4tens hat es ☿al Wasser der Sonne, sein Feuer und seine Seele, welche alle in liquorischer, mercurialischer

appears, rising from the vapors, but after the creation of the earth, GOD did not let it rain upon the earth straight away, but a stream came out to water the garden. One must also follow this: if the rain comes too early it will be the deluge and the earth will be spoiled, so this river must now moisten the earth. Finally, through frequent moistening there are cruel storm winds, which crack, crash, thunder, and flash. This stirs the earth and separates the soul from the body, since one is killed by the woman and flees with it; and the clouds began to take on colors: black, yellow, and red, bitter and inflamed. But if one works so hastily and clumsily that the clouds permeate each other too thickly and darkly, as in heavy hail and cloudburst, such that one hears crashing and cracking in the glass, there is risk of a thunderstrike, whereby the spirit escapes in the roaring of the wind, and the work is irretrievably lost. These two repelling ones are called the two dragons: one has wings, the other has none; this one is the earth, the winged one is the water, the ☿. They are now like the eagle and the toad connected with a chain: the eagle always flees upwards, the toad pulls it back down to earth through the chain. Take what you trample on,[78] if you try to climb without ladders I assure you that you will fall on your head. The ladder is the help from the water with so many rungs, the many parts of the impregnations is the marriage of brother to sister, to whom a cup of love is drunk. Often they must be moistened until they grow in the inner body, which is the hermaphrodite, masculine and feminine in nature, and can marry itself with everything and is the luminous ☿al water. Nor does anything in the whole world have a duplicate sulphur within itself like this, possessing two natures, volatile and fixed. I am not saying anything wrong, as it has four ☿. First it has the simple virgin ☿, secondly it has the fixed [☿], which is the virgin earth; thirdly it has the *spiritum* ☿ which is its oil; and fourthly it has the ☿al water of the sun, its fire and its soul, which all appear in liquid mercurial form,

Gestalt erscheinen, sehr schwer. Wenn nun Wasser und Erde, Mann und Weib, in einem Leib gewachsen, sind von den Elementen noch 2 übrig, der Leib ist nun der doppelte ☿, die Grundfeuchtigkeit so im Feuer bleibt, diese widerwärtigen Dinge zusammen zu setzen, dass sie in einem Leib wachsen, braucht große Kunst und geschieht nicht bald, sondern durch lange Zeit und viele *Sublimationes*, dass der Mann in das Weib und das Weib in den Mann verkehrt werde, das fixe volatilisch und das volatilische fix werde, alsdann sind sie nicht wieder zu scheiden, wo das eine hin will, reißt es das andere mit, sie müssen beide aufs höchste gereinigt sein, dann werden sie in einem Leib wachsen, das Obere dem Unteren und das Untere dem Oberen gleich werden, die Erde zum Himmel und der Himmel zur Erde, welches die verkehrten *Elementa* sind, die Materie aber spiritualisch gewesen, in Sausen und Brausen eines gewaltigen Windes, darinnen wächst diese Erde und wird darinnen der Stein geboren, denn im Bauche des Windes muss er getragen werden, die Erde muss prudeln und sieden, wodurch das Wasser aufwallt und Blasen bekommt, welche aufreißen und Dünste geben, tut man hier zu viel, so ist der Geist nicht zu erhalten und schlägt das Gefäß in 1000 Stücke, es tut einen Donnerschlag, es sein die widerwärtigen *Elementa*, welche müssen vereinigt werden, bis sie alle ausgestritten, das Wasser in Erde verwandelt, die Erde in Luft, die Luft in das Feuer. Wenn das Feuer coaguliert, ist es eine blinkende blutrote Erde, schwerer als Blei, ehe sich aber das Wasser ganz in Erde verwandelt, welches die Begrabung des Leibes in die Erde ist, darinnen der Leib verfaulen muss, geschehen viele Wunder bei den *Conjunctionibus*, ehe sich diese Quell-Geister alle bilden, einer den andern zeugt und in den 7ten als in einem ausgeht, in die Begreiflichkeit, welches ist das *Corpus*, so aus den 6 Geistern geboren wird, darinnen alle Figuren stehen, sich alles bildet und formiert, darinnen alle Farben, Schönheit und Freude aufgeht, der rechte Geist der Natur, ja die Natur selber, ein geschickter Geist dringt durch alle Geister, sieht, fühlt, riecht und schmeckt sie. Der erste Geist, wenn er zur Erde kommt, zieht zusammen, weil er herbe wird vom Salz, astringiert das süße Wasser in der

which is very heavy. Now when water and earth, man and woman, have grown in one body, there are still two of the elements left. The body is now double ☿, and the basic moisture that remains in the fire to put these repelling things together so that they grow in one body requires great art and does not happen quickly, but rather over a long period of time and many *sublimationes*, so that the man is transformed into the woman and the woman into the man, and the fixed becomes volatile and the volatile becomes fixed. Then they cannot be separated again, for where one wants to go, the other rises with it, and they both have to be purified to the highest degree. Then they will grow into one body, and the upper will be like the lower and the lower will be like the upper, the earth like heaven and the heaven like earth, which are the wrong *elementa*; but the matter was spiritual, in the rushing and roaring of a mighty wind, in which this earth grows and the stone is born. Since it has to be carried in the belly of the wind, the earth must simmer and boil, causing the water to swell and produce bubbles which rip open and give off vapors. If you do too much here, the spirit cannot be preserved and breaks the vessel into a thousand pieces with a thunderous noise. These are the mutually hostile *elementa*, which must be united until these hostilities have been exhausted, turning the water into earth, the earth into air, the air into fire. When the fire coagulates, it is a flashing blood-red earth, heavier than lead; but before the water is all turned to earth, which is the burial of the body in the earth, wherein the body must rot, many miracles occur in the *conjunctionibus*; and before these source-spirits are all formed, one begets the other until in the seventh they finally go out as one in the form of a single tangible body which has been born from the six spirits, in which all figures stand, everything is shaped and formed, in which all colors, beauty, and joy arise, the proper spirit of nature, indeed nature herself, a skillful spirit penetrates all spirits, sees, feels, smells, and tastes them. The first spirit, when it comes to earth, contracts because it becomes harsh from the salt, astringes the sweet water in the earth so that it becomes

Erde, dass es natürlich begreiflich wird zusammengezogen, sieht gleich dem Himmel blau aus, wenn es zu sehr ertrocknet. Die andere Qualität sänftigt, weil es süße wird, wenn das Licht oder Leuchten darinnen aufgeht, so sieht es gelb aus, dem Edelgestein Jaspis gleich. Die dritte, der Geist der Bitterkeit, der bitteren Qualität entsteht aus den anderen und ersten, wenn das Salz aufgelöst und sie sich entzünden, so geht an die Grimmigkeit im Feuer und wird selbstständig, im süßen wird es erweicht, im harten corporalisch und steht das Licht mitten, weil die Geister lauter werden und das Licht in sich fassen. Wenn nun das Licht mitten in die Finsternis scheint, so steigt alles auf, als wenn es lebte und formt sich in grünliche Gestalt und geht aller 4 Geister Kräfte auf, weil sie das Licht beseelt, lebendig und schwanger macht und derselbe fasst die Liebe des Lebens, das ist der fünfte Geist, wenn die Hitze darinnen aufgeht, so formt sich die grünliche Gestalt in eine ganz rötliche. Nun stehen sie alle in großer Liebe und Freude, schmekken und empfinden einander, weil alle diese Geister ineinander gegangen und aufgelöst und geht aus diesen auf der sechste Geist, darinnen es erhärtet, pochend und klingend, wird zur Begreiflichkeit. Wenn nun das helle reine Licht darin aufgeht und darin scheinend wird, bekommt es seine gelbliche und weißliche Farbe, welche mit nichts verglichen werden kann, weil aus diesem ☿ oder Natur geht aus der siebente Geist, in Blüten, Formungen, Früchten, Farben, schönem Geruch und ganzen Kräften, Bildungen und Vollkommenheit. Ehe aber sich alle 7 Geister bilden, sieht die Erde im Wasser rund, bald oval, bald als ein ☿, viel schöner als der gemeine, bis sich die beiden widerwärtigen Fechter, welche mit ihrem Realgar nun gesetzt, zum vollkommenen Metall kochen, auch beide zur süßen Medizin werden wie Milch und Honig. Wenn diese beiden Drachen miteinander verfault, ist es das Kraut, welches eine schwarze Wurzel und weiße Blüte hat, weil es in der Erde schwarz liegt und fault, wenn es aber reif wird, so ist es die weiße glänzende *Terra Foliata*, der doppelte ☿, die *Lunaria*, ja gar die Sonne und Mond, welche sich aus der Finsternis und Schatten der Nacht hervor gemacht und nun ohne Flecken erscheinen und durch

tangible naturally, contracted, and appears sky-blue when it dries too much. The second quality is softening, because it becomes sweet; when the light or luminosity rises within it, it appears yellow, like the precious stone jasper. In the third, the spirit of bitterness, the bitter quality arises from the second and the first, when the salt is dissolved and they ignite; thus the fierceness enters the fire and becomes self-sufficient; in the sweetness it is softened, in the hardness [it becomes] corporeal, and the light stays in the middle, because the spirits become raucous and absorb the light. When the light shines in the middle of the darkness, everything rises as if it were alive and takes on a greenish appearance and the powers of all four spirits grow, because the light inspires them, enlivens them, and impregnates them, and they are seized by the love of life, which is the fifth spirit. When the heat rises within it, the greenish appearance becomes extremely reddish. Now they all stand in great love and joy, taste and feel each other, because all these spirits have merged and dissolved and the sixth spirit emerges from them, in which it hardens, throbbing and resounding, to become tangible. When the bright, pure light now rises and shines within, it receives its yellowish and whitish color, which cannot be compared with anything, because out of this ☿ or nature comes the seventh spirit, in blossoms, shapes, fruits, colors, in beautiful fragrance and full strength, formation, and perfection. But before all seven spirits are formed, the earth looks round in the water, sometimes oval, sometimes as a ☿, much more beautiful than the common one, until the two repelling fencers, which are now seated with their realgar, boil themselves into perfect metal, both also become sweet medicine like milk and honey. When these two dragons putrefy together, it is the herb which has a black root and white flower, because it lies black and rots in the earth, but when it ripens, it is the white lustrous *terra foliata*, the double ☿, the *Lunaria*, even the sun and the moon, which emerged from the darkness and shadow of the night and now appear spotless and can give health because previously they were a stinking earth. But

den Grund die Gesundheit geben können, da sie zuvor eine stinkende Erde waren. Wenn aber das Wasser, als das flüchtige Weib, einen Ausgang findet, so geht es fort und lässt eine tote Erde hinter sich, so zu nichts zu gebrauchen, auch unmöglich die Tinctur zu machen ist und die ganze Kunst in Kochung des Wassers besteht, koche, koche, koche, bis es fix wird bei der Erde und die Erde der ☿ wird, auch im Himmel wie dieser gestiegen und die Erde Flügel bekommen, der Himmel die Erde und die Erde der Himmel geworden, das Oberste das Unterste und das Unterste das Oberste. Diese giftige Schlange vermählt sich mit keinem anderen als mit ihrem eigenen Ehegatten, darum heißt es, diese Schlange hat sich selbst gebissen, sich selber getötet, sich auch selber wieder lebendig gemacht. Wenn dieser Drache seinen eigenen Zorn und Gift getrunken, der Basilisk sich im Spiegel beschaut, so tötet er nicht mehr. Dieser Stein nun hat Seele, Geist und Leib, wie ein Mensch und ist das glühende weiße Feuer, dadurch das Silber 7-mal geläutert worden und einen Stein geboren und im glühenden roten Feuer das Gold, welches die rechten Eheleute sind, auch der Mond und die Sonne, welche sich rein gewaschen und das lebendige Wasser geben, darinnen die hellen Gestirne ihre Gesichter waschen, da sie hernach unzerstörlich sind, weil sie alle ihre Feinde überwunden, zuvor waren sie Tod und Gift, jetzt geben sie die Gesundheit und Leben, zuvor waren sie bitter, sauer, nagend, beißend, herbe und stinkend, jetzt süß, lieblich und wohlriechend, durchdringen alle Adern, Nerven in allen äußersten Gliedern, macht den Menschen jung und neu geschaffen, vertreibt durch seines Lichtes Strahlen alle dunklen finsteren Geister in dem Menschen, die Krankheiten mögen herkommen von bösen Geistern oder Hexerei, von *Astris* oder den Elementen, wird es doch durch seine Ausstrahlungskraft wie ein Blitz die Krankheit verzehren, erstlich durch das Herz, Lunge und Leber gehen, auch sich im ganzen Leib in allen zerteilen, weil es alles geistlich und spiritualisch ist, wegen seiner Reinigkeit, göttliche und himmlische Tugenden hat, ja der Geruch allein kann gesund machen. Es ist nicht allein des Goldes *Astrum*, sondern auch aller Metalle und wenn es zur Tinctur

when the water, as the fleeing woman, finds an exit, it goes away and leaves dead earth behind, so useless that it is impossible to make the tincture from it. The whole art consists in boiling the water, and one must cook, cook, cook, until the earth is fixed and becomes the ☿, risen to heaven like this [☿], and the earth receives wings. Heaven has become earth and the earth has become heaven, the highest the lowest and the lowest the highest. This venomous serpent mates with none other than its own mate, therefore it is said that this serpent bit itself, killed itself, and also made itself alive again. When this dragon has drunk its own anger and poison, if the basilisk has seen itself in the mirror, it kills no more. This stone now has soul, spirit, and body like a human being and is the glowing white fire, whereby the silver was refined seven times and a stone was born, and in the glowing red fire the gold [was born], which are the true marriage pair, also the moon and the sun who have washed themselves clean and have given the living water in which the bright stars wash their faces, since afterwards they are indestructible, because they have conquered all their enemies. Before they were death and poison, but now they give health and life. Before they were bitter, sour, nagging, acrid, astringent, and stinking but now they are sweet, lovely and fragrant. They pervade all veins, nerves, in all outermost limbs, and they make people youthful and regenerated, driving all dark spirits out of them by its luminous rays. The illnesses may come from evil spirits, or witchcraft, from *astris,* or from the elements. Through its irradiating power it consumes the sickness like lightning, moving first through the heart, lungs, and liver, thereby distributing its power throughout the whole body; and because everything is subtle and spiritual, and due to its purity has divine and heavenly virtues, the very smell alone can make one healthy. It is not only the *astrum* of gold, but also of all metals, and when it has become a tincture, it is far more powerful still, competely angelic and

worden, ist es noch weit höher an Kraft, ganz engelhaft und himmlisch, ja der Baum des Lebens, wenn GOTT nicht den Tod zur Strafe der Sünden gesetzt, so würde der Mensch dadurch ewig leben und keiner Krankheit noch Schmerzen unterworfen sein. Hier möchte nun jemand fragen, wie doch aus Gift und Tod eine solche heilsame Medizin und Lebens-Balsam zu bereiten? So dient zur Antwort: Dass alles erst von GOTT gut und heilsam erschaffen, aber durch die Sünde verlosch alles reine, himmlische Wesen, trat inwendig zurück ins *Centrum*, blieb äußerlich ein grobes irdisches *Subjectum*, welches mehr Gift als heilsame Arznei zeugt, denn der Fluch hat es verschlossen und ist gleichsam tot, wenn es aber in die Arbeit kommt, so wird es noch giftiger, weil es erstlich recht umgewendet, eröffnet, spiritualisch und flüchtig durchdringend ist, so ist es die giftige Schlange, wie die Schlange *Thyrus*, aus welcher der Theriak gemacht wird, so hernach wider alles Gift dient: Also mit dieser roten feurigen Schlange auch. Wenn das *Confusum Chaos* wohl eröffnet, wird der ganze Körper der Schlangen übersteigen und alle Farben in Gestalt eines Regenbogens zu sehen sein, wie wenn das Chaos in der großen Welt eröffnet, sich alles in die Feuchte resolviert, lässt sich der Regenbogen sehen, welches ein Zeichen des Regens und der Nässe, aber diese Elementa sind bitter, sauer, herb, grimmig, giftig und tödlich, müssen also dieselben verlassen werden, weil sie nur Mütter der Metalle sind, doch sind sie unser Anfang und der Geist unser Welt nach der Poeten Vers:

> Der Geist der alle Ding belebte,
> Im Anfang auf dem Wasser schwebte,
> Den unzerteilten Klumpen schied,
> Aus einer Form viel Formen zoge,
> Das Schwere sank, das Leichte floge,
> Das Licht auch aus den Finstern blüht,
> Das Schwere macht er kugelrund,
> Und macht es, dass es stunde feste
> Mit Luft und Wasser auf das beste,

heavenly, indeed the tree of life; and if GOD had not made death the punishment for sins, human beings would thereby live forever and be subject to no disease or pain. Now would someone like to ask how such a healing medicine and balm of life can be prepared from poison and death? This serves as an answer: because everything was first created good and salubrious by GOD, but through sin, every pure, heavenly essence faded and receded inwardly into the *centrum*, while on the outside a coarse, earthly *subjectum* remained, which engenders more poison than healing medicine, because the malediction has closed it off and it is as dead. Yet when it comes into the work it becomes even more poisonous, because once it is truly inverted, open, and [has become] spiritually and volatilely pervasive, then it is the poisonous serpent, like the serpent *Thyrus* from which is made the theriac[79] that subsequently serves against all poison. So too with this red fiery serpent. When the *confusum chaos* is sufficiently opened, the whole body of the serpent will rise and all the colors will be seen in the form of a rainbow, just as when the chaos opens in the macrocosm and everything dissolves into a moisture, the rainbow can be seen, which is a sign of rain and wetness; but these elements are bitter, sour, acrid, fierce, poisonous, and deadly so they must be abandoned because they are only mothers of the metals. Yet they are our beginning and the spirit of our world, according to the poet's verse:

> The spirit which made all things alive
> In the beginning it did above the water glide
> Separated the undivided lump
> Draw many shapes from one shape
> The heavy sank, the light flew to escape
> The light blooms from the darkness sump
> It makes the heavy round as a ball,
> And make it solid for the rest
> With air and water for the best

Dass es ohne Fall nicht wanken kunt,
Er goss ihm Seel und Samen ein.
Der Sternen abzufließen pfleget,
Dadurch das Feuer soll geheget,
Der Himmel auch erhalten sein.

Wenn aber die Schlange in unseren Elementen ihre vielen Farben nicht spiritualisch zeigt, als gelb, rot, blitz-blau, wie der Regenbogen, so wird es keine rechte Scheidung sein, *purum ab impuro*, wird auch das Paradies-Wasser nicht recht zu machen sein und wird sich solches annoch mit der unreinen Erde nicht vermischen, weder prudeln und sieden, auch keine Winde noch Sturm auf der See geben und die Arbeit ganz unglücklich und umsonst sein, weil kein *Agens* noch *Patiens* da ist, dass sich mit bei dem anderen erhitzen kann, daher werden sich solche nicht vereinigen, das Wasser wird bei der Erde nicht gerinnen, auch solches dieselbe nicht solvieren. Ist also des Jasons seine Schifffahrt umsonst nach der Insel Kolchis, das güldene Vlies zu holen, welches der Drache bewahrt, wenn er keinen Wind hat, so kann das Schiff nicht fortgehen, wird also nichts besseres sein, als die Schifffahrt einzustellen, bis man sich besser zur Reise geschickt, die Dinge dazu besser eingerichtet und seine Mängel verbessert, denn an dieser Schifffahrt ist alles gelegen. Man muss erst über das Meer und das Toben der Wellen und Brausen des Windes nicht scheuen, welches das Schiff zuletzt an Port bringt und in das herrliche gute Land, dem Drachen das güldene Vlies wegzunehmen, wenn das vielköpfige Tier bestritten und die zugerichtete Suppe auf sie gegossen, dann pflügt Jason mit feurigen Ochsen, bricht den Acker, dass er zubereitet wird, die Schlangen-Zähne in sich zu fassen, so darein gesät werden, woraus Kriegs-Leute erwachsen, starke Riesen, die sich alle untereinander erwürgen, da dann durch diesen Sieg und Triumph sie in eben die Herrlichkeit verwandelt werden und die Ausbeute des Schatzes von den Stärksten, so den Sieg erhalten, davon gebracht, wird der natürliche Schatz, die erste Ausbeute, welches man halb ausnehmen kann, mit der anderen

That it could not totter without a fall
It poured soul and seed into it.
To flow away to the stars much higher
This also helped to contain the fire
The sky was also preserved a bit.[80]

But if the serpent in our elements does not show its many colors spiritually—such as yellow, red, lightning blue, like the rainbow—then there will be no real separation, *purum ab impuro*. Moreover, it will not be possible to prepare the water of paradise in the right way, and it will not mix with the impure earth, will neither bubble nor boil, nor will it give winds or storms on the sea, and the work will be quite unhappy and in vain, because there is no *agens* and nor *patiens* that are able to heat each other. Accordingly they will not unite, the water will not coagulate here on earth, and will not dissolve it. Thus Jason voyages in vain to the island of Colchis to fetch the golden fleece which the dragon guards, because if there is no wind, the ship cannot depart; and so there is nothing better than to suspend the voyage until one is better prepared for it and things are better arranged and the shortcomings improved, for everything depends on this voyage. One must first sail across the sea and not be afraid of the raging waves and roaring wind, which eventually bring the ship to the port and into the wonderful good country, where one can take away the golden fleece from the dragon when the many-headed beast is challenged and the prepared broth cast on it. For Jason plows with fiery oxen, breaks up the field so that it is ready to take in the serpent's teeth which will be sown in it. From these seeds men of war grow, robust giants who all strangle one another, since through this victory and triumph they are transformed into the same magnificence and the spoils of the treasure are taken away by the strongest, who received the victory, the natural treasure, the first spoils, which can be half fleeced. In the other half of the work

Hälfte das Werk *in infinitum* vermehren und größere Leute, große Riesen schicken, die Stadt zu bestürmen, da wohl ♂ ihr Führer sein wird, welchen sie mit Liebe überwinden, wenn ihm die ♀ schön geputzt in Liebe entgegen geht, vor welcher Schönheit ♂ erstarrt und sein Schwert zu Boden sinken lässt: Die Göttin ♀ hat ihren Bruder ♂ gefunden, mit dem sie vorher als Zwilling in einem Leibe gelegen, so werden sich diese beiden aufs neue unzertrennlich vereinigen und sich dieser durch seine Streitbarkeit auf den höchsten Thron setzen. Es hat viel Zeit und Mühe ehe sich die Planeten alle ausfechten, einer den anderen von seiner Herrschaft stößt und ist sonderlich der ☿ wie ein nackendes kleines Kind, ehe es von seiner Feuchtigkeit zu seinen männlichen Jahren kommt, hitzig und trocken wird, doch liebt er alle seine Brüder, die Planeten, mit welchem er sich einmal hat eingelassen, aus dessen Hause ist er nicht wieder zu bringen und wird sich der unsrige, ob er schon der nackende und unbekleidete, sich erst bei dem rechten martialischen Salz in ♄ coagulieren und wenn diese wieder in ☿ gekocht wird, es unser ♃, wenn solcher wieder gekocht, ist abermals ein herrlicher Leib gewachsen, bis der königliche ☿ erscheint, welcher die Schlangen nicht in Händen, sondern in Zepter und Kronen trägt, nicht mehr giftig, sondern wohlriechend, zur heilsamen Arznei geworden, zuvor war es Realgar, jetzt ist er wider alles Gift, männlicher und weiblicher Natur, vertritt bald Mannes, bald Weibes Statt, er ist Gift und heilt den Aussatz, er ist der Drache, der in dem verschlossenen Glase tödliches Gift ausspeit, er ist das angenehme Ei der Natur, der grüne ☊ und der rote ☊ lieben ihn, er ist die Mutter in des Kindes Leib, so sie geboren, er ist das *Astrum* aller Metalle, er ist der ♄, ♃, ♂, ☉, ♀, ☿, ☽, welche er alle in seinem Leibe verborgen und in Gestalt schöner klarer Leinwand, klarer als der zarteste Flor, das Buch durch 7 *Sigul* verschlossen, das gesalzene Meer-Wasser und *Sal pontilum*, Salz der Weisen, ist auswendig weiß, inwendig grün, gelb und rot, welches auch die vielfarbigen Dünste in seiner Arbeit gezeigt, bis endlich der Mond durch seinen 7-fachen Umlauf voll geworden. Dieser Hermaphrodit ist nun der Leib, der sich so gern gradieren lässt, wenn dieser Drache seinen eigenen

it can be increased *ad infinitum*, sending forth larger men, great giants, to storm the city, since ♂ will be their leader, which they will overcome with love when the beautifully groomed ♀ goes against him in love, and before whose beauty ♂ freezes and lets his sword sink to the ground. The goddess ♀ has found her brother ♂, with whom she had previously laid in one body as twins, so these two will be inseparably reunited, and he will sit on the highest throne through his stridency. It takes a lot of time and effort before the planets all fight one other. One pushes the other from its rule, and ☿ in particular is like a naked little child before it reaches its masculine years, and from its dampness becomes hot and dry. But he loves all his planetary brothers, with whom he once became engaged, and from whose house he cannot be brought back, and which will become ours, whether he is already naked and undressed, and will only coagulate into ♄ with the right martial salt, and if these are boiled again in ☿, it will be our ♃, and if this is boiled again, a glorious body will have regrown until the royal ☿ appears, which no longer carries the serpents in its hands, but carries scepters and crowns. It is no longer poisonous but fragrant, for it has become a healing medicine. Before it was realgar, but now it is against all poisons. It is of male and female nature, takes the place of men and women, it is poison and heals leprosy, it is the dragon that spits out deadly poison in the sealed glass. It is nature's pleasing egg, the green ☊ and the red ☊ love it, it is the mother in the child's womb when it is born, it is the *astrum* of all metals, it is the ♄, ♃, ♂, ☉, ♀, ☿, ☽, which are all hidden in its body in the form of a beautiful canvas clearer than the tenderest gauze, the book closed by seven seals. It is the salted sea water and *sal pontilum*, salt of the wise, white on the outside, green, yellow, and red on the inside, which the multicolored vapors also show in its work, until finally the moon has become full through its sevenfold revolution. This hermaphrodite is now the body that likes to be gradated. When this dragon has drunk its own wrath and poison and stained the

Zorn und Gift getrunken und das Blatt mit Gift gefärbt und durch ihren eigenen ♄ so hoch gradieren wird und solchen mit großer Lust in sich zieht, wegen seiner hitzigen und trockenen Natur, sich seiner Natur erfreut, zur hohen Farbe und unser Zinnober-Erz wird, welches eine hohe färbende Tinctur ist, wenn diese geblätterte Erde ihren Durst gelöscht und sich satt getrunken bis sie nichts mehr zu sich nehmen will, dann tingiert ein Teil viele 1000 Teile und dieser ☿ und kein anderer ist der, so die Transmutation annimmt, sich veredeln und aus einer Stadt und Ort in den anderen versetzen lässt, bis er der gekrönte rote König wird, der seinen anderen Brüdern güldene Kronen aufsetzen kann; der gemeine ☿ tut solches nimmermehr, ob er schon noch so sehr sublimiert; so er doch wieder zu revificieren, auch wenn er gleich zum Zinnober gemacht wird und sind alle seine Teile, sie mögen noch so klein sein als sie wollen, nicht die geblätterte Erde, weil sie keine doppelte Natur, *fixum & volatile*, in sich haben, als wie der unsere, so durch seine 7-fache Sublimation all sein Wasser und Feuchtigkeit in sich getrunken und eingetrocknet zum trockenen Wasser, dass die Hand nicht nass macht und wenn solcher zum Zinnober, danach zum Rubin und dunkel gemacht worden, kann er endlich zum leuchtenden Karfunkel-Stein gebracht werden. Dieser gibt endlich himmlisches Licht, welches ein Bild des klaren Himmels und die engelhafte himmlische Welt ist, da vorige Röte und gestocktes Blut nichts dagegen zu rechnen, auch ein Fünklein solchen Feuers viele 1000 Teile tingiert, aber ehe dieses alles gemacht werden kann, so muss eine Auflösung und Scheidung der Elemente vorhergehen, damit sich Geist, Leib und Seele scheiden kann, alsdann wird der Leib in die Erde vergraben, auch wenn er nach dem Tode wieder auferstanden, muss er mit Geist und Seele wieder vereinigt werden, welches die andere Schöpfung und Erzeugung des paradiesischen und himmlischen Leibes ist, der unsterblich und keiner Corruption unterworfen. Endlich durch noch höhere Aufführungen engelhafte, himmlische und göttliche Tugenden bekommt, da dieses das allergeringste, die geringen Metalle in ☉ zu verwandeln, es sind wohl andere höhere Dinge darinnen zu sehen, die

leaf with poison and will be gradated so highly by its own ♁, drawing it into itself with great lust due to its dry and fiery nature, delighting in its nature, it becomes the high color and our cinnabar ore, which is a high coloring tincture. After this leafed earth has quenched its thirst and drunk its fill until it will absorb no more, one part tinges many thousand parts; and it is this ☿ and no other which accepts the transmutation and allows itself be ennobled and transferred from one city and place to another, until it becomes the crowned red king who can set golden crowns on its other brothers. The common ☿ never does such a thing, no matter how highly sublimated; if it is revived again, or even if it is made like unto cinnabar, all of its parts, no matter how small they may be, are not the leafed earth, because they do not have a double nature, fixed and volatile, as does our [☿]. Through its sevenfold sublimation it has drunk all its water and moisture and dried it up to a dry water that does not wet the hand and when such has been made into cinnabar, then into a ruby and made to become dark, it can finally be made into a shining carbuncle. This finally gives heavenly light, which is a picture of the clear sky and the angelic heavenly world, since previous reddening and stagnant blood count nothing against it. Even a little spark of such fire tinges many thousands of parts, but before all this can be done a dissolution and separation of the elements has to happen so that spirit, body, and soul can separate. Then the body is buried in the earth. Even if it rose again after death, it must be reunited with spirit and soul, which is the other creation and begetting of the paradisiacal and heavenly body, being immortal and not subject to corruption. Finally, through still higher performances, one gets angelic, heavenly and divine virtues; and since transforming the inferior metals into ☉ is the least of these virtues, there are probably other higher things to be seen in them which must not be written down at all.

gar nicht geschrieben werden dürfen, die Unverständigen würden ansonsten die *Adeptos* für die ärgsten Zauberer und Teufels-Banner halten, da doch bei diesem reinen Licht kein Teufel bleiben kann, wenn von diesem reinen *Electro* nach seiner Arbeit und Reinigung Spiegel, Glocken, Bilder und dergleichen gemacht werden, welches die rechten *Instrumenta* zur göttlichen heiligen *Magia* sind, dabei sich kein Teufel einmischen kann, den Menschen zu betrügen und in einen Engel des Lichts zu verstellen, so können dadurch die Wundertaten GOTTES ausgeführt werden, welche die heiligen Propheten und Männer GOTTES sich gebraucht, auch die Opfer damit angezündet und bei der Kirche Alten Testamentes das *Urim* und *Thumim* im Leib-Rock des Hohen-Priesters gewesen, dadurch die Israeliten GOTT um Rat gefragt. Als Kisch, der Vater von Saul, seine Eselin verloren und Saul solche suchen ging, ging er zum Seher zu fragen: Dort sagte Saul, als das kluge Weib zu Endor den Samuel hervor brachte, zu Samuel: Der HERR ist von mir gewichen und antwortet mir weder durch Träume noch durchs Licht. David sagte vor dem Priester, als er vor Saul fliehen und zuvor GOTT fragen wollte: Lange den Leib-Rock her. Wenn sich nun das *Urim* und *Thumim* mit einem lichten Glanz und Licht verklärt, wenn die Gemeinde GOTTES um Rat gefragt, so fiel die ganze Gemeinde auf die Knie, dankten und lobten GOTT, denn solches war eine Verheißung des Sieges wider ihre Feinde; Wo sich aber durch Besprengung des Wassers das *Urim* und *Thumim* verdunkelte, so fasteten, weinten und beteten sie, forschten, wurfen auch das Los an wem die Stunde wäre: Als dort durchs Los Jonathan getroffen war, da er hatte Honig gekostet und Saul durch das *Urim* und *Thumim* GOTT fragte und derselbe ihm nicht antwortete. Als der gottlose König Ahasia durchs Gitter fiel in seinem Saal zu Samaria, schickte er hin zu fragen Baal Zebub, den Gott zu Eckron, dies empfand GOTT übel, schickte einen Engel zu Elia, der den Abgeordneten sagen musste, ist denn kein GOTT in Israel das du hingehst zu fragen Baal Zebub, den Gott zu Eckron? Darum sollst du nicht von dem Bette kommen, darauf du dich gelegt hast, sondern sollst des Todes sterben. Dieses Licht und

Otherwise the ignorant would take the adepts for the worst sorcerers and exorcists, though no devil can remain with this pure light, when mirrors, bells, pictures, and the like are made of this pure *electro* after its work and purification, which are the proper instruments for the divine holy *magia* in which no devil can interfere, being unable to deceive people and disguise themselves as angels of light. Thus the miracles of GOD which the holy prophets and men of GOD used, can now be carried out; and the sacrifices are ignited with it, and in the churches of the Old Testament the *Urim* and *Thumim* were in the vestments of the High Priest, through which the Israelites asked GOD for advice. When Kish, Saul's father, lost his asses and Saul went looking for them, he went to ask the seer:[81] "When the woman with the familiar spirit at Endor brought up Samuel, Saul said to Samuel: "the LORD has departed from me and does not answer me by dreams or by light".[82] David said to the priest, when he wanted to flee from Saul and asked GOD beforehand: bring on the vestment. Now when the *Urim* and *Thumim* are transfigured with a bright splendor and light, when the congregation asked GOD for advice, the whole congregation fell on their knees, gave thanks, and praised GOD, for such was a promise of victory against their enemies; but when the *Urim* and *Thumim* were darkened by the sprinkling of water, they fasted, wept, and prayed, searched, and cast their lot to whom it was the hour: when the lot fell on Jonathan, he had tasted honey, and Saul asked GOD by the *Urim* and *Thumim*, and the same did not answer him. When the wicked king Ahasia fell through the bars in his hall in Samaria, he sent to ask Baal-zebub the god of Ekron, which GOD felt bad about, and sent an angel to Elijah, who had to tell the deputies, is there no GOD in Israel that you go to ask Baal-zebub, the god of Ekron? Therefore you shall not get off the bed on which you lie down, but you shall surely die. This light and heavenly fire, with which they lit the sacrifices, was hidden by the Jews

himmlische Feuer dadurch sie die Opfer anzündeten, hatten die Juden in ihrer Flucht in einem Berg versteckt nebst der Lade des Bundes und den Tafeln Moses, wie nun GOTT ihre Gefängnis gewendet und Nehemias die Opfer wieder verrichten lassen wollte, schickte er die Priester hin, solch Feuer zu holen, sie kamen wieder, brachten kein Feuer, sondern sagten, sie hätten ein dickes Wasser gefunden, das hieß er sie bringen, welches auch die Opfer angezündet. Nun fragt es sich, welches Feuer sich in einem Berg verstecken, durch lange Zeit sich in ein dickes Wasser resolvieren und wieder von der Sonne anzünden lässt? Ich sage, unser coaguliertes Feuer und künstliches *Electrum*. Dieses hohe und große Werk hat nun so einen geringen und schlechten Anfang und wird durch die Arbeit so hoch gebracht, dass es aus der äußeren groben finsteren Welt zum Licht und Glanz des Paradieses gelangt und aus dieser in die engelhafte und himmlische Welt, dass es lauter Feuer und Glanz ist. Die Arbeit nach ihrer rechten Ordnung ist, dass der Körper, die rote Erde, durch sein eigenes Wasser in einen Liquor gebracht werde, welches Chaos Grund wird, die rote Erde hält unser ☉ in sich, aus dem Chaos müssen die *Elementa* separiert werden, die gereinigten *Elementa* wieder zusammensetzen, das geborene Kind mit Milch erhalten, speisen und tränken, bis es zum vollen Alter kommt. Die erste Zusammensetzung der roten Erde ist keine Conjunction im rohen Werk, denn da wird Wasser und ☿ zusammengesetzt, dieses ist wieder zu scheiden. Die 2. ist die rechte Composition, Wasser und Erde, dies ist Mann und Weib. Die 3. Composition in Geist, Seele und Leib. Die 4., wenn das Wasser in der Erde eintrocknet und durch die Luft wieder belebt, auch solche die Erde wohl durchgangen, so kann der ☿ sich darin vermengen lassen, dass es zum roten Blut wird. Die erste Haupt-Solution ist die Reduction des unvollkommenen Körpers im Liquor oder Chaos, die 2. ist eine Absonderung vom Chaos durch die Destillation in die *Elementa*. Die 3. Solution ist des vollkommenen reinen Körpers, Auflösung des Mannes und Coagulierung des Weibes. Die 4. Solution des Ferments Seele, Leib und Geist zu vereinigen. 5. Solution den weißen ☿ mit dem roten solvieren und denselben darin-

during their flight in a mountain together with the ark of the covenant and the tablets of Moses, now when GOD ended their prison time and Nehemiah wanted to perform the sacrifices again, he sent the priests to bring such fire, they came back, brought no fire, but said they had found a thick liquid, which he told them to bring, which also lit the sacrifices.[83] Now the question is, what fire can be hidden in a mountain, changing over a long time into a thick water, and can be set on fire by the sun again? I say, our coagulated fire and artificial *electrum*. Now this high and great work has such a lowly and poor beginning and is brought so high through the work that although starting from the outer gross dark world it reaches the light and splendor of paradise and from there the angelic and heavenly world, so that it is nothing but fire and brilliance. The work according to its proper order is that the body, the red earth, is brought into a liquor through its own water, which becomes the basis of chaos, the red earth contains our ☉, the *elementa* have to be separated from the chaos, the purified *elementa* put back together, nurturing the born child with milk, feeding and watering it until it comes to full age. The first composition of the red earth is not a conjunction in the raw work, because water and ♄ are put together, this has to be separated again. The second is the proper composition, water and earth, this is man and woman. The third composition is in spirit, soul, and body. The fourth, when the water dries up in the earth and is revived by the air, the air even passed through the earth, the ♄ can be mixed with it so that it becomes red blood. The first major solution is the reduction of the imperfect body into liquor or chaos, the second is a separation of chaos through the distillation into the *elementa*. The third solution is the dissolution of the completely pure body, the man and coagulation of the woman. The fourth solution is to unite the soul, body, and spirit of the ferment. Fifth: dissolve the white ♄ with the red one and redden the same inside it. First we assemble

nen zu röten. Erstlich setzen wir zusammen, lassen es verfaulen, das verfaulte lösen wir auf, das aufgelöste teilen wir, das geteilte reinigen wir, das gereinigte vereinigen wir, das vereinigte figieren wir. In dieser Arbeit sehen wir die Schöpfung des Menschen, sein Leben im Paradies, seinen Fall, die Wiederbringung durch den Erlöser, durch seine Geburt, Tod und Leiden, auch wie der Mensch gleichfalls durch Tod und Leiden muss seinem Erlöser nachfolgen, will er anders hier und dort ewig mit ihm vereinigt bleiben. Es ist ein Gleichnis des himmlischen und irdischen Adams, der irdische Adam kam aus dem roten Ton oder Letten, von dem der ganzen Welt Menschen herstammen und von der unreinen Eva, welche diese unreine Erde liebte, aber der himmlische Adam war rein, geistlich und himmlisch, vermengte sich mit der unreinen Erde nicht, sondern wurde empfangen von der reinen Maria, welche sich mit der Erde, dem Adam, nicht vermischte, so war auch die Empfängnis rein und ist die jungfräuliche Geburt, welche nicht den sündhaften unreinen Menschen im tierischen Fleisch und Blut oder Erde und irdisch gesinnt, sondern das jungfräuliche Paradies-Bild, den inneren Menschen, in reinem Fleisch und Blut, ohne Sünde an sich nahm. Also unser reiner ☿ vermengt sich mit der unreinen Erde im geringsten nicht, sondern mit dem reinen jungfräulichen Salz, darinnen er empfangen, nach 9 Monden geboren, sichtbarlich und begreiflich in seiner Mutter Leib wird, da ihn sonsten niemand fassen noch halten konnte. Wenn er nun geboren, übertrifft er an Schönheit alle seine Brüder und Schwestern, die anderen Metalle, er aber muss leiden, sterben und sein Blut vergießen, damit er seine Brüder von Fluch und Tod erlösen und sie ihm alle an Klarheit gleich machen kann, an Schönheit und Beständigkeit, so muss nun diese reine Geburt den Stich der Schlange fühlen, welche ihn gleichsam kreuzigt und tötet, weil er aber himmlischer Natur, so geht er durch des Todes und der Höllen Reich durch und steht verherrlicht und verklärt wieder auf, tingiert seine armen presshaften Brüder und Schwestern, dass sie alle verklärt und beständige Leiber bekommen und in der Feuer-Probe bestehen, weil er ihr Fleisch und Blut, kann er sie ihm gleich

[it], let it putrefy, we dissolve the putrefied, separate the dissolved, purify the separated, combine the purified, and fix the combined. In this work, we see the creation of the human being, its life in paradise, its fall, the restoration by the redeemer through his birth, death, and suffering, also how the human being must follow his redeemer through death and suffering if he wants to be united with him forever here and in the hereafter. It is a parable of the heavenly and earthly Adam. The earthly Adam came from the red clay or loam, from which all the world's people descended and from the impure Eve who loved this impure earth, but the heavenly Adam was pure, spiritual, and heavenly; he did not mingle with the impure earth, but was conceived by the pure Mary, who did not mingle with the earth, the Adam, and so the conception was pure and is the virgin birth which did not take on the sinful impure human being of animal flesh and blood, or of earth, or the earthly-minded [human being], but rather the virgin image of paradise, the inner human being of pure flesh and blood without sin. So our pure ☿ does not mix with the impure earth at all, but with the pure, virgin salt, in which it was conceived, born after nine months, becoming visible and tangible in its mother's womb, since otherwise nobody could grasp or hold it. Now when it is born, it surpasses in beauty all its brothers and sisters, the other metals, but it must suffer, die, and shed its blood that it may redeem its brothers from malediction and death and make them all equal in beauty to itself in clarity and constancy, so this pure birth must now feel the sting of the serpents, which crucify and kill it, but because it is of heavenly nature, it goes through death and the kingdom of hell and rises glorified and transfigured again, tingeing its poor pressed brothers and sisters, so that they become transfigured, solid bodies which can pass the test by fire; and because it is [of] their flesh and blood, it can make them like itself.

machen. Wer sich dieses Werkes untersteht, der soll wissen, dass er die schwerste Sache der ganzen Welt vornimmt, darinnen lauter Geheimnisse des göttlichen Wesens, der Natur und der Creatur, Himmel und Paradies, Engel und Menschen darinnen zu finden, in welchen wir die 7 Geister GOTTES erblicken, die 7 Quell-Geister der großen Natur, der 7 Planeten, die 7 Quell-Geister in unserem Werk, welche uns die 7 Metalle gebären und wie diese Metalle durch unsere reinen Luft-Geister aufs neue glorifiziert, in unsere hell leuchtenden Planeten verkehrt und wenn diese Sonne und Mond ihre große Finsternis ausgestanden, leuchten sie in ewigem Glanz und können nicht mehr verdunkelt werden noch untergehen. Es ist die siebente Zahl eine geheime Zahl, in der Woche 7 Tage, im Werk 7 Siegel, welches in der Offenbarung Johannis in 7 darinnen abgebildet, die ihre Stimmen geredet, im Vater Unser 7 Bitten, im Menschen 7 Quell-Geister, auch in der ganzen Natur und allen Creaturen, weil nichts ohne diese sein kann, die Himmel und Natur ihre Ausgeburt aus den 7 Geistern GOTTES genommen, im siebenten begreiflich worden, wie GOTT die Himmel, Engel und die ganze Natur aus diesen, als aus sich selbst erschaffen, durchs Wort, als das Licht den Sohn und Herz GOTTES, sich auch in allen seinen Geschäften abgebildet und eingedrückt, ist auch weder im Himmel noch auf Erden, noch unter der Erden kein Ort wo GOTT nicht ist, es heißt:

> Führe ich im Himmel so bist du da, bettet ich mir die Hölle, siehe so bist du auch da, nehme ich Flügel der Morgenröte und bliebe am äußersten Meer, so würdest du mich doch finden und deine Rechte mich halten.

Wenn nun in unserem Werk unsere 7 *Systemata*, Grade oder Gestalten erfüllt und solche alle zu einem hungrigen und weißen Kalk reverberiert, haben sie ihren Sabbat gefunden, so sehnt sich der Leib nach der Seele und die Seele begehrt des verklärten Leibes, darinnen sie nun ewig wohnen will und zum himmlischen Geist und Engel beisammen

Anyone who undertakes this work should know that he is undertaking the most difficult task in the whole world, for he will be seeking after the sheer mysteries of the divine being, nature, and creatures, heaven and paradise, angels and humans, searching therein for the seven spirits of GOD which are the seven source-spirits of great nature, the seven planets, the seven source-spirits in our work, spirits which give birth to the seven metals and show how these metals, glorified anew by our pure air spirits, turn into our brightly shining planets. When this sun and moon endured their great darkness, they shine in everlasting splendor and can no longer be darkened nor do they set. The seventh number is a secret number: seven days in the week, seven seals in the work, which in the Revelation of John[84] are depicted as seven [seals] that spoke their voices, seven petitions in the LORD's Prayer, in people seven source-spirits; in the whole of nature and among all creatures, because nothing can be without them, the heavens and nature took their birth from the seven spirits of GOD. In the seventh it became comprehensible how GOD created the heavens, angels, and all of nature from these as from himself, through the word as the light, the son, and the heart of GOD, also depicted and imprinted in all his dealings. There is no place either in heaven or on earth, nor under the earth where GOD is not, it says:

> If I ascend up into heaven, you are there: if I make my bed in hell, behold, you are there. If I take the wings of the morning and dwell on the uttermost parts of the sea; Even there shall your hand lead me, and your right hand shall hold me.[85]

If now in our work, our seven *systemata*, grades, or forms are fulfilled and all of them reverberate to a hungry and white lime, they have found their Sabbath, the body longs for the soul and the soul longs for the transfigured body in which it now wants to dwell forever and become the heavenly spirit and angel together. So also humans, when

werden. Also auch der Mensch, wenn die Posaune ruft: Steht auf ihr Toten und kommt zum Gericht. Die 4 *Elementa* geben uns 3 Anfänge, ☉, ♃, ☿, diese 3 Leib, Seele und Geist und diese 3 eins, sowohl in GOTT, Vater, Sohn und Heiligen Geist und diese 3 ist ein GOTT, also auch in allen Creaturen und Geschöpfen. Wenn nun der Künstler diese 3 wieder in eins gebracht, ist es ein unzertrennliches Wesen und das himmlische Geschöpf, so große Tugenden hat, denn darinnen kommt man wieder zu der X Zahl, worin das Werk beschlossen wird, von dieser Zahl geht man zu der Ewigkeit, weil keine Creatur über diese Zahl schreiten kann, man muss denn wieder an dem einen anheben, so auch, wenn man aufs Kreuz kommt, muss man zugreifen und nicht darüber schreiten, man hebe denn wieder an dem einen an, alsdann geht es zu der Ewigkeit, da immer größere und größere Wunder erzeugt und geboren werden. Und also glaube ich, wird es GOTT auch mit uns halten, wenn wir erstlich durch Tod und Leiden mit *Christo* auferstanden, uns nach dem natürlichen Tode verklären und neu schaffen und noch immer höher und höher, von einer Gestalt in die andere verklären wird, bis wir dem verklärten Leib unseres Erlösers gleich werden, nachdem wir uns hier als gute Streiter *Christi* gehalten, mit unsern 7 Quell-Geistern nicht in der Welt Geist, sondern im Geist GOTTES inqualiert, da dann die reinen Lehrer werden leuchten wie des Himmels Glanz und die viele zur Gerechtigkeit weisen, wie die Sterne immer und ewiglich, nur dass wir uns hier durch wahre Buße und reinen Tugendwandel im neuen Menschen müssen reinigen und geschickt machen, durch Tötung des alten Adams, damit die groben sündlichen irdischen Schlakken von uns wegfallen, welche verhindern, dass die göttliche Tinctur nicht in uns gehen und tingieren und ihm gleich machen kann, denn die Sünden scheiden uns und GOTT voneinander, der Geist GOTTES kommt nicht in eine boshafte Seele und wohnt nicht in einem Leibe der Sünden unterworfen, GOTT und der Satan stehen nicht beisammen und wie stimmt *Christus* mit Belial, oder was hat die Gerechtigkeit für Lohn mit der Ungerechtigkeit? Waschet euch, reiniget euch, tut euer böses Wesen von meinen Augen, alsdann kommt und lasst

the trumpet calls: Rise up, you dead, and come to judgment. The four *elementa* give us three beginnings, ☉,[86] ♃, ☿, these three—body, soul, and spirit—these three are one, both in GOD: father, son and holy spirit, and these three are one GOD, so too in all creatures and creations. Now when the artist brings these three back into one, it is an inseparable being and the heavenly creature that has such great virtues, because in it one comes again to the number X [ten][87] in which the work is concluded. From this number one goes to eternity, because no creature can step over this number, we must start again at one, so even if we reach the cross, we have to seize it and not step over it. We start at one again, and so it goes to eternity, as greater and greater miracles are being generated and born. And so I believe that GOD will also do [so] with us, when we first have risen through death and suffering with Christ, have transfigured ourselves after natural death and [been] created anew, and continue to transfigure ourselves higher and higher, from one form to the other, until we become like the transfigured body of our redeemer, having maintained ourselves here as good warriors of Christ with our seven source-spirits infused not in the spirit of the world but in the spirit of GOD, since the pure teachers will shine like the splendor of heaven and point many to righteousness, like the stars always and forever, only that we have to cleanse ourselves here through true repentance and pure virtue into new people and make ourselves fit, by the killing of the old Adam, so that the coarse sinful earthly slag falls away from us, [the slag] which prevents the divine tincture from entering us and tingeing us and making us like him, for the sins separate us from GOD. The Spirit of GOD does not come into a malicious soul and does not dwell in a body subject to sin. GOD and Satan do not stand together, and how does Christ agree with Belial, or what reward has righteousness with injustice? "Wash and make yourself clean; put away the evil of your doings from before mine eyes. Come now, and

uns miteinander Rechten, spricht der HERR, wenn eure Sünde gleich blutrot ist, soll sie doch schneeweiß werden und wenn sie ist wie Rosinenfarbe, soll sie doch wie Wolle werden. Wenn unsere Geister mit GOTT inqualieren, so befinden sie einen steten Hunger nach GOTT, wenn sich das göttliche Feuer anfängt zu einem hungrigen und weißen Kalk zu reverberieren, da unsere Seele nach GOTT dürstet, wie ein dürres Land, so sollen wir diesem göttlichen Zug folgen, wieder in GOTT eingehen, weil die Seele sonsten keine Ruhe findet als in GOTT, da sie wieder in ihr *Centrum* eingeht.

Aber der meiste Teil der Menschen sucht diesen Zug oder Hunger der Seele, diese göttliche Traurigkeit, welche wirkt eine heilsame Reue zur Seligkeit, in der Welt zu sättigen oder die Melancholie zu vertreiben mit lustiger Gesellschaft irdisch gesinnter Menschen, beim Trunk, Tanzen, Gastereien und Welt-Freude, da doch *Christus* vor der Tür steht und anklopft, so jemand seine Stimme hört und die Tür auftut, zu dem will er eingehen und das Abendmahl mit ihm halten und er mit ihm. *Christus* will in des Menschen Seele kommen und die Seele wieder in GOTT als in ihre Ruhe, diese himmlische Gesellschaft, diese Gasterei, da sich *Christus* den Seelen anbietet, versäume niemand, sobald er den Zug des Vaters fühlt, der sie durchs Gesetz und Zorn-Feuer schreckt und sich die Traurigkeit findet, so suche er keine Gesellschaft, gehe in seine Seele, schaffe weg die bösen Geister, räume auf, damit *Christus* eingehen kann, höre auch nicht auf mit weinen und klagen, bis *Christus* sich mit Trost den Seelen zeigt, so wird das Herz erfreut, frisch und fröhlich, als wenn viele Centner Stein vom Herzen wären und der himmlische Trost sich gefunden, dass man vor Freuden und Liebe zu GOTT und *Christo* sich und der ganzen Welt vergisst und dieses so oft als man diesen Zug empfindet, welches, je öfter sich dieses zuträgt, je besser und ist ein Vorgeschmack des ewigen Lebens, so geht man getrost wieder in seinem Beruf fort, weil man die rechte Ruhe für seine Seele gefunden, dies ist die rechte Gesellschaft, so uns die Melancholie vertreiben kann. Auch hat man in diesem reinen und

let us reason together",⁸⁸ says the LORD. Even though your sins are blood-red, they shall become snow-white, and if they are rose-colored, they shall become like wool. When our spirits fuse with GOD, they find a constant hunger for GOD, when the divine fire begins to reverberate to a hungry, white lime, as our soul thirsts for GOD like a barren land, we are to follow that divine pull and go back into GOD, because the soul finds no other rest than in GOD, since it goes back into its *centrum*.

But most people are seeking to satiate this pull or hunger of the soul, this divine sadness which brings about a healing repentence leading to bliss, in the world; seeking to banish melancholy with the merry company of earthly-minded people: drinking, dancing, banqueting—through worldly joy; and yet Christ is standing at the door knocking; if anyone hears his voice and opens the door, he will come in and take communion with him, and he with him. Christ wants to come into the soul of human beings and the soul again into GOD as unto its rest. This heavenly company, this banquet where Christ offers himself to souls, nobody should miss this. As soon as he feels the pull of the father, who frightens them by law and wrathful fire, and sadness is found, so let him seek no company, go into his soul, get rid of evil spirits, clean up so that Christ can enter, also do not stop weeping and lamenting until Christ shows comfort to souls. The heart will be happy, fresh and cheerful, as if many hundred weights of stone were lifted from the heart and heavenly consolation has been found, that one forgets oneself and the whole world for joy and love for GOD and Christ, and this as often as one feels this pull, which, the more often this happens, the better it is and is a foretaste of eternal life, so one goes confidently back to one's profession, because one has found the right peace of mind before one's soul, this is the right company, which can drive melancholy away from us. One also has to look at this pure and divine

göttlichen *Mysterio* zu sehen, gleich wie ihm in der ganzen Natur nichts zu vergleichen an Reinigkeit, Schönheit, Herrlichkeit, Kraft, Wirkung, Geruch und Geschmack, so ist doch das einzige, der Mensch, noch drüber, der nach GOTTES Bild geschaffen, dem er selber seinen göttlichen Odem eingeblasen und die Menschen mehr als die Engel liebte, denn da diese fielen, gab er ihnen nicht seinen Sohn als Erlöser, als wie den Menschen, welchen er auch die Engel zu dienen verordnet, damit sie vom bösen Feind nicht sollen beschädigt werden. Ja wenn wir unserem Erlöser durch Kreuz und Tod nachgefolgt, hier in seinem Geist gelebt, so sollen wir zu ihm als seine Brüder und Kinder GOTTES kommen wo er ist, nämlich zur rechten Hand GOTTES, noch einen Thron höher als die Engel, ja ewig mit unserem Bräutigam vereinigt bleiben, welcher göttlichen Vereinigung kein Engel gewürdigt worden, welche Vereinigung die Engel gelüstet anzuschauen, es liegt nur an uns, dass wir uns in acht nehmen und wohl bedenken, da wir als Kinder GOTTES aus dem Schoß GOTTES, in die tiefste Hölle gefallen, hat uns GOTT wieder durch *Christum* heraus gerissen. Wo wir aber sein Leiden und Tod gering achten, so sind wir die groben unreinen Metalle, welche nichts als Schlacken, darin die göttliche Tinctur nicht eingehen und tingieren und zum verklärten Leib bringen kann, sondern gehören als verworfene Schlacken und die *Terra Damnata* in das höllische Feuer. Es ist ganz gewiss, dass ein Mensch durch Verfertigung des Steins himmlischen und engelhaften Verstand bekommt, GOTTES Herrlichkeit, Majestät, Allmacht und Weisheit erkennen lernt, GOTT, Natur und Creatur, den Schöpfer und das Geschöpf, sieht die Liebe GOTTES zu seinen Geschöpfen, wie er seine Kraft in sie gesenkt, wie er für alles sorgt und es erhält, diese auch alle seinen Willen tun, nur der vernünftige Mensch nicht, welcher GOTT zuwider, den er doch so herrlich erschaffen, ihm so viel Gutes getan, alles unter seine Füße geworfen und ihn zum Herren darüber gemacht. Dies alles sollen wir wohl bedenken, täglich GOTT innigst Dank sagen und in steter Buße leben. Wer aber dieses Geheimnis ohne allen Verstand und Nachsinnen findet, der brauche es nicht wie ein Tier, lebe auch nicht wie ein

mysterio, for just as nothing in nature can compare to it in terms of cleanliness, beauty, splendor, power, effect, smell, and taste, so the only thing that is still beyond that is the human being, created as an image of GOD, into which he himself blew his divine breath. He loved humans more than angels, for when they fell, he did not give them his son, the redeemer, as he had done to humans, whom he also commanded angels to serve, that they might not be damaged by the evil enemy. Indeed, if we followed our redeemer through crucifixion and death, lived here in his spirit, then we should come to him as his brothers and children of GOD, where he is, namely at the right hand of GOD, one throne higher than the angels. Indeed, we shall remain forever united with our bridegroom, though no angel has been honored by such a divine union, a union which angels long to behold. It is up to us alone to take heed and consider well, as we as GOD's children have fallen from the bosom of GOD into the deepest hell, GOD has torn us out again through *Christum*. But when we despise his suffering and death, we are the coarse, impure metals, which are nothing but slag, into which the divine tincture cannot enter and tinge and bring to a transfigured body, but belongs, as discarded slag and the *terra damnata*, into the hell fire. It is quite certain that through the manufacture of the stone a human being acquires a heavenly and angelic mind and learns to recognize GOD's glory, majesty, omnipotence, and wisdom. He sees GOD, nature, and creature, the creator and the creature, sees the love of GOD for his creatures as he imbued his strength into them, sees how he takes care of everything and preserves it. They all do his will, except for the human being capable of reasoning, a refusal which is contrary to GOD, who created man so gloriously. God has done so much good and thrown everything under humanity's feet and made them master over it. We should think about all this, give GOD heartfelt thanks every day, and live in constant penance. But whoever finds this secret without any understanding or reflection should not use it

Vieh und mache es nicht gemein, sondern preise und lobe GOTT und diene damit seinem armen Nächsten und missbrauche es nicht, sonst ist er ein Brecher des himmlischen Siegels und wird die Rute vielfach über ihn gebunden werden und das Feuer viel heißer fühlen müssen.

Noch ein Geheimnis, welches uneröffnet nicht lassen kann, wenn die 7 Gestalten der Natur vorbei gegangen und im 7ten Grad seine Endschaft erreicht, so ist es der reine paradiesische Leib, welcher durch das Öl-Meer, den Geist der Tinctur, neu versinkt, alsdann muss es durchs rote Feuer-Meer seiner Seele, so vereinigen sich dann diese 3 zusammen, erhärten und werden unzertrennlich, der herrliche blutige Stein und gläsernes Meer mit Feuer gemenget, der sich von der äußeren Welt entrissen durchs Paradies ins engelhafte innere *Principium* versetzen lassen, da es keiner Verfinsterung unterworfen, sondern ewig unsterblich bleibt: Also auch der Mensch muss seinen 7 Quell-Geistern, die bisher in dem äußeren *Principio* mit der Schlange und seinem eigenen bösen Willen und unreinen irdischen Begierden inqualiert, nach und nach ertöten, dem 7-köpfigen Tier einen Kopf nach dem anderen weghauen, wie *Christus* sagt: Ärgert dich den Auge, reiß es aus, ärgert dich deine Hand oder Fuß, haue es ab und wirf es vor dir. Wenn nun die bösen Begierden, Wort und Werke ausgeschafft, so werden sie wieder in GOTT inqualieren und vom Sünden-Tod aufstehen, so ist der paradiesische Leib wieder gefunden, alsdann kann er durch das Öl-Meer, den Geist GOTTES gesalbt werden, welcher nun in diesen reinen Leib gehen kann und ihn vollends zubereiten, dass die steinernen Herzen wie Wachs schmelzen, damit das Feuer-Meer, die blutrote Tinctur, ihn durchdringen kann und zum himmlischen, ja dem Leib des Erlösers gleich machen, dass er mit ihm ein herrschender König wird. Wenn wir unsere Seelen von toten Werken gereinigt haben, so kommt der Heilige Geist zu uns, macht Wohnung in uns, alsdann steht *Christus* in der Seele auf, erst ist GOTT Mensch geworden, nun ist der Mensch in GOTT, wer in GOTT lebt, sündigt nicht, denn GOTT lebt in ihm, dies ist die Wiedergeburt durch Wasser und Geist, wer aus GOTT geboren ist, überwindet die Welt und unser Glaube ist der Sieg, der die

like an animal, nor live like cattle. He should not make it public knowledge, but glorify and praise GOD and serve his poor neighbor with it, abstaining from its abuse, for otherwise he is a breaker of the heavenly seal; and the staff will be bound many times over him and the fire will feel much hotter.

Another secret that I cannot leave unopened is that when the seven forms of nature have passed and it has reached its final stage in the seventh degree, it is the pure paradisiacal body, which sinks anew through the sea of oil, the spirit of the tincture. Then it has to go through the red sea of fire of its soul, so that these three unite together, harden, and become inseparable. The glorious bloody stone and sea of glass mingled with fire, which snatched itself from the outer world, moved through paradise into the angelic inner *principium*, since it is not subject to eclipse there, but remains eternally immortal: so a person must also gradually kill his seven source-spirits which in the outer *principio* have hitherto been imbued by the serpent and his own evil will and impure earthly desires; must cut off one head after the other from the seven-headed beast, as Christ says: "If your eye offends you, pluck it out,[89] if your hand or your foot offend you, cut them off, and cast them from you."[90] If the evil desires, words, and deeds are now cast out, they will again unite with GOD and rise from the death of sin, then the paradisiacal body will be found again, since if it can be anointed by the sea of oil, the spirit of GOD, which now can go into this pure body and prepare it completely, so that the stony hearts melt like wax, so that the sea of fire, the blood-red tincture, can penetrate it and make it equal to the heavenly, even the body of the redeemer, so that it becomes a ruling king with him. When we have purified our souls from dead works, the holy spirit comes to us and dwells in us, then Christ rises in the soul, first GOD became human, now the human is in GOD, whoever lives in GOD does not sin, because GOD lives in him, this is rebirth through water and spirit. "He who is born of GOD conquers the world, and our faith is the victory that overcomes the world",[91]

Welt überwindet, welches ist die erste Auferstehung, über die hat der andere Tod keine Macht. Wache auf der du schläfst und stehe auf von den Toten, so wird dich *Christus* erleuchten. Selig ist der, so Teil hat an der ersten Auferstehung, über die hat der andere Tod keine Macht; Ja wenn der Mensch nur einen ersten Vorsatz hat, sein Fleisch und Blut zu töten und sich zu schwach befindet, indem er die Welt so sehr liebt, so sucht ihn *Christus*, als sein verlorenes Schaf, schickt ihm Kreuz und Unglück zu, macht ihm die Welt bitter und verhasst, welche ihn plagt, kreuzigt und tötet, alsdann geht er in sich mit dem verlorenen Sohn, sagte Vater, ich habe gesündigt im Himmel und vor dir und bin fort nicht wert, dass ich dein Kind heiße, so küsst ihn der Vater, *Christus* in der Seele, schenkt ihm das neue Kleid seiner Unschuld, steckt ihm den Vermählungsring an und ist Freude im Himmel über einen Sünder der Buße tut. Viele hat GOTT durch Kreuz und Unglück zu sich gezogen und gefunden, die sich durch Glück und Ehre, auch Liebe zur Welt vor ihm verlaufen. Ach, dankt GOTT für das Kreuz und küsst die Rute. Unsere Trübsal, die zeitlich und leicht ist, schafft eine ewige und über alle Maßen wichtige Herrlichkeit in uns, die wir nicht sehen auf das Sichtbare, sondern auf das Unsichtbare, denn was sichtbar ist, das ist zeitlich, was aber unsichtbar ist, das ist ewig. GOTT begehrt vom Sünder nicht mehr als Reue und Leid, die Wiederkehrung zu GOTT, sobald der Sünder von Herzen sagt: GOTT bist mir Sünder gnädig, so erschallt die Gegenantwort: Sei getrost mein Sohn, deine Sünden sind dir vergeben, *Christus* nimmt die Sünder an, auch die größten, ja wenn die Teufel, die gefallenen Engel, zu tiefer Reue und Leid kämen, würde ihnen GOTTES Liebes-Erbarmen auch verzeihen, weil GOTT nichts erschaffen zum Zorn, sondern die Seligkeit zu besitzen, sie sind ja alle sein, Engel und Menschen, dem Herrscher und Liebhaber des Lebens. Auch muss kein Mensch dem Satan verstatten, dass er stets in unserer Seele einen Eingang hat und unserer Sünden wegen vor GOTT verklagt, wodurch alles kindliche Vertrauen und Liebe zu GOTT aufhört, dass wir uns mit Adam vor GOTT verstecken, da wir ihn eben am meisten suchen sollten, uns ihm als dem Hohen-Priester zeigen,

which is the first resurrection, over which the second death has no power. "Awaken you who sleeps, and arise from the dead, and Christ shall enlighten you".[92] "Blessed is he that has part in the first resurrection: on such the second death has no power".[93] Indeed, if a human has only one first resolution, to kill his flesh and blood, and finds himself too weak because he loves the world too much, so Christ seeks him as his lost sheep, sends him the cross, sends him misfortune, makes the world bitter and hateful for him, which plagues, crucifies, and kills him. Then he goes into himself with the prodigal son, said father, I have sinned in heaven. I am not worthy any more to be called your child. Then the father, and Christ within his soul, kisses him, giving him the new robe of his innocence. He puts on his wedding ring and there is joy in heaven over a sinner who repents. Through the bearing of the cross and misfortune, GOD has drawn and found many who stray from him because of success and esteem, as well as love for the world. Oh, thank GOD for the cross, and kiss the rod. Our affliction, which is light and temporary, creates an eternal and exceedingly important glory, for us who do not look to the visible but to the invisible, since what is visible is temporary, but what is invisible is eternal. GOD desires no more from the sinner than remorse and suffering, the return to GOD, as soon as the sinner says from the heart: "GOD have mercy upon me, a sinner",[94] then the answer resounds: "Son, be of good cheer, your sins are forgiven".[95] Christ accepts sinners, even the greatest; indeed, if the devils, the fallen angels, came to deep remorse and suffering, GOD's loving mercy would also forgive them because GOD did not create anything for anger, but to possess blissfulness. After all, they are all his, angels and humans, the ruler and lover of life. Also, no one has to allow Satan to have constant ingress into our soul where he accuses us of our sins before GOD, whereby all childlike trust and love for GOD ceases so that we hide from GOD with Adam, since we should seek him most and show ourselves to him as we would to the high

dass wir von unserem Sünden-Aussatz rein würden, es muss uns nicht leid sein, dass wir mit unseren Sünden die Hölle verdient, denn das haben wir billig verdient, sondern das muss uns Leid sein, dass wir von unserem großen Schöpfer abgefallen, ihn als unseren lieben Vater erzürnt und zuwider gelebt, dies ist die rechte Buße und nicht die Furcht vor der Hölle, auch die Sünde lernen hassen, fliehen und meiden, so wird uns GOTT um *Christi* Willen gnädig sein.

Nun ist noch übrig von der Zeit zu melden, da man dieses unser Werk von dem äußeren *Principio* der äußeren Welt in das innere *Principium* der Licht-Welt, in das Paradies und zuletzt in die engelhafte und himmlische Welt aufführen kann, so bedarf es freilich viel und lange Zeit, die Wasser abzukochen und wäre sehr gut, wenn einer gefunden würde, der das Wasser auf einmal zur Erden eintrocknen könnte und die 7 Grade und Gestalten der Natur in einer Arbeit so exaltieren, auch auf den höchsten Thron zu steigen machen könnte, dass es auf der 7ten Stufe des Königs Sitz einnehmen und den reinen paradiesischen verklärten Leib anziehen könnte. Ich sage, es ist möglich, obschon die Künstler sehr rar, doch ist ein Weg dazu, die meisten aber müssen sich mit Abkochung des Wassers plagen. Im Menschen ist es auch also, die meisten so wieder in GOTT kehren wollen, sind so unkräftig, dass sie viele Jahre streiten, ehe sie sich einem von diesen Quell-Geistern entreißen, von der Welt und dem äußeren *Principio* bis sie mit allem wieder in GOTT inqualieren, weil viele Jahre ihre Seelen als der Tempel GOTTES wüste gelegen, weder Opfer noch Brand-Opfer darinnen getan, sondern ist den Heiden eingegeben zum Greuel der Verwüstung, da die heilige Stadt GOTTES zertreten, da weder Tempel noch Altar gesäubert, die falschen Götzen und eigene Bilder, eigene Liebe und der Welt Liebe nicht ausgeworfen, wie kann nun *Christus* und der Heilige Geist zu ihnen kommen und Wohnung bei ihnen machen? Wie kann der Heilige Geist die Seele erleuchten, heiligen und verklären, dass *Christus* in der Seele aufstehen kann? Doch sind ihrer etliche, die hurtig und geschwind mit ihrer großen ernstlichen Buße ihre Seele, ihre 7 Quell-Geister, so bisher in der Welt inqualiert, mit Gewalt von der

priest when we wish to show that we are purified from our sin, from our leprosy. We do not have to be sorry that we deserve hell because of our sins, because we earned that cheaply; but we have to be sorry that we fell away from our great creator, angered him as our dear father and lived contrary to him. This is the right penance and not the fear of hell. One must then also learn to hate, flee, and avoid sin, so that GOD will be merciful to us for Christ's sake.

Now it remains to report the time when this work of ours can be carried out from the outer *principio* of the outer world into the inner *principium*, the world of light, into paradise, and finally into the angelic and heavenly world. Of course it takes a long time to boil the waters, and it would be very good if someone were found who could dry the water down to earth at once and exalt the seven grades and forms of nature in one work, so exalted that they could also ascend to the highest throne, that it could sit on the seventh step of the king's seat and put on the pure paradisiacal transfigured body. I say it is possible, even if the artists are very rare, but there is a way to do it, though most of them struggle to decoct the water. It is the same with humans, for most of those who want to return to GOD are so weak that they fight for many years before they break away from one of these source-spirits, from the world and the outer *principio*, until they unite again with everyone in GOD; because for many years their souls, as the temple of GOD, lay desolate; neither sacrifices nor burnt offerings were made therein, but it was given to the pagans for the abomination of desolation, to enter the holy city of GOD, since neither temple nor altar were cleansed, false idols and their own images, their own love and the love of the world, were not cast out. How can Christ and the holy spirit come to them and dwell with them? How can the holy spirit enlighten, sanctify, and glorify the soul so that Christ may arise therein? But some of them, which are nimble and quick with the great, serious penance of their souls, their seven source-spirits, until now infused into the world, are snatched from the world by force, purified and cleansed,

Welt entrissen, geläutert und gereinigt und mit großer Buße und Glauben in GOTT gekehrt. Solches haben wir am Schächer der sagt: HERR, gedenke an mich, wenn du in dein Reich kommst, bekam darauf die trostvolle Antwort: Heute wirst du mit mir im Paradies sein. Wenn er die rechte wahre Buße und den starken Glauben in *Christo* nicht gehabt, würde er die trostvollen Worte nicht bekommen haben, hätte ihnen auch nicht geglaubt. 3 Dinge im Himmel sind die Zeugen, der Vater, das Wort und der Heilige Geist, der Geist, das Wasser und das Blut und diese 3 sind eins. Also auch in unserem Werk, Seele, Geist und Leib, ☉, ♃, ☿, diese 3 sind auch eins, auch alles in einem verborgen. Dieses *Subjectum* accordiert mit der Natur und Creatur, man sehe an dem Menschen, Adam aus der roten Erde, aus diesem das Weib, alle beide nach GOTTES Bilde, rein, vollkommen, als unser ☿, geziert mit dem jungfräulichen Kranz der Gerechtigkeit und Heiligkeit, diese sollten Engels-Brot essen und durch den Mund GOTTES gespeist werden, sie sollten Kinder zeugen nach GOTTES Bild, reine paradiesische Geburten, Kinder GOTTES, so nicht auf tierische Art, wie jetzt in Unzucht, durch Schmerzen und Zerreißung ihrer Leiber. Exempel haben wir an Vögeln und Fischen, welche keine solche Glieder haben, sich aber doch vermehren, sie waren aber lüstern nach der Welt, fingen an, diese zu lieben und sich daran zu vergaffen, da sie doch nichts als GOTT lieben sollten, in Einfalt und Demut bleiben, so kosteten sie dieser Welt Früchte, selbigen Augenblicks figurierte sich die Welt, zog ihnen das tierische Bildnis an, Magen und Därme, mit den fressenden Zähnen und tierischen geilen Gliedern, deren Frucht sie gekostet haben und eben im selben Augenblick sturben sie ab des göttlichen Bildes, ihre schöne Krone und Kranz entfiel ihnen, wussten nicht von GOTT und Paradies, wurden ihres tierischen Leibes gewahr, dass sie nackend waren, machten ihnen Schürzen und versteckten sich aus Scham und Furcht und muss nun zur Strafe seiner Sünden den Tod und alles Unglück der Welt dulden. Wenn er aber durch Kreuz und Unglück wohl gereinigt, den alten sündigen Adam täglich tötet, ersäufen und untergehen lässt, dass der Mensch zum verlorenen Bild GOT-

and turned back to GOD with great penance and faith. We have this in the malefactor, who said: "LORD, remember me when you come into your kingdom",[96] he got the comforting answer: "Today you shall be with me in paradise".[97] If he did not have true repentance and strong faith in Christ he would not have received the comforting words, nor would he have believed them. Three things in heaven are the witnesses, the father, the word, and the holy spirit, the spirit, the water, and the blood; and these three are one. So also in our work: soul, spirit, and body, ☉,[98] ♁, ☿, these three are also one, and everything is hidden in one. This *subjectum* accords with nature and creature. Look at the human being, Adam, created from the red earth. From him the woman was formed, both of them in GOD's image, pure and perfect like our ☿, being adorned with the virginal wreath of righteousness and holiness. They were meant to eat angel's bread and to be fed through the mouth of GOD; and they were meant to father children in GOD's image, pure heavenly births, children of GOD, not procreate like animals who fornicate through the pain and tearing of their bodies. Examples are birds and fish, which have no such limbs, but nevertheless reproduce, but they were lusting after the world, began to love it and fell in love with it, although they were supposed to love nothing but GOD, to remain in simplicity and humility. Thus they tasted the fruits of this world, and at the same moment the world shaped them: they were clothed in the animal image, stomach and intestines, with gnawing teeth and beastly, horned limbs, whose fruit they tasted. At the same moment the divine image died in them, and their beautiful crown and wreath fell from them. They no longer knew of GOD and paradise, but became aware of their animal body; and realizing that they were naked, they made aprons and hid themselves out of shame and fear. Now as punishment for their sins they have to tolerate death and every misery in the world. But when he, well purified by the cross and misfortune, kills, drowns, and destroys the old sinful Adam each day, so that humanity is renewed again to the lost image of GOD, through faith and a pious life,

TES wieder erneuert wird, durch Glauben und ein frommes Leben, welches wir hier doch nicht vollkommen haben, weil uns die Erbsünde immer anklebt, bis durch den Tod unser tierischer Leib abgeschieden, welcher uns stets verhindert und die groben Schlacken sind, wohl verfault und von uns abgefallen, denn Fleisch und Blut kann das Reich GOTTES nicht erben, sondern der erste reine Mensch, das Bild GOTTES, welches wir nach dem Tode vollkommen werden wieder haben, wenn wir erwachen nach seinem Bild und schauen GOTTES Angesicht in Gerechtigkeit und in unserem Fleische, dem paradiesischen Leibe, GOTT sehen, dann werden wir in Engel verklärt und gar dem Leib unseres Erlösers gleich werden. Wie nun der Mensch durch Leiden und Tod hindurch muss, ehe er zur Herrlichkeit gelangt, so auch unser ☿, welcher auch sterben, in der Erde verfaulen und zur Klarheit auferstehen muss, dann wird er ins Paradies versetzt, leuchtet und funkelt, worin kein Gift noch Tod, auch kein Schwinden mehr, weil er durch das Wasser in dem reinen Element gewaschen, schneeweiß worden, nun kann er durch den himmlischen Geist in einen Engel verklärt werden, welches des Geistes Taufe ist, so schadet ihm das Feuer-Gericht nichts, er dringt durch, isst Engels-Brot, welches er vorher nicht konnte, weil er keinen reinen Mund dazu hatte, nun aber kann ihn die reine Tinctur in einen größeren Engel verwandeln; aus der paradiesischen Licht-Welt in die himmlische Welt versetzen, weil nichts corporalisches mehr da, sondern lauter reine Licht-Geister, durchsichtig, leuchtend, brennend und funkelnd, alles wie es vor dem Fall war. Vor dem Fall bei dem Menschen war es auch also, aber nach dem Fall bleibt dem Menschen nichts als der tierische Leib, Erde und Kot; sollte nun dieses, also der Mensch, wieder mit GOTT vereinigt werden, musste eine doppelte Natur da sein, göttlich und menschlich, als der Sohn GOTTES ein Mediator, welcher die 2 widerwärtigen Naturen vereinigt: Also in diesen Werk ist der ☿ das himmlische Wasser, welches sich mit der unreinen Erde nicht vermengt, sondern mit der reinen sauberen *Materia*, welches, weil es selbst himmlisch und geistlich ist, solche Erde ihm gleich macht, so ist die Erde bei der Natur,

which we do not have perfectly here, because the original sin always clings to us until our animal body is separated from our souls through death; for this corruptible body is a constant hindrance, the coarse slag that has putrefied and fallen away from us, because flesh and blood cannot inherit the kingdom of GOD, but only the first pure human being, the image of GOD, which we will have again perfectly after death, when we wake up in his image and see the face of GOD in righteousness and see GOD in our flesh, the paradisiacal body, when we will be transfigured into angels and even become like unto the body of our redeemer. As a human being must now go through suffering and death before it attains glory, so too our ☿ must die, putrefy in the earth, and rise again in an illuminated form which will be transferred to paradise. This heavenly body will be shining and sparkling, and in it there will be no poison nor death, nor will it ever diminish, because it has become snow-white through the water, washed in the pure element. Now it can be transfigured into an angel by the heavenly spirit, which is the baptism of the spirit, so the judgment by fire will not harm it. It penetrates, eats angel's bread, which it could not do before because it had no clean mouth to do it, but now the pure tincture can transform it into a greater angel, and transfer it from the paradisiacal world of light into the heavenly world, because there is nothing corporal anymore, there is nothing but pure spirits of light, transparent, luminous, burning, and sparkling, everything as it was before the fall. It was so with humans before the fall, but after the fall humans are left with nothing but the animal body, dirt, and excrement. For this fallen human being to be reunited with GOD, a double nature needed to be introduced, a person both divine and human, i.e., the son of GOD, who acts as a mediator who unites the two repulsive natures. In like fashion, in this work, ☿ is the heavenly water which does not mix with the impure earth, but with the pure, clean *materia*, which, because it is heavenly and spiritual, makes such earth like it, so that the earth becomes

irdisch und himmlisch, ohne welches Eingang oder Mittel der reine Geist sich mit dem Leibe nicht vereinigen kann oder einen Eingang darein hat, das Wasser aber der ☿, ist das *Medium conjugendi*. Auch hätte GOTT nicht leiden können, wenn er nicht menschliche Natur an sich genommen, denn er war nicht begreiflich, hatte keinen Leib, als den er von der Maria annahm und wenn dieses nicht geschehen wäre durch *Christum*, so hätten wir Menschen stinkende Höllen-Brände bleiben müssen, wären nimmermehr mit GOTT vereinigt worden, so ist *Christus* unser Mittler und Gnaden-Thron, der uns durch sein Blut tingieren und ihm gleich machen kann; und wie die Tinctur die metallischen Leiber tingiert und zu ☉ macht, also auch *Christus*, da er unser Fleisch und Blut an sich genommen, uns am Fleisch gleich geworden, sind wir dadurch fähig geworden, seine Tinctur, sein Blut, an uns zu nehmen, dadurch wir rein gewaschen von Sünden und zum anderen und neuen Leben erneuert, dass wir im neuen Leben sagen können: *Christus* Fleisch ist unser Fleisch und unser Fleisch ist *Christus* Fleisch. Er ist die edle Rebe in uns gepfropft und wir seine Reben an ihm, als an dem rechten edlen Weinstock. Soll ich mich nun nicht freuen, dass mein Fleisch mit im Himmel sitzt zur rechten Hand GOTTES, in dem Leibe des HERRN *Christi* und ob ich schon noch hier auf der Erde bin, so wird er mich doch als das Glied seines Leibes nicht zurücklassen, sondern fest halten und gewiss nach sich ziehen, dass mich die Welt, Tod und Höllen-Reich ihm nicht nehmen kann. Wenn wir das Verwesliche liegen lassen und das Unverwesliche anziehen, dann werden wir mit seinem göttlichen Geist vereinigt, dass eine Union geschieht, eher kann uns seine Tinctur nicht tingieren und ihm gleich machen. Erst ist GOTT Mensch geworden, nun ist der Mensch GOTT, die Himmels-Tür ist uns angelweit aufgetan, dass heißt recht, er wird ein Knecht und ich ein Herr, das mag ein Wechsel sein, wie könnte es doch sein freundlicher, das herzige Jesulein. Diese Vereinigung ist nun die Hochzeit des Lammes, da wir uns hier als sein Weib zubereiten sollen. Durch die Arbeit des philosophischen Werks bekommen wir engelhaften und himmlischen Verstand, wir sehen, wie GOTT die Welt geschaf-

earthly and heavenly by nature. Without this entrance or means, the pure spirit cannot unite with the body or obtain entrance into it; but the water, the ☿, is the *medium conjugendi*. Nor could GOD have suffered if he had not taken human nature unto himself, for he was untouchable, having no body, which he took on from Mary; and if this had not happened through Christ, we humans would have had to remain stinking hell fires, could never have been reunited with GOD, therefore Christ is our mediator and throne of grace, who can tinge us with his blood and make us like him; and just as the tincture stains the metallic bodies and turns them into ☉, so also Christ, since he took our flesh and blood to himself and became like us in the flesh; we have thereby become capable of his tincture, taking his blood upon us, we were thereby washed clean from sins and renewed to the next and new life, that in the new life we may say that Christ's flesh is our flesh, and our flesh is Christ's flesh. He is the noble branch grafted onto us, and we his branches on him as on the true noble vine. Shall I not then rejoice that my flesh is seated in heaven at the right hand of GOD, in the body of the LORD Christ, and though I am still here on earth, yet as a member of his body, he will not leave me, but will hold tight and certainly make sure that the world, death, and the realm of hell cannot take me from him. If we leave the perishable and put on the imperishable, then we are united with his divine spirit, a union that enables his tincture to tinge us and make us like him. First GOD became human, now the human is GOD, the door of heaven is wide open to us, that is correct, he becomes a servant and I, a master. That may be a change, and how could it be friendlier, that dear little Jesus. Now this union is the marriage of the lamb, for here we are to prepare ourselves as his wife. Through the labor of the philosophical work we gain angelic and heavenly understanding, we see how GOD created the world, how the

fen, wie die *Elementa* geschieden, wie sie noch wirken, erhalten und fruchtbar machen durch den Geist der großen Welt, wir sehen auch, wie die erste Welt durch die Sintflut gestraft, die andere durchs Feuer gerichtet wird, wenn sie durch ihr eigenes Feuer brennen muss, wie GOTT die reinen Wasser der reinen Erde wieder geben wird und eine neue Erde machen, auch solche himmlisch schaffen. Er sagt: Ich schaffe einen neuen Himmel und eine neue Erde, in welcher Gerechtigkeit wohnt. GOTT hat aus den reinsten Elementen die Planeten erschaffen, wir machen sie aus unseren auch, diese clarificieren wir, darinnen sie ihre Finsternis erdulden, hernach geht die Sonne nicht mehr unter, sondern scheint von Ewigkeit zu Ewigkeit und sind nicht mehr die irrenden Planeten, sondern sind selber der Thron und Himmel, das innere *Principium* der engelhaften Welt, weil es alles lauter reine, trockene Feuer-Geister sind, da alle Erde verschwunden und alles zum Himmel worden, keines Mondes noch Sonnenschein bedarf, denn dieses engelhafte *Principium* ist selbst mehr als der Sonnen Glanz. Also wird auch GOTT den Himmel, die 7 Planeten, so diese Erde erleuchten, wieder clarificieren und renovieren und alles in himmlisches Licht und Glanz verwandeln, weil die Erde keiner Sonne nicht mehr bedarf und GOTT selber ihr Licht und Glanz ist, welche die Auserwählten bewohnen werden. Siehe ich schaffe alles neu. Siehe da, eine Hütte GOTTES bei den Menschen. GOTT sagt zu Abraham: Hebe deine Augen auf gegen Morgen, gegen Abend und Mittag, das alles will ich deinem Samen ewig zu besitzen geben, welches auch geschehen wird, wenn sie renoviert. Wir sehen auch aus dem philosophischen Werk die Zeit, wann die Welt ihren Sabbat und Ruhe-Tag halten wird, wenn sie ihre Tage-Werke vollendet, die 6000 Jahre in Mühe und Arbeit gelebt, in 6 Quell-Geistern im ringenden Rade, im 7ten im Schauen steht alles zur Begreiflichkeit und Erkenntnis, ihre Ruhe oder großen Sabbat einmal halten und der Spruch erfüllt: Ich will meinen Geist ausgießen über alles Fleisch, eure Söhne und Töchter sollen weissagen, eure Ältesten sollen Träume haben, eure Jünglinge sollen Gesichte sehen, auch will ich zur selben Zeit beides über Knechte und

elementa are separated, how they still act, are sustained, and made fruitful through the spirit of the macrocosm. We also see how the first world was punished by the deluge, the second by the fire, when they must burn by their own fire, as GOD will restore the pure waters of the pure earths, and make a new earth, and also make such heavenly. He said: "I create new heavens and a new earth,[99] wherein righteousness dwells".[100] GOD created the planets from the purest elements, we make them from ours too, we clarify these, while they endure their darkness, afterwards the sun no longer sets, but shines from eternity to eternity; and they are no longer the wandering planets, but are themselves throne and heaven, the inner *principium* of the angelic world, because they are all pure, dry, fire spirits, since all earth has disappeared and everything has become heaven. Neither moon nor sunshine is needed, for this angelic *principium* is itself more than the sun's splendor. So GOD will also clarify and renovate the sky, the seven planets that illuminate this earth, and transform everything into heavenly light and splendor, because the earth, which will be inhabited by the chosen ones, no longer needs a sun, and GOD himself is its light and luster. "Behold, I make all things new.[101] Behold, the tabernacle of GOD is with men".[102] GOD said to Abraham: "Lift up now your eyes, and look from the place where you are northward, and southward, and eastward, and westward:[103] all this I will give to your seed forever",[104] which will also happen when it is renovated. We also see from the philosophical work the time when the world will keep its Sabbath and rest day, after it has completed its daily work, after living for six-thousand years in toil and work, in six source-spirits through the wheel of struggle, in the seventh through seeing that everything stands for comprehensibility and knowledge, to keep their rest or long Sabbath once and the saying is fulfilled: "I will pour out my spirit upon all flesh; and your sons and your daughters shall prophesy, your old men shall dream dreams, your young men shall see visions: and also upon the servants and upon the handmaids in those days will I pour out my

Mägde meinen Geist ausgießen. Danach wird sie erneuert und geht das ewige Reich an und die *Terra Damnata* wird abgeworfen, den Teufeln und Verdammten zu ihrer ewigen Wohnung eingeräumt, welche alle in ihrem eigenen Feuer brennen und GOTTES Zorn-Feuer ihre Quell-Geister entzündet, in Heulen und Zähneklappern, kalt, herb, hitzig, brennend, bitter als Galle, reißend wie die hitzige Pestilenz, hitziger als brennender ♃, stetiger Feindschaft und Zank, Hass und grimmiger Zorn, stetiger Donnerschlag und Zittern, welche alle 7 die *Turba* anzünden, in lauter Wüten und Toben, da ihr Wurm nicht stirbt, ihr Feuer nicht verlischt und sie alle in Gestalt schändlicher gräulicher Würmer sind. Das Ausgehen im siebenten Quell-Geist ist sonsten das Haus des Lebens, im Höllen-Reich aber des Todes und Trauer-Haus, stetig Ach und Weh, auf ihrem Stuhl oder Thron sitzt ein anderer König und hält ein ewiges Gericht, welches ist GOTTES grimmiger Zorn, ihre Speise ist Greuel und böse Taten, welche sie in sich fressen und wieder ausspeien. Ob sich nun GOTT nach so vielen Ewigkeiten, wenn der Gerechtigkeit GOTTES völlige Genüge geschehen und von den Gottlosen aller Schaum abgebrannt sein wird, ihrer erbarmen möchte und sie endlich durch *Christum* im großen Jubel- oder Erlass-Jahr wiedergebracht werden, wissen wir nicht. Es erscheint wohl ein Licht, weil *Christus* ja für der ganzen Welt Sünde gestorben und sein Blut so kostbar, dass es nicht eine Welt, sondern viele tausend Welten erlösen könnte; und wie durch eines Menschen Sünden-Fall alle Sünder geworden, also sind durch eines Gerechtigkeit alle gerecht geworden und ist der Gerechtmacher größer als der Sünder, welcher nur ein Mensch, der aber, so gerecht macht, ist GOTT, welcher teuer und überflüssig Lösegeld für die Sünder bezahlt. Er sagt: Ich will sie erlösen aus der Hölle und vom Tod erretten, Tod ich will dir ein Gift sein, Hölle ich will dir eine Pestilenz sein. Welches er auch nach seinem Tode erwiesen, da er die Höllen-Fahrt gehalten und den Geistern im Gefängnis gepredigt, die zu den Zeiten Noahs waren sicher gewesen, welchen er sich als ihren Erlöser gezeigt. Wenn nun die Menschen aus dem Höllenreich in bessere Wohnungen versetzt, wird sich erst Reue und Leid

spirit".[105] After that it is renewed and goes to the eternal kingdom and the *terra damnata* is thrown off, given to the devils and the damned as their eternal dwelling place, which all burn in their own fire and GOD's wrath-fire ignites their source-spirits, in the weeping and gnashing of teeth, cold, harsh, hot, burning, bitter as bile, tearing like the heated pestilence, hotter than burning ♀, with constant enmity and quarrels, hate, and grim anger, constant thunderstrikes and trembling, these all seven ignite the *turba*, in sheer fury and rage, since their worm does not die, their fire is not quenched, and they are all in the form of vile abominable worms. Going out in the seventh source-spirit is otherwise the house of life, but in the kingdom of hell it is [the house of] death and the house of mourning, constant sorrow and woe. Another king sits on their chair or throne and holds eternal judgment, which is GOD's fierce anger, their food is abominations and evil deeds, which they devour and spit out again. We do not know if they will finally be brought back through Christ in the great year of jubilee or remission, after so many eternities, when GOD's justice has been fully satisfied and all scum from the ungodly will have burned off. A light will certainly appear, because Christ died for the sin of the whole world and his blood is so precious that it could redeem not one world but many thousands of worlds; and just as through the fall of one human all became sinners, so through one alone all were justified and the one who acts justly is greater than the sinner, who is only a human, but whoever acts righteously is GOD, who paid costly and unnecessary ransom for sinners. He says: "I will ransom them from the power of the grave; I will redeem them from death: O death, I will be your plagues; O grave, I will be your destruction."[106] This he also proved after his death, when he made the descent into hell and preached to the spirits in prison, which in the days of Noah had been safe, to whom he showed himself to be their redeemer. If the humans are now transferred from the kingdom of hell to better homes, the devils will only feel remorse and

bei den Teufeln finden, wenn er sieht, dass sein Höllen-Reich geräumt und er allein der verstoßene von GOTT ist, welches ihm ein tiefes Nagen und Beißen erregen, auch große Reue und Leid erwecken wird, so könnte wohl kommen, dass sich GOTTES Liebes-Erbarmen, welches unendlich, auch über ihm ausgösse und als einem wohlgestäupten Knecht Gnade wiederfahren ließe und das Wunder aller Wunder erfüllt, seine Göttliche Allmacht dadurch groß zu machen, diese aus dem höllischen Kerker zu erheben und als Reuende wieder annehmen und vorige Klarheit geben, ihre Hölle in eine klare leuchtende *Sphæra* oder Himmel verwandeln, weil er auch ihr Schöpfer und hasst nicht, was er gemacht hat. Welche Liebe, Güte und Gnade GOTTES das größte Wunder sein wird, ja größer als da er seinen Sohn zur ganzen Welt Heiland ins Fleisch kommen ließ, welches von den wenigsten erkannt worden. Aber dieses Wunder aller Wunder würde offenbar in allen Thronen und Herrschaften aller Engel und auserwählten Menschen, für welche Wiederbringung sich alle Engel und Menschen höchst verwundern und GOTT ewig loben, ehren, rühmen und preisen werden, wenn durch *Christum* alles wieder gebracht worden, was im Himmel und auf Erden gefallen. Aus der Schrift und aus dem Werk sehe ich wohl eine endliche Errettung für die Menschen, weil der Sohn GOTTES der ganzen Welt Heiland ist und für alle Menschen gelitten und gestorben, auch aller Menschen Fleisch und Blut an sich genommen. Für die gefallenen Engel aber sehe keinen Rat, als GOTTES Liebes-Erbarmen durch *Christum*, welcher den Zorn GOTTES versöhnt und sich seine Liebe und Erbarmung über alle Runde ausbreitet und größer als Himmel und Erde ist und wenn er alles wiedergebracht, was im Himmel und Erden gefallen, wird er das Reich seinem himmlischen Vater wieder überantworten, aber nach seiner menschlichen Natur bleibt er in Ewigkeit unser König und Hoher-Priester. Jetzt steht die Erde im 6ten Alter der Welt, in der 6ten Gestalt der Natur, in der siebenten wird sie clarificiert als in ihrem Sabbat, in Licht und Erkenntnis GOTTES erleuchtet, damit sie *Christus* beherrschen und in ihr wohnen kann. Andere Geheimnisse dürfen nicht geschrieben wer-

suffering when they see that their kingdom of hell has been cleared and that they are the only ones cast out by GOD, which causes them to gnaw and bite deeply, and will also arouse great remorse and sorrow; and it could well happen that GOD's merciful love, which is infinite, also pours out over them and, as well-punished servants, allows grace to come and the miracle of all miracles is fulfilled, thereby making his divine omnipotence great enough to raise them out of the hellish dungeon and accept them again as repentant and to restore their former glory by transforming their hell into a clear shining *sphæra* or heaven, because he is also their creator and does not hate what he has made. This love, goodness, and grace of GOD will be the greatest miracle, even greater than when he let his son come to the whole world as a savior in the flesh, which was recognized by only very few. But this wonder of all wonders would be revealed, in all the thrones and dominions of all angels and chosen humans, for which restoration all angels and humans will marvel and laud, honor, glorify, and praise GOD forever, when everything that has fallen in heaven and on earth has been restored through Christ. From the scriptures and from the work I see a final salvation for the people, because the son of GOD is the savior of the whole world and suffered and died for all people, also taking the flesh and blood of all people. But I see no other advice for the fallen angels but GOD's merciful love through Christ, who atones for GOD's wrath and spreads his love and mercy all round and all over and is greater than heaven and earth and when he has brought back everything that has fallen in heaven and earth, he will hand over the kingdom to his heavenly father again, but according to his human nature he remains our king and high priest forever. Now the earth is in the sixth age of the world, in the sixth form of nature, in the seventh it will be clarified as in its Sabbath, enlightened in the light and knowledge of GOD that Christ may rule and dwell in it. Other mysteries must not be written, the world does not believe what form it will re-

den, die Welt glaubt nicht, was sie für Gestalt bekommen wird, wenn
der Geist der Wunder wird offenbar werden, noch mehr, wenn die
Wunder der ganzen neuen Schöpfung erfüllt und das lichte Feuer der
Heiligen Dreifaltigkeit seine Wunder wird offenbaren, welches alles
auch in unserem Werk vorgeht, welches doch nur ein Schatten gegen
ihnen ist oder als ein Tropfen gegen das große Meer. Wer von GOTT
so glückselig ist, dass er ihm den Schlüssel Davids verleiht, der versteht die Schrift vollkommen und sieht das philosophische Werk aus
der Heiligen Schrift, wie sie miteinander concordieren, er sieht im
Göttlichen, *Christum* den Eckstein und Fundament, aus welchem und
auf welchem wir alle gegründet und gebaut und wie er in der Zeit aus
dem menschlichen Geschlecht müssen geboren werden. Wer aber den
Schlüssel Davids nicht hat, GOTT gibt ihm aber die Gnade dies Werk
zu finden, der sieht aus diesem Werk erstlich den Schöpfer, auch wie
dieser seinen Sohn dargibt in die Natur, solche zu erlösen, seine Geburt, Leiden und Tod, Auferstehung, Höllen- und Himmelfahrt, auch
die Erlösung aller seiner Brüder, die er durch sein Blut tingiert und
sich gleich macht, weil er ihr Fleisch und Blut ist. Durch dieses Werk
sind die Heiden Christen geworden, weil sie durch dieses Licht das
göttliche Licht gesehen, auch wie alles durch Mittel des himmlischen
Geistes renoviert und sich neu gebiert, welches auch bei dem Alter der
Welt geschehen wird, da sich schon die Creaturen ängstigen über der
Eitelkeit der sie unterworfen sind und gerne himmlische Früchte gebären wollten, daher sie sich sehnen los zu sein von den Banden und befreit, dass sie im Licht vor GOTT erscheinen sollen. Ist nun durch den
Willen GOTTES dem Menschen zugelassen, aus der äußeren finsteren
Welt die Paradies- und Licht-Welt zu schaffen, diese in das himmlische
und engelhafte Licht zu verwandeln, was meint der Mensch wohl, was
GOTT für Wunder und große herrliche Taten tun kann durch sein
göttliches Wort und Geist, der der rechte Schöpfer und Neugebärer ist,
wodurch er so viele wunderschöne Throne und Herrschaften der Engel geschaffen, die Engel selbsten so viele Millionen, da immer einer
schöner als der andere, wie kein Mensch dem anderen ähnlich sieht,

ceive when the spirit of miracles will be revealed, even more, when the miracles of the whole new creation will be fulfilled and the bright fire of the holy trinity will reveal its miracles, all of which also takes place in our work, which is only a shadow against them, or as a drop against the great sea. Whoever is so blessed by GOD that he gives him the key of David understands the scriptures perfectly and sees the philosophical work from the holy scriptures, how they concord with one another, and sees in the divine, Christ, the cornerstone and foundation from which and upon which we are all founded and built, and how in time he is to be born of the human race. But who does not have the key of David, but GOD gives him the grace to find this work, sees the creator from this work first, also how he gives his son into nature to redeem such, his birth, suffering, and death, resurrection, descent into hell, ascension, and also the redemption of all his brothers, whom he tinges with his blood and makes like himself because he is their flesh and blood. Through this work the heathen became Christian, because they saw the divine light through this light, and like everything else, renovated and born again through the means of the heavenly spirit, which will also happen in the old age of the world, when the creatures are already afraid of the vanity to which they are subjected and eager to bear heavenly fruits, hence they long to be freed from bondage and freed to appear in the light before GOD. If GOD's will now allows humans to create the world of paradise and light out of the outer, dark world, to transform it into the heavenly and angelic light, what miracles and great deeds of glory do humans think GOD can achieve through his divine word and spirit, who is the true creator and parturient, through which he created so many beautiful thrones and angelic dominions, the angels themselves so many millions, because one is always more beautiful than the other, as no human being resembles the other,

immer eines anders als das andere, so auch die Engel, der unermessliche Glanz der Sterne, ihre Größe, ihre unzählbare Menge, welche so hell leuchten, die vielen Himmel und Sphären, da sich doch kein *Circul* mit den anderen vermengt, sondern alle in ihrer Ordnung laufen? Wie viel 1000 mal schöner muss GOTT selber leuchten und durch sein Licht und Glanz alles in Wundern und Schönheit bis in Ewigkeit unendlich aufführen, dass wir alles hier nicht begreifen können, auch unsere Gedanken nur über die Sterne bis an GOTT reichen, was über den Sternen in der Höhe ist, da wissen wir nichts von, da doch GOTT wohl noch einen größeren Schatz-Kasten hat, von vielen engelhaften Kronen. Wir können seine Allmacht und Wunder nicht begreifen, so wenig, als wir das Meer mit einem Löffel ausschöpfen können, weil solches immer wieder voll läuft. Also ist auch die große Allmacht, Majestät und Herrlichkeit GOTTES nicht auszugründen und obwohl die Engel und Erzengel GOTTES Angesicht schauen, so wissen sie doch seine Allmacht und Herrlichkeit nicht zu ergründen und die Tiefe seiner Gottheit zu erforschen, sie sind einfältige, fromme, gerechte, heilige Geisterlein und warten täglich auf GOTTES Befehl, den sie mit Freuden verrichten, loben, rühmen, preisen und danken GOTT ohne Aufhören, denn darinnen besteht ihr Leben und Freude. Wir wollen diesen heiligen Geisterlein nachfolgen in Lob und Dank GOTTES, hier mit unserer schweren Zunge, bis wir dort mit Engels-Zungen sein Lob ausbreiten und das 3-mal *Sanctus* singen, hier aber als Ritter und Streiter des HERRN *Christi* mit der Sünde, unserem eigenen Fleisch und Blut, den Teufel und der gottlosen Welt tapfer streiten, bis wir alle Feinde des HERRN *Christi* durch *Christum* in uns überwinden und durch Not und Tod zu ihm gedrungen, weil er uns durch sein schmerzliches Leiden und Tod der Hölle und dem Tod aus dem Rachen gerissen, da sonsten keine Errettung gewesen und uns durch sein Blut erkaufen müssen, uns Pfand und Siegel gegeben, die *Sacramenta*, dass wir uns darauf, als auf seine ewige Treue und Gnade verlassen und nimmermehr seiner großen Liebe vergessen, ihn wieder lieben sollen, so lange wir im Leibe leben, damit wir dort in seiner Liebe ewig mögen

always one different from the other, so also the angels, the immeasurable splendor of the stars, their size, their innumerable multitude, which shine so brightly, the many heavens and spheres, since no *circul* mixes with the others, but all run in their order? How many thousand times more beautifully must GOD himself shine and, through his light and splendor, perform everything in wonder and beauty for eternity, such that we cannot understand everything here, even our thoughts only reach GOD via the stars, and what is in the heights above the stars, we know nothing, maybe GOD has a larger treasure chest there with many angelic crowns. We cannot comprehend his omnipotence and wonders any more than we can scoop up the sea with a spoon, because such things always keep filling up again. So the great omnipotence, majesty, and glory of GOD cannot be fathomed, and although the angels and archangels look into the face of GOD, they still do not know how to fathom his omnipotence and glory and to explore the depth of his divinity, they are simple, pious, righteous, holy little spirits and daily await GOD's command, which they carry out with joy, praising, extolling, glorifying, and thanking GOD without cease, for in this consists their life and joy. We want to follow these little holy spirits in praise and gratitude to GOD, here with our heavy tongues, until we spread his praise there with angelic tongues and sing the three-times *Sanctus*,[107] but here as knights and warriors of the LORD Christ we are bravely fighting against sin, and struggling with our own flesh and blood, with the devil and the godless world, until we have overcome all enemies of the LORD Christ through Christ in us and have reached him through hardship and death. For he has torn us from the throat of hell and death through his painful suffering and death, as otherwise there was no salvation. We had to buy it with his blood, and he gave us the *Sacramenta* as a pledge and seal, that we may rely on them as on his eternal faithfulness and grace and never forget his great love, and should also love him, as long as we live in the body, so that we may

eingeschlossen bleiben, dass helfe uns GOTT, Vater, Sohn und Heiliger Geist, hochgeliebt und gelobt in Ewigkeit, Amen.

BESCHLUSS DIESES ERSTEN TRAKTATS

Im Himmel ist die Sonne der Gerechtigkeit, der Sohn GOTTES, *Christus* JESUS, welcher alle himmlischen Kronen, Engel und heilige Menschen erleuchtet, davon sie ihr Leben, Licht und Kraft haben, ohne dieses göttliche Licht wäre alles finster und tot, ohne Licht, Liebe und Gnade. In der großen Welt ist die Sonne, welche alle Gestirne, auch die ganze Welt erleuchtet, denn sie ist der Geist der großen Welt, davon sie ihre Kraft, Licht, Leben, Wachstum und Fruchtbarkeit, als von ihrem lebendig-machenden Geist hat, denn ohne die Sonne oder ihren Schein wäre alles tot, finster und erstarrt. In unserer kleinen Welt ist auch unsere Sonne, die alles erleuchtet, lebendig macht, Leben und Wachstum gibt, ohne dieselbe oder ihre Wärme, Feuer und Licht, wäre alles tot, ohne Kraft, Leben und Tinctur. Die erste Sonne, der Sohn GOTTES, das Herz GOTTES ist geboren aus GOTT von Ewigkeit und ist ausgangen in himmlischen Thronen, sonderlich in den Thron des Lucifers, welches der schönste Engel und aus dem Licht des Sohnes GOTTES erschaffen. Die Sonne der großen Welt ist aus den Quell-Geistern der großen Welt am 4ten Tage im Mittel der 7 Geister ausgangen als das Licht, dass die ganze Erde erleuchten soll, dieses Licht hat GOTT aus den 4 Elementen, als aus der Quintessenz der reinen Elemente erschaffen. In unserer kleinen Welt ist unsere Sonne auch aus dem Mittel der Quell-Geister als das Licht und Herz, Quintessenz aus den 4 Elementen erkoren, rein, lauter, himmlisch, Licht, Kraft und Leben. Das Paradies war ein Mittel-Ort, darin Adam gesetzt war, zu versuchen, ob er bestehen oder fallen würde und ist noch, denn GOTT hat den Cherub davor gelegt, zu bewahren den Weg zu dem Baum des Lebens; auch sagt unser Erlöser vor dem Schächer: Heute wirst du mit mir im Paradies sein. Paulus ward im 3ten Himmel ins Paradies ent-

remain locked up there in his love for ever, so help us GOD, the father, son and holy spirit, beloved and praised for eternal time, Amen.

CONCLUSION OF THIS FIRST TREATISE

In heaven is the sun of righteousness, the son of GOD Christ JESUS, who enlightens all heavenly crowns, angels, and holy people, from which they derive their life, light, and strength, without this divine light everything would be dark and dead, without light, love, and grace. In the macrocosm is the sun, which illuminates all the stars, including the whole world, for it is the spirit of the macrocosm, from which it draws its power, light, life, growth, and fertility, as from its life-giving spirit, because without the sun or its luminosity all would be dead, dark, and frozen. In our microcosm there is also our sun, which illuminates everything, makes everything alive, gives life and growth; without this sun, or its warmth, fire, and light, everything would be dead, without power, life, and tincture. The first sun is the son of GOD, the heart of GOD is born of GOD from eternity and has gone forth in the heavenly thrones, especially in the throne of Lucifer, which is the most beautiful angel and was created from the light of the son of GOD. The sun of the macrocosm came out of the source-spirits of the macrocosm on the fourth day from the midst of the seven spirits, as the light that should illuminate the whole earth, GOD created this light from the four elements, as the quintessence of the pure elements. In our microcosm, our sun also emerged from the midst of the source-spirits as the light and the heart, the quintessence of the four elements, clean, pure, heavenly; light, strength, and life. Paradise was a middle place in which Adam was set to test whether he would stand or fall, and still is, for GOD set the cherub to guard the way to the tree of life. Our redeemer also said to the malefactor: "Today you shall be with me in paradise."[108] Paul was swept up to the third heaven, up into

zückt und hört unaussprechliche Worte, nur dass wir es mit unseren tierischen Augen nicht sehen können, wie dorten des Propheten Diener, da er sah, dass sein Herr und er belagert waren, sprach er: Ach wie wollen wir ihn tun? Der Prophet aber betet zu GOTT: HERR öffne ihnen die Augen, da sah er lauter feurige Wagen und Rosse, ein Heer der heiligen Engel um sich und seinen HERRN. Also sind uns die Augen auch nicht geöffnet, das Paradies zu sehen, bis unser tierischer Leib durch den Tod abgeschnitten, dann können wir das Paradies und den Baum des Lebens JESUM *Christum* finden, davon essen und ewiglich leben. Warum aber GOTT Adam in das Paradies gesetzt, als in den Mittel-Ort, geschah darum, dass ihm wieder konnte geholfen werden, weil GOTT den Neid des Teufels wohl sah, dass er das neue Geschöpf anfeinden würde und zum Abfall bewegen, so erschuf GOTT aus der Quintessenz der Erde einen roten Letten oder Ton. Dieser Klumpen war angefeuchtet mit dem geseelten reinen, süßen Licht-Wasser von den *Astris* zu einem Kloß. Dieses Geschöpf, der Mensch, wäre nun der paradiesische Leib, das Bild nach GOTT geschaffen, ein reiner unbefleckter Spiegel, darinnen sich GOTT beschauen wollte als ein Vater in seinem Sohn. Wäre er nun bestanden, so wäre er in einen Engel wie Lucifer, durch das Licht des Sohnes GOTT verklärt worden, welches ihn erleuchtet und seinen paradiesischen Leib alsobald in einen Engel verklärt, wäre er hernach gefallen, so wäre ihm unmöglich wieder zu helfen gewesen, wie dem Lucifer. Aber GOTT behielt diesmal seine Hilfe zurück, er wollte erweisen, seine göttliche Gerechtigkeit und Allmacht, dass er die bösen Engel stürzen, den armen bußfertigen Menschen aber erretten und befreien könnte. Dieser Adam sollte nun herrschen über die Gestirne und die große Welt, ein engelhaftes Heer aus sich erzeugen, lauter heilige Kinder GOTTES, er sollte die *Astra* oder äußere Welt nicht lieben, weder die Natur noch Creatur, auch nicht sich selbst, nichts begehren, wissen, wollen und lieben als GOTT: Er solle nicht dieser Welt irdische Früchte essen, sondern Engels-Brot, Früchte des Paradieses aus den reinen und heiligen Elementen, aber er ward lüstern nach der großen Welt ihren Früchten und

paradise and heard unspeakable words, but we cannot see this with our animal eyes; for there the prophet's servant, when he saw that he and his master were besieged, said: "Alas, how will we do it?"[109] But the prophet prayed to GOD: "LORD, open his eyes, and he saw: the mountain was full of horses and chariots of fire,"[110] an army of holy angels around himself and his LORD. So our eyes are not open to see paradise until our animal body is cut off by death, then we can find paradise and the tree of life, JESUS Christ, eat from it and live forever. But why GOD put Adam in paradise, as in the middle place, this happened because he could help him later and because GOD saw the envy of the devil, that he would be hostile to the new creature and cause it to fall, so GOD created a red loam or clay from the quintessence of the earth. This lump was moistened into a clod with the blessed pure, sweet luminous waters of the *astris*. This creature, the human being, would now be the paradisiacal body, the image created after GOD, a pure immaculate mirror in which GOD wanted to look at himself, as a father at his son. If he had passed, he would have been transfigured into an angel like Lucifer, through the light of the son of GOD, which would have enlightened him and immediately transfigured his paradisiacal body into an angel; but had he fallen afterwards, it would have been impossible to help him again like Lucifer. But GOD withheld his help this time, he wanted to prove his divine justice and omnipotence and show that he could overthrow the evil angels, while saving and liberating the poor penitent human. This Adam should now rule over the stars and the macrocosm, create an angelic army from himself, all holy children of GOD. He was not to love the *astra* or the outer world, neither nature nor creatures, not even himself, but to desire, know, want, and love nothing else but GOD: He was not supposed to eat earthly fruits of this world, but angel's bread, fruits of paradise from the pure and holy elements, but he became lustful for the macrocosm and its

ging wieder zurück in der Mutter Leib, daraus er gangen war ins Paradies, aus welchem er in die engelhafte Welt sollte versetzt werden, ging er zurück, aß Erde und irdische Früchte, sobald als er diese kostete, ward er ein Tier, denn er bekam die irdische Gestalt anstatt des Engels. Der Teufel, als er sah, dass der Mensch durch Luft halb überwunden, brauchte er auch seine List, er wusste wohl, dass Hoffart und GOTT gleich sein wollen, die größte Sünde wäre, denn dadurch war er gefallen, sah den Versuchungs-Baum mit den schönen Früchten der Welt, machte den Menschen weis, wenn sie davon äßen, würden ihre Augen aufgetan und sein wie GOTT. Dieses glaubten unsere ersten Eltern, welche noch zur selben Zeit in lauter reiner Einfalt und Unschuld waren, aßen von der Welt ihren Früchten, da doch GOTT geboten, welches Tages du davon isst, sollst du des Todes sterben. Der Lügen-Teufel aber sagte: Ihr werdet nicht des Todes sterben. Nun kannten sie den Teufel nicht, er erschien ihnen auch nicht in Teufels-Gestalt, sondern redete aus der Schlange, welche vor der Verfluchung ein schönes Geschöpf gewesen, aßen davon, zur Stunde figurierte sie der große Welt-Geist, nachdem sie lüstern geworden und seiner Frucht gegessen hatten, zog ihnen das tierische Bild an, sie sturben ab des göttlichen Bildes, ihr paradiesischer Leib war mit dem tierischen plumpen schweren Körper bedeckt und angefüllt, mussten auch als Tiere von dieser Welt Früchten essen, sich von den Tieren kleiden. Das Gestirn mit seiner bösen Influenz bekam den Menschen auch in seine Gewalt, es wäre GOTT nicht um einen Apfelbiss zu tun, jedweder *Circul* oder Welt sollte sich nicht in den anderen vermischen, GOTT ist ein GOTT der Ordnung, es sollte alles im Gehorsam und Willen GOTTES verbleiben, es sollte Adam Engels-Brot aus dem Munde GOTTES essen, so aß er von der tödlichen vergänglichen Speise und aß daran den Tod. Nun war es um ihn und alle seine Nachkommen geschehen, konnten nun keine Kinder zeugen nach GOTTES Bilde, sondern nach dem sündlichen Adams Bilde, hier war weder Hilfe noch Rat, niemand in Himmel noch auf Erden konnte diese erlösen, der Gerechtigkeit GOTTES genug tun und den Zorn der Liebe versöhnen, außer der Sohn

fruits and went back into the mother's womb, from which he had gone to paradise, and from which he was to be transferred into the angelic world. He went back, and he ate earth and the fruits of the earth, and as soon as he tasted them he became a beast, for he received the earthly form instead of the angelic one. The devil, when he saw that the human was half conquered by air, also needed his cunning; and he knew well that ambitious pride and the desire to be like GOD would be the greatest sin, because that was what caused him to fall. He saw the tree of temptation with the beautiful fruits of the world, and he told the humans that if they ate of it, their eyes would be opened and be like GOD. This is what our first parents believed, who then ate the fruits of the world while they were still in pure simplicity and sheer innocence, although GOD had commanded: "In the day that you eat thereof you shall surely die".[111] But the lying devil said: "You shall not surely die".[112] Now they did not know the devil, and he did not appear to them in the form of a devil either, but spoke out of the serpent, which had been a beautiful creature before the malediction; they ate and at that very hour the great world-spirit shaped them after they had become lustful and had eaten of its fruit, imposing on the humans the animal image as the divine image died in them: their paradisiacal body was covered and filled with the clumsy, heavy, animal body; and as animals of this world they had to eat fruit and clothe themselves with animals. The celestial body with its evil influence also brought the human under its power. GOD would not be concerned with the biting of an apple, any *circul* or world should not mix with the others. GOD is a GOD of order, everything should be done in constant obedience to the will of GOD. Adam was to eat angel's bread from the mouth of GOD, but he ate of the deadly perishable food and ate death from it. It was now over for him and for all his descendants, for he could not father children in GOD's image anymore, but only in the sinful Adam's image. Here was neither help nor counsel, nobody in heaven or on earth could redeem them or do enough of GOD's justice to reconcile the wrath with love, except the son of GOD, who now had to come into

GOTTES, der musste nun in der Zeit ins Fleisch kommen, menschliche Natur, unser Fleisch und Blut an sich nehmen und uns der göttlichen reinen Natur wiederum teilhaftig machen, dass er mit seiner reinen Empfängnis und Geburt auch reinigte, das wir durch die neue Geburt wieder einen Eingang ins Paradies hätten, so musste der helle Stern aus Jacob, unsere Sonne, die uns erleuchten sollte, aus des Weibes Samen geboren werden, der Funken des paradiesischen und himmlischen sollte durch dieses göttliche Feuer wieder anglimmen, doch musste diese Sonne der Gerechtigkeit und Aufgang aus der Hölle erst ihre große Finsternis leiden durch Kreuz und Tod, alle unsere Schmach und Schande tragen, damit unser Fleisch ihm konnte verklärt werden und wieder GOTT sehen, denn in ihm sind die so weit abgelegenen *Centra*, GOTT und der Mensch wieder vereinigt, durch ihn muss das Paradies wieder eröffnet, wenn durch den Tod der tierische Leib wieder von uns geschieden, werden wir den paradiesischen Leib wieder haben und durch unsere Sonne, den Fürsten des Lebens, durch sein Licht und Klarheit in Engel GOTTES verwandelt werden, danach wird uns in Ewigkeit weder Tod noch Finsternis mehr treffen. Wie der Mensch im Paradies, so ist auch unser paradiesischer Stein, unser reiner ☿, wenn dieser sich wollte wieder zurück wenden in die grobe unreine metallische Erde und mit solchen Schlacken vermengen und besudeln, so würde sein ganzer reiner Leib zu einer hässlichen Schlacke und könnte derselbe in Ewigkeit nicht wieder gereinigt noch mit dem reinen Geist erleuchtet, in Engels-Gestalt verklärt werden, das aus diesem paradiesischen Leib ein Geist würde, woraus der Himmel oder Thron, der steinerne Palast könnte gebaut und in das gläserne Meer gebracht, welches mit Feuer gemengt werden kann und zu lauter reinen himmlischen, glänzenden, feurigen, lichten Geiste werden, sondern bliebe in Ewigkeit unrein und verworfen, untüchtig zur Tinctur. Es ist Adams Fall nicht ein geringes Ding, ein bloßer Apfelbiss, dadurch er sich zu einem schändlichen garstigen Wurm gemacht, wenn ihn GOTTES Gnade nicht erhalten durch Verheißung des Messias, des Weibes Samen: Also in unserem Werk unserer Jungfrau Sohn,

the flesh, into human nature, taking upon himself our flesh and blood and to make us partakers of the divine pure nature again. And with his pure conception and birth he also purified us so that through the new birth we would again have an entrance into paradise, then the bright star of Jacob, our sun, which should illuminate us, had to be born from the woman's seed, that the spark of the paradisiacal and heavenly should glow again through this divine fire, but this sun of justice and rising from hell[113] first had to suffer its great darkness, through the crucifixion and death, wearing all our disgrace and shame, so that our flesh could be transfigured for him and become capable of seeing GOD again. For in him there are the remote *centra*, [in which] GOD and humans are reunited, through him paradise must be opened again, when the animal body is separated from us again through death, we will have the paradisiacal body again and will be transformed into GOD's angels by our sun, the prince of life, through its light and clarity, after which neither death nor darkness will come to us in eternity. Like unto a human in paradise is our paradisiacal stone, our pure ☿; and if it wanted to turn back into the coarse, impure, metallic earth and mix and soil it with such slag, then its whole pure body would become an ugly slag and could then neither be purified again in eternity nor be transfigured with the pure spirit in the form of angels, so that this paradisiacal body would become a spirit from which the heaven or throne, the stone palace, could be built and brought into the sea of glass, which could be mingled with fire and become nothing but pure, heavenly, shining, fiery, light spirit. Instead it would remain impure and rejected for eternity, unfit for tincture. It is not a trifling thing in Adam's case, a mere bite of an apple, whereby he is made a shameful nasty worm, unless GOD's mercy is preserved for him, by the promise of the messiah, the woman's seed: so in our work, our virgin's son, who unites with

der sich mit der reinen Maria vereinigt, von ihr geboren, dieser muss sich lassen töten, in die Erde begraben und diese Sonne auch ihre große Finsternis ausstehen, nach welcher sie nicht mehr untergeht und allen ihren Brüdern güldene Kronen aufsetzen kann, das ganze metallische Reich in ☉ tingieren, dass sie eben wie die Sonne leuchten und jedweder Himmel seine sonderliche Sonne hat, in dieser unserer kleinen Welt nach allen 3 Reichen in der äußeren Welt ist es der solarische ☿, das feurig leuchtende Wasser, der brennende *Rubicundus*, in der paradiesischen Welt ist es unser erstandener Sieges-König, leuchtet nach dem Tod und Finsternis, welche nicht mehr verdunkelt werden kann. In der engelhaften Welt ist es das leuchtende himmlische Licht-Feuer, die hohe Tinctur. Aus diesem allen sehen wir, dass die große Welt mit den 7 Planeten nach dem Fall der Engel geschaffen, als Lucifer in seinem Thron-Himmel auch in sich alle 7 Quell-Geister anzündete und sich alles in einen wüsten Klumpen reducirte, aus diesem schaffte GOTT die Welt und wurden die gefallenen Engel ausgestoßen zwischen dem Mond und der Erde, darinnen erhalten werden zum Tage des Gerichts und sind Geister, so in der Luft herrschen, der Fürst dieser Welt, in der angezündeten Grimmigkeit, in den siderischen Geistern, woraus des Menschen Geister im animalischen Geist auch bestehen und durch diesen hat er einen Eingriff in unser Herz: Wenn aber unsere Geister mit GOTT inqualieren, unseren animalischen Geist erleuchten, so werden die siderischen Geister ganz inbrünstig und des göttlichen Lichts begierig, da wütet und tobt der Teufel, welcher in der äußeren Geburt des Menschen seinen Sitz hat, denn er kann dieses göttliche Licht nicht vertragen, es war sein und war ihm entzogen, das gönnt er den Menschen nicht, wütet und tobt, will alle Quell-Geister in Zorn entzünden, dass sie im Zorn GOTTES brennen sollen, welches der neue Mensch in *Christo* nicht leiden will, löscht im alten Adam, im Zorn-Reiche, dann geht der Streit an, Michael und seine Engel streiten mit den Drachen, dieser Streit währt bis in unseren Tod. Überwinden wir in der neuen Geburt, welche in GOTT steht im Paradies, so sind wir nach dem Tode auch drinnen, überwindet uns

the pure Mary, born of her, he must allow himself to be killed, buried in the earth, and this sun also endures its great darkness, after which it no longer sets and can put crowns on all its brothers, tingeing the whole metallic kingdom into ☉, so that they shine just like the sun and every sky has its own special sun, in this our microcosm after all three kingdoms in the outer world, it is the solar ☿, the fiery shining water, the burning *rubicundus*. In the paradisiacal world it is our risen victorious king which shines after death and darkness, which can no longer be darkened; in the angelic world it is the luminous celestial light-fire, the high tincture. From all this we see that the macrocosm with the seven planets was created after the fall of the angels, when Lucifer in his heavenly throne also ignited all seven source-spirits in himself and everything was reduced to a desolate lump, from which GOD created the world. The fallen angels were cast out between the moon and the earth, in which they are preserved for the Day of Judgment, and are spirits reigning in the air with the prince of this world, in incendiary wrath, in the sidereal spirit,[114] of which human spirits in animal spirits also consist; and through this he has an intervention in our heart. But when our spirits fuse with GOD, who enlightens our animal spirit, the sidereal spirits become extremely fervent and greedy for the divine light. At this point the devil, who has his seat in the outer birth of the human, rages and rampages because he cannot endure this divine light. It was once his but was withdrawn from him and he begrudges it to the humans, rages and rampages, wants to ignite all source-spirits in anger so that they shall burn in the wrath of GOD, which the new human in Christ does not want to suffer, extinguishing the old Adam in the kingdom of anger. Then the battle begins, Michael and his angels clash with the dragons, and this battle lasts until our death. If we overcome it in the new birth, which is in GOD, in paradise, then we are also inside [paradise] after death, whereas if the old Adam overcomes us,

der alte Adam, die sündliche Geburt, mit der wir im Zorn und des Teufels Reich stehen, dessen sind wir auch nach dem Tode und stehen wir hier im ganzen Leben mit einem Fuß in der paradiesischen Licht-Welt, mit dem anderen in der finsteren Zorn-Welt des Teufels und der Höllen Reich. Wenn nun der Mensch durch *Christum* siegt und überwindet, so sieht er die Gefahr, worin er gesteckt und lernt sich vor dem Teufel hüten und wird im Tode ganz von ihm geschieden; überwindet aber der Teufel die Seele, ergibt sich der Welt und ihrem Laster-Leben, so ist der Mensch nach dem Tode im Höllen-Reich und des Teufels Leibeigener, daher kommt auch die leibliche Besitzung des Teufels, wenn er sein Raub-Schloss im Menschen macht. Hieraus sehen wir, dass Paradies und Hölle auch in uns ist, in welcher wir hier leben, darinnen sind wir im Tode; leben wir in der neuen Geburt, im neuen Menschen, so dürfen wir das Paradies nicht weit suchen, denn das Reich GOTTES ist inwendig in uns, leben wir aber in der alten sündlichen Geburt, so dürfen wir auch die Hölle nicht weit suchen, denn wir leben schon drinnen und sobald die gottlose Seele den Leib verlässt, hat sie der Teufel in Händen. *Christus* hat das meiste an der Seele, denn erstlich hat sie GOTT erschaffen, da sie durch Sünde fiel in des Todes und der Höllen Reich, ward der Sohn GOTTES Mensch, sie zu erlösen, litt den bittern Tod, fuhr in die Hölle, nahm den Gefangenen, dem Teufel die gefangenen Seelen, das heißt, er hat die Gefangenen aus dem Gefängnis gefangen geführt und den Raub ausgeteilt, er errettete sie aus der Grube, da kein Wasser war, dessen nun, der uns durch den Tod, Streit, Kampf und Sieg erobert und erworben, auch aus Liebe für uns sein Leben gelassen, dessen sind wir. Des Todes und der Höllen Reich wird uns nicht halten können, *Christus* hat den Tod verschlungen in den Sieg, Tod wo ist dein Stachel, Hölle wo ist dein Sieg? GOTT sei Dank, der uns den Sieg gegeben hat durch unsern HERRN JESUM *Christum!*

the sinful birth, with which we stand in wrath's and devil's kingdom, we are also inside this after that death, and we stand here throughout life with one foot in the paradisiacal world of light and with the other in the dark world of anger, of the devil, and the kingdom of hell. Now when the human triumphs and overcomes through Christ, he sees the danger in which he is stuck and learns to guard himself against the devil and is completely separated from him in death; but if the devil overcomes the soul, which surrenders to the world and its life of vice, then after death the human being is in the kingdom of hell and is the devil's serf, hence he is the physical possession of the devil, when he puts his thieve's castle in the human being. From this we see that paradise and hell are also in us, and depending upon which of the two we abide in during life, therein we are to remain also in death. If we live in the new birth, in the new human being, we must not look far for paradise, because the kingdom of GOD is within us; but if we live in the old sinful birth, we must not look far for hell either, because we already live inside [hell], and as soon as the godless soul leaves the body, the devil has it in his hands. Christ has most of the soul, because first GOD created it, and when it fell through sin into the kingdom of death and hell, the son of GOD became human to redeem it. He suffered a bitter death, descended to hell, took away from the devil the imprisoned soul, that is, he led the captive prisoner out of prison and distributed the spoils. He rescued them from the pit in which there was no water. We belong to him who conquered and won us through death, strife, battle, and victory, and who also laid down his life for us out of love; to him we belong. The kingdom of death and hell will not be able to hold us, Christ has swallowed up death in victory. "Death where is your sting, hell where is your victory?"[115] Thank GOD who gave us the victory through our LORD JESUS Christ!

DER ANDERE TRAKTAT

*Den paradiesischen Stein betreffend, von der inneren und Licht-Welt,
worin sich dieser Stein versetzen lässt, dies heißt das Paradies
bauen und bewohnen, hier muss nichts
unreines eingehen*

Dieser Stein hat nur 2 Geburten; der Sohn GOTTES hat auch 2 Geburten, eine von Ewigkeit, die andere in der Zeit, eine göttliche und eine menschliche, in dem Leib Marias, der Mensch hat auch 2 Geburten, eine sündliche, die andere durch die Taufe, Wasser und Geist. Soll nun der Mensch im neuen Leben wandeln, muss er durch den Geist wiedergeboren werden, ob schon der äußere Mensch mit der Wasser-Taufe getauft wird, der innere Mensch aber mit dem heiligen Element, so zeugt er doch nach der Taufe keinen neuen Menschen, sondern den alten Adam, zu allem gottlosen Leben behende und geschickt, zum Guten faul und träge von Kindheit an bis ins Alter; soll nun dieser wieder umwenden und den neuen Menschen anziehen, so muss er durch den Geist wiedergeboren werden, den alten Adam kreuzigen und töten, dass er täglich ersäuft wird mit allen Sünden und bösen Lüsten und herauf komme ein neuer Mensch der in Gerechtigkeit und Reinigkeit für GOTT ewig lebe. Das Wasser wäscht ab die Erbsünde, der Geist aber durchdringt den ganzen neuen Menschen, nimmt alle Flecken weg, erneuert den Menschen in Geist, Seele und Leib, da der Apostel sagt: Der GOTT des Friedens heilige euch durch und durch und euer Geist ganz, samt Seele und Leib, müsse behalten werden unsträflich, bis auf die Zukunft unseres HERRN Jesu *Christi*. Da meint er die Feuer-Taufe durch den Heiligen Geist, welche unsere Erleuchtung und Verklärung ist: Exempel haben wir an den Jüngern des HERRN *Christi*, ob sie wohl im neuen Menschen als Kinder GOTTES lebten,

THE SECOND TREATISE

Regarding the paradisiacal stone from the inner and luminous world, wherein this stone is set down, which means to build and inhabit paradise; here nothing impure must enter

This stone has only two births; the son of GOD also has two births, one from eternity, the other in time, one divine and one human, in the womb of Mary. A human also has two births, one sinful, and the other through baptism, water and spirit. If the human being is to walk in the new life, they must be born again through the spirit. While the outer human is baptized with water, the inner human [is baptized] with the holy element. After the baptism [the outer human] does not beget a new human, but the old Adam, nimble and fit for all ungodly life, indolent to all good from infancy to old age. If [the outer human] is to transform and put on the new human, they must be born again through the spirit, crucifying and killing the old Adam, drowning him daily with all his sins and evil lusts so that a new human will arise who will live forever in righteousness and purity for GOD. The water washes away the original sin, but the spirit penetrates the new and complete human being, removing all stains, renewing the human in spirit, soul, and body, as the apostle says: "And the very GOD of peace sanctify you wholly; your whole spirit and soul and body be preserved blameless unto the coming of our LORD Jesus Christ".[116] He means the baptism of fire by the holy spirit, which is our enlightenment and transfiguration. We have an example for that in the disciples of the LORD Christ, although they lived in the new human as children of GOD, having

Christum versöhnlich bei sich hatten, auch an ihn glaubten, so waren sie doch in göttlicher Erkenntnis Kinder und musste ihnen der HERR *Christus* das Verständnis öffnen nach seiner Auferstehung, hatten aber doch Zweifel und Unglauben, lauter Unvollkommenheit, fielen noch manchmal in die Sünde, als wie Petrus, konnten noch keine Wunder tun, bis nach der Himmelfahrt des HERRN *Christi* der Heilige Geist über sie ausgegossen ward, danach waren sie die Wunder-Leute und Heiligen Apostel durch den Heiligen Geist, taten alle die Wunder, als der Sohn GOTTES selbst, bestätigten ihre reine Lehre durch ihr Blut, welches sie als freudige Märtyrer vergossen. Von dieser Taufe sagt Johannes: Ich taufe mit Wasser, der nach mir kommt, wird euch mit dem Heiligen Geist und mit Feuer taufen. Diese Taufe sollen alle gläubige fromme Kinder GOTTES haben, denn es ist der Geist unserer Verklärung, dadurch wir in Engel verwandelt werden, welches uns im Paradies vorbehalten war. Wäre Adam und Eva im paradiesischen Leibe mit dem Heiligen Geist erleuchtet gewesen und wäre hernach gefallen, so wäre es ihm gegangen wie dem Lucifer und wäre der Abtrünnige von GOTT geblieben, aber GOTT wollte den Menschen erhalten zu Trotz dem Teufel, ließ den Menschen in dem paradiesischen unverklärten Leibe versuchen, da er als ein lauter reiner Mensch in Unschuld, gerecht, rein, keusch, einfältig und fromm war, nichts wusste als von GOTT, nichts sah und kannte als GOTT, bis sich ihnen die große Welt mit ihren Früchten der Versuchungs-Baum präsentierte, da wurde er lüstern nach den Früchten der äußeren Welt, welche doch tödlich waren und er dadurch in Sünde und Tod fiel. Es wollen alle Menschen gern den Heiligen Geist haben und dadurch erleuchtet sein, GOTT will auch den Heiligen Geist geben denen, die ihn darum bitten, aber wir müssen zuvor beten: Schaff in mir GOTT ein reines Herz und gib mir einen neuen gewissen Geist; Es ist die größte Gnade von GOTT, die ein sterblicher Mensch erlangen kann, wenn er sich hier von Sünden los reißt, im Geist *Christi* lebt und durch den Heiligen Geist hier erleuchtet wird, aber es ist auch die größte Gefahr dabei, kehrt ein solcher wieder um, so hat er seine Hand an den Pflug

Christ with them in reconciliation and also believing in him, they were still children in the knowledge of GOD; and LORD Christ had to open their understanding after his resurrection. And yet [they] still had doubts and unbelief, every imperfection, and [they] still sometimes fell into sin like Peter. Nor could they perform miracles until after the ascension of the LORD Christ, when the holy spirit was poured out on them, after which they were the miracle-people and holy apostles through the holy spirit. All worked miracles like the son of GOD himself, confirming their pure teaching by their blood, which they shed as joyful martyrs. Of this baptism John says: "I baptize you with water, but he that comes after me, he shall baptize you with the holy spirit and with fire".[117] All believing, pious children of GOD should have this baptism, for it is the spirit of our transfiguration, through which we are transformed into angels, which was reserved for us in paradise. If Adam and Eve had been enlightened with the holy spirit in the paradisiacal body and had fallen afterwards, they would have fared like Lucifer and would have remained apostates from GOD, but GOD wanted to keep humans in defiance of the devil, and so allowed him to tempt humans in the paradisiacal, untransfigured body, when he was an untarnished, pure human being in innocence, being just, pure, chaste, simple-minded, and pious, knowing nothing but GOD, seeing nothing but GOD, until the macrocosm with its fruits of the tree of temptation presented itself to them. Then they became lustful for the fruits of the outer world, which were deadly and thereby they fell into sin and death. All humans want to have the holy spirit and be enlightened by it; GOD also wants to give the holy spirit to those who ask for it, but we must first pray: "Create in me a pure heart, O GOD, and renew a steadfast spirit within me".[118] It is the greatest mercy of GOD that a mortal human can obtain, when they break free from sin here, live in the spirit of Christ, and are enlightened by the holy spirit here, but there is also the greatest danger should they return. If they had put their hand on the plow and pulled it back again, then they would not

gelegt und wieder zurückgezogen, der ist nicht geschickt zum Reich
GOTTES und wäre besser, dass er den rechten Weg nie erkannt hätte
und werden wieder in die vorigen Sünden geflochten, ist mit ihnen
das Letzte ärger worden als das Erste. Es ist *Christus* für die Sünder gestorben zu einem mal, wenn nun solcher mit *Christo* auferstanden vom
Tod der Sünden, mit *Christo* in einem neuen Leben gewandelt und mit
dem Heiligen Geist gesalbt und erleuchtet, davon abermals der Apostel sagt: Denn es ist unmöglich, dass die, so einmal erleuchtet sind
und geschmeckt haben die himmlischen Gaben und teilhaftig sind des
Heiligen Geistes und geschmeckt haben das gütige Wort und die Kräfte der zukünftigen Welt, wo sie abfallen und wiederum ihn selbst, den
Sohn GOTTES kreuzigen und für Spott halten, dass sie wieder sollten
erneuert werden zur Buße. Das ist die Sünde im Heiligen Geist, denn
der Geist der Gnaden wird geschmäht, mit dem er doch versiegelt ist,
diese Sünde wird nicht vergeben, weder in dieser noch in jener Welt,
deswegen muss gebetet werden: Nimm deinen Heiligen Geist nicht
von mir. Exempel haben wir an dem erleuchteten Salomo, wie er gefallen, wir lesen aber nicht, dass er den Geist GOTTES wieder bekommen, denn wo derselbe einmal seine Wohnung aufgeschlagen und aus-
gejagt wird, da kommt er nimmermehr wieder. Das sehen wir am Fall
Lucifers, wäre er nicht durch das Licht GOTTES erleuchtet gewesen,
hätte er so tief nicht fallen können, weil aber der Mensch noch nicht
dadurch erleuchtet war und fiel, so konnte ihm wieder geholfen werden durch den Sohn GOTTES, durch Annehmung unserer Natur, welche durch seine göttliche Natur wieder gebracht, unser neuer Mensch
gereinigt, damit er kann durch den Heiligen Geist erleuchtet werden,
doch konnte diese Erleuchtung und Ausgießung des Geistes über alles
Fleisch nicht eher erfüllt werden, bis nach seinem Hingange zum Vater, da unser Erlöser selber fragt: Wo ich nicht hingehe, so kommt der
Tröster nicht zu euch, so ich aber hingehe, will ich ihn zu euch senden,
derselbe wird euch in alle Wahrheit leiten. Was war aber die Ursache,
dass *Christus* erst musste zu seinem und unserem himmlischen Vater
gehen? Antwort: Er musste unser Fleisch und Blut, welches er nach

be suitable for the kingdom of GOD; and it would be better if they had never recognized the right way, for in this case they would be braided again into the previous sins, the last of which, for them, have become worse than the first. Christ died for sinners at one time, when now such [a sinner] rose with Christ from the death of sins, walked with Christ in a new life and was anointed and enlightened with the holy spirit, of which the apostle again says: "For it is impossible for those who were once enlightened, and have tasted of the heavenly gift, and were made partakers of the holy spirit, and have tasted the good word of GOD, and the powers of the world to come—should they fall away—to renew them again unto repentance; seeing them crucify the son of GOD afresh and put him to open shame".[119] This is the sin in the holy spirit, because the spirit of graces is reviled, with which it is sealed. This sin will not be forgiven, neither in this world nor in the next, therefore one must pray: "Take not your holy spirit from me."[120] We have an example of the enlightened Salomon, how he fell, but we do not read that he received the spirit of GOD again, for where the same once opened his dwelling and was chased out, he will never come back. We see that in the fall of Lucifer, if he had not been enlightened by the light of GOD, he could not have fallen so far; but because the human was not yet enlightened by it, and fell, they could be helped again by the son of GOD, by acceptance of our nature, restored by his Divine nature, our new human [existence] purified to be enlightened by the holy spirit, but this enlightenment and outpouring of the spirit upon all flesh could not be accomplished until after his departure to the father, there our redeemer himself asks: "for if I go not away, the Comforter will not come unto you; but if I depart, I will send him unto you,[121] he will guide you into all truth".[122] But what was the reason that Christ first had to go to his and our heavenly father? Answer: He had to transfigure our flesh and blood, which he took with him into heaven after

seiner Auferstehung mit in den Himmel nahm, in seiner göttlichen Person verklären und durch den Heiligen Geist erleuchten lassen, dass unser ganzer Leib in seiner Person, in Seele, Geist und Leib verklärt, Paradies und Himmelreich dem Menschen wieder geöffnet und wir alle fähig, den neuen Menschen im Geist *Christi* anzuziehen und noch hier im sterblichen Leib durch den Heiligen Geist können erleuchtet werden, welches sich auch noch zuträgt bei den rechtgläubigen Kindern GOTTES und wird auch noch der Heilige Geist über alle Menschen ausgegossen werden, in dem großen Sabbat der Welt, wenn der Geist der Wunder wird offenbar werden, welcher als ein heller Blitz sich über die ganze Welt offenbaren wird, ob solchem schon der Anti-Christ widerspricht, so wird sich doch die 3te Person der Gottheit den Menschen auch offenbaren. GOTT der Vater offenbarte sich durch Moses, durch Gesetz, der Sohn GOTTES offenbarte sich in unserem Fleisch durch Mariam, der Heilige Geist, der Geist der Wunder, der von Vater und Sohn ausgeht, wird sich offenbaren in der Gläubigen Herzen, wenn wir werden alle von GOTT gelehrt sein, wenn der Anti-Christ entdeckt, das Kind des Verderbens, das Tier und der falsche Prophet, dann ist der Heilige Geist doppelt, ja dreifach in uns, denn er geht aus vom Vater, Sohn und Heiligen Geist und sehen ihn nicht mehr als den zorneifrigen GOTT, als er das Gesetz gab mit Donner und Blitz, sondern wie dort bei den Propheten, da der Wind vorüber war, kam ein stilles sanftes Sausen, dieses alles kann und wird mir niemand glauben, der nicht durch den Heiligen Geist manchmal einen kurzen und kleinen Blick in die göttlichen Geheimnisse gehabt, da auf einmal sein Herz durch das göttliche Licht erleuchtet und dieses reine, sanfte, entzündete Feuer in seiner Seele empfunden, welches mit keiner Feder zu beschreiben und die Freude dieses göttlichen Lichts in der Seele nicht auszusprechen ist, die erleuchtet alle Menschen zum ewigen Leben.

Nun will ich unseren Paradies-Stein, seine Geburten auch ausführen, die ist nun wie der Mensch aus der roten Erde, daraus der Mensch zu einem paradiesischen Leibe erschaffen, aus der Quintessenz der

his resurrection, in his divine person, and let it be enlightened by the holy spirit, so that our whole body was transfigured in his person, in soul, spirit, and body, paradise and the kingdom of heaven opened up again for humans and we are all capable of putting on the new human being in the spirit of Christ and can still be enlightened here in the mortal body by the holy spirit, which also happens with the orthodox children of GOD. And the holy spirit will also still be poured out over all people, in the great Sabbath of the world, when the spirit of miracles will be revealed, which will reveal itself as a bright flash over the whole world, even if this is contradicted by the Antichrist; and the third person of the deity will reveal itself to the people. GOD the father revealed himself through Moses, through law; the son of GOD revealed himself in our flesh, through Maryam; the holy spirit, the spirit of miracles, proceeding from the father and son, will reveal itself in believers' hearts when we will all be taught by GOD, when the Antichrist discovers the child of perdition, the beast, and the false prophet, then the holy spirit is twice, indeed thrice in us, for it emanates from the father, son, and holy spirit and we no longer see it as a GOD who raged in wrath when he gave the law with thunder and lightning, but as with the prophets, when the wind had passed, there came a quiet, gentle hissing, no one can and will believe me in all this who has not sometimes had a brief, small glimpse into the divine mystery through the holy spirit, because suddenly their heart was illuminated by the divine light and this pure, gentle, enkindled fire was felt in their soul, which cannot be described with any pen; and the joy of this divine light in the soul, which enlightens all people to eternal life, cannot be expressed.

Now I want to carry out our stone of paradise and its birth, which is now like the human being from the red earth, from which the human being was created as a paradisiacal body from the quintessence of the

Elemente, dieser unser paradiesischer Leib oder Stein ist unser Erz, aussätzig wie Naeman, muss sich 7-mal im Jordan taufen und durch die Taufe perlweiß gewaschen werden, nicht durch gemeines Wasser, sondern durch das Paradies-Wasser im reinen Element. Ob er nun wohl durchs Wasser rein und schneeweiß ist, so hat er doch noch Schiedligkeit und Unvollkommenheiten, er ist noch kein Engel, dass er kann durch verschlossene Türen gehen und so, glaube ich, ist Adam geschaffen und aus eben der Materie, wenn er nun im Paradies bestanden wäre, so hätte ihn GOTT bald in einen Engel verklärt, aber er ward lüstern nach der Speise der irdischen Welt und als er die kostete, so fiel er aus der Licht-Welt in die finstere Welt, in den Zorn GOTTES, die irdische figurierte ihn mit dem tierischen Leib, fressende Zähne, Magen und Gedärme, als auch tierischen geilen Gliedern, dass er nun sein Geschlecht nach tierischer Art fortpflanzen muss, er starb augenblicklich ab des göttlichen Bildes, er wusste nichts mehr von GOTT, Engel noch Paradies, er sah seine tierische Gestalt, welcher er sich schämte, die Engel verließen ihn nach dem Sünden-Fall, die Teufel ließen sich sehen, die er nun wohl kannte, denn nun wusste er was böse war, zuvor aber kannte er nur das Gute. Die Teufel spotteten ihrer, denn sie waren nun in seinem Reich, ihre Rede kann gewesen sein: Du schöner Engel! Bist du nun GOTT gleich, willst unser Fürstentum beherrschen, woraus wir verstoßen, du bist auch gefallen, wir sind Herren der Welt und Herrscher über dich, du bist unser Gefangener im Höllen-Reich. Aber die Gerechtigkeit GOTTES erbarmte sich des armen Menschen, weil er war durch Verleitung und Hass des Teufels von ihm abgefallen, rief ihm wieder zu: Adam wo bist du? Wie hast du dich durch Sünde von mir verlaufen? Verhieß ihm des Weibes Samen zum Erlöser. Durch dies Wort, als sie dieses in ihrer Seele fassten, bildete sich wieder der neue Mensch in ihnen im Geist *Christi* und ist kein Mensch unter der Sonne, er sei Türke, Jude oder Heide, der nicht etwas von dem Geist GOTTES in sich hat, denn er wird stets von demselben zum Guten angeregt, obschon der wenigste Teil dem Geist GOTTES folgt, wie GOTT schon vor der Sintflut klagte, die Menschen wollen sich meinen Geist

elements. This, our paradisiacal body or stone, is our ore, leprous like Naaman,[123] and it must be baptized seven times in the Jordan and washed pearly white by baptism, not by common water, but by the water of paradise, in the pure element. Though it may be water which is clear and snow white, it still has divisiveness and imperfections; it is not yet an angel which can walk through closed doors, and so I believe Adam is created of the same matter. If he had persisted in paradise, GOD would have transfigured him into an angel soon enough, but he became lustful for the food of the earthly world and when he tasted it, he fell from the luminous world into the dark world, in GOD's wrath, the earthly [world] clothed him with an animal body with teeth, stomach, and intestines for eating food, as well as horned animal limbs, so that he now had to reproduce his lineage in a beastly way. He died immediately from the divine image, he knew nothing more of GOD, nor the angels' paradise. He saw his own animal form, which he was ashamed of. The angels left him after the fall, and the devils showed up, whom he now knew well, because now he knew what evil was, but before that he only knew the good. The devils mocked him, for they were now in their kingdom, their speech might have been: "You beautiful angel! You are now equal to GOD, wanting to rule our principality, from which we were expelled, you fell too, we are masters of the world and rulers over you, you are our prisoner in the kingdom of hell." But the justice of GOD had mercy upon the poor human, because he had fallen away from him through the temptation and hatred of the devil, and [GOD] shouted out to him again: "Adam, where are you?"[124] How did you stray from me through sin? And he promised him the woman's seed as a redeemer. Through this word, when it took hold in their souls, the new human being was formed again in the spirit of Christ; and there is no human being under the sun, be it Turk, Jew, or heathen, who does not have something of the spirit of GOD in himself, because he is always stimulated by the same [spirit] towards the good, although the least part [of humans] follow the spirit of GOD, as GOD already complained before the flood: "My spirit shall not always strive

nicht mehr strafen lassen. Als nun unsere ersten Eltern den Glauben an Messiam fassten, wichen die Teufel. Die Menschen müssen sich aber in dem Leibe dieses Todes gedulden, bis der zeitliche Tod den tierischen sündlichen Leib abschneidet und der Mensch nach dem Tode den vorigen reinen Leib im reinen Element anzieht, dann wird er von dem Geist *Christi* verklärt und gleich den Engeln GOTTES, denn der tierische Leib kann GOTT nicht sehen, Fleisch und Blut können das Reich GOTTES nicht ererben, der neue Mensch in Geist und Seele, der kennt GOTT, wenn er durch den Geist GOTTES ein Licht bekommt, so dauert dieses göttliche Anschauen nicht lange, sondern unser Wissen ist Stück-Werk, weil wir im Fleisch und Blut GOTTES nicht begreifen können.

Ich komme wieder zu unserem ☿, unserem paradiesischen Leib. Im Paradies ist nichts tödliches, nichts unreines, die Erde ist die *Terra Sancta*, weil die *Terra Damnata* abgeworfen, die Luft ist rein und helle, die Wasser gesund, welche ☉ und Edelgesteine führen, die schönsten wohlriechenden Kräuter wachsen aus der reinen und glänzenden Erde, wie Perlmutt, die Erde ist silbern, die Bäche gülden, in den Bächen schwimmt das Fischlein *Æscheneis* mit silbernen Schuppen und roten Floßfedern, welche Bächlein endlich ganz voll, zu lauter Fischlein werden, sie ergießen sich zu 7 unterschiedenen Malen in die Erde, worin sie sich verlieren und wieder hervorkommen, welches endlich zu einem himmlischen Wasser wird, unzerstörlich und unvergänglich, welches unsterblich macht und alles neu gebärt und über die Elemente herrscht. Jene Wasser wuschen nur ab, als wenn aus einem Tuche die grobe Unsauberkeit abgewaschen wird, da die grobe Erde in der Solution gestanden, ist aber in ihrem vorigen Wesen der Corruption und Unvermögenheit noch unterworfen geblieben und die 7 Quell-Geister noch in ihrem *Confuso Chao* gesteckt, noch nicht in Schiedligkeit gestanden, noch in eins, ins Licht getreten, darum ist dieses Paradies-Wasser sehr unterschieden vom vorigen groben elementischen Wasser, darum ist es klar, leuchtend, unbefleckt, von seiner Impurität geschieden. Aber was bringt uns das Paradies für Früchte, Lust und Ergötz-

with man".¹²⁵ Now when our first parents started to believe in the messiah, the devils departed. But people must be patient in this perishable body until timely death separates the soul from the sinful animal body; and after death man puts on the previous pure body in the pure element, at which point he is transfigured by the spirit of Christ and like unto the angels of GOD. For the animal body cannot see GOD, and flesh and blood cannot inherit the kingdom of GOD, whereas the new human being in spirit and soul knows GOD when he receives light through the spirit of GOD. Although this divine contemplation does not last long, our knowledge is patchwork, because we cannot understand GOD while we remain in flesh and blood.

I come back to our ☿, our paradisiacal body. In paradise there is nothing deadly, nothing impure. The earth is *terra sancta* because the *terra damnata* has been thrown off. The air is pure and bright; the water which carries ☉ and precious stones is healthy; the most beautiful, fragrant herbs grow from the pure and shiny earth like mother of pearl; the earth is silver; the brooks are golden; the little fish *echeneis* swims in the brooks with silver scales and red fins, until these brooks finally become completely full, becoming nothing but a stream of little fishes. At seven different times they pour forth into the earth, wherein they are lost before emerging again, finally becoming an indestructible and imperishable celestial water, which imparts immortality and gives birth to everything anew and reigns over the elements. These waters only washed away, as if the coarse uncleanliness were being washed off a cloth, since the coarse earth stood in the solution, but in its previous [form of] being it was still subject to corruption and incapacity and the seven source-spirits, which were still contained in their *confuso chao* and were not yet in divisiveness but were still unified, entered into the light. That is why this water of paradise is very different from the previous coarse, elemental water and why it is clear, luminous, undefiled, and separated from its impurity. But what sort of fruit, pleasure, and

lichkeit? Es bringt uns den süßen Balsam oder Zucker ♄, den güldenen Regen des ♃, das heilsame Kraut, so alles Leid und Trauern vertreibt, welches eine schwarze Wurzel und weiße Blüte hat, ☿ genannt, der ♀ ihre Liebe, ihren Honig und Balsam, ihren grünen Wald, da Stier und Widder weiden und die Tauben der Diana fliegen, der *Luna* ihr Zukker und Himmels-Brot, der Marcipan und Trank-Nectar, so von Honig und Milch gebraut, Ambrosia, die Speise der Götter, das himmlische Lebens-Wasser, das vegetabilische, darinnen die Metalle wachsen und große *Circulat, Circulatum minus*, ist das vorhergehende, da der Schwan auf dem philosophischen Meer geschwommen, nun aber ist der gebundene Schwan eine Speise des Königs und *Circulatum maius*, die Juno mit ihrem Pfau, Jason mit dem güldenen Vlies, alle Götter in ihrer hochgüldigen Freude, ☿ ist ihrer aller Bote, bringt ihnen Briefe von zarter dünner Leinwand oder dünnem Flor, dass nun die Zeit der Hochzeit vorhanden, das Paradies ist erbaut aus der kleinen Erde, so aus der vorigen roten Erde, aus ihrem *Centro*, als das Weib vom Mann genommen und mit dem mittleren Wasser zur kristallischen Klarheit angefüllt, aus welchem ein Brunnen entsprungen, daraus Lebens-Wasser quillt, ist dem Strom gleich, so aus dem Tempel geflossen und gemessen worden, bis er zuletzt so tief worden, dass er nicht zu ergründen und die Ewigkeit bedeutet, weil es durch dieses Wasser zur Ewigkeit fortgeht und *in infinitum* verherrlicht wird, welches ein Mann gemessen in Erzgestalt im Propheten Hesekiel, da aller Menschen Verstand verstummen muss in diesem großen Werk. Hieraus haben die Menschen den Sohn GOTTES kennen lernen, wie er hat müssen geboren werden von einer Jungfrau, wie er hat müssen leiden und der Schlangen Stich fühlen, wie er die Schlange, den Teufel, überwunden und sein Reich zerstört. Die ganze Heilige Schrift weist uns dies Werk mit Fingern. Wer von GOTT den Schlüssel Davids hat, versteht es, wer aber solche Gnade von GOTT nicht hat, aber von ihm in dem natürlichen Licht erleuchtet wird, dass er dieses Werk findet, der versteht danach die Heilige Schrift besser und sieht das göttliche Wesen viel klarer als wir in einem Spiegel. Im Alten Testament wussten dieses nur

delight does paradise bring us? It brings us the sweet balm or sugar ♄, the golden rain of ♃; the healing herb that drives away all suffering and sorrow, which has a black root and white blossom, called ☿; the love of ♀, its honey and balm, its green forest, where bull and ram graze and the doves of Diana fly; and the sugar of *Luna* and heavenly bread; the marcipan and potable nectar brewed from honey and milk, ambrosia, the food of the gods; the heavenly water of life, the vegetabilic [water], in which the metals grow and the great *circulat, circulatum minus*[126] being the previous one, since the swan swam on the philosophical sea, but now the bound swan is a food of the king and the *circulatum maius*, Juno with her peacock, Jason with the golden fleece, all gods in their most golden joy, ☿ is their messenger, bringing them letters made of thin, delicate canvas or gauze, so that the time of marriage is now at hand, paradise is built from the small earth, which was taken from the former red earth, from its *centro*, as the woman is taken from the man, and filled with the middle waters to crystalline clarity, from which a fountain sprung, from which the water of life springs, like the river that flowed out of the temple and was measured until at last it has become so deep that it cannot be fathomed and thus signified eternity, because it proceeds through this water to eternity and is glorified *ad infinitum*, which was measured by a man in the form of ore, in the prophet Ezekiel, since all human understanding must be silent in this great work. From this the humans have come to know the son of GOD, how he had to be born of a virgin, how he had to suffer and feel the sting of the serpent, and how he overcame the serpent, i.e., the devil, and destroyed his kingdom. All of holy scripture is pointing at this work with its fingers. Whoever has the key of David from GOD understands it, but whoever does not have such grace from GOD, yet is enlightened by him in the natural light to find this work, will thereafter understand the holy scriptures better, seeing the divine essence much more clearly as though looking in a mirror. In the Old Testament only the high priests and the prophets knew this,

die Hohen-Priester und die Propheten, hernach kamen diese *Studia* auf die Könige, und sollten es heutigen Tages die Priester und Regenten auch wissen, so würden ganz andere Leute aus ihnen werden.

Ich komme wieder zu unserem Wasser, solches ist nun ein reiner Geist, lauter Licht und Leben, daher bringt es tausendfältige Frucht, macht unsterblich, weil von diesem Wasser unsterbliche Geschlechter gezeugt werden, die nun engelhafte Leiber bekommen. Aus den vorigen elementarischen reinen Wassern wurden unsere 7 Planeten geschaffen oder unsere 7 Metalle, diese aber müssen vergehen in diesem lebendigen Wasser, welches nun ein Geist ist, darinnen sie engelhafte und himmlische Leiber überkommen und in Edelgesteine verwandelt, welche tingieren, die auch den Himmel, die Ewigkeit besitzen, weil sie nicht untergehen und keine Nacht mehr sehen. Hier ist noch der Cherub mit dem hauenden Schwert, er ist ein Engel, das ist flüchtig, ich möchte ihn nennen den Engel ♄, der mit seinem Schwert austreibt, die darinnen erschaffen worden, der ♄ hat nun über seine Kinder Gewalt, welches sind unsere reinen Metalle, welche alle aus seinen Lenden kommen, die er nun als der oberste Richter richtet und probiert, ob sie auch bestehen. Es ist das hermetische Siegel, womit er ihnen das Auge blendet, dass sie das Paradies nicht mehr finden können, daher sie erstarren und zum Glas werden, zum Stein der ersten Ordnung, welches der natürliche Schatz genannt wird, weil dieser paradiesische Stein 7-mal ertötet und wieder lebendig worden, alle 7 Planeten sich ausgefochten und hierinnen ruhen, sich alle vor Sonne und Mond demütigen, welche jenen die Macht und Gewalt genommen. Diese herrschen nun und regieren bis ihrer aller Schöpfer sich über sie erbarmt, sie zu herrlicheren und edleren Geistern macht, dass sie in die engelhafte und himmlische Welt versetzt, zu leuchtenden, tingierenden Edelgesteinen und zur Ewigkeit gelangen, da sie nicht mehr wie der Mond und die Sonne untergehen und verfinstert werden, sondern stets in Licht und Glanz leuchten und funkeln. Also wird GOTT uns Menschen auch verklären: Eine andere Klarheit hat die Sonne, eine andere der Mond, eine andere die Sterne. Also werden wir auch von

afterwards these *studia* came to the kings; and if today's priests and regents would know this, they would become completely different people.

I come back to our water, such is now a pure spirit, sheer light and life, therefore it produces a thousandfold fruit and imparts immortality, because immortal generations which now acquire angelic bodies are begotten from this water. Our seven planets or our seven metals were created from the previous pure elemental waters, but these must perish in this living water, which is now a spirit, in which they conquer angelic and heavenly bodies and are transformed into precious stones which tinge, and which also possess heaven and eternity, because they never perish and see no more night. Here is still the cherub with the cutting sword, he is an angel, which means volatile, I would like to call it the angel ♄ who drives out those which were created within with his sword, ♄ now has power over his children, which are our pure metals, which all come from his loins, which he now judges as the supreme judge and tests whether they also pass. It is the Hermetic seal with which he blinds their eyes so that they can no longer find paradise, hence they solidify and become glass, the stone of the first order, which is called the natural treasure. Because this stone of paradise was killed and came to life again seven times, all seven planets fought each other and rest here, all humble themselves before the sun and moon, which took away their power and violence. These now rule and reign until their creator will have mercy on them, make them more glorious and noble spirits, so that they are transported into the angelic and heavenly world, into shining, tingeing jewels, attaining to eternity, since they are no longer like the moon and the sun, which decline into darkness, but always shine and sparkle in light and luster. So G O D will also glorify us humans: there is one glory of the sun, and another glory of the moon, and another glory of the stars.[127] So we will also advance

einer Klarheit zu der anderen vordringen, bis wir endlich dem verklärten Leib unseres Erlösers gleich werden, da wir erstlich durch Leiden und Tod ihm nachfolgen müssen und wenn wir in der Erde verfault, in unserem paradiesischen Leib wieder auferstehen und immer mehr und mehr verherrlicht werden, bis wir zum Berg Zion kommen und endlich in das neue Jerusalem aufgenommen werden, worin die Vermählung unseres Bräutigams vollzogen und wir mit ihm herrschen werden, worauf er sich hier mit uns verlobt, Brief und Siegel, auch gesagt: In meines Vaters Hause sind viel Wohnungen und ich gehe hin, euch die Stätte zu bereiten, da die Hochzeit des Lamms gehalten wird, da die Seele von dem himmlischen Gewächs des Weinstocks, *Christo Jesu* gesättigt wird, als mit einem Strom und er das Abendmahl mit uns neu halten wird, in seines Vaters Reich, da Freude, die Fülle und liebliches Wesen, zur rechten GOTTES immer und ewiglich sein wird:

Ubi sunt Gaudia?
Nirgends mehr denn da,
da die Engel singen:
Nova cantica,
und die Schellen klingen,
in Regis curia,
Eya wär ich da, Eya wär ich da!

Glückselig ist, der in Ewigkeit also die göttlichen Geheimnisse erforscht, glückselig ist der hier also das große Geheimnis der Natur findet, der Allerglückseligste ist der alle drei, GOTT, die Zeit und Ewigkeit daraus erkennen lernt und in diesem Geheimnis erblickt. Die Heiden haben aus dem natürlichen Werk, aus diesem Geheimnis GOTT lernen erkennen die Heilige Dreifaltigkeit, wie das Wort ist Fleisch geworden, wie er von einer Jungfrau geboren, wie er leiden und sterben müssen, ehe er zur Herrlichkeit eingegangen und erhöht worden und seinen Stuhl über die Engel gesetzt nach seiner Menschheit, nämlich zur rechten GOTTES und andere Geheimnisse mehr. Hat

from one glory to the other, until we finally become like the glorified body of our redeemer, since we must first follow him through suffering and death, putrefying in the earth and rising again in our paradisiacal body that is always more and more glorified, until we come to Mount Zion, and are finally taken up into the new Jerusalem, wherein the marriage of our bridegroom shall be consummated, and we shall reign with him, whereupon he is here betrothed to us, letter and seal, also saying: "In my father's house are many mansions: I go to prepare a place for you",[128] where the marriage of the lamb will be held, where the soul will be satisfied by the heavenly crop of the vine, *Christo* JESU, as with a river, and he will dine with us anew, in his father's kingdom, where joy, abundance and sweetness will be forever and ever at the right hand of GOD:

> *Ubi sunt gaudia?*[129]
> In any place but there?
> There the Angels singing
> *Nova cantica,*[130]
> And there the bells are ringing
> *In Regis curia.*[131]
> O that we were there![132]

Happy in eternity is he who explores the divine secrets, happy is he who finds the great secret of nature here, and most happy of all is he who learns to recognize from it GOD, time, and eternity, seeing all three in this secret. The heathen have learned from the natural work, from this mystery of GOD, to know the holy trinity, how the word became flesh, how he was born of a virgin, how he had to suffer and die before he entered into glory and exaltation, when his Chair was set over the angels according to his humanity, namely, on the right hand of GOD — and other mysteries besides these. If GOD has not withheld such

GOTT solche Wissenschaft den Heiden nicht versagt, so wird er sie uns Christen auch nicht versagen. Der Geist forscht alle Dinge, auch die Tiefe der Gottheit. Wenn nun unsere Erde recht hell und rein in klaren Wassern in der Solution gestanden und zu ihrer ersten Diaphanität gebracht, ist es die jungfräuliche Geburt, wenn es in der siebenten Gestalt zur begreiflichen Selbstständigkeit ausgegangen und beider Natur irdisch und himmlisch, der doppelte Leib, der doch ein flüchtiger Geist ist, aber noch kein Engel, weil er noch nicht durch verschlossene Türen gehen kann, weil er noch nicht im Geist verklärt und mit seiner ewigen Seele vereinigt und noch nicht engelhafter Natur ist, welches dem Wasser gleich vom Propheten Hesekiel 47. Kapitel gemessen worden, so unter der Schwelle des Tempels ausgeflossen, da es ihm an die Knöchel, hernach an die Schenkel und Gürtel gegangen, endlich so tief, dass man darüber schwimmen müssen und unergründlich worden. Dieses Wasser sollte von einem Meer ins andere fließen, so sollen dieselben Wasser gesund werden.

> Ja alles was darinnen webt und lebt, wohin diese Ströme kommen, das soll leben und sehr viel Fische haben und soll alles gesund werden, wo dieser Strom hinkommt und an demselben Strom und Ufer werden fruchtbare Bäume wachsen und ihre Blätter nicht verwelken, auch ihre Früchte nicht verfaulen und werden alle Monde neue Früchte bringen, denn ihr Wasser fließt aus dem Heiligtum, ihre Frucht wird zur Speise dienen und ihre Blätter zur Arznei.

Zuvor aber erschien ein Mann in Erz-Gestalt, der hatte eine Mess-Rute, der ist der Anfang und das Ende des Werks und der Schützer dieses Wunder-Gebäudes, welches er erst abmisst, weil er der Baumeister, wiewohl dieses auch auf die christliche Kirche zu deuten ist, da der GOTTES-Dienst des Neuen Testaments vorgebildet, der innere Tempel und das verschlossene Tor, da niemand dadurch gehen soll, als al-

knowledge from the heathen, he, Christ, will not withhold it from us either. The spirit explores all things, including the depths of deity. If now our earth stood quite bright and pure in clear waters in the solution it would be brought to its first diaphaneity. It is the virgin birth. If it ended in the seventh form to tangible independence and is of both natures, earthly and heavenly. It is the double body, which is yet a fleeting spirit, but not yet an angel, because it cannot yet pass through closed doors; and because it is not yet transfigured in spirit and united with its eternal soul. Nor is it yet of an angelic nature, which is like the water measured by the prophet Ezekiel in chapter 47, which flowed out from under the threshold of the temple, as it went to his ankles, then to his knees and loins, finally becoming so deep that one had to swim across it once it became unfathomable. This water should flow from one sea to another, so the waters shall be healed.

> And it shall come to pass, that everything that lives, which moves, whithersoever the rivers shall come, shall live: and there shall be a very great multitude of fish, because these waters shall come thither: for they shall be healed,[133] and by the river upon the bank thereof, shall grow fruitful trees, whose leaf shall not fade, neither shall the fruit thereof spoil: for it shall bring forth new fruit according to his months, because the waters from the sanctuary flow to them: and the fruit thereof shall be for meals, and the leaf thereof for medicine.[134]

Before that, however, a man appeared in the form of brass, who had a measuring rod, who is the beginning and the end of the work and the protector of this miracle building, which he measures first because he is the master builder, although this can be interpreted to mean the Christian church, because the service of GOD of the New Testament is prefigured here, the inner temple and the closed gate, through which

lein der HERR der GOTT Israel, welches die Jungfrau Maria bedeutet und die unsichtbare Kirche mit ihrem Fürsten, dem Messias, als auch das ganze Neue Testament und den GOTTES-Dienst. Aber weil in allen Bildungen und Gesichten sich GOTT, die Natur und Creatur dem Menschen abbilden wollen, auch wir solches abermals beisammen finden, so habe dieses nicht vorbei gehen können. Dies Wasser nun, die reine und saubere Maria, hat sich mit dem unreinen Adam, der unreinen Erden nicht vermengt. Daher auch eine reine Geburt folgt, auch vermengt sich dieses mit nichts unreinem, kann auch nichts unreines darein gehen und was darinnen geht, wird neu geboren, die *Elementa* sind weggetan, sie sind nur Mütter, daraus die paradiesischen Leiber wachsen, unsere Metalle. Also wachsen aus diesen oberen und himmlischen Wassern, aus der Quintessenz, Tincturen der Metalle oder unsere Planeten, werden mit Jungfer-Milch, als mit dem Paradies-Wasser gespeist, wenn sie aber hier ihre Geburt und Auferziehung erhalten, erwachsen und zur vollkommenen Reife und Größe gelangen, Sonne und Mond die Herrschaft vollkommen, so hat das Paradies sein Ende und werden diese ausgelassen, dass sie in die engelhafte Welt unter die Zahl der Götter aufgenommen werden, weil unsere Sonne und Mond ein unsterbliches Geschlecht der Götter zeugen soll und hat dieser philosophische Tag sein Ende, weil ein neuer Himmel und eine neue Erde geschaffen wird. Diese paradiesische Sonne in der Licht-Welt wird nun zeugen den Sonnen-Sohn, der die engelhafte Welt besitzen soll.

no one is to go through except the LORD, the GOD of Israel, which means the Virgin Mary and the invisible church with its prince the messiah, as well as the whole New Testament and the service of GOD. Yet because GOD, nature, and creatures want to portray themselves to humans in all such formations and visions, we too find such things together again, so I could not ignore this one. Now this water, the pure and clean Mary, did not mix with the unclean Adam of the unclean earths. Therefore a pure birth follows, and it also does not mix with anything impure. Nothing impure can enter it either and what goes in is reborn, the *elementa* are gone, they are only the mothers from which grow the paradisiacal bodies, our metals. So from these upper and celestial waters, from this the quintessence, grow the tinctures of the metals or our planets. They are fed with virgin milk as with the water of paradise, but if they receive their birth and rearing here, they grow up and reach perfect maturity and greatness. Sun and moon complete the reign, then paradise has its end and if these are left out, they are accepted into the angelic world and counted among the gods, because our sun and moon are supposed to beget an immortal race of gods and so this philosophical day has its end because a new heaven and a new earth will be created. This paradisiacal sun in the world of light will now beget the son of the sun who is to possess the angelic world.

DER DRITTE TRAKTAT
DES STEINS

Seine Beschreibung und Vergleichung in der engelhaften und himmlischen Welt, auch von ihren Früchten

Unsere Planeten, welche durch X Kreuz und Feuer-Proben durchgegangen, darinnen sie gesäubert und gereinigt, in Seele, Geist und Leib, dass kein Flecken mehr an ihnen erscheint, ist nun die weiße Tinctur, das Weib mit der Sonne bekleidet, der Mond unter ihren Füßen, sie ist schwanger von der Sonne, lauter solarische Früchte, diese sollen verklärt werden in Edelgesteine, welche im Brust-Schildlein Aaronis gestanden, mitten aber das *Urim* und *Thumim*, Licht und Recht, geleuchtet, in diese Edelgesteine muss erst die Tinctur, der Smaragd, von einem Glanz zu dem anderen verwandelt werden, bis es durch wunder-würdige Stufen auf den höchst-glänzenden Thron sich setzt. Erst war es in seiner Verherrlichung der glasförmige *Azoth* des *Lullij*, der reine Kristall, den er vom Engel Raphael empfangen, nun ist es das gläserne Meer, sieht grünlich und leuchtend. Zuvor als sich der Kristall in die geblätterte Erde begab, schmeckte dieses Büchlein süß wie Honig, in der Feuer-Probe, da dieses gerichtet und zum Stein der ersten Ordnung geschmolzen, schmeckt es etwas herbe, welches den *Lapidem* bedeutet, wird aber in der himmlischen und engelhaften Welt durch die himmlische und engelhafte Speise und Trank wieder in eine Honig-Süße verkehrt, der Trank ist himmlischer Nectar, das Wasser darinnen sie baden und ihre Gesichter waschen, ist das Wasser, so keine Hand nass macht, aus dem Brunnen des lebendigen Wassers geholt, ihre Speise ist Himmels-Brot, gebacken von dem weißen Blut des grünen ♌, so von dem Apollo, dem gestirnten Adler gereicht wird, des

THE THIRD TREATISE
OF THE STONE

*Its description and comparison in the angelic and
heavenly worlds, also of their fruits*

Our planets, which have gone through ten tests of crucification and fire, where they have been cleansed and purified in soul, spirit, and body, so that no stain appears on them anymore, is now the white tincture. The woman is clothed with the sun, and the moon is under her feet. She is pregnant from the sun, bearing nothing but solar fruits which are to be transfigured into the precious stones that were set in the breast-shield of Aaron; but in the middle *Urim* and *Thumim*, light and justice, shone. In these precious stones the tincture, the emerald, must first be transformed from one splendor to the other, until through a series of steps worthy of marvel it sits upon the most-splendid throne. First it was in its glorification the glassy *azoth* of Lullius, the pure crystal he received from the angel Raphael;[135] now it is the sea of glass, which appears greenish and luminous. Before, when the crystal went into the leafed earth, this booklet tasted as sweet as honey in the crucible. When it was judged and melted into the stone of the first order, it tasted a little tart, signifying the *lapidem*, but in the heavenly and angelic worlds it changed again into a sweet honey by the heavenly and angelic food and drink. The drink is heavenly nectar, the water in which they bathe and wash their faces is the water that wets no hand; it is taken from the wellspring of living water, their food is heavenly bread, baked from the white blood of the green ♌ that is handed out by Apollo, the starry eagle, Phœbe's draft of pearl is served, our sun and

Phöbi Perlen-Trank wird aufgetragen, unsere Sonne und Mond besteigen ihr Ehe-Bett und zeugen den Sonnen-Sohn, die rote Tinctur. Ehe aber dieses alles geschieht, müssen die 7 Gestalten und Grade der 7 Quell-Geister in der engelhaften Welt auch das ihre vollenden, bis sie aus einem funkelnden Edelgestein oder Himmel von einer erhöhten Farbe zu der anderen gelangt, bis zur gewünschten Röte des Rubins, davon dieser Sonnen-Sohn eine Krone hat, der himmlische Thron, die Geister und Luft verändern sich in allen Himmeln, woraus man des erhöhten Königs Gestalt, Farbe und Krone abnehmen kann, will dieser König sich abermals multiplicieren, so hat er eine Krone von Karfunkel-Stein. Will sich dieser abermals multiplicieren, so hat er eine Krone, welche ein himmlisches Licht zeigt, da der Karfunkel-Stein dunkel dagegen. Dieses ist das Licht und Recht und ist ein König der Könige, der große Gewalt, Kraft und Reichtum besitzt, aus der Erhöhung des Steins, welches *in infinitum* geschieht und durch dieses Licht viel 1000 Lichter können angezündet werden, dadurch er immer mehr und mehr verherrlicht wird an Kraft und Macht, auch solches in Ewigkeit seinen Fortgang hat, nicht veraltet, sich immer neuer und schöner gebärt aus sich selbst, welche seine große Kraft er dem ganzen metallischen Reiche austeilt, auch dem animalischen und vegetabilischen Reiche, welche er verändert, sonderlich im mineralischen Reiche, da er sie in ☉ verkehrt. Bekommen diese Metalle aber viel von seiner Kraft, so werden Edelgesteine daraus, auch macht er alle Kristalle zu Edelgesteinen, wodurch ich abermals einen Blick ins *Mysterium magnum*, in die heilige göttliche Dreifaltigkeit bekomme, wie dieser unwandelbare GOTT nicht veraltet, sondern immer seinen Sohn, als das Licht und Herz GOTTES gebiert und solches in alle Ewigkeit, auch solches so zu reden alle Augenblicke in seiner göttlichen Kraft und Allmacht erneuert, erhöht und nimmermehr abnimmt, noch veraltet und ob er gleich durch seinen Geist und Wort die Himmel und Engel geschaffen, auch den Menschen und sie alle durch seinen Geist und Odem erhält, auch von seiner göttlichen Kraft gespeist und getränkt werden und GOTT sich so vielfältig allen seinen Geschöpfen und der ganzen Natur mitge-

moon climb into their marriage bed, and show the sun's son the red tincture. But before all this happens, the seven forms and grades of the seven source-spirits in the angelic world must also complete theirs, until from one sparkling jewel or sky, from one elevated color to another, they reach the desired red of the ruby, from which this son of the sun has its crown, the heavenly throne; the spirits and air change in all heavens, from which one can take the form, color, and crown of the exalted king. This king wants to multiply itself again, so it has a crown of the carbuncle stone. If it wants to multiply again, it has a crown that shows a heavenly light, compared to which the carbuncle stone is dark. This is the light and the justice and it is a king of kings, which possesses great power, strength, and wealth from the elevation of the stone, which happens *ad infinitum* and through this light many thousands of lights can be lit, thereby glorifying it more and more in strength and power, which proceeds in eternity and never becomes obsolete, ever giving birth to itself newly and more beautifully, which distributes its great power to the entire metallic kingdom and also to the animal and vegetabilic kingdoms, which it transforms, and especially the mineral kingdom, where it turns them into ☉. But if these metals receive a lot of its power, they become precious stones and it also turns all crystals into precious stones, which gives me another glimpse into the *mysterium magnum* into the holy divine trinity, how this unchangeable GOD does not become obsolete, but always gives birth to his son as the light and heart of GOD, and such to all eternity, so to speak, every moment in its divine power and omnipotence renewed, increased, and never decreases, nor becomes obsolete, even though he created the heavens and the angels, and humans too, through his spirit and word, and keeps them all alive through his spirit and breath, and feeds and waters them with his divine power; and GOD communicates himself in so many ways to all his creatures and to the whole of nature, through

teilt, dadurch alles erhalten wird, absonderlich die Himmel und reinen Geister, welche ihr ganzes Wesen aus GOTT haben und solche nicht zu zählen sind und in Ewigkeit von ihm erhalten und aus ihm gespeist werden, so nimmt er doch nicht ab und wird geringer, sondern er erzeigt sich immer in unendlichen Wundern, Schönheit und Herrlichkeit, Glanz und Klarheit und schönen Farben, dass es weder Engel noch Mensch aussprechen kann, sich auch der Schönheit, Lust und Freude keine Creatur satt sehen kann, welche alle unendlich sind, immer in größeren Wundern, Schönheit, Licht, Glanz und Herrlichkeit erneuern, welches alle reine Geister, Engel und Menschen empfinden, riechen, sehen, schmecken, fühlen, hören und empfinden, wodurch sie in allen ihren Quell-Geistern als im göttlichen Licht entzündet und verherrlicht, worin das große Freuden-Reich besteht, wenn die 7 Geister GOTTES mit unseren Geistern vereinigt und die Kraft GOTTES uns durchdringt, welches die rechte Hochzeit des Lammes ist, da wir die göttliche Kraft, Liebe, Licht und Leben in unseren Geistern schmecken, riechen, fühlen, sehen, hören und empfinden, dass wir vor Freuden aus unseren Geistern ausgehen in die Geister GOTTES, da wir nichts als von GOTT wissen, welche Freude und himmlische Wollust uns als ein Strom in Ewigkeit erfüllen und wir dadurch getränkt werden, auch solche Freude in Ewigkeit dauern wird. Es nehme ein rechter Christ ein Exempel, der von Herzen sucht, GOTT in einem bußfertigen heiligen Leben zu dienen, wenn er von dem Geist GOTTES besucht wird und sich die himmlische Sophia der Seelen zeigt, wie er plötzlich mit einer Freude und Licht in GOTTES Erkenntnis überfallen wird, dass er vor Liebe zu GOTT sich und der ganzen Welt vergisst, nichts sieht, hört, noch weiß, als GOTT, auch sich nichts anderes einbildet, als er stehe schon im Paradies, aber diese göttliche Anschauung währt nicht lange im tierischen Leibe, sondern ist ein Vorgeschmack des ewigen Lebens, welches uns dort vorbehalten wird, der Schatz, so uns im Himmel beigelegt, da unser Erlöser sagt: Sammelt euch Schätze im Himmel, die die Motten nicht fressen und die Diebe nicht nachgraben, kein Teufel noch Feind uns solche rauben

which everything is preserved, especially the heavens and pure spirits, which have their whole beings from GOD; such [beings] are countless and are preserved by him and fed from him for eternity. He does not decrease and become less, but always displays himself in infinite wonders, beauty and glory, luster and clarity, with beautiful colors that neither angels nor humans can express, and no creature can be satiated by this beauty, desire, and joy, all of which are infinite, ever renewing in greater wonders, beauty, light, splendor, and glory, which all pure spirits, angels, and humans sense, smell, see, taste, touch, hear, and feel, whereby they are ignited and glorified in all their source-spirits with divine light, wherein the great kingdom of joy consists, when the seven spirits of GOD unite with our spirits and the power of GOD penetrates us, which is the true marriage of the lamb, since we taste, smell, feel, see, hear, and perceive the divine power, love, light, and life in our spirits, that we go with joy out of our spirits into the spirits of GOD; and because we know nothing but from GOD, this joy and heavenly pleasure fills us like a river for eternity and we are watered by it, and such joy will last for eternity. Take as an example a true Christian who sincerely seeks to serve GOD in a penitent holy life: when they are visited by the spirit of GOD, and the heavenly Sophia of the souls shows herself, they [the true Christian] are suddenly filled with joy and light and GOD's knowledge such that they lose themselves and the whole world for love of GOD, they see, hear, and know nothing but GOD and imagine only that they are already in paradise; but this divine vision does not last long in the animal body, but is a foretaste of eternal life, which is reserved for us there, the treasure set aside for us in heaven, when our redeemer says: "But lay up for yourselves treasures in heaven, where moth does not corrupt, and where thieves do not break through nor steal:"[136] neither devil nor enemy can rob us of such,

kann, das ewige beständige Gut, wo unser Schatz ist, da soll auch unser Herz sein. Auch sehe ich in der philosophischen Tinctur noch ein Geheimnis in der Heiligen Dreifaltigkeit. Wenn die rote Tinctur fertig und viele 100 Teile tingiert, so kann ihre Kraft und Gewalt durch keine Hilfe wieder zur Multiplication gebracht werden. Dieser himmlische Stein und Tinctur vermische sich denn mit dem gemeinen Golde und nehme diesen Leib aus dem mineralischen Reiche an sich, wodurch seine Kraft und Macht erst recht offenbar und sich wieder aufs neue erhöht und in seiner Unendlichkeit zunimmt, sich in viele 1000 multipliciert. Daraus wir abermals sehen, wenn der Sohn GOTTES nach seiner göttlichen Natur nicht hätte sich mit unserer menschlichen Natur vermischt, so wären so viele 1000 Millionen Seelen nicht erleuchtet noch zu GOTT gekommen und aus der Hölle erlöst worden und wäre die Kraft und die Macht und das Reich GOTTES nicht offenbar worden, welches doch in dieser und der zukünftigen Welt unter dem menschlichen Geschlecht sollte offenbar werden, welches das Reich GOTTES war, wie uns unser Erlöser im Vater Unser beten lernt: Zu uns komme dein Reich und abermals: Denn dein ist das Reich und die Kraft und die Macht und die Herrlichkeit in Ewigkeit. Der Dank für diese Offenbarung geschieht noch von allen Heiligen hier und im ewigen Leben. Offenbarung Johannes Kap. 12. Nun ist das Heil und die Kraft und die Macht und das Reich unseres GOTTES seines *Christus* worden, weil der verworfen ist, der sie verklagt Tag und Nacht vor GOTT und seinem Stuhl. Welches herrliche Reich GOTTES und *Christi* durch die Erlösung des Menschen hat müssen offenbar werden, dadurch GOTT in vielen 1000 Millionen Zeugen hier und dort ewig gelobt und gepreist wird, denn sagt Johannes: Ich sahe eine Zahl welche niemand zählen konnte, aus allen Völkern und Heiden vor dem Stuhl und vor dem Lamm, angetan mit weißen Kleidern und Palmen in ihren Händen. Dieses sind die seligen Einwohner dieses Reichs. Ich wende mich wieder zu unserem Himmel. Wenn dieser König der Könige, welcher die funkelnde leuchtende Krone trägt, sich abermals multiplicieren wollte, sein Geschlecht zu vermehren, so wäre zu besorgen, dass

the everlasting treasure. For where your treasure is, there will your heart be also".[137] I also see another mystery in the philosophical tincture, in the holy trinity. When the red tincture is finished and tinges a hundred parts, no help can bring its strength and power back to multiplication. Let this heavenly stone and tincture mingle with the common gold and take this body from the mineral kingdom, whereby its strength and power becomes genuinely evident and increases again and increases in its infinity, multiplied by thousands. From this we see again, if the son of GOD with his divine nature had not mingled with our human nature, so many thousand millions of souls would not have come to GOD and been enlightened and redeemed from hell; and the strength and the power and the kingdom of GOD would not have been revealed, which was to be revealed in this world and the world to come among the human race, which was the kingdom of GOD, as our redeemer teaches us to pray in the LORD's Prayer: "Thy kingdom come[138] to us", and again: "For yours is the kingdom and the power and the glory for ever".[139] Gratitude for this revelation is yet to come from all the saints here and in eternal life. Revelation 12: "Now is come salvation, and strength, and the kingdom of our GOD, and the power of his Christ: for the accuser of our brethren is cast down, which accused them before our GOD day and night".[140] This glorious kingdom of GOD and Christ had to be revealed through the redemption of those humans through whom GOD is eternally glorified and praised in many thousand-million witnesses both here and there, because John says: "I saw a great multitude, which no man could number, of all nations and heathen, stood before the throne, and before the lamb, clothed with white robes, and palms in their hands".[141] These are the blessed inhabitants of this realm. I turn back to our heaven. If this king of kings, which wears the sparkling shining crown, should multiply itself again in order to increase its lineage, it is to be feared that its

sein ganzes Wesen sich in ein himmlisches und leuchtendes Feuer resolvieren möchte, gleich dem lauteren und himmlischen Licht und er als ein Geist gar in seinem Thron verschwinden möchte, daher ihm besser zu raten, dass er sich mit dem gemeinsamen Golde vermählt, damit sein geistlicher und engelhafter Leib zur Erde, zur Corporalität gebracht wird, damit er nicht gar verschwindet und zum lauteren Feuer wird, denn in der engelhaften Welt ist keine Erde mehr, der Grund von durchsichtigen Edelgesteinen, das Wasser so in diesem Climate zu finden, ist das lebendige Wasser, von den Oberen, Mittleren und Unteren geschieden, von aller Feuchtigkeit entbunden und ein reiner Luft- und Feuer-Geist ist, doch ist es ihr Bad, dadurch diese geistlichen, durchsichtigen und reinen Leiber wieder aufs neue renoviert und immer herrlicher und herrlicher glorifiziert worden. Vorher war es nur Gold in Glas-Gestalt, vergleicht sich mit dem Grund und Pflaster des neuen Jerusalems, da die Tore von Perlen, welche Engel bewachen, die Stadt keiner Sonne bedarf, weil sie selber vom himmlischen Licht erleuchtet, mitten ist der Tempel und der Thron, daran die Wappen unserer Helden, Götter und Göttinnen aufgehängt, es darf nichts unreines herein gehen, sondern lauter reine Feuer-Geister. Die Schönheit und Wunder sind unendlich, auch nicht zu beschreiben, aller Götter und Göttinnen Name steht im Buch des Lebens, ihr Gedächtnis wird nicht untergehen, hier ist alle Kraft der Obersten und der Untersten, weil man durch den Himmel und die Erde sehen kann, die Erde auch zum Himmel worden und verklärt, zu lauter himmlischem Geist, die Sonne geht nun nicht mehr unter, sondern scheint von Ewigkeit zu Ewigkeit, weil sie keine Finsternis mehr leiden darf. Das heißt die Welt erschaffen, die Erde auf den himmlischen Thron der Freundschaft setzen, der Mensch sitzt hier an GOTTES statt, kann den Himmel und der Erde gebieten, allen Geistern, alles ist dem Menschen unterworfen, so weit dass nichts wider GOTTES Willen vorgenommen wird, alles was unter dem Himmel und auf Erden ist, kann ein solcher Besitzer erlangen, viele 1000 unzählbare Wunder damit verrichten, aller Verstand verstummt hier bei diesem Werk in der Multiplication, es ist bei

whole being should resolve itself into a heavenly and luminous fire like the pure and heavenly light and it would like to disappear as a spirit into its throne. Therefore it is better to advise that it marries itself with the common gold so that its spiritual and angelic body is brought to earth, to corporeality, so that it does not disappear at all and becomes pure fire, for in the angelic world there is no more earth, the ground is of transparent jewels, the water to be found in this climate is the living water, separated from the upper, middle, and lower [waters], devoid of all moisture and a pure spirit of air and fire, but it is their bath, whereby these spiritual, transparent, and pure bodies are renewed ever again and are ever more gloriously ennobled. Before it was only gold in glass form, comparable with the ground and pavement of the new Jerusalem, where the gates which are guarded by angels are made of pearls. The city needs no sun, because it is itself illuminated with heavenly light, and in the midst of [this city] is the temple and the throne on which hang the coats of arms of our heroes, gods, and goddesses. Nothing impure may enter, only pure spirits of fire. The beauty and wonders are infinite, and thus impossible to describe. The names of all gods and goddesses are in the book of life, and their memory will not perish. Here resides all the power of the highest and lowest, because one can see through heaven and earth. The earth has also become heaven and has been transfigured to pure heavenly spirit; the sun no longer sets, but shines from eternity to eternity, because it is no longer allowed to suffer darkness. This signifies the creation of the world, setting the earth upon the heavenly throne of friendship. Here humanity sits in GOD's place, and can command heaven and earth and all spirits. Everything is subject to humanity insofar as nothing is done against GOD's will; everything that is under heaven and upon earth can be attained by one who holds this position, and many thousands of innumerable miracles can be performed through this. All intellect falls silent here in this work of multiplication, for there is something divine

diesem Werk etwas Göttliches, es ist das Geheimnis der Gottheit, der Natur und Creatur. Sündiger Mensch, erkenne hier deine Niedrigkeit, dass dich GOTT zu seinem Bild geschaffen und da du so tief von ihm gefallen, dich doch solches wissen lassen, auch den Verstand dazu gegeben, dir in der äußeren Welt ein solches Licht gezeigt, dass du das Paradies gefunden, ins *Mysterium magnum* eingedrungen, in diesem Spiegel GOTT, das Geheimnis der Heiligen Dreifaltigkeit gesehen, Zeit und Ewigkeit, den Schöpfer und das Geschöpf lernen erkennen. Demütige dich vor GOTT, sage ihm mit gebogenen Knien des Herzens und Leibes innigsten, unendlichen Dank, dass er dich hier erleuchtet und dich solches Licht sehen lassen und bitte ihn ganz demütiglich, dass er dich auch zu seinem göttlichen Licht wolle lassen gelangen, denn in seinem Licht sehen wir das Licht? Und ich sah ein gläsernes Meer mit Feuer gemengt und die den Sieg behalten an dem Tier und seinem Bilde und sein Malzeichen und seines Namens Zahl, dass sie stunden an dem gläsernen Meer und hatten Harfen und sungen das Lied Moses, des Knechts GOTT und das Lied des Lammes und sprachen: Groß und wundersam sind deine Werke, HERR Allmächtiger GOTT, gerecht und wahrhaft sind deine Wege, du König der Heiligen, wer soll dich nicht fürchten und deinen Namen preisen, denn du bist allein heilig, denn alle Heiden werden kommen und anbeten vor dir, denn deine Urteile sind offenbar geworden.

in this operation. It is the secret of divinity, of nature and creature. Sinful human, recognize your lowliness here, and know that GOD created you in his image although you fell so far from him, and yet [he] let you know such things, and has also given you intelligence. He showed you such a light in the outer world that you found paradise and penetrated the *mysterium magnum*. Seeing GOD in this mirror and witnessing the mystery of the holy trinity, time and eternity, you learned to recognize the creator and the creature. Humble yourself before GOD, give him heartfelt, infinite thanks with bowed knees of heart and body so that he enlightens you here and lets you see such light, and ask him with complete humility to allow you to reach his divine light, for it is through his light that we see the light. "And I saw a sea of glass mingled with fire: and those that had attained victory over the beast, and over his image, and over his mark, and over the number of his name, stand on the sea of glass, having the harps of GOD. And they sing the song of Moses the servant of GOD, and the song of the lamb, saying, great and marvelous are your works, LORD GOD Almighty; just and true are your ways, you king of saints. Who shall not fear you and glorify your name? For only you are holy: for all nations shall come and worship before you, for your judgments are made manifest".[142]

DER VIERTE TRAKTAT

Vom Universal-Stein oder der Universal-Tinctur

Wer dieses will zubereiten, der suche das Licht, dieses wohnt im Menschen, im Verstand, auch in allen Geschöpfen GOTTES. Der Mensch bitte GOTT, so wird er ihn durch sein göttliches Licht erleuchten und seinen finsteren Verstand eröffnen, dass er erstlich in der Erde als in sich selbst den neuen Menschen finde und in demselben das Reich GOTTES, die teure Perle, alsdann suche er in der reinen metallischen Erde die teure schneeweiße Perle des *Universals*, so wird er sehen, fühlen, riechen, schmecken und begreifen, was für herrliche Früchte die Elemente geben, so viele wunderbare Gewächse von schönen Farben, Geruch und Geschmack. Solches sehen wir an den Elementen der großen Welt. Was hat die Erde für schöne Früchte an Blumen, Kräutern, Bäumen, Weinstock, Korn, Öl, Metallen, Mineralien, Salzen, unzählig vielerlei Früchte, auch Tiere? Das Wasser, so vielerlei Arten Fische; auch die Vögel haben ihren Ursprung aus dem Wasser und sind daraus geschaffen. Die Luft ist angefüllt mit Geistern und alle animalischen Leiber haben ihr Leben aus der Luft. Ja nichts kann ohne Luft wachsen, weil alles aus der Luft seinen Geist nimmt und erhält. Im Feuer sind weit größere Wunder, weil es die Seele der Welt, ist in der Schöpfung geschieden, dass das Feuer den obersten Ort über Luft und Wasser angenommen, doch ist noch Feuer genug in der Erde, auch in allen Dingen. Dieses sind die *Elementa* der großen Welt und wer kann alle ihre Früchte erzählen? Aber unsere *Elementa* geben viel 1000 mal herrlichere und beständigere Früchte, weil dieselben von keinem Win-

THE FOURTH TREATISE

Of the Universal Stone or the Universal Tincture

Anyone who wants to prepare this should look for the light, this dwells in humans, in the mind, also in all of GOD's creatures. Humans should ask GOD, he will enlighten them with his divine light and open their dark mind, so that they first find the new human in the earth as in themselves and in the same the kingdom of GOD, the precious pearl; then they should search for the precious snow-white pearl of the *universal* in the pure metallic earth, so that they will see, feel, smell, taste, and understand what wonderful fruits the elements give, all the many wonderful plants of beautiful colors, smells, and tastes. We see this in the elements of the macrocosm. What beautiful fruits does the earth have of flowers, herbs, trees, vines, grain, oil, metals, minerals, salts, innumerably diverse fruits, and also animals? As for the water, it has so many kinds of fish. Birds also have their origin in water and are created from it. The air is filled with spirits and all animal bodies derive their life from the air. Indeed, nothing can grow without air, because everything takes and maintains its spirit from the air. In fire there are far greater miracles, because it is the soul of the world, it was separated during creation, so that fire has taken the highest place above air and water, but there is still enough fire in the earth, also in all things. These are the *elementa* of the macrocosm, and who can tell all their fruits? But our *elementa* give fruits that are a thousand times more noble and more enduring, because they are not

ter berührt, sondern stets Sommer haben, wenn sie reif, werden sie abgebrochen, aufs neue umgepflanzt und in bessere Erde versetzt, absonderlich unser Perlen-Baum, bis er die Paradies-Blumen, die weiße Lilie und die rote Rose bringt. Endlich schließen sich diese Blüten zu und werden die reifen Früchte der philosophischen Sonne und Mond. Dann wird dieser Perlen-Baum, das Gewächs der drei Principien, ins engelhafte und himmlische Licht und Glanz verwandelt, dass er das Holz des Lebens wird und die Speise der Götter trägt, welche alle Monde reif werden. Es hat GOTT allen Geschöpfen sein göttliches Licht, Kraft und Leben eindrücken wollen, absonderlich in dem mineralischen Reiche, worin die beständigen Geister, welche kein Feuer verbrennen kann, darinnen ein himmlisches Licht enthalten, solches ist auch von GOTT mit Fruchtbarkeit begabt, dass es sich in viele 1000 multipliciert. Wie aber solches Licht und Kraft durch Hände-Arbeit zu scheiden vom finsteren toten Körper und solchen ins Leben und Wachstum zu bringen, folgt in diesem kurzen Bericht. Unsere *Materia* hat eine fünffache Solution, erstlich die Haupt-Solution. Die andere der Elementen. Die dritte die philosophische Solution, Wasser und Erde. Die vierte durchs weiße Ferment, den Geist. Die fünfte Solution durch Feuer, als die Seele. Die erste Haupt-Solution ist eine Reduction des vollkommenen Körpers in einen Liquor. Die andere Solution der Elemente ist eine Absonderung der Teile von Chaos durch die Destillation in Feuer, Wasser, Luft und Erde. Die dritte Solution ist eine Auflösung Mannes und Weibes. Die vierte Solution des Ferments ist eine Reduction des weißen Schwefels in sein weißes Öl. Die fünfte Solution des weißen Steins in sein rotes Öl, dadurch der Stein gerötet, welches ist der Seele ihr Wagen, da die Sonne auf 4 Rädern fährt. Hierinnen sind nun enthalten dreierlei Vereinigungen:

1. Das Wasser mit der Erde.
2. Des Geists mit dem Leibe.
3. Der Seele mit Geist und Leib.

Die Arbeit zu solchem ist, die Körper in einen Liquor zu bringen, der

touched by winter, but only have summer, and when they are ripe, they are plucked, replanted, and placed in better soil, especially our pearl tree, until it brings the flowers of paradise, the white lily and the red rose. At last these blossoms close and become the ripe fruits of the philosophical sun and moon. Then this pearl tree, the growth of the three principles, is transmuted into angelic and celestial light and splendor that it becomes the wood of life, and bears the food of the gods, which ripens in every month. GOD wanted to impress his divine light, power, and life on all creatures, especially in the mineral kingdom, in which the enduring spirits, which no fire can burn, bear a heavenly light; it is also granted by GOD with fertility, so that it multiplies into many thousands. But how one separates such light and power through manual work from the dark, dead body and brings it into life and growth follows in this short account. Our *materia* has a fivefold solution: first, the main solution; second, the [solution] of the elements; third, the philosophical solution, water and earth; fourth, [the solution] through the white ferment, the spirit; the fifth solution, through fire, like the soul. The first major solution is a reduction of the perfect body into a liquor. The second solution of the elements is a separation of the parts of chaos through the distillation in fire, water, air, and earth. The third solution is a dissolution of man and woman. The fourth solution of the ferment is a reduction of the white sulphur to its white oil. The fifth solution is of the white stone into its red oil, thereby reddening the stone, which is the chariot of souls, since the sun rides on four wheels. Herein are contained three kinds of unifications:

1. The water with the earth.
2. The spirit with the body.
3. The soul with spirit and body.

The work towards such [unifications] is to bring the bodies into a

Hyle oder Chaos genannt, aus dem Chaos die *Elementa* separieren, die gereinigten *Elementa* zusammen zu setzen, das Kind mit Milch ernähren, speisen und tränken, dass es zum vollkommenen Alter kommt und solches zu fermentieren. Mit der Zusammensetzung verhält es sich so: Die erste ist keine Conjunction im rohen Werk, denn da wird dieses Wasser und ♁ zusammen gesetzt. Diese sind wieder zu scheiden. Die andere ist die rechte Composition, Wasser und Erde, dies ist Mann und Weib. Die 3. ist die Composition mit ♁, wenn kein Wasser mehr, sondern alles in die Erde vertrocknet, dann kann sich der ♁ mit dem ☿ vereinigen. 4te Composition mit dem Feuer, wenn es heißt: Gib dem Feuer Feuer, dem ☿ den ☿, alsdann wirst du die Güter des Glücks haben. Aus den 4 Elementen kommt die Quintessenz im Glase. Erstlich setzen wir zusammen, lassen es verfaulen, das Verfaulte lösen wir auf, solches teilen wir, das Geteilte reinigen wir, das Gereinigte vereinigen wir, das Vereinigte figieren wir. Wenn nach der Tötung der Leib auferstanden und mit der Seele vereinigt wird, so ist es der verklärte paradiesische Leib. Praxis aus dem Wasser. Aus unserem einfachen ☿ und seiner eigenen Erden wachsen unsere 7 Metalle. Dieser jungfräuliche ☿ wird 7-mal mit seiner Erde sublimiert, bis er von seiner Wässrigkeit entbunden, aus seiner feuchten Natur zur Trockne und Reife gekommen, zu einem weißen Stein, als Kristall. Dieser Wasser- und Salz-Stein wird in die *Terram Foliatam* sublimiert. Den doppelten ☿ calciniert man mit dem allergrößten Feuer, weil man sich auf den sublimierten ☿ nicht zu verlassen, sondern auf den figierten, in seinem eigenen Feuer, seinem *Oleo*, welches ihn figiert und calciniert, so ist der Stein fertig, das gläserne Meer. Will man ihn nun multiplicieren in *Quantitate & Qualitate*, muss man die gekrönte weiße Königin in dem fliegenden gekrönten Adler solvieren und in diesen Brunnen unsere rote Sonne steigen lassen, so wird die Sonne durch ihre Hitze den Brunnen austrocknen und ist die weiße Königin mit der roten Sonne in einem Leibe gewachsen, unseren roten gekrönten König. Die erste ausgekochte figierte Tinctur auf rot, vereinigen wir als unsere Sonne mit unserer roten flüssigen Sonne, dieses gibt eine sehr große Röte. *Sapienti sat!*

liquor called *hyle* or chaos, separated from the chaos of the *elementa*, to put together the purified *elementa*, to nurture, feed, and water the child with milk, so that it may come to perfect old age and ferment it. The composition is like this: the first is not a conjunction in the crude work, because this water and ♁ are brought together. These are to be separated again. The second is the right composition of water and earth, this being male and female. The third is the composition with ♁ when there is no more water, but everything is dried up in the earth, so that then the ♁ can unite with the ☿. The fourth composition with fire, when it says: give the fire to the fire, the ☿ to the ☿, and then you will have the goods of happiness. The quintessence in the glass comes from the four elements. First we bring it together and let it putrefy, then we dissolve what is rotten, separate it, purify what has been separated, unite what has been purified, and then fix what is united. If after having been killed, the body is resurrected and united with the soul, then it is the transfigured paradisiacal body. The practice from water: our seven metals grow from our common ☿ and its own earth. This virgin ☿ is sublimated seven times with its earth, until it is relieved of its wateriness, has come from its wet nature to dryness and maturity to [form] a white stone like a crystal. This stone of water and salt is sublimated into the *terram foliatam*. The double ☿ is calcined with the greatest fire, because one does not rely on the sublimated ☿ but on the fixed one, in its own fire, its *oleo*, which fixes and calcinates it, and then the stone is ready, the sea of glass. If you now want to multiply it in *quantitate & qualitate*, you have to dissolve the crowned white queen in the flying, crowned eagle and let our red sun rise into this well, so the sun will dry up the well with its heat and [this] is the white queen with the red sun grown in one body, our red crowned king. We unite the first cooked, fixed tincture in red, as our sun, with our red liquid sun. This gives a magnificent reddening. *Sapienti sat!*[143]

FÜNFTER UND LETZTER TRAKTAT

DES PHILOSOPHISCHEN PERL-BAUMS VOM GROSSEN UNIVERSAL-STEIN

Worin sechs Figuren aus der Offenbarung Johannis ausgelegt, erstlich geistlicher Weise und dann auch auf das natürliche Werk

Die Offenbarung Johannis auszulegen erfordert einen Menschen, der sowohl die göttliche *Cabbalam*, als auch die himmlische und engelhafte *Magiam*, und die natürliche *Philosophiam* nebst der Natur-Sprache versteht, weil die ganze Offenbarung, absonderlich viele magische Bilder und Figuren hat, welche vorstellen die 3 Reiche oder 3 *Principia*: 1. Das *Mysterium magnum*, die göttliche, himmlische und engelhafte Welt. 2. Die paradiesische Licht-Welt. 3. Die äußere finstere Welt, die alle 3 in GOTT, in einem Zirkel verfasst, doch keiner den anderen begreift und doch GOTT in allen ist, auch alle von ihm ausgehen, die Engel, Natur und Creatur, auch der Mensch, welcher von GOTT und aus der paradiesischen Welt gefallen in die äußere finstere Welt, da er sich den Tieren gleich machen muss, aber durch sein Ausgehen von GOTT ist dem zeitlichen Tode heimgefallen, wenn er aber aus dem alten Adam ausgeht und im neuen Menschen lebt, so kann er nach dem Tode, wenn der tierische Leib verfault, wieder in die paradiesische Licht-Welt versetzt werden und wieder in GOTT einkehren, da er gleich den Engeln GOTTES von GOTT erleuchtet wird, weil nun die Natur und Creatur mit dem Oberen, Mittleren und Untersten in gleicher Concordanz steht, GOTT auch seine Dreifaltigkeit in die Natur und Creatur eindrücken wollen, weil seine Gegenwart alle Wege, nur dass wir ihn mit unsern tierischen Augen nicht sehen noch erkennen, so sollen wir ihn in der Natur und Creatur, in allen Geschöpfen, absonderlich in uns sehen und finden, worin ihn auch die Heiden gefun-

FIFTH AND FINAL TREATISE

OF THE PHILOSOPHICAL PEARL TREE FROM
THE GREAT UNIVERSAL STONE

In which six figures from the Revelation of John are interpreted, first in a spiritual way and then also in the natural work

Interpreting the Revelation of John requires one who understands the divine *cabbalam*, as well as the heavenly and angelic *magiam*, and the natural *philosophiam* along with the natural language, because strangely enough, the whole of Revelation has many magical images and figures representing the three realms or three *principia*: (1) The *mysterium magnum*, the divine, heavenly, angelic world. (2) The paradisiacal world of light. (3) The outer, dark world, all three in GOD, composed in a circle, yet none apprehends the other and yet GOD is in all, while all emanate from him: angels, nature, and creatures. Included among these creatures is the human being, who comes from GOD and has fallen from the paradisiacal world into the outer dark world, where it must make itself like the beasts. It has fallen victim to temporal death through its falling away from GOD, but if it comes out of the old Adam and lives in the new human, then it can be transferred back to the paradisiacal world of light and return to GOD after the death of its animal body. Thereupon it is enlightened by GOD like the angels of GOD, because now nature and creatures stand with the upper, middle, and lowest ones in the same concordance. GOD also wants to impress his trinity into nature and into creatures, for he is present in every way, yet we do not see or recognize him with our animal eyes, so we should see him in nature and creature and all created things, and strangely enough, see and discover him in ourselves, wherein the heathen also found and recognized him,

den und erkannt, in unser kleinen Welt, welche mit allen accordiert. Weil ich aber die *Cabbalam, Magiam* oder *Philosophiam* nicht verstehe, so befinde mich viel zu ungeschickt, solche auszulegen, nachdem ich aber durch GOTTES Gnade vom natürlichen Licht einen Schein habe und etliche Figuren, welche das göttliche Geheimnis vorstellen, erkannt, worunter Johannes zugleich das *Universal* auch mit abgebildet, so habe diese Figuren in etwas erklären wollen, jedem seine Meinung unbenommen. Es möchte sich aber mancher verwundern, warum doch Moses und alle Propheten, als welche die *Cabbalam, Magiam,* und *Philosophiam* verstanden, in dunklen Worten, Bildern, Figuren, Träumen, auch dunkel geredet und geschrieben, ja der Sohn GOTTES selbst hat die himmlische Lehre in Gleichnissen ausgeredet, sagte auch vor seinen Jüngern: Euch ists gegeben zu wissen das Geheimnis des Reichs GOTTES, den anderen aber in Gleichnissen. Warum aber dies alles? Antwort: Darum, weil die gottlosen Verächter solches nicht wert wären und nur verspotteten, so sollten die Perlen und das Heiligtum nicht vor die Hunde und Säue geworfen werden, auch musste solches der Teufel nicht sehen und wissen, damit er nicht vor der Zeit sein Unkraut darein streuen möchte. Die ganze Heilige Schrift, *Altes* und *Neues Testament,* ist voll von göttlichen und natürlichen Geheimnissen und liegt nur an uns, dass wir solches nicht verstehen, wir dürfen das Licht nicht bei den Heiden in ihren Büchern suchen, sondern nur GOTT um Weisheit bitten mit Salomo: Gib mir die Weisheit, die stets in deinem Thron ist, dass sie mit mir sei und mit mir arbeite, dass ich erkennen möge, was dir wohl gefällt und durch die Weisheit selig werde. Denn wer will GOTTES Rat erfahren, es sei denn, dass er Weisheit gebe und sende seinen Heiligen Geist aus der Höhe. Die Weisheit, so von GOTT kommt, ist besser denn Königreiche und Fürstentum und Reichtum ist nichts gegen sie, kein Edelgestein gleicht ihr und Gold ist gegen ihr wir Sand, sie ist besser denn gesunder und schöner Leib, das Licht und Glanz, so von ihr geht, verlöscht nicht, in ihrer Hand ist Reichtum und Ehre, sie ist den Menschen ein unendlicher Schatz, wer GOTTES Freund ist, liebt sie, denn in ihr ist der Geist der ver-

in our microcosm, which accords with all. But because I do not understand the *cabbalam, magiam,* or *philosophiam,* I find myself much too inept to interpret this, but since I have, by the grace of GOD, a semblance of natural light and have recognized several figures representing the divine mystery, under which John simultaneously envisioned the *universal,* I wanted to explain these figures a little, everyone being free to express their opinion. But some might wonder why Moses and all prophets who understood the *cabbalam, magiam,* and *philosophiam,* spoke and wrote in obscure words, pictures, figures, and dreams. In fact, the son of GOD himself expatiated the heavenly teaching and lessons in parables, and he spoke before his disciples: "Unto you it is given to know the mysteries of the kingdom of GOD: but to others in parables".[144] But why all this? Answer: because the ungodly despisers were not worthy of such things and only mocked them; pearls and the sacred should not be cast before dogs and swine, neither should the devil see or know such things, so that he would not sow his weeds in them prematurely. All holy scripture, Old and New Testament, is full of divine and natural mysteries and it is only up to us that we do not understand such things, we must not seek the light from the heathen in their books, but, with Solomon, only ask GOD for wisdom: "Give me the wisdom that sits by your throne",[145] that it may be with me and work with me, that I may know what pleases you and that this wisdom may make me happy. For who will know the advice of GOD unless he gives wisdom and sends his holy spirit from high above. The wisdom that comes from GOD is better than kingdoms and principalities, and wealth is nothing compared to it; no precious stone is like it and in comparison gold is like sand. It is better than a healthy and beautiful body; the light and splendor which proceed from it does not fade; wealth and honor are in its hand, it is an infinite treasure for humanity; whoever is GOD's friend loves it, for within it is the spirit that is understanding, reasonable, unique,

ständig, billig, einzig, mannigfaltig, behend, beredt, rein, klar, sanft, freundlich, ernst, frei, wohltätig, leutselig, fest, gewiss, sicher, vermag alles, sieht alles und geht durch alle Geister, wie verständig, lauter und scharf sie sind, sie ist das Hauchen der göttlichen Kraft und ein Strahl der Herrlichkeit des Allmächtigen, sie ist ein Glanz des ewigen Lichts und ein unbefleckter Spiegel seiner Gütigkeit und der göttlichen Kraft und ein Strahl der Herrlichkeit des Allmächtigen, darum kann nichts unreines zu ihr kommen und gibt sich für und für in die heiligen Seelen und macht GOTTES Freunde und Propheten, denn GOTT liebt niemand, er bleibe denn bei der Weisheit, sie geht einher, herrlicher denn die Sonne und alle Sterne gegen das Licht gerechnet, geht sie weit vor, sie ist herrlichen Adels, denn ihr Wesen ist bei GOTT, und der HERR aller Dinge hat sie lieb, sie ist der heimliche Rat in Erkenntnis GOTTES und ein Angeber seiner Werke. Begehrt einer viele Dinge zu wissen, so kann sie erraten beides, was vergangen und zukünftig ist, sie versteht sich auf alle verdeckten Worte und weiß die Rätsel aufzulösen, Zeichen und Wunder weiß sie zuvor, wie es zu Zeiten und Stunden ergehen soll, ein Jüngling hat durch sie Herrlichkeit bei dem Volk und Ehre bei den Alten. Man wird einen unsterblichen Namen durch sie bekommen und ein ewiges Gedächtnis bei den Nachkommen lassen, es ist kein Verdruss mit ihr umzugehen, noch Unlust um sie zu sein, sondern Lust und Freude und kommt unendlicher Reichtum durch die Arbeit ihrer Hände und Klugheit durch ihre Gesellschaft und ein guter Ruhm durch ihre Gemeinschaft und Rede, es kommt alles gutes von ihr und unzähliger Reichtum von ihrer Hand, sie ist einzig und tut doch alles, sie bleibt, das sie ist und erneuert doch alles für und für, ihr Lob ist unendlich.

diverse, agile, eloquent, pure, clear, gentle, kind, earnest, free, benevolent, affable, firm, certain, sure; it can do everything, see everything, and pass through all spirits; but however intelligent, pure, and sharp these [things] are, it is the breath of divine power and a ray of the almighty's glory, it is a splendor of the eternal light and a spotless mirror of its goodness and therefore nothing impure can approach it and gradually it enters the holy souls and makes them into GOD's friends and prophets; for GOD loves nobody unless he adheres to wisdom. It accompanies, more glorious, for the sun and all the stars are counted towards the light, it proceeds and expands, it is of glorious nobility because its essence is with GOD, and the LORD of all things loves it. It is the secret counselor in the knowledge of GOD and an informer of his works. If someone desires to know many things, [this wisdom] can divine both the past and the future, it understands all hidden words and knows how to solve riddles; it knows beforhand [by] signs and marvels how things should fare at all times and hours. Through [this wisdom] a youth has glory among the people and honor among the elders. An immortal name will be obtained by [this wisdom], and an everlasting memory will be left among the offspring. There is no vexation in dealing with it, nor displeasure in being around it, but rather pleasure and joy; and infinite riches come from the work of its hands, as well as wisdom from its company and good glory through its fellowship and speech. In sum, all good things come from it, including innumerable riches from its hand, for it is unique and yet it does everything. It remains what it is and yet renews everything through and through, its praise is infinite.

DIE ERSTE FIGUR
OFFENBARUNG JOHANNIS AM 1. KAPITEL

Die Stimme, ich bin das A und das O, der Erste und Letzte. Das Gesicht, 7 güldene Leuchter und mitten unter den 7 Leuchtern einer, der war eines Menschen Sohn gleich, angetan mit einem Kittel und begürtet um die Brust mit einem güldenen Gürtel. Sein Haupt aber und sein Haar waren weiß wie weiße Wolle, als der Schnee und seine Augen wie Feuer-Flammen und seine Füße gleich dem Messing das im Ofen glüht und seine Stimme wie groß Wasser rauscht und hatte 7 Sterne in seiner rechten Hand, aus seinem Munde ging ein scharfes zweischneidiges Schwert und sein Angesicht leuchtet wie die Sonne und er sagte: Ich bin der Erste und der Letzte und der Lebendige, ich war tot und siehe, ich bin lebendig von Ewigkeit zu Ewigkeit und habe die Schlüssel der Hölle und des Todes.

GEISTLICHE AUSLEGUNG DIESER FIGUR

Diese Figur bedeutet den Sohn GOTTES, das *Alpha & Omega*, Er ist von Ewigkeit und bleibt in Ewigkeit, der eingeborene Sohn GOTTES, das Licht der Welt und der ganzen Welt Heiland: Die 7 güldenen Leuchter sind die 7 Gemeinden, welche in allen 7 Quell-Geistern in GOTT imaginieren sollten, in den 7 Gnaden-Gaben des Heiligen Geistes leuchten und brennen, im Glauben, Liebe, Demut, Geduld, Keuschheit, Hoffnung, Beständigkeit, so will er mitten unter ihnen wohnen und ihr Hoher-Priester sein. Er war eines Menschen Sohn gleich, angetan mit einem Kittel, &c. Unsere arme menschliche Natur, darein er sich verkleidet. Umgürtet um die Brust mit einen güldenen Gürtel, &c. Seine Gottheit, welche er fest mit der Menschheit verbunden. Sein Haupt aber und Haar wie weiße Wolle, &c. Seine Reinigkeit und Unschuld, wie er als unser Haupt von den Sündern abgesondert, rein und schneeweiß ist. Seine Augen wie Feuer-Flammen, &c. Aus welcher das Licht der Gottheit leuchtet, mit welchen er, wenn er die Sünder erblickt, ihre steinernen Herzen als Wachs schmelzen kann, dass sie mit Petro bit-

THE FIRST FIGURE
REVELATION OF ST. JOHN, FIRST CHAPTER

The voice: I am Alpha and Omega, the first and the last.[146] *The vision: seven golden candlesticks, and in the midst of the seven candlesticks one like unto the son of man, clothed with a garment down to the foot, and girt about the breast with a golden girdle.*[147] *His head and his hair were white like wool, as white as snow; and his eyes were as a flame of fire. And his feet like unto fine brass, as if they burned in a furnace; and his voice like the sound of many waters. And he had in his right hand seven stars: and out of his mouth went a sharp two-edged sword: and his countenance was as the sun shineth in his strength*[148] *and he said: I am the first and the last: I am he that liveth, and was dead; and, behold, I am alive forevermore and have the keys of hell and death.*[149]

SPIRITUAL INTERPRETATION OF THIS FIGURE

This figure means the son of GOD, the *Alpha* and the *Omega*, he is from eternity and remains for eternity, the only begotten son of GOD, the light of the world and the savior of the entire universe. The seven golden candlesticks are the seven churches, which should be imagined in all the seven source-spirits in GOD, [which] shine and burn in the seven gifts of grace from the holy spirit, in faith, love, humility, patience, chastity, hope, and firmness; so he will dwell among them and be their high priest. "He was like the son of man, clothed with a garment down to the foot etc." signifies our poor human nature in which he disguises himself. "Girt about the breast with a golden girdle etc." signifies his divinity, which he firmly conjoined with humankind. "His head and his hair were white like wool etc." signifies his purity and innocence, how he, as our head, is set apart from the sinners, being pure and snow-white. "His eyes were as flames of fire etc." From these shine the light of divinity, and when he beholds sinners with them, he can melt their stony hearts like wax so that they weep bitterly with Peter over their sins; but in doing this he will also burn the hardened sinners like straw and stubble

terlich weinen über ihre Sünde, die verstockten Sünder aber als Stroh und Stoppeln damit verbrennt, wenn sie nicht Buße tun. Seine Füße gleich wie Messing, dass im Ofen glüht, ist seine Niedrigkeit, da er mit uns, mit der unreinen Erde oder Erz vermischt und deswegen so viel leiden müssen und im Ofen des Elends geprüft werden durch Versuchung, Armut, Leiden, Marter, Pein und bitteren Tod. Seine Stimme wie groß Wasserrauschen, &c. Sein Wort und Evangelium, mit dem Wasser-Bad der Taufe und Abendmahl, wenn wir in seinem Namen getauft werden in dem großen rauschenden Gnaden-Wasser, in dem heiligen Element, welches uns von der schwarzen Erbsünde abwäscht. Und hat 7 Sterne in seiner rechten Hand, &c. Das sind die 7 Geister GOTTES, durch welche er alles geschaffen, die engelhaften Fürstentümer und Fürsten-Engel, die 7 Planeten, welches alles in dem siebenten Geist seinen Ausgang genommen, nämlich die Natur und Creatur und darinnen sichtbarlich und begreiflich geworden und alle Geheimnisse und Schöpfungen im Himmel und auf Erden aus den 7 Geistern GOTTES ihren Ursprung haben. In dem Menschen sind auch 7 Quell-Geister, auch in allen Creaturen, in der Natur 7 *Gradus* oder *Systemata*, in der Erde 7 Metalle, in der Woche 7 Tage, der siebente der Ruhe-Tag, wenn die Welt 6000 Jahre gestanden, wird sie ihren großen Sabbat halten, 7 Siegel, 7 Zeiten, 7 Bitten im Vater Unser und ist in der siebenten Zahl lauter Geheimnis zu finden, davon anderswo weiter. Die 7 Sterne werden hier gezeigt als Engel der Gemeinde, die reinen göttlichen Lehrer, die durch des Heiligen Geistes 7 Gnaden erleuchtet sind. Aus seinem Munde ging ein scharfes zweischneidiges Schwert, &c. Ist das Gesetz und Evangelium, damit die Sünder geschreckt und die Frommen getröstet werden, es ist auch das scharfe zweischneidige Schwert, welches den alten Adam abschneidet vom neuen Menschen, damit der neue Mensch im Geist *Christi* leben kann. Sein Angesicht leuchtet wie die Sonne, &c. Seine Gottheit, Licht und Glanz, mit welchen er uns zum ewigen Leben erleuchtet. Er sagt: Ich bin der Erste und der Letzte, &c. Er hat keinen Anfang und Ende, er ist in Ewigkeit und auch heute gezeugt und bleibt auch in Ewigkeit der eingeborene Sohn des Vaters

if they do not repent. His "feet like fine brass" glowing in a furnace is his lowliness, since he mingles with us, with the unclean earth or ore, and therefore must suffer greatly and be tested in the furnace of misery through temptation, poverty, suffering, torture, torment, and bitter death. "His voice like the sound of many waters etc." means that his word and gospel are bound up with the water of baptism and holy communion, so that when we are baptized in his name with the great rushing waters of grace, this holy element washes away from us the black original sin. "And he had in his right hand seven stars etc." These are the seven spirits of GOD, through which he created everything, the angelic principalities and princely angels, the seven planets, all of which originated in the seventh spirit, namely nature and creatures, and therein became visible and comprehensible, and all mysteries and creations in heaven and on earth have their origin from the seven spirits of GOD. In the human being there are also seven source-spirits, just as there are in all creatures: in nature seven *gradus* or *systemata*, in the earth seven metals, in the week seven days, the seventh being the day of rest when the world has stood for six-thousand years and will have its great Sabbath; seven seals, seven times, seven petitions in the LORD's Prayer; sheer mystery can be found in the seventh number, more in which elsewhere. The seven stars are shown here as angels of the church, the pure divine teachers illuminated by the seven degrees of the holy spirit. "Out of his mouth went a sharp two-edged sword etc." This signifies the law and gospel by which sinners are terrified and the godly are comforted; and it is also the sharp two-edged sword that cuts off the old Adam from the new human, so that the new being may live in the spirit of Christ. "His countenance was as the sun shines etc.": his divinity's light and splendor, with which he enlightens us to eternal life. "He says: I am the first and the last, etc." He has no beginning and no end, he was begotten in eternity and also today and remains the only begotten son of the father forever; "I am he that lives, and was

und der Lebendige, ich war tot und siehe ich bin lebendig von Ewigkeit zu Ewigkeit. Er war tot durch der Schlangen Stich, er musste leiden und sterben, aber der Tod konnte ihn nicht halten und wurde lebendig in drei Tagen und lebt nun von Ewigkeit zu Ewigkeit, das ist von einer Ewigkeit zu der anderen und hat die Schlüssel der Hölle und des Todes, &c. Durch ihn werden wir von Sünden, Tod und Hölle erlöst, GOTT, sein und unser himmlischer Vater, hat ihm das Gericht über uns gegeben, weil er uns durch sein Blut erworben und von GOTT zu unserem Thron-Fürsten gesetzt, da er unser ewiger König sein und uns beherrschen soll. Er sagt: Ich will sie erlösen aus der Hölle und vom Tod erretten, Tod ich will dir ein Gift sein, Hölle ich will dir eine Pestilenz sein.

AUSLEGUNG DIESER FIGUR
AUF DAS GROSSE NATÜRLICHE WERK

Aus welchem die Heiden haben GOTT lernen erkennen und sich GOTT in der Natur abbilden wollen, weil GOTT nur einzig, die Natur auch, welche von GOTT geschaffen und hat solche mit allen Kräften, lebendigen Geistern und Wachstum gesegnet, dass alles fruchtbar sein und sich vermehren soll in allen 3 Reichen, im animalischen, vegetabilischen und mineralischen, in welchem letzten solche Beständigkeit, reine Geister, Balsam und Öle eingesenkt, auch alle Schönheiten der Himmel und Engel. Solches sehen wir an den Edelgesteinen, wie solche leuchten, aber noch mehr im philosophischen Werk, wenn die 7 philosophischen Metalle in hell leuchtende durchsichtige Edelgesteine gebracht und diese in der Multiplication in engelhafte Leiber, das ist tingierende Geister verkehrt werden.

UNSERE MATERIA

Dieses ist auch nur eine einzige *Materia*, daraus unser Anfang und Ende geht. Die 7 Leuchter sind die 7 Geister aus dieser *Materia*, unse-

dead and, behold, I am alive forevermore". He was dead by the sting of the serpent, he had to suffer and die, but death could not hold him and he came alive again in three days; and he now lives from eternity to eternity, that is, from one eternity to the next, and has the keys of hell and death, etc. Through him we will be redeemed from sin, death, and hell. GOD, his and our heavenly father, has given him judgment over us, because he acquired us through his blood and was made our throne prince by GOD, since he shall be our eternal king and rule us. He says: "I will ransom them from the power of the grave; I will redeem them from death: O death, I will be your plagues; O grave I will be your destruction".[150]

INTERPRETATION OF THIS FIGURE ON THE GREAT NATURAL WORK

From which the heathen have learned to recognize GOD and wanted to depict GOD in nature, because GOD is only unique, nature too, which GOD created and has blessed it with all powers, living spirits, and growth, so that everything is fruitful and multiplies in all three kingdoms, in the animal, vegetabilic, and mineral. In the last of these lie latent durability, pure spirits, balsams, and oils, as well as all the heavenly and angelic beauties. We see this in the precious stones, how they shine, but even more so in the philosophical work, when the seven philosophical metals are brought [i.e., transformed] into brightly shining transparent jewels and these are transmuted by multiplication into angelic bodies, that is into tingeing spirits.

OUR MATERIA

This is also just a single *materia* from which stems our beginning and end. The seven candlesticks are the seven spirits from this *materia*, our

re 7 Metalle, welche erstlich leuchten und funkeln in ihrer Wachsung und Witterung, wenn sich der Schimmer in Glimmer begibt und sich läutert, wenn es die 7 *Systemata* erreicht, der Blicker sich sehen lässt, mitten unter den 7 Leuchtern oder den 7 Gestalten der Natur, wird das *Subjectum* beschrieben, welches alles dieser Wunder Anfang und Ende ist. Er war angetan mit einem Kittel, umgürtet um die Brust mit einem güldenen Gürtel, &c. Das *Subjectum* ist von außen armseliger Gestalt, hat ein graues Kittelchen um, in seinem Innersten ist es lauter ☉, welches noch roh und unreif. Es ist der ☉ Baum und seine Wurzel. Sein Haupt und Haar waren weiß wie Wolle, &c. Sein oberster Teil, als sein flüchtiger ☿, ist weißer als der Schnee. Seine Augen wie Feuer-Flammen, etc. Wenn es gereinigt, ist es das leuchtende ☿al Wasser, die funkelnden Fisch-Augen. Seine Füße gleich wie Messing, &c. Sein fixes Teil, das *Electrum* ♀, welches hernach in ☉ verkehrt wird. Seine Stimme wie großes Wasserrauschen, &c. Sein ☿al Wasser ist das rauschende und tobende Meer, wenn es mit seinem fixen Teil soll vereinigt werden, so geschehen die Sturm-Winde, welche das Meer zu sehr bewegen, dass es prudelt und siedet und hat 7 Sterne in seiner rechten Hand. Wenn die 7 philosophischen Metalle in funkelnde Sterne, als in 7 Planeten, verwandelt werden, dass sie als funkelnde Sterne leuchten, wenn sie in dem renovierten Lebens-Wasser aufgelöst und darinnen glorifiziert werden. Aus seinem Munde ging ein scharfes zweischneidiges Schwert, &c. Dieses sind die 2 feurigen *Menstrua*, in welchen ersten sein siebenfacher ☿ præcipitiert, calciniert und fix zur weißen Tinctur wird, im anderen wird die Welt durchs Feuer gerichtet und der Stein rot und fix gemacht, wenn das gläserne Meer mit Feuer gemengt wird. Sein Angesicht leuchtet wie die Sonne, &c. Danach ist es der Sonnen-Sohn, welcher alles in ☉ verkehrt, wo er sein leuchtendes Angesicht hin wendet, worauf die Tinctur geworfen wird. Er ist der Erste und der Letzte, &c. Die erste *Materia* und die letzte *Materia* ist eine *Materia*, nur die erste war unreif, diese aber reif und beständig, reif und gekocht. Er war tot und ist nun lebendig und lebt von Ewigkeit zu Ewigkeit, etc. Diese *Materia* als sein ☿ musste leiden, sterben und in

seven metals, which first shine and sparkle in their growth and weathering, and then when the shimmer turns into a glimmer, and becomes clear when it reaches the seven *systemata*, the gazer lets itself be seen; and in the midst of the seven candlesticks or the seven forms of nature, the *subjectum* is described, which is the beginning and end of all these miracles. "He was clothed with a garment down to the foot, and girt about the breast with a golden girdle etc." The *subjectum* is poor on the outside, clothed with a gray coat, inside it is pure ☉, which is still raw and immature. It is the ☉ tree and its root. "His head and his hair were white like wool etc." Its uppermost part, its volatile ☿, is whiter than the snow. "His eyes were as flames of fire etc." When purified, it is the luminous ☿al water, the sparkling fish-eyes. "His feet like fine brass etc." This refers to its fixed part, the *electrum* ♀, which is then inverted into ☉. "His voice like the sound of many waters etc." This signifies its ☿al water, and it is the rushing and raging sea, and at the point when it is to be united with its fixed part, the storm winds arise, which disturb the sea too much, so that it bubbles and boils. "And he had in his right hand seven stars." This signifies that when the seven philosophical metals are transformed into sparkling stars, that is, into seven planets, they are dissolved and glorified in the renovated water of life. "Out of his mouth went a sharp two-edged sword etc." This signifies the two fiery *menstrua*, in the first of which its sevenfold ☿ precipitates, calcinates, and becomes a white tincture; and in the second of which the world is judged by fire and the stone is made red and fixed when the sea of glass is mingled with fire. "His countenance was as the sun shines etc." This signifies that afterwards it is the son of the sun who turns everything into ☉ where he turns his shining face, whereupon the tincture is thrown. "He is the first and the last etc." This means that the first *materia* and the last *materia* is one *materia*, only the first was immature; but this last one was ripe and permanent, ripe and cooked. "He was dead and is now alive and lives forevermore etc." This signifies that *materia* as his ☿ had to suffer, die, and be buried

die Erde begraben werden, alsdann lebendig mit seiner Seele vereinigt zum glorifizierten Leib wieder auferstehen, der nun ewig lebend ist und nicht mehr sterben kann und hat die Schlüssel der Hölle und des Todes. Er hat Macht, die Metalle zu töten und auch wieder lebendig zu machen, weil er ihr Bruder, metallischer Natur, ihr Fleisch und Blut, so erweckt er sie zum anderen und neuen Leben. Hier ist nun generaliter in einer *Summa* das ganze Werk als in einer Vorrede angezeigt, die andern Auslegungen werden nun ordentlich folgen.

ANDERE FIGUR
OFFENBARUNG JOHANNIS AM 4. KAPITEL

Eine Tür wurde aufgetan und ein Stuhl gesetzt im Himmel und auf dem Stuhl saß einer, war gleich anzusehen wie der Stein Jaspis und Sardis und ein Regenbogen um den Stuhl, gleich anzusehen als ein Smaragd, um den Stuhl die Ältesten mit weißen Kleidern und güldenen Kronen auf den Häuptern. Vom Stuhl aus ging Blitz, Donner und Stimmen und 7 Fackeln brannten vor dem Stuhl, welches sind die 7 Geister GOTTES *und um den Stuhl und vor dem Stuhl ein gläsernes Meer, gleich dem Kristall und mitten im Stuhl und um den Stuhl 4 Tiere, voller Augen vorn und hinten. Das erste Tier gleich einem Löwen, das andere einem Kalbe, das dritte hatte ein Angesicht wie ein Mensch, das vierte einem fliegenden Adler. Und ein jegliches der 4 Tiere hatte 6 Flügel und waren voller Augen vorn und hinten und hatten keine Ruhe Tag und Nacht und gaben Preis, Ehre und Dank dem, der auf dem Stuhl saß und die Ältesten warfen ihre Kronen vor den Stuhl und sprachen: Du bist würdig zu nehmen Ehre und Kraft, denn du hast alle Dinge geschaffen und durch deinen Willen haben sie das Wesen und sind geschaffen.*

GEISTLICHE AUSLEGUNG DER ANDEREN FIGUR, WELCHE
MUSSTE VORHERGEHEN, EHE DIE NACHFOLGENDE DRITTE
KONNTE GEZEIGT WERDEN

in the earth, and was then resurrected, his soul reunited to the glorified body, which is now eternally alive and can no longer die and has the keys of hell and death. He has the power to kill metals and also to bring them to life again, because he is their brother, being metallic in nature, he can restore their flesh and blood, awakening them to a new life. Here, more generally, the whole work is presented in a *summa* as indicated in the preface, and the other interpretations will now follow in an orderly manner.

SECOND FIGURE
REVELATION OF ST. JOHN IN THE FOURTH CHAPTER

A door was opened[151] *and a throne was set in heaven, and one sat on the throne.*[152] *He was to look upon like a jasper and a sardine stone and there was a rainbow around about the throne in sight like unto an emerald.*[153] *Round about the throne were elders sitting, clothed in white raiment; and they had on their heads crowns of gold.*[154] *And out of the throne proceeded lightnings and thunderings and voices: and there were seven lamps of fire burning before the throne, which are the seven spirits of* GOD, *and before the throne there was a sea of glass like unto crystal: and in the midst of the throne, and round about the throne, were four beasts full of eyes before and behind. The first beast was like a lion, and the second beast like a calf, and the third beast had a face as a man, and the fourth beast was like a flying eagle. And the four beasts had each of them six wings about him; and they were full of eyes within: and they rest not day and night,*[155] *and give glory and honour and thanks to him that sat on the throne,*[156] *the elders cast their crowns before the throne, saying: You are worthy to receive honour and power, for you have created all things, and for your pleasure they are and were created.*[157]

SPIRITUAL INTERPRETATION OF THE SECOND FIGURE,
WHICH MUST GO BEFORE THE THIRD
COULD BE SHOWN

Die offene Tür ist der Leib der Jungfrau Maria, da *Christus* musste durchgehen, unser Fleisch annehmen und uns erlösen, dass er unser König wäre, uns ewig zu beherrschen, &c. Der Stuhl ist der Stuhl, der zubereitete Thron des ganzen *Principii*, und fürstliche Thron, für den Fürsten-Engel, den Sohn GOTTES, in dem ausgestoßenen Reiche des Lucifers, welches wir durch *Christum* einnehmen, da er ewig unser König ist. Auf dem Stuhl saß einer, sah wie der Stein Jaspis und Sardis, bedeutet den Sohn GOTTES in seiner göttlichen und menschlichen Natur, der Stein Jaspis seine Gottheit, der rote Stein Sardis seine Menschheit, durch sein Leiden und Blut. Der Regenbogen um den Stuhl zeigt, dass dieser Himmel, Fürstentum und Thron aus den 7 Quell-Geistern GOTTES und aus dem himmlischen Licht erschaffen, da alle die schönen himmlischen Farben sich bilden und zeigen, da alles lebt und sich formt und das himmlische Freudenreich darinnen aufgeht, wenn es sich als ein heller grüner Blitz entzündet, leuchtet, glänzt und scheint. Vom Stuhl aus gingen Blitz, Donner und Stimmen &c. In welchen sich diese 7 Geister gebären, auch die Gnaden-Gaben des Heiligen Geistes bei der triumphierenden Kirche im Himmel und bei der streitenden hier auf Erden. 7 Fackeln mit Feuer brannten vor dem Stuhl, &c. Die Geister GOTTES und Gaben des Heiligen Geistes, durch die heilige Lehre und sind die 7 Quell-Geister allenthalben zu finden, in allen himmlischen Thronen, Engeln und Menschen und in der Natur und Creatur, denn durch sie lebt alles und ist geschaffen, wird auch durch sie erhalten und ist durch den 7ten als in dem einzigen durchs Wort des HERRN gemacht und durch den Geist seines Mundes ausgegangen. Diese 7 Fackeln, welche vor dem Stuhl brennen, zeigen auch die 7 Quell-Geister der Kinder GOTTES an, welche in allen ihren Geistern in der Liebe GOTTES entzündet, vor seinem Stuhl durch heilige reine Lehre leben und wandeln, brennen und leuchten. Die Ältesten mit weißen Kleidern und güldenen Kronen auf ihren Häuptern, &c. Das sind die reinen Lehrer und durch den Geist GOTTES erleuchtete Menschen, welche ihr geistliches Priestertum in der Welt in acht genommen, nun aber ihre Kronen, ihr Amt und

The open door is the body of the virgin Mary, where Christ had to go through, take on our flesh, and redeem us to be our king to rule over us forever etc. The throne is the chair, the prepared throne of the whole *principii*, and the princely throne, for the princely angel, the son of GOD, in the outcast realm of Lucifer, which we occupy through Christ, since he is our king forever. The one who sat on the throne appeared like jasper and a sardine stone, signifies the son of GOD in his divine and human nature: the jasper stone represents his divinity, and the red sardine stone his humanity through his passion and blood. The rainbow around about the throne shows that this heaven, principality, and throne are created from the seven source-spirits of GOD and from the heavenly light, where all the beautiful heavenly colors form and show themselves, where everything lives and forms and where the heavenly realm of joy rises within, when it ignites, glows, sparkles, and shines as a bright flash of green. "Out of the throne proceeded lightnings and thunderings and voices etc.", in which these seven spirits give birth, also the gifts of grace of the holy spirit in the triumphant church in heaven and in the fighting [churches] here on earth. "Seven lamps of fire burning before the throne etc." This signifies the spirits of GOD and the gifts of the holy spirit through the holy teaching. The seven source-spirits can be found everywhere, in all heavenly thrones, angels, and people as well as in nature and creatures, because through them everything lives and is created, and will also be maintained through them and through the seventh, which is the only [spirit] that is made through the word of the LORD and which departed through the spirit of his mouth. These seven lamps of fire which are burning before the throne also indicate the seven source-spirits of the children of GOD, which in all their spirits are ignited in the love of GOD, live and walk, burn and shine before his throne through the pure, holy teaching. "The elders with white raiment and crowns of gold on their heads etc." represent the pure teachers and humans enlightened by the Spirit of GOD, who took heed of their spiritual priesthood in the world, but who now throw their crowns, their office, and adornment

Zierde vor den Stuhl werfen, vor den König der Könige und HERR aller HERREN, dem Sohn GOTTES, als dem ewigen Hohen-Priester überantworten. Und vor dem Stuhl ein gläsernes Meer gleich dem Kristall, &c. Das ist das durchsichtige Welt-Gebäude, welches in der verklärten Gestalt ewig vor dem Sohn GOTTES, vor seinem Thron, stehen wird, auch sind wir Menschen solches selber, wenn wir mit unseren 7 Quell-Geistern in GOTT inqualieren, in einem heiligen Leben, im neuen Menschen, darinnen er in unser reines und neues Herz sehen kann, wenn wir alle schwarze Erde und stinkende Sünden ausgefegt haben, rein und lauter in GOTTES Willen leben. Vom Stuhl gingen aus Blitz, Donner und Stimmen, &c. Die Gaben des Heiligen Geistes, dadurch wir berufen, erleuchtet, zu Engeln und Untertanen dieses herrlichen himmlischen Reiches erwählt und bestätigt werden und wenn die Zahl der gefallenen Engel erfüllt, wir dieses Reich ewig besitzen, da *Christus* JESUS unser ewiger König ist, auf dem Thron Majestät und Herrlichkeit, alles beherrscht, bei der triumphierenden Kirche im Himmel und bei der streitenden hier auf Erden, die 4 Tiere, die 4 Stände auf Erden, der Löwe die Obrigkeit, der Ochse den Lehrerstand; der Adler der Schülerstand; der Mensch den Hausstand. Diese Tiere hatten 6 Flügel und waren voller Augen vorn und hinten und hatten keine Ruhe Tag und Nacht, &c. Diese alle sollen nicht ruhen oder faul sein in dem Stand darein sie GOTT gesetzt hat, die Flügel zeigen an, dass sie die Sünden und unreine Welt sollen fliehen und meiden und sich nach GOTT und dem Himmel schwingen: die vielen Augen die Vorsichtigkeit, stets Wachen und zum Lob und Preis GOTTES geschickt sein. Die 4 Tiere sind auch des HERRN CHRISTI Menschwerdung, sein Opfer, Himmelfahrt und Auferstehung. *Item*, die Lehre der 4 Evangelisten.

AUSLEGUNG DER ANDEREN FIGUR
AUF DAS GROSSE NATÜRLICHE WERK

Die aufgetane Tür ist die aufgesperrte Pforte aller Elemente, ohne die-

before the throne, before the king of kings and LORD of LORDS, the son of GOD, as the eternal high priest. "And before the throne there was a sea of glass like unto crystal etc." This is the transparent world-edifice, which will stand in the transfigured form forever before the son of GOD, before his throne. Also we humans will be like that ourselves when our seven source-spirits unite with GOD in a holy life, in the new human being in which he can see into our pure and new heart, when we have swept out all black earth and stinking sins, living purely and solely in GOD's will. "Out of the throne proceeded lightnings and thunderings and voices etc." This represents the gifts of the holy spirit, by which we are called, enlightened, chosen and confirmed as angels and subjects of this glorious heavenly kingdom; and when the number of the fallen angels is fulfilled, we possess that kingdom forever, since Christ Jesus is our everlasting king, radiating majesty and glory from his throne, ruling everything through the triumphant church in heaven and through the churches still engaged in spiritual combat here on earth. The four beasts represent the four estates on earth: the lion is the ruling class; the calf the teacher's estate; the eagle the estate of the disciples; and the human represents the estate of the commons. "The beasts had each of them six wings and they were full of eyes before and behind and they did not rest day and night etc." All of these should not rest or be lazy in the estate in which GOD has set them. The wings indicate that they should flee and avoid sin and the impure world and swing to GOD and heaven: the many eyes are caution and awareness, always awake and prepared to praise and glorify GOD. The four beasts are also the LORD Christ's incarnation, his sacrifice, ascension, and resurrection. They also symbolize the four Gospels.

INTERPRETATION OF THE SECOND FIGURE
ON THE GREAT NATURAL WORK

The opened door is the unlocked gate of all the elements, nothing can

selben ist nichts zu tun, der Stuhl bedeutet, dass nun der steinerne Palast soll erbaut werden. Der Sitz oder Stuhl ist das hermetische Erz, *Hyle*, *Vas Viride Saturni*, der sich darauf setzen soll und als künftiger König herrschen, ist der reine jungfräuliche ☿, sobald er diesen Stuhl oder Sitz einnimmt, bekommt er die Farben von Jaspis und Sardis, die güldischen und martialischen Farben. Um den Stuhl ein Regenbogen, &c. Wenn sich die 7 Quell-Geister bilden und formen, erscheinen die Farben, absonderlich die glückselige und gesegnete Grüne, bei unserem grünenden Erz, durch so viele Abkochungen. Um den Stuhl die Ältesten mit weißen Kleidern und güldenen Kronen, welche sie alle vor den Stuhl werfen und dem ihre Herrschaft lassen, welcher den Thron eingenommen, &c. Das sind die reinen metallischen Geister der 7 Planeten, welche sie alle ihrem König aufopfern, ihm zu Füßen fallen und selber als nichts werden und vergehen, dieser aber groß und mächtig wird, je mehr er Untertanen und Könige beherrscht, so wird er König der Könige. Vom Stuhl aus ging Blitz, Donner und Stimmen und 7 Fackeln mit Feuer brannten vor dem Stuhl, &c. Die Blitze, Donner und Stimmen sind die vielen Farben und Witterungen, welche auch bedeuten das Jubel-Geschrei der Seelen, die Unruhungen der vielen widerwärtigen Geister, welches dem Blitzen und Wetterleuchten gleich ist. 7 Fackeln mit Feuer brannten vor dem Stuhl, &c. Sind die 7 philosophischen Metalle oder unsere Planeten, welche durch die 7 Quell-Geister gebildet und begreiflich worden. Vor dem Stuhl ein gläsernes Meer gleich dem Kristall, &c. Wenn die ersten 7-fachen magischen Zahlen erfüllt, wenn alle die 7 Gestalten geboren, sichtbarlich und begreiflich geworden und zum Stein erhärtet, welches der glasförmige *Azoth* und Kristall des ☿ ist. Die 4 Tiere mit Flügeln voller Augen, &c. Das sind nicht die rechten *Elementa*, weil Feuer und Luft in diesen Kristall nicht kommen und jene 2 *Elementa* noch zurück sind, derhalben sie auch nicht hier als sonstigen die *Elementa* beschrieben werden, weil es nicht die ganz vollkommenen *Elementa* sind, sondern nur vermischt, weil sie unmöglich ganz rein zu scheiden, denn bei diesem jungfräulichen ☿ ist viel von der öligen Luft geblieben, bei der Erde viel vom Feuer, so

be done without it. The throne signifies that now the stone palace is to be built. The seat or throne is the Hermetic ore or *hyle* and the *vas viride Saturni* which will set itself upon it and rule as the future king; it is the pure virgin ☿ and as soon as it takes this seat or throne, it acquires the colors of jasper and sardis, the golden and martial colors. "Around about the throne was a rainbow etc." When the seven source-spirits shape and form themselves, the colors—especially the blissful and blessed green—appear on our verdant ore through so many decoctions. "Round about the throne the elders with white raiment and crowns of gold, who throw them all before the chair and let him reign who has occupied the throne etc." These represent the pure metallic spirits of the seven planets which all sacrifice to their king. They fall at his feet and become nothing and perish, but he becomes great and powerful the more he dominates his subjects and kings, then he becomes king of kings. "Out of the throne proceeded lightnings and thunderings and voices and seven lamps of fire burning before the throne etc." The lightnings, thunderings, and voices are the many colors and atmospheres, which also signify the jubilant cries of the souls, the commotion of the many spirits who are present, which are like sheets and bolts of lightning. "Seven lamps of fire burning before the throne etc." These are the seven philosophical metals (or our planets) which were formed and made tangible through the seven source-spirits. "Before the throne a sea of glass like unto crystal etc." When the first seven-fold magic numbers are fulfilled, when all the seven forms are born, they become visible and tangible and harden into stone, which is the glassy *azoth* and crystal of ☿. "The four beasts with wings full of eyes etc." These are not the proper *elementa*, because fire and air do not come into this crystal and those two *elementa* remain behind, which is why they are not described here as the usual *elementa*, because they are not the completely perfect *elementa*, but only mixed ones. They are mixed because they are impossible to separate into completely pure forms, because in this virgin ☿ much of the oily air has remained, and with the earth much of fire, so that the *elementa* are of course still there and have no

sind freilich die *Elementa* noch da und haben wegen widerwärtiger Natur keine Ruhe, der vielerlei Arten der flüchtigen Geister in den vielen Aufkochungen und Farben, wenn sie stets ihre Flügel erheben und im Glase stüben bis sie endlich ihren Ruhe-Tag und Sabbat gefunden und fix liegen bleiben. Die Ältesten mit weißen Kleidern um den Stuhl, &c. Dass die *Materia*, wenn sie fix werden will, sich von außen anfängt zu weißen und ob es schon weiße Kleider an hat, außen in weißer Gestalt erscheint, sind diese Farben so alle in Schiedligkeit gestanden und in eine eingekehrt doch güldisch, welches die güldene Kronen anzeigt, dass solche Leiber güldisch sind. Der Blitz und Donner so vom Stuhl ausgeht, die vielen Erzitterungen und unruhigen Bewegungen, vielerhand Farben wie ein Blitz und Wetterleuchten, Sturm und Winden, je unruhiger, desto besser unsere Metalle wachsen und geben die flüchtigen Geister, die funkelnden Augen, welche anheben zu leuchten, wenn dieser wütend ist und Castor und Pollux sich sehen lässt, so wird es wieder stille und eine geistliche Schifffahrt.

DRITTE FIGUR

OFFENBARUNG JOHANNIS AM 5. KAPITEL

Und ich sah in der rechten Hand dessen, der auf dem Stuhl saß, ein Buch geschrieben, in und auswendig versiegelt mit 7 Siegeln und niemand im Himmel noch auf Erden, noch unter der Erden konnte das Buch auftun und darein sehen. Und ich sah mitten im Stuhl und der 4 Tiere und mitten unter den Ältesten stand ein Lamm wie es erwürgt wäre und hatte 7 Hörner und 7 Augen, &c. Welches sind die 7 Geister GOTTES. *Und es kam und nahm das Buch von der Hand des der auf dem Stuhl saß und da es das Buch nahm, fielen die 4 Tiere und die 24 Ältesten vor das Lamm und hatten Harfen und güldene Schalen voll Räucher-Werks und sungen ein neues Lied: Du bist würdig zu nehmen das Buch und aufzutun seine Siegel, denn du bist erwürgt und hast uns erkauft mit deinem Blute.*

rest because of the recurrent nature. The many types of volatile spirits, in the many boilings and colors when they always raise their wings and linger in the glass until they finally find their day of rest and Sabbath and stay put. "The elder with white raiment round about the throne etc." That the *materia*, when it wants to become fixed, begins to turn white from the outside and appears white on the outside even though it is already wearing white clothes. These colors, which were standing in divisiveness, have turned into one [color], yet [are] gold-bearing; by the golden crown is indicated that such bodies are golden. The lightnings and thunderings proceed from the throne, along with many tremors and restless movements, accompanied by many different colors like bolts and sheets of lightning, storms, and winds. The more restless [the throne is], the better our metals grow and give forth volatile spirits, the sparkling eyes, which begin to shine. When it is raging, and Castor and Pollux[158] allow themselves be seen, it becomes quiet again, a spiritual ship voyage.

THIRD FIGURE:
REVELATION OF ST. JOHN IN THE FIFTH CHAPTER

And I saw in the right hand of him that sat on the throne a book with writing on both sides, sealed with seven seals[159] *and no human in heaven, neither on earth nor under the earth, was able to open the book, nor to look upon it.*[160] *And I saw in the midst of the throne encircled by the four beasts and the elders, where stood a lamb looking as if it had been slain, having seven horns and seven eyes etc. These are the seven spirits of* GOD.[161] *And he came and took the book out of the right hand of him that sat upon the throne and when he had taken the book, the four beasts and four and twenty elders fell down before the lamb, each bearing harps and golden vials full of incense,*[162] *and they sung a new song: you are worthy to take the book, and to open the seals thereof: for you were slain, and have redeemed us by your blood.*[163]

GEISTLICHE AUSLEGUNG DER DRITTEN FIGUR

Und ich sah in der rechten Hand des der auf dem Stuhl saß ein Buch, in- und auswendig beschrieben, &c. Der auf dem Stuhl saß bedeutet den Sohn GOTTES, den herrschenden König in diesem Thron: Das Buch in seiner rechten Hand in- und auswendig beschrieben ist die Natur und Creatur nach allen 7 Quell-Geistern und Ausgeburten: In- und auswendig beschrieben, &c. Ist die göttliche Regierung und Vorsehung, sonderlich bei den Menschen durchs Wort und Evangelium; weil der Mensch aber durch die 7 Quell-Geister nicht in GOTT inqualiert und in die Schiedligkeit fiel, so war der Mensch im Zorn und Gerichten GOTTES, welche durch 7 Siegel versiegelt mit Gerechtigkeit und Wahrheit, das ihm niemand widersprechen darf, auch dem Menschen verborgen sein die göttlichen Gerichte, es sei denn, dass er ihm solche selbst eröffne, seinen Willen und er wieder von der Schiedligkeit einwendet und diese Siegel eins nach dem anderen erbrochen werden und also wieder im neuen Menschen eingehen, mit GOTT inqualieren und aus dem Zorn wieder in das Licht und die Gnade GOTTES verwandelt, doch anders nicht als durch *Christum*. Und niemand im Himmel noch auf Erden, noch unter der Erden, konnte das Buch auftun und darein sehen. Und ich sah mitten im Stuhl und der 4 Tiere und mitten unter den Ältesten stand ein Lamm, wie es erwürgt wäre und hatte 7 Hörner und 7 Augen, welche sind die 7 Geister GOTTES und es kam und nahm das Buch von der Hand des der auf dem Stuhl saß. Niemand im Himmel noch auf Erden, noch unter der Erden, weder Engel noch Erz-Engel, konnte den gefallenen Menschen, der mit allen 7 Geistern hart in GOTTES Zorn verschlossen und versiegelt war, wieder eröffnen und befreien, auch durch göttliche strenge Zorn-Gerichte brechen und solches durchdringen, den Zorn des gerechten GOTTES wieder versöhnen, in Sanftmut und Liebe verwandeln und sich als Mediator darstellen, die menschliche Natur wieder reinigen, erneuern und mit GOTT vereinigen, er müsste denn beide Naturen haben, die göttliche und menschliche, welches das erwürgte

SPIRITUAL INTERPRETATION OF THE THIRD FIGURE

"And I saw in the right hand of him that sat on the throne a book with writing on both sides etc." He who sat on the seat is the son of GOD, the reigning king on this throne: the book in his right hand with writing on both sides is nature and all its creatures, according to all seven source-spirits and monstrosities. "With writing on both sides etc." The writing is the divine government and providence, especially with respect to humans as expressed through the word and gospel. But because humans were not imbued in GOD by the seven source-spirits and fell into divisiveness, they became subject to the wrath and judgments of GOD, which were sealed by seven seals with justice and truth, which no one is allowed to contradict. Moreover, the divine judgments are hidden from humans, unless he reveals his will to them himself. He objects again to the divisiveness and these seals are broken one after the other and thus enter again into the new human, [who is] infused with GOD and converted again from the wrath into the light and grace of GOD, but not otherwise than through Christ. "And no human in heaven, neither on earth nor under the earth, was able to open the book, nor to look upon it. And I saw in the midst of the throne encircled by the four beasts and the elders, where stood a lamb looking as if it had been slain, having seven horns and seven eyes. These are the seven Spirits of GOD. And He came and took the book out of the right hand of him that sat upon the throne." No one in heaven, on earth, or under the earth, neither angels nor archangels, could reopen and liberate the fallen human, who was firmly locked and sealed with all seven spirits in GOD's wrath. No one could break through the severity of the divine wrathful judgments and penetrate them in order to reconcile the wrath of the righteous GOD and transform it into gentleness and love by presenting themselves as a mediator who would purify human nature again, renewing it and uniting it with GOD. To do so they would have to have both natures, the divine and the human, and this

Lamm GOTTES ist, das der Welt Sünde getragen, mit 7 Augen, die 7 Geister GOTTES in seiner göttlichen Natur, die 7 Hörner, die Macht und Gewalt in seiner göttlichen und menschlichen Natur, in unsere 7 Quell-Geister, die er all erneuern, reinigen, gewiss und beständig machen kann. Dieses Lamm stand mitten im Stuhl und der 4 Tiere und mitten unter den Ältesten. Da er nach seiner angenommenen Menschheit, Marter und Leiden, in den herrschenden Thron erhoben, dass nun GOTT und Mensch in einer Person herrscht und regiert, doch also, dass die göttliche Natur, das Buch der Regierung in der rechten Hand gehabt, von Ewigkeit, die menschliche aber dasselbe empfangen in der Person vermittelst des Stuhls und ist nun mit unermesslicher Gewalt und unendlicher Gnade des Heiligen Geistes geziert. Und die 4 Tiere und die Ältesten, da es mitten unter ihnen, sind die 4 Stände unter den Menschen, die Ältesten die reinen Lehrer und Prediger, da er mitten unter uns sein will in seiner göttlichen und menschlichen Natur, unser König und Hoher-Priester, in der triumphierenden Kirche im Himmel und der streitenden hier auf Erden, welches wundergroße Geheimnis Engel und Menschen in Ewigkeit preisen und GOTT loben werden, auch das neue unerhörte Lied singen, dass er uns durch sein Blut erkauft und aus Höllen-Bränden zu Königen, Priestern und Kindern GOTTES gemacht, welches Dank und Lob teils hier abgestattet, dort aber in der Ewigkeit vollkommener geschieht.

AUSLEGUNG DIESER FIGUR
AUF DAS NATÜRLICHE GROSSE WERK

Das Buch mit 7 Siegeln ist die geblätterte Erde, so aus vorhergegangenen 7 Geistern zur Ausgeburt kommen und aus solchem Kristall und glasförmigen *Azoth* sich sublimiert, welches mit 7 Siegeln verschlossen, durch 7 Eintränkungen zubereitet, welches die 7 Gestalten der Natur, hier aber unsere 7 Metalle sind, so durch 7 *Systemata* in höchsten Grad gebracht, in weiß ☉, in die geblätterte Erde oder *Luna fixa*, sieht schön, weiß, gelb. Nichts kann nun dieses Buch aufschließen oder sei-

is precisely the slain lamb of GOD that bears the sins of the world, with seven eyes, the seven spirits of GOD in his divine nature, the seven horns, the might and power in his divine and human nature, in our seven source-spirits, all of which he can renew, purify, and make sure and permanent. This Lamb stood "in the midst of the throne encircled by the four beasts and the elders". Because he was raised to the ruling throne after he accepted human incarnation, torture, and suffering, GOD and humanity now rule and reign in one person, and the divine nature had the book of government in his right hand from eternity, but humanity received the same in person by means of the throne, and is now adorned with the immeasurable power and infinite grace of the holy spirit. And the four beasts and the elders, since it [the lamb] was in the midst of them, are the four estates among humans, the elders, the pure teachers, and the preachers, since he wants to be among us, in his divine and human nature, being our king and high priest, in the triumphant church in heaven and the struggling [churches] here on earth, this great miracle and mystery that angels and humans will praise for eternity and glorify GOD. They will also sing the new unheard song that he bought us with his blood and which made kings, priests, and children of GOD out of hell fires; thanks and praise for this is partly accomplished here, but is more perfectly completed there, in eternity.

INTERPRETATION OF THIS FIGURE
ON THE NATURAL GREAT WORK

The book with seven seals is the leafed earth, which came to birth from the previous seven spirits and sublimated itself from its crystal and glass-shaped *azoth*, which is closed with seven seals and prepared by seven infusions which are the seven forms of nature. Yet here are our seven metals, thus brought to the highest degree by seven *systemata*, in white ☉, in the leafed earth or *Luna fixa*. It appears as a beautiful shade of white and yellow. Nothing can now unlock this book or break

ne 7 Siegel erbrechen oder resolvieren, er sei denn eben von diesem Geblüt entsprungen oder solchem teilhaftig, doppelter Natur, geistlicher und leiblicher: Die 7 Hörner zeigen an, dass das Auflösende und Wiederbringende auch von diesen 7 Gestalten und aus diesen Graden entsprungen. Die 7 Augen seine geistliche doppelte Natur mit der lebendig-machenden Seele. Dieser doppelte Geist kann die 7 Siegel durchbrechen. Bei jeder Siegel-Eröffnung geschieht eine große Veränderung im Werk, in der Exaltation, und wäre viel davon zu schreiben geistlicher Weise und auch in diesem Werk, befleißige mich aber jetzt der Kürze, bis zuletzt das Erdbeben geschieht bei den 6 Siegeln, die Sonne schwarz und der Mond in Blut verwandelt wird. Bei dem 7ten Siegel geschehen die größten Wunder.

VIERTE FIGUR:
OFFENBARUNG JOHANNIS AM 10. KAPITEL

Und ich sah einen anderen starken Engel vom Himmel herab kommen, der war mit einer Wolke bekleidet und ein Regenbogen auf seinem Haupt und sein Antlitz wie Sonne und seine Füße wie Feuer-Pfeiler und er hatte in seiner Hand ein Büchlein aufgetan und er setzt seinen rechten Fuß auf das Meer und den linken auf die Erde und er schrie mit großer Stimme als ein Löwe brüllt und da er schrie, redeten 7 Donner ihre Stimmen und da die 7 Donner ihre Stimmen geredet hatten, wollte ich sie schreiben, da hörte ich eine Stimme vom Himmel sagen zu mir: Versiegle, was die Donner geredet haben, dasselbe schreibe nicht. Und der Engel, den ich sah stehen auf der Erde, hub seine Hand auf gen Himmel und schwor bei dem lebendigen GOTT *von Ewigkeit zu Ewigkeit, der den Himmel geschaffen hat und was darinnen ist, das hinfort keine Zeit mehr sein soll, sondern in den Tagen der Stimme des 7ten Engels, wenn er posaunen wird, soll vollendet werden das Geheimnis* GOTTES. *Und ich hörte eine Stimme mit mir reden und sagen: Gehe hin und nimm das Büchlein von der Hand des Engels der auf dem Meer und auf der Erden steht und ich ging hin zum Engel und sprach: Gib mir das Büchlein und er sprach: Nimm es hin und verschling es und es wird dich in deinem*

or unseal its seven seals, unless it is something that has just sprung from this lineage or has such a thing as a double nature: spiritual and corporeal. The seven horns indicate that which dissolves and brings back, and it also arose from these seven forms and from these grades. The seven eyes represent its spiritual dual nature with the life-giving soul. This double spirit can break through the seven seals. At each opening of the seals, a great change occurs in the work and in the exaltation; and there would be much to write about, both spiritually and also in this work, but I will now take heed of brevity, until at last the earthquake of the sixth seal occurs, and the sun is turned black and the moon into blood. The greatest miracles happen at the seventh seal.

FOURTH FIGURE:
REVELATION OF ST. JOHN IN THE TENTH CHAPTER

And I saw another mighty angel come down from heaven, clothed with a cloud: and a rainbow was upon his head, and his face was as it were the sun, and his feet like pillars of fire: and he had in his hand a little book open: and he set his right foot upon the sea, and his left foot on the earth, and cried with a loud voice, as when a lion roars: and when he had cried, seven thunders uttered their voices, and when the seven thunders had uttered their voices, I was about to write, and I heard a voice from heaven saying unto me: seal up those things which the thunders uttered, and write them not. And the angel which I saw stand upon the earth lifted up his hand to heaven, and swore by him that lives for ever and ever, who created heaven, and the things that therein are, that there should be time no longer: But in the days of the voice of the seventh angel, when he shall begin to sound his trumpet, the mystery of GOD *should be finished.*[164] *And the voice which I heard spoke to me and said: Go and take the little book from the hand of the angel which stands upon the sea and upon the earth, and I went unto the angel, and said: Give me the little book, and he said: take it, and eat it up; and it shall make your belly bitter, but shall be in your mouth sweet as honey,*[165] *and as soon as I had eaten it, my*

Bauche grimmen, aber am Ende wird es süße sein wie Honig und da ich es gegessen hatte, grimmt es mich im Bauche und er sprach: Du musst abermals weissagen den Völkern und Heiden und Sprachen. Und es ward mir gegeben ein Rohr, einem Stecken gleich und sprach: Stehe auf und miss den Tempel GOTTES und Altar und die darinnen anbeten, aber den inneren Chor des Tempels wirf hinaus und miss ihn nicht, denn er ist den Heiden gegeben und die heilige Stadt werden sie zertreten, 42 Monde.

GEISTLICHE AUSLEGUNG DIESER VIERTEN FIGUR

Dieser Engel vom Himmel bedeutet abermals den Sohn GOTTES in göttlicher und menschlicher Natur, verklärt durch die 7 Quell-Geister GOTTES, welche die menschliche durchdrungen. Er war mit der Wolke bekleidet, &c. Das ist seine Majestät und Glanz in der 7ten Ausgeburt. Und ein Regenbogen auf seinem Haupt, &c. Ist mit den Geistern der ganzen Natur umgeben und mit unseren menschlichen Quell-Geistern, die in so vielen Schiedligkeiten und Farben stehen, vereinigt, durch Annehmung unseres Fleisches, welche Schiedligkeiten und Ausgehung von ihm er doch in Gnaden und Barmherzigkeit wieder zu sich nimmt durch unsere Buße und mit seinem reinen göttlichen Geist vereinigt und unsere Geister seinem Geist gleich macht. Sein Angesicht leuchtet wie die Sonne und seine Füße wie Feuer-Pfeiler, &c. Nach seiner Auferstehung ist seine menschliche Natur in die göttliche verklärt, dass er in der menschlichen Natur auch GOTT ist. Die Füße wie Feuer-Pfeiler, bedeutet die Menschheit, welche in der ersten Figur als Messing oder Erz erschienen, nun aber nach seinem Leiden, Sterben und Auferstehen in ihm verklärt und geläutert worden, wie das Silber durchs Feuer siebenmal. Er setzt seinen rechten Fuß auf das Meer, &c. Seine Gottheit, Macht und Gewalt, welche er im Himmel und auf Erden hat und als ein wahrer GOTT herrscht. Den linken auf die Erde, &c. Welche er in seiner angenommenen menschlichen Natur beherrscht und über die Geister und Menschen, Lebendige und Tote ein HERR ist. Seine Stimme da er wie ein Löwe brüllt, &c. Sein Wort

belly was bitter, and he said: you must prophesy again before many people, and nations, and tongues, and kings.[166] *And there was given me a reed like unto a rod, and [the angel] said: rise, and measure the temple of* GOD, *and the altar, and those that worship therein, but the court which is within the temple leave out, and measure it not; for it is given unto the gentiles: and the holy city shall they tread under foot forty and two months.*[167]

SPIRITUAL INTERPRETATION OF THIS FOURTH FIGURE

This angel from heaven again means the son of GOD in divine and human nature, transfigured by the seven source-spirits of GOD, which penetrated human [nature]. "He was clothed with a cloud etc." [means] his majesty and splendor in the seventh birth. "And a rainbow was upon his head etc.", [means he is] surrounded by spirits of every nature and is united with our human source-spirits, which stand before him in so many divisivenesses and colors. Through the acceptance of our flesh he reconciles our alienation from him, taking us again to himself in grace and mercy through our penance, uniting us with his pure divine spirit and making our spirits like his spirit. "His face was as it were the sun, and his feet like pillars of fire etc." After his resurrection his human nature is transfigured into the divine [nature], so that in human nature he is also GOD. The feet like pillars of fire signifies humanity, which in the first figure appeared as brass or ore; but now after his suffering, death, and resurrection has been transfigured and purified within him, like silver seven times through fire. "He set his right foot upon the sea etc." [This is] his deity, power, and authority, which he has in heaven and on earth, which he reigns as a true GOD. "His left [foot] on the earth etc.", which he rules in his assumed human nature and is LORD over spirits and humans, living and dead. "His voice when he roars like a lion etc." [is] his word and gospel when

und Evangelium, so es in der ganzen Welt erschallt und gepredigt wird. Er schwor bei dem lebendigen GOTT, bei seinem Vater und sich selbst, dass nach der 7ten Posaune keine Zeit mehr sein soll, sondern soll vollendet werden das Geheimnis GOTTES. Die 7te Posaune ist die Vollendung der Wunder und geht in den 7ten Geist aus, welches der Natur Ende, der großen Welt ihr Sabbat und Ruhe-Tag und soll das Toben und Wüten der Feinde CHRISTI und seiner Kirche ein Ende nehmen. Das offene Büchlein ist die Heilige Schrift, die soll uns nicht mehr versiegelt sein, sondern wir werden GOTT in seinem Wort ganz klar erkennen, welches uns süßen Geschmack wie Honig geben wird, aber dennoch auch im Trost, Kreuz und Leiden erregen und geben wird in uns selber, weil wir doch stets mit unseren eigenen Quell-Geistern streiten müssen und ihnen, wenn sie sich in Liebe der Welt, in Zorn und dergleichen Lastern entzünden wollen, Einhalt tun und solche löschen und uns täglich das Büchlein das Wort GOTTES zu Nutz machen, welches uns dieser Engel, der Sohn GOTTES, unser Thron und Fürstenengel, gegeben, so soll sein Wort unsere stetige Speise sein, dadurch wir unsere Seelen zum ewigen Leben erhalten. Das Rohr, einem Stecken gleich, &c. Das ist die Richt-Schnur der Heiligen Schrift, mit welcher wir den Tempel GOTTES abmessen und erweisen, welches die rechte Kirche sei, den inneren Chor, da sich unsere falschen Quell-Geister, falsche Lehre und Meinungen haben an GOTTES Statt gesetzt, wo sich die falschen Priester aufhalten, die sollen wir auswerfen und als heidnische Greuel verdammen, welche die christliche Kirche 42 Monde anfechten wird, eine recht gar lange Zeit.

AUSLEGUNG DIESER VIERTEN FIGUR AUF
DAS GROSSE NATÜRLICHE WERK

Der Engel kommt mit einer Wolke bekleidet, den Regenbogen auf seinem Haupt, sein Angesicht wie die Sonne, seine Füße wie Feuer-Pfeiler, &c. Der ist unserer Jungfrau Sohn, der durch die 7 Gestalten der Natur und durch die vielen Farben des Regenbogens seine Geburt genom-

it resounds and is preached throughout the world. He swore by the living GOD, by his father and by himself that after the seventh trumpet there should be no more time, but the mystery of GOD should be accomplished. The seventh trumpet is the fulfillment of the miracle and proceeds in the seventh spirit, which is the end of nature, of the macrocosm, its Sabbath and day of rest, and it shall bring to an end the raging and rampaging of the enemies of Christ and their churches. The open book is the holy scripture, which should no longer be sealed to us, but we will recognize GOD very clearly in his word, which will give us a sweet taste like honey, yet will nevertheless also arouse within us crucifixion and suffering and give us consolation, because we must always fight with our own source-spirits to stop them when they want to be kindled in love of the world, in anger, and similar vices. We must eliminate such [things] and use the little book, the word of GOD, to our daily advantage, which this angel, the son of GOD, our throne and princely angel, has given to us. Then his word shall be our constant food, whereby we maintain our souls in eternal life. "The reed like unto a rod etc." is the guiding line of the holy scriptures, with which we measure the temple of GOD and prove which is the true church. It is the inner court, where our false source-spirits, false teachings, and false opinions have taken GOD's place, where are found the false priests whom we are to cast out and condemn as pagan abominations, who will contest the Christian church for forty-two months, quite a long time.

INTERPRETATION OF THIS FOURTH FIGURE ON THE GREAT NATURAL WORK

"The angel comes clothed with a cloud, a rainbow upon his head, his face as it were the sun, his feet like pillars of fire etc." He is our Virgin's son, who was born through the seven forms of nature and through the many colors of the rainbow and is made of water and earth upon

men und besteht aus Wasser und Erde, darauf er seine beiden Füße setzt, welches seine doppelte Natur anzeigt und nun in einer trockenen Wolke übersteigt, als das trockene lebendige Wasser, wegen seiner Reinigkeit und himmlischen Natur in Seele, Geist und Leib und unser Erz, siebenmal durchs weiße Feuer geläutert worden, so sieht das Fixe die Erde, als seine Füße, wie Feuer-Pfeiler, sein Angesicht leuchtet wie die Sonne, &c. Es ist in seiner ganzen Natur güldisch, welches über Erde und Meer einen güldischen Spiegel, einen Gold-Glanz macht, welches der wahre lebendig-machende Geist ist und besteht aus dreien, Seele, Geist und Leib, das göttliche Tugenden hat, das Wunder der Natur und der Kunst, der Brunnen des lebendigen Wassers, so keine Hand nass macht. Die große Stimme, &c. ist sein Lob und Ruhm. Wenn er in seiner Stärke wie ein Löwe brüllt, von den 7 Donnern ihre Stimme, &c. Wenn er durch abermalige 7 Eintränkungen seine Brüder, die philosophischen Metalle, vom Tode erweckt, solche neu gebiert zu schönen funkelnden Sternen, zur Tinctur verklärt und erleuchtet. Welche 7 Donner die andere siebenfache magische Zahl bedeutet und dieses Geheimnis aller Geheimnis soll verschwiegen bleiben und nicht geschrieben werden. Wenn nun der siebente Engel posaunt, so soll vollendet werden das Geheimnis GOTTES, welches der Natur Ende. Die andere Umkehrung des Rades, da die philosophische, paradiesische Welt in der Beschaulichkeit steht, ihren großen Sabbat-Tag hält, zum gläsernen Meer erhärtet, welches nun mit Feuer gemengt werden kann, das Büchlein aber sollen wir essen. Es soll erstlich unsere Arznei sein, weil es der ☿ vitæ, der rechte Lebens ☿ ist, welcher unseren Geist erneuern und stärken wird, davon wir gesund werden, auch weissagen lernen, das ist klug werden, andere lehren und unterrichten. Das Rohr, einem Stecken gleich, ist unser güldener Mess-Stab und Wunder-Stekken, damit wir das neue Gebäude abmessen, den Tempel dieser Götter, daran ihre Wappen zum ewigen Gedächtnis sollen aufgehängt werden.

which he sets his two feet, showing his double nature, now surmounting a dry cloud: as the dry, living water, due to its purity and heavenly nature in soul, spirit, and body; and as our ore, purified seven times through the white fire, so that the fixed sees the earth as its feet, like pillars of fire. "His face was as it were the sun etc.". It is golden in its whole nature, which makes a golden mirror over earth and sea, a golden luster, which is the true life-giving spirit; and it is threefold, consisting of soul, spirit, and body, having divine virtues, the miracle of nature and art, the fountain of living water which does not wet the hands. "The loud voice etc." [is] his praise and glory. "When in his strength he roars like a lion, from the seventh thunder its voice etc." When he awakens from the dead his brothers, the philosophical metals, through seven repeated impregnations, they are reborn into beautiful sparkling stars and are enlightened and transfigured into a tincture. The seventh thunder means the other sevenfold magic number, and this secret of all secrets shall be kept secret and shall not be written down. When the seventh angel sounds the trumpet, so shall the mystery of GOD be completed, which is the end of nature. The other reversal of the wheel, since the philosophical, paradisiacal world stands in tranquility, keeps its great Sabbath day, and hardens into the sea of glass, which can now be mingled with fire; but we are to eat the little book. First of all, it should be our medicine, because it is the ☿ vitæ, the right ☿ of life, which will renew and strengthen our spirit. We will become healthy through it, and we shall also learn to prophesy; that is to say, we shall become wise so that we can teach and instruct others. The reed like unto a rod is our golden measuring stick and miraculous rod, with which we measure the new building, the temple of these gods, on which their coats of arms are to be hung for everlasting remembrance.

FÜNFTE FIGUR:
OFFENBARUNG JOHANNIS AM 12. KAPITEL
DIE SIEBENTE POSAUNE, VOLLENDUNG DES GEHEIMNISSES

Und es erschien ein großes Zeichen im Himmel, ein Weib mit der Sonne bekleidet und der Mond unter ihren Füßen, auf ihrem Haupt eine Krone von 12 Sternen und sie war schwanger und schrie und war in Kindes-Nöten und hatte große Qual zur Geburt und es erschien ein anderes Zeichen im Himmel und siehe ein großer roter Drache, der hatte 7 Häupter und 10 Hörner und auf seinen Häuptern 7 Kronen und sein Schwanz zog den dritten Teil der Sterne und warf sie auf die Erde und der Drache trat vor das Weib, wenn sie geboren hätte, er ihr Kind fräße und sie gebar einen Sohn, ein Knäblein, der alle Heiden sollte weiden mit der eisernen Rute und ihr Kind ward entzückt zu GOTT und seinem Stuhl und es wurden dem Weib 2 Flügel gegeben, wie eines großen Adlers, dass sie in die Wüsten flöhe vor das Angesicht der Schlange und die Schlange schoss nach dem Weib aus ihrem Munde ein Wasser wie einen Strom, dass er sie ersäufte, aber die Erde half dem Weib und tat ihren Mund auf und verschlang den Strom, den der Drache aus seinem Munde schoss und der Drache ward zornig auf das Weib und ging hin zu streiten mit den übrigen von ihrem Samen, die GOTTES Gebot halten.

GEISTLICHE AUSLEGUNG DIESER FÜNFTEN FIGUR

Das Weib mit der Sonnen bekleidet, &c. Das ist die christliche Kirche, die reine Menschheit des neuen Menschen im Leben JESU CHRISTI, mit der Sonne bekleidet mit CHRISTO, den sie in der Taufe angenommen. Hat unter den Füßen den Mond, &c. Ist alles Zeitliche, Vergängliche, &c. Sind die vielen Schiedligkeiten und Quell-Geister, welche sie nicht über sich herrschen lässt und solche mit Füßen tritt. Die Kronen von 12 Sternen, &c. Sind die heiligen Lehrer der Apostel und Propheten. Sie hat alle 7 Gestalten der Natur und Creatur unter sich getreten, die böse Influenz und Reizung der Planeten und ihren eigenen Quell-Geistern und bösen Reizungen hat sie sich widersetzt

FIFTH FIGURE:
REVELATION OF ST. JOHN IN THE TWELFTH CHAPTER
THE SEVENTH TRUMPET, FULFILLMENT OF THE SECRET

And there appeared a great wonder in heaven, a woman clothed with the sun, and the moon under her feet, and upon her head a crown of twelve stars: and she being with child cried, travailing in birth, and pained to be delivered and there appeared another wonder in heaven; and behold a great red dragon, having seven heads and ten horns, and seven crowns upon his heads, and its tail swept a third of the stars [from the sky] and cast them to the earth: and the dragon stood before the woman in order to devour her child as soon as it was born, and she brought forth a male child, who was to rule all nations with a staff of iron: and her child was snatched up unto GOD and to his throne.[168] *And to the woman were given two wings of a great eagle, that she might fly into the wilderness, from the face of the serpent, and the serpent cast out of its mouth water as a flood after the woman, that he might cause her to be carried away by the flood, but the earth helped the woman, and the earth opened her mouth, and swallowed up the flood which the dragon cast out of its mouth, and the dragon was wroth with the woman, and went to make war with the remnant of her seed, who keep the commandments of GOD.*[169]

SPIRITUAL INTERPRETATION OF THIS FIFTH FIGURE

"The woman clothed with the sun etc." is the Christian church, the pure humanity of the new human being in the life of JESUS CHRIST, clothed with the sun, with CHRIST, whom they accepted in baptism. "Has the moon under her feet etc." [is] everything temporal, ephemeral, and so forth, the many divisivenesses and source-spirits, which she does not submit to, but tramples under her feet. "The crowns of twelve stars etc." [are] the holy teachers of the apostles and prophets. She has trampled under her feet all seven forms of nature and creature, and she has resisted the evil influence and irritation of the planets, her own source-spirits and evil irritations—and she has not

und nicht über sich herrschen lassen. Sie ist schwanger mit dem neuen Menschen in CHRISTO und hat große Qual zur Geburt, ehe sie sich der Welt und ihren eigenen Reizungen entreißt und *Christus* eine Gestalt in uns gewinnt, daher muss die Kirche, die Braut *Christi*, viel Leiden und Dulden, der innere Mensch, endlich wird das Kind, der innere Mensch, durch Schmerzen und Angst geboren. Dies Kind war entzückt zu GOTT und seinem Stuhl, &c. Dieser neue Mensch ist das paradiesische Bild, das Kind GOTTES, des HERRN CHRISTI liebes Brüderlein und seine Braut, der Feind der Kirche ist der höllische Drache mit seinem Anhang, der gottlosen Welt und ihren Gewaltigen, welche die Kirche GOTTES, die frommen Kinder GOTTES verfolgen, rot, die Blutgierigkeit, die 7 Häupter, die 7 Quell-Geister im Zorn und Höllen-Reich, die Kronen auf den Häuptern, die 7 Quell-Geister der Welt, in Augen-Lust, Fleisches-Lust und hoffärtiges Leben, aufgeputzt und aufgeschmückt mit einem Ansehen, als etwas herrliches durch Ehre und Hoheit, als groß, herrlich und majestätisch. Die 10 Hörner, dass der Teufel auch die Großen und Gewaltigen zu seinen Gehilfen braucht, wenn er seine Gewalt will an den Frommen ausüben. Sein Schwanz zog den dritten Teil der Sterne und warf sie auf die Erde, &c. Er verführt auch die Lehrer, welche als Sterne der Kirche leuchten sollten, steckt sie mit irriger Lehre an, verfinstert GOTTES Wort, riß sie vom Licht GOTTES, warf sie auf die Erde in die irdischen Schiedligkeiten, darinnen sie GOTT wollten suchen und verloren ihn gar, scheut sich nicht vor der Kirche, die doch CHRISTUS durch sein Blut erworben, welche er verfolgt, dass sie muss in Wüsten fliehen und müssen sich die armen Christen so im neuen Menschen, im Geist *Christi*, als in der rechten Kirche leben wollen, vor dem Drachen, dem Teufel und seinem Anhang, seinem großen Schwanz verstecken und verdecken und sich wohl verborgen halten, so lange der Anti-Christ und seine Tyrannen wüten und toben. Aber Michael mit seinen Engeln stritten mit dem Teufel und erretten die Kirche und ist dieser Streit zu verstehen von der letzten Offenbarung des *Evangelii*, so der Engel, welcher durch den Himmel geflogen, ein ewiges Evangelium hatte, welches

allowed them to rule over her. She is pregnant with the new human in Christ and has great pain in delivering before she tears herself away from the world and her own irritations. Then Christ gains a form in us, therefore the church, the bride of Christ, has to suffer and endure much. The inner human, then finally the child of the inner human, is born through pain and fear. "This child was snatched up unto GOD and to his throne etc." This new human is the paradisiacal image, the child of GOD, the LORD CHRIST's dear little brother and his bride, the enemy of the churches is the infernal dragon with its followers of the godless world and its mighty ones, who persecute the church of GOD and the pious children of GOD. Its redness signifies that it is bloodthirsty, while its seven heads represent the seven source-spirits in wrath and the kingdom of hell. As for the crowns on the heads, these represent the seven source-spirits of the world, in lust for the eyes, lust for the flesh and the arrogant life, being decked out and adorned with a reputation as glorious by honor and majesty, as great, glorious, and majestic. The ten horns mean that the devil also needs the great and mighty as his assistants if he wants to exercise his power over the pious. "His tail swept the third of the stars [from the sky] and cast them to the earth etc." He also seduces the teachers who are supposed to shine as the stars of the churches, infecting them with erroneous teachings, darkening GOD's word, tearing them from GOD's light and throwing them to earth in earthly divisivenesses where they try to seek GOD but lose him altogether. He does not shy away from the church, which CHRIST acquired through his blood. Instead he persecutes it so that Christians have to flee into the deserts and the poor believers who want to live in the new humanity, in the spirit of Christ as in the true church, must hide, cover up, and keep well hidden before the dragon, the devil, and his followers, with his great tail, as long as the Antichrist and his tyrants rage and rampage. Yet Michael and his angels fought with the devil and saved the church. This combat is to be understood from the last revelation of the gospel in such wise that the angel who flew through heaven had an everlasting gospel, a gospel which will spread

sich durch die ganze Welt ausbreiten wird. Der Drache wird ausgeworfen und gebunden 1000 Jahre, dass er nicht mehr verführen darf die Heiden und hat die Kirche, die Braut *Christi* überwunden durch des Lammes Blut, drum freut euch ihr Himmel und die darinnen wohnen, &c. alle Kinder GOTTES, &c. Ein großes Frohlocken der Auserwählten, eine nochmalige Verfolgung der Kirche, so aber nicht lange. Der Drache schoss aus seinem Munde einen Strom, &c. Das Wasser der Wollust und Eitelkeit, aber die Erde hilft dem Weibe, welche den Strom in sich gesoffen: Die irdisch gesinnt sind, saufen das Unrecht in sich wie Wasser dadurch der Drache abermals dem Weibe, der Kirche GOTTES, nichts anhaben kann. Er wird aber zornig und geht hin mit den übrigen von ihrem Samen zu streiten, die GOTTES Gebot halten und haben das Zeugnis JESU CHRISTI mit dem kleinen Häuflein der Kinder GOTTES, die im Geist *Christi* im neuen Menschen in der neuen Geburt leben.

AUSLEGUNG DER FÜNFTEN FIGUR IM GROSSEN NATÜRLICHEN WERK

Nach der Posaune des 7ten Engels erscheint der Natur dieses Werks Ende, die weiße Tinctur, das Weib mit der Sonne bekleidet, den Mond unter ihren Füßen, &c. Sie ist wahrhaftig die weiße Sonne oder das weiße ☉, welches mit ☉ fermentiert, alle ☉ Proben aushalten wird und ist die keusche Königin, welche alle die 7 Quell-Geister und Schiedligkeiten mit den vielen Farben, welche alle darinnen adulteriert in des Mondes unbeständigen Schein und die 7 Grade oder *Systemata* unter ihre Füße getreten und sie alle beherrscht, sich zur Regentin darüber gemacht, die Sonne ruht darinnen in der weißen Tinctur. Es ist die Sonne noch hinter dem Mond verborgen, welches man vor dem Schein des Mondes nicht wohl sehen kann, sie ist mit der solarischen Frucht schwanger. Das Kind, so sie gebären wird, wird unter die Götter aufgenommen, welches der Sonnen-Sohn werden soll, dem Weib werden Flügel eines großen Adlers gegeben, dass sie in die Wüsten flieht und

through the whole world. The dragon will be cast out and bound for a thousand years so that it can no longer deceive the heathen, for the church, the bride of Christ, has prevailed through the blood of the lamb, and therefore there is rejoicing in heaven amidst those who dwell in it etc. "All the children of GOD etc." This refers to a great rejoicing of the chosen ones, and a renewed persecution of the churches, but not for long. "The dragon cast out of his mouth water as a flood etc." This is the water of lust and vanity, but the earth which has drunk the river unto itself helps the woman. Those who are worldly minded drink injustice unto themselves like water, whereby the dragon again cannot harm the woman, the church of GOD. But it [the dragon] becomes angry and goes to fight with the remnant of her seed, who keep the commandments of GOD and have the testimony of Jesus Christ with a small group of GOD's children, who live in the spirit of Christ, in the new humanity, in the new birth.

INTERPRETATION OF THE FIFTH FIGURE IN THE GREAT NATURAL WORK

After the trumpet of the seventh angel, the nature of this work's end appears: the white tincture, or the woman clothed with the sun, the moon under her feet etc. She is truly the white sun or the white ☉, which fermented with ☉, will endure all ☉ assays. And she is the chaste queen which adulterates all the seven source-spirits and the many colored divisivenesses in the moon's fluctuating light, and the seven grades or *systemata* are trampled under her feet and ruled by her, for she made herself regent over them. The sun now rests within the white tincture. The sun is still hidden behind the moon, which cannot be seen clearly before the moon shines, but she is pregnant with the solar fruit. The child that she shall bear shall be received among the gods, and shall be the son of the sun. "The woman was given two wings of a great eagle so that she might fly into the wilderness and be

daselbst ernährt wird, sie wird mit dem fliegenden gekrönten Adler aufgelöst, dass sie flüchtig. Der rote Drache schießt einen Strom nach ihr sie zu ersäufen, aber die Erde hilft dem Weibe und vertrocknet den Strom, sie macht sich wieder trocken, ihre solarische Frucht hat nun mit diesem Drachen zu streiten, welcher sie in allen Teilen ihres Leibes mit Blut bespritzt, weil sie aber Götter oder engelhafter Natur, schadet ihr das Blut des Drachens nicht, sondern sie erhitzen sich durch diesen Streit, als durch welchen ihr Feuer entzündet und sie immer mehr und mehr in hellere und leuchtendere Sterne verwandelt werden, das ist zu größeren und herrlicheren Tincturen, durch den siebenköpfigen gekrönten Drachen, welcher durch so oftern Streit erregt worden, bis der Drache getötet und diese den Sieg erhalten. Der Drache mit den 7 Häuptern und Kronen ist die blutige Tinctur und gekrönte rote Löwe, das Feuer des Steins oder das glühende rote Feuer. Durch siebenfache rote Einträngung diese Könige ihre Kronen nehmen, 10 Hörner ist die Beschließung in der zehnten Zahl.

SECHSTE FIGUR:
OFFENBARUNG JOHANNIS AM 15. KAPITEL

Und ich sah ein anderes Zeichen am Himmel das war groß und wundersam, 7 Engel, die hatten die letzten 7 Plagen, denn mit demselben ist vollendet der Zorn GOTTES und ich sah ein gläsernes Meer mit Feuer gemengt und die den Sieg behalten hatten an dem Tier und seinem Bilde und ein Malzeichen seiner Namens-Zahl, dass sie stunden an dem gläsernen Meer und hatten GOTTES Harfen und sungen das Lied Moses des Knechts GOTTES und das Lied des Lammes und sprachen: Groß und wundersam sind deine Werke, HERR Allmächtiger GOTT, gerecht und wahrhaftig sind deine Wege, du König der Heiligen, wer soll dich nicht fürchten HERR und deinen Namen preisen.

nourished there" means that she shall be dissolved with the flying, crowned eagle, and that she becomes volatile. The red dragon casts out water as a flood after her to drown her, but the earth helps the woman and dries up the stream. She dries up again, her solar fruit now has to fight with these dragons, which spatters her with blood on all parts of her body; but because she is of godly or angelic nature the blood of the dragon does not harm her; but through this quarrel they heat each other up and her fire is kindled and she is increasingly transformed into ever-brighter and more brilliant stars—that is to say, into greater and more glorious tinctures—by the crowned seven-headed dragon, who was stimulated by this frequent strife until it was slain; and by this the victory is obtained. The dragon with the seven heads and crowns is the bloody tincture and crowned red lion, the fire of the stone or the blazing red fire. By sevenfold red impregnation these kings take their crowns; the ten horns signifies the conclusion in the tenth number.

SIXTH FIGURE:
REVELATION OF ST. JOHN IN THE FIFTEENTH CHAPTER

And I saw another sign in heaven, great and marvelous, seven angels having the seven last plagues; for in them is filled up the wrath of GOD, and I saw a sea of glass mingled with fire: and those that had gained the victory over the beast, and over his image, and over his mark, and over the number of his name, stand on the sea of glass, having the harps of GOD, and they sing the song of Moses the servant of GOD, and the song of the lamb, saying: great and marvelous are your works, LORD GOD almighty; just and true are your ways, you king of saints; who shall not fear you, O LORD, and glorify your name?[170]

GEISTLICHE AUSLEGUNG DIESER SECHSTEN FIGUR

Erstlich die Ausrüstung der 7 Engel, wer sie berufen. GOTT der hat aufgetan die Hütten des Zeugnis GOTTES, da geweissagt worden von der herrlichen Offenbarung des *Evangelii*, in den letzten Zeiten, zu welcher GOTT seine Diener abgefertigt, so wider den Anti-Christ predigen müssen, was GOTT für Mittel dazu gebraucht. Eins der 4 Tiere gab den 7 Engeln Schalen voll Zorn GOTTES, &c. Das ist der Hausstand, als da GOTT mehrenteils seine Kirche hat, welche den Anti-Christ sehen, sonderlich unter dem rechtgläubigen Häuflein und GOTT darum anrufen, dass er das Kind des Verderbens soll offenbar machen, durch die Predigt der Wahrheit, deren Kraft und Nachdruck die Herzen der Menschen erfüllen, dass sie voll von Rauch und göttlicher Erkenntnis, dass kein Welt-Geist noch Geist aus der Hölle darein gehen und in diesem Tempel herrschen kann. Das gläserne Meer mit Feuer gemengt, &c. In der schnöden Welt da sich das Feuer der Trübsal immer mit einmischt und wir in lauter Zerbrechlichkeit unserer 7 Quell-Geister stehen, wenn sich einer entzündet, so sind sie alle im brennen und wir haben genug zu löschen, dass sich solche nicht gar anzünden und uns zu Höllen-Bränden machen, wie beim Lucifer. Die gläubigen Kinder GOTTES sind verwundert über GOTTES großer Allmacht, der sie durch so viel Kreuz und Unglück der Welt, der Hölle, des Todes und des Teufels Reich erlöst, welches nun alles in durchscheinender Figur ewig vor ihm steht und sie augenblicklich sehen, wie wunderbar sie GOTT erhalten, beschützt und heraus gerissen, so brachen sie vor Freuden aus und sungen das Lied Moses und das Lied des Lammes, sie loben GOTT in Ewigkeit zu Ewigkeit.

AUSLEGUNG DIESER SECHSTEN FIGUR AUF DAS NATÜRLICHE GROSSE WERK

Dies Zeichen ist groß und wundersam, &c. der Smaragd oder grüne ♌, das gläserne Meer, &c. die weiße Tinctur ist mit Feuer gemengt, &c.

SPIRITUAL INTERPRETATION OF THIS SIXTH FIGURE

First to be described here is the equipment of the seven angels, as well as who called them. GOD has opened the tabernacle of GOD's testimony, since it was prophesied by the glorious revelation of the gospels in the last times, when GOD dispatched his servants to preach against the Antichrist, what GOD needed as a means to achieve this. "One of the four beasts gave the seven angels vials full of the wrath of GOD etc." refers to the family estate, where GOD mostly has his church, and which sees the Antichrist (especially among the orthodox few) and calls on GOD, praying that he should make the child of perdition manifest through the preaching of the truth. The strength and emphasis of this fills the hearts of humans so that they are filled with smoke and divine knowledge, with the result that no worldly spirit or hell spirit can enter and reign in this temple. "The sea of glass mingled with fire, etc." In the despicable world, in which the fire of affliction is always mixed, we stand in the sheer fragility of our seven source-spirits; and if one is ignited, they all burn; and we have enough [fires] to extinguish to stop them from igniting [further] and causing us to burn in the flames of hell with Lucifer. The faithful children of GOD are amazed at GOD's great omnipotence, which redeems them through so much crucifixion and worldly misfortune, through hell, death, and the devil's kingdom, which now all stands before him in transparent form forever; and they see immediately how wonderful they were preserved by GOD, who has protected and saved them, so they errupted in joy and sang the song of Moses and the song of the lamb, praising GOD forever and ever.

INTERPRETATION OF THIS SIXTH FIGURE
ON THE NATURAL GREAT WORK

"This sign is great and marvelous etc." refers to the emerald or green ♌, while "the sea of glass etc." refers to the white tincture; "is mingled

die Schalen des Zorns haben die 7 Engel ausgegossen, die dreimal siebenfachen magischen Zahlen sind erfüllt durch Wasser, Luft und Feuer, dadurch es geplagt und schwere Pein ausgestanden. Eins der 4 Tiere gab den Engeln 7 Schalen voll Zorn GOTTES, die letzten Plagen. Eines der 4 Tiere ist eines von den Elementen, als das Feuer, wodurch es geplagt und gebrannt wird, wenn es seinem eigenen Zorn und Gift trinken muss, durch diese letzte wird es ganz feurig rot, als eine glühende Kohle, der Streit hat sich nun gelegt, alle Wunder der Kunst und der Natur haben ihr Ende gefunden, das Oberste ist dem Unteren gleich, das Untere ist dem Oberen gleich. Der Anfang hat das Ende gefunden und das Ende den Anfang. Die vielen Geister und Schiedligkeiten so von ihm ausgegangen, sind alle wieder in eines gewendet, aus dem Tod und Gift ist die herrliche Arznei worden und ist nun selber der Arzt Raphael. Es sind darinnen die allerklärsten Wasser der allerlautersten Erde, die teure weiße Perle, so in des Phöbi Perlen-Trank resolviert worden, es ist darinnen der Strom des lebendigen Wassers, klar wie ein Kristall, so aus dem Heiligtum geflossen, es sind darinnen alle 7 Quell-Geister und unsere 7 Metalle, welche alle in Gold verwandelt, in ihr Wesen eingegangen, es sind darinnen die 7 Geister, welche sind die Geister und Kräfte der Planeten, es ist darinnen die Quintessenz der irdischen, paradiesischen und engelhaften Welt, weil es sich aus einer in die andere durch neue Gestalten versetzen lassen und ist das rechte Gewächs der 3 Principien, so sich mit allen 3 Welten vergleichen lässt und es ist selber diese drei. Es hat die Kräfte der engelhaften Welt, welches lauter reine, trockene Feuer-Geister sind, es hat die Kräfte der paradiesischen Licht-Welt, weil kein Tod, Krankheit und Gebrechlichkeit darinnen, sondern lauter Licht und Klarheit, welche aus unserer kleinen Welt aus dem *Confuso Chao* geschaffen, welche ihren Anfang von GOTT hat. Johannes kann es nicht genug beschreiben, sondern sagt: Ich sah ein Zeichen das war groß und wundersam. Es kann endlich diese Figur noch dieses Geheimnis haben, wenn nach so vielen Ewigkeiten und Zeiten durch den Sohn GOTTES alles wird wiedergebracht werden und das große Jubel-Jahr, das Wunder aller Wunder vergehen

with fire etc." refers to the vials full of wrath poured out by the seven angels, and the thrice sevenfold magic numbers are filled with water, air, and fire, causing it to be afflicted and endure severe torment. One of the four beasts gave unto the seven angels seven vials full of the wrath of GOD, the last plagues. One of the four beasts is one of the elements, being the fire by which it is plagued and burned when it has to drink its own anger and poison; and by this it becomes all fiery red like a glowing coal. The struggle has now settled, all wonders of art and nature have come to an end. The upper is equal to the lower, the lower is equal to the upper. The beginning has found the end and the end has found the beginning. The many spirits and divisivenesses that emanated from it have all become one again, death and poison have become the marvellous medicine, and it is now the physician Raphael himself. Within it are the clearest waters of the purest earths; the precious white pearl having been dissolved in Phœbi's pearl potion. Within it is the stream of living water, flowing out of the sanctuary, clear as a crystal; and all seven source-spirits are in it, as are our seven metals, which are all transmuted into gold. The seven spirits are in there, have entered into its being, and these are the spirits and forces of the planets; and there is the quintessence of the earthly, paradisiacal, and angelic world, for it can be transferred from one [world] to the other through new forms, and is the proper growth of the three principles, so it allows itself to be compared with all three worlds, and it is itself these three [worlds]. It has the powers of the angelic world, which are nothing but pure, dry, fire spirits; it has the powers of the paradisiacal world of light, because there is no death, sickness, or frailty in it, but only light and clarity, which come from our microcosm created by the *confuso chao*, which in turn has its beginning from GOD. St. John cannot describe it well enough, but says: "I saw a sign that was great and marvelous". Finally, this figure can still have this secret when, after so many eternities and times, everything will be brought back through the son of GOD, and the great year of jubilee, the miracle of miracles, will arrive. Then everything will be reunited with GOD; angels and

wird, da alles wieder mit GOTT vereinigt, Engel und Menschen wieder zur Ruhe in GOTT werden eingegangen sein, die sich von GOTT in so viele Schiedligkeiten ausgeteilt und nirgends Ruhe gefunden, bis sie wieder in ihr *Centrum* eingekehrt und GOTT alles in allem ist. Welches wohl in diesen Werk als in einem Spiegel der 3 Welten zu sehen. Der große GOTT, Schöpfer und Erhalter Himmels und der Erde, gebe uns, dass wir in seinem Licht als in dem erleuchteten Gnaden-Schein des Heiligen Geistes, ihn als das wahre vollkommene Licht sehen und hier als Kinder des Lichts im Licht wandeln mögen, damit wir dort mit seinem Licht ewig vereinigt sein und bleiben mögen! Das helfe uns GOTT, Vater, Sohn und Heiliger Geist vollbringen, hochgelobt und gepreist in alle Ewigkeit!
Amen.

humans will be entered into GOD again to rest; that which has been dealt so many divisivenesses by GOD has found no rest anywhere until it has returned to its *centrum* and GOD [where] all is in everything. It can be seen in this work as in a mirror of the three worlds. May the great GOD, creator and preserver of heaven and earth, grant us that we may see him in his light as in the enlightened glow of grace from the holy spirit, see him as the true perfect light so that we may walk here as children of light in the light, and so that we may be and remain
united there with his light forever! Help us to accomplish
this, O GOD, father, son, and holy spirit,
highly praised and glorified
for all eternity!
Amen.

Key to the Cabinet of Nature's Secret Treasure Chamber

SCHLÜSSEL ZU DEM CABINET DER

Geheimen Schatz-Kammer der Natur

ZUR SUCHE UND FINDUNG DES
STEINS DER WEISEN

Durch Fragen und Antwort gestellt

VERFERTIGT UND DER WELT GEZEIGT
DURCH

D. I. W.

VON WEIMAR AUS THÜRINGEN

LEIPZIG
Verlegt von Johann Heinrichs Witwe
1706

KEY TO THE CABINET OF

Nature's Secret Treasure Chamber

FOR SEEKING AND FINDING
THE STONE OF THE WISE

Presented Through a Series of Questions and Answers

PREPARED AND REVEALED TO THE WORLD
BY

D.I.W.

FROM WEIMAR IN THÜRINGIA

LEIPZIG
Published by Johann Heinich's Widow
1706

VORREDE

BELIEBTER UND NACH STANDESGEBÜHR GEEHRTER LESER. Unter allen Autoren, so jemals von der geheimen chymischen Wissenschaft geschrieben, hat sich, ohne Ruhm zu melden, keiner so deutlich, klar und hell, weder in der Vor- noch Nach-Arbeit, heraus gelassen, als ich in meinen beiden Traktätlein, dem *Mineralischen Gluten* und dem *Philosophischen Perl-Baum*, dem *Gewächs der 3 Principien*, und dann hier in diesem Traktat, darinnen der Welt die Geheimnisse derselben ganz bloß und entdeckt vorgestellt, welches doch alle *Philosophi* zugedeckt und nur bisweilen, als wie schlummernd und schlafend, ein Wort fahren lassen und sich gleichsam belustigt, der Nach-Welt solche süßen Träume zu erzählen, wodurch es niemand erraten können. Daher viele unglückselige Arbeiter ihr Geld, Mühe und Zeit vergebens angewendet und ins Unglück geraten. Weilen ich nun gesehen, dass unser Leben von Jugend an bis ins Alter eine beschwerliche Reise, da wir, so zu reden, auf dem Schiff vom ungestümen Meere und widrigen Winden zu dem tiefsten und finstersten Abgrund des Todes hinunter gestoßen werden, indem wir von Kindheit auf, innerliche Feinde, nämlich ganze Haufen beschwerlicher Krankheiten, welche nicht zu zählen, von außen aber ein erschreckliches Heer vieler Ungelegenheit und Unfälle, deren Heerführerin die Armut ist, haben. Wider diese beiden Widersacher nun, welche sich gleichsam wider unser Glück verschworen und einen heimlichen Schluss wider uns gemacht, ist bisher nichts dafür gefunden worden, als die Geduld: Jedoch wider-

PREFACE

DEAR READER, HONOURED ACCORDING TO YOUR STATUS, if I may say so without bragging then I do maintain that of all the authors who have ever written about the secret chymical sciences, none have expressed themselves so explicitly, clearly, and brightly, either in the preliminary work or in the after-work, as I have done in my two little treatises, *The Mineral Gluten* and *The Philosophical Pearl Tree and the Growth of the Three Principles*. I do so now again in this treatise, in which the secrets of the world are presented to the public in a completely open and disclosed way, secrets which all *philosophi* had covered up and who had only occasionally allowed a word out, as if they were slumbering and sleeping, and, as it were, amused to tell posterity such sweet dreams that no one could guess. Hence many unfortunate workers have expended their money, labor, and time in vain and fallen into misfortune. For now I have seen that our life from youth to old age is an arduous journey, as we are, so to speak, tossed in the ship by stormy seas and adverse winds to the deepest and darkest abyss of death, having from infancy onward internal enemies, namely entire masses of grave illnesses, which cannot be enumerated. From the outside we face a terrible army of multiple inconveniences and accidents, the chief of which is poverty. Against these two adversaries,[171] which, as it were, conspired against our happiness and made a secret agreement against us, nothing has been found so far but patience. However,

setzt sich diesem Unglück der tapfere und unermüdete Geist des Menschen, lässt sich durch so viel Unglück nicht zu Boden schlagen noch überwinden, sondern rüstet sich aus mit den herrlichen Waffen der unüberwindlichen Weisheit und durchsucht das Herz der Welt und der Natur, ob er etwas finden könne, dass diese Feinde erlegen und ihnen Abbruch tun möchte. Er bittet und ruft in diesen Nöten zu GOTT um Erleuchtung und um den Geist der Weisheit, dass er seinen Weg erkennen und seine Werke ihm angenehm sein mögen, darauf folgt er getrost dem Weg der Natur, weil GOTT darinnen sehr große Schatz-Kästen hat. Wenn er nun einig mit GOTT, so lässt ihm GOTT zu, sich seiner Schatz-Kästen zu gebrauchen, denn er vertraut ihm den Schlüssel der Gesundheit und des Reichtums. Diese Schlüssel soll er gebrauchen zu GOTTES Ehren und dem Nächsten zu Nutz und Dienst. Daher ist mir durch GOTTES Gnade vergönnt worden, meinem Neben-Christen diese Schlüssel zu reichen, damit derselbe durch Liebe und Wohltat möge zu GOTT gezogen und aus dem Angst-Meer der Krankheit und Armut gerissen werden, auch an dem sicheren Port des guten Landes anlanden möge und nicht mehr klagen dürfe: Ich kann vor Schmerzen des Leibes nicht beten, ich kann vor Arbeit, Hunger und Kummer, mich und die Meinigen zu erhalten, nicht in die Kirche gehen, habe nicht Zeit, an GOTT zu gedenken, muss wegen der Bauch-Sorgen noch gar zur Hölle fahren: Und dergleichen ungeduldige und unchristliche Reden und Gedanken mehr. Dieses alles, lieber Leser, hat mich bewogen, so offenbar und klar zu schreiben, weil mir des Nächsten Not zu Herzen gegangen, deswegen habe ich ihn in allem und jedem unterrichtet, die *Materia* genannt und des groben Körpers Solution, und der Elemente Separation gezeigt, welches sie alle verschwiegen. Ingleichen die Conjunction des Wassers und der Erde, des Geistes mit dem Leibe, des Leibes mit der Seele, Geist und Leib, die Figierung des weißen und roten ♃, auch solche zu incerieren und zum Öl zu machen vorgeschrieben, den weißen und roten ♃ zu conjugieren, solchen noch mehr zu röten und zu färben, zu illuminieren und zu multiplicieren, auch etliche Abkürzungen und wie es in der Medi-

the brave and indefatigable spirit of humanity resists this misfortune, does not let itself be knocked down nor overcome by so many disasters, but equips itself with the glorious weapons of invincible wisdom, and searches the heart of the world and of nature to see if it could find anything that might slay these enemies and to break free of them. Amidst these hardships [humanity] pleads and calls out to GOD for enlightenment and the spirit of wisdom so that it may recognize his way and that his works may be pleasant to him, after which it confidently follows the path of nature, for herein GOD has very large treasure chests. When it is in agreement with GOD, GOD allows it to use his treasure chest, for he entrusts it with the key to health and wealth. It should use these keys for the honor of GOD and for the benefit and service of its neighbor. Therefore, through GOD's grace, I have been granted the privilege of handing these keys to my fellow Christians, so that through love and kindness they may be drawn to GOD and torn from the sea of fear of illness and poverty, and that they may land at the safe port of the good country and may no longer complain: "I cannot pray because of physical pain, I cannot go to church because of work, hunger, and sorrow to support myself and my family, I do not have time to think of GOD because of my stomach's worries, and I even have to go to hell", and more such impatient and unchristian speeches and thoughts. All of this, dear reader, has moved me to write so openly and clearly, because the need of my neighbor has moved my heart. This is why I have instructed them in each and every [detail], named them the *materia*, and showed them the dissolution of the coarse body and the separation of the elements, which they all concealed. Likewise, I have given instruction on the conjunction of water and earth; of spirit with body; body with soul; and the fixation of the white and red ☿. I have also given the prescription of how to incinerate and make an oil, and how to conjoin the white and red ☿, to redden and color such even more, to

zin gebraucht werden solle, dargestellt. Dieses, lieber Leser, ist noch nie so aufrichtig und alles und jedes entdeckt worden, als in diesem meinen dritten Traktätlein zu finden, sonnenklar mit allen Umständen gewiesen. Ich hätte zwar wohl auch schweigen können, denn die Wissenschaft hätte mir das Herz nicht eingedrückt, in Maßen auch weder Ehre noch Lob dadurch gesucht, sonst würde meinen Namen ausgeschrieben haben, wenn ich damit hätte vor der Welt bekannt sein wollen: Allein es ist geschehen aus mitleidender und erbarmender Liebe gegen meinen Nächsten, weil das menschliche Leben kurz, voller Mühe, Jammer und Not ist, damit es durch diesen Zucker sollte versüßt werden. Und wird GOTT seine Arbeiter schon dazu ausersehen haben. Die unwürdigen aber gleichwohl in Finsternis tappen und ihnen das helle Licht, so ich in meinem dritten Traktätlein angezündet, nicht sehen lassen, daher den glückseligen Arbeiter dieser Dinge bitte, dass er solches GOTT zu Ehren und dem armen Nächsten zum Besten anwenden wolle. Er mache sich nicht durch Kirchen-Aufbauen einen großen Namen und suche keine Ehre, sondern gebe den Privat-Armen, schicke Witwen und Waisen Geld in die Häuser, versorge die Kranken, richte Armen-Apotheken auf und versorge die Hospitäler, ohne dass er von den Leuten Dank dafür einnimmt, dadurch wird er diesen Schatz, so er an die Armen wendet, im Himmel 1000-fältig wiederfinden und von CHRISTO an jenem Tage dies Lob hören: Ei, du frommer und getreuer Knecht, du bist über wenige getreu gewesen, ich will dich über viele setzen, gehe ein zu deines HERRN Freude. Dem fleißigen Leser und Arbeiter wünsche von GOTT Glück, Segen und Gedeihen dazu! Ich verbleibe dem GOTT-liebenden Leser weiter christlich zu dienen verpflichtet.

GEGEBEN WEIMAR, DEN 13. JULI 1705

illuminate and to multiply; in addition, quite a few shortcuts, and how they should be used in medicine, are shown. This, dear reader, has never been disclosed so honestly in each and every detail as found in this, my third little treatise, as clear as day with all the instances. I might as well have remained silent, because science would not have impressed my heart, nor would it have sought praise for every honor, otherwise I would have written out my name if I had wanted to be known to the world: it happened out of compassionate and merciful love towards my neighbor, because human life is short, full of trouble, misery, and distress, so it should be sweetened with this sugar. And GOD will already have chosen his workers for this. The unworthy, however, still grope in darkness; and I do not let them see the bright light that I ignited in my third little treatise, so I bid the blessed ones who work on these things to use such things to honor GOD and to use them to benefit their poor neighbors. They shall not make a name for themselves by building churches and shall not seek honor, but give privately to the poor, send money to the houses of widows and orphans, care for the sick, set up apothecaries and support hospitals for the poor, without receiving thanks from the people. Thereby they will find this treasure that they gave to the poor again in heaven a thousandfold, and they will hear this praise from Christ on that day: O, you pious and faithful servant, you have been faithful about a few [people], I want to set you above the many, enter into to the joy of your LORD. To the diligent reader and worker, GOD wishes you good luck, blessings, and prosperity! I remain committed to provide Christian service to the GOD-loving reader.

PRESENTED IN WEIMAR, 13 JULY 1705

SCHLÜSSEL ZU DEM CABINET DER
GEHEIMEN SCHATZ-KAMMER DER NATUR

1. *Erste Frage. Was ist die Schatz-Kammer der Natur?*

Antwort: Es ist die ganze Natur selber. Als 1. nach der feurigen finsteren Welt. 2. Nach der wässrigen elementarischen Welt. 3. Nach der astralischen Luft-Welt, worin die Sterne und Planeten. 4. Nach unserer Welt, welches die sterbliche vergängliche Welt genannt wird, als wie es jetzt in der Scheidung steht, da es GOTT in die 4 *Elementa* zerteilt, als in Luft, Wasser, Feuer und Erde. Das Feuer hat den höchsten Ort inne, danach das Wasser, in der Luft ist das Licht die Sterne und diese *Meteora* wirken auf die Erde, welche als ein Magnet die Einflüsse aus der Luft an sich zieht. Die ganze Natur aber oder alle ihre Kräfte und Wirkungen zu beschreiben, ist keinem Menschen möglich, wenn er auch 1000 Jahre lebte, so würde er es doch nicht alles erforschen noch erfahren. Denn in dem Feuer-Himmel oder der feurigen finstern Welt über den Sternen ist kein Mensch gewesen, obschon dieses Chaos sich oft eröffnet und Feuer vom Himmel gefallen, als bei der Umkehrung Sodom und Gomorra, ingleichen da der Prophet ließ das Feuer GOTTES vom Himmel fallen, welches die Haupt-Leute mit ihren 50 Männern musste verzehren. Auch noch zum öftern mal Feuer vom Himmel gefallen zu unseren Zeiten. In der wässrigen elementarischen Welt sind wir auch nicht gewesen. Viele halten solches für die sogenannte Milch-Straße am Himmel, darinnen so viele Millionen Sterne sind, welches ich aber nicht glaube, sondern vielmehr dieses, dass es der klare lautere Himmel von dem allerlautersten geseelten Luft-Wasser ist, welches der feu-

KEY TO THE CABINET OF
NATURE'S SECRET TREASURE CHAMBER

1. *First Question: What is nature's treasure-chamber?*

Answer: it is the whole of nature herself. This includes: (1) The fiery dark world. (2) The watery elemental world. (3) The astral world of air, in which the stars and planets reside. (4) Our world, which is called the mortal, perishable world, as it is now in separation, since G OD divides it into the four *elementa*, that is, air, water, fire, and earth. Fire occupies the highest place, then water, in air is the light of the stars and these *meteora* act upon the earth, which draws the influences from the air to itself like a magnet. However, it is not possible for any human being to describe the whole of nature or all its powers and effects, and even if they lived for a thousand years, they would still not explore or experience everything. For there was no human in the fiery heaven or the fiery dark world above the stars, although this chaos often opens up and fire falls from heaven, as in the overthrow of Sodom and Gomorrah,[172] just as the prophet let the fire of G OD fall from heaven, which consumed the captain and his fifty men.[173] Fire has also often fallen from heaven in our times. We have also not been to the watery elemental world. Many consider this to be the so-called Milky Way in the sky, in which there are so many millions of stars, but I do not believe this; rather, I believe that it is the clear, pure sky of the very purest, ensouled air-water, which stands against the wall, so to speak,

rigen *Sphæræ* so lange zur Mauer, so zu reden, steht, dass sich solches nicht mit dem Licht der Sterne vermischen kann. In der astralischen Luft-Welt ist auch niemand gewesen, jedoch weil die Menschen solche mit Augen gesehen, hat ihr Geist nicht nachgelassen zu forschen der Sterne Lauf, ihre Wirkung, auch ihre Größe zu messen, welches auch, was ihre Wirkung und Lauf anbelangt, ziemlich zutrifft, aber die Größe solcher Körper zu messen dünkt mich, wird nicht zutreffen. Woraus aber solche Körper bestehen, hat noch bis dato kein Mensch erforschen können, als dass man sagt, es sind Lichter, welche leuchten. Dass sie aber brennen sollen, ist nicht also, sie geben ja keine Flamme und Rauch, wiewohl die Sonne durch und durch Feuer ist, hitzt, erwärmt, auch erleuchtet alles und ist der Geist der großen Welt, doch ist solches kein verbrennliches Feuer, sondern das Feuer und Licht, welches vorher den ganzen Himmel erleuchtete, klar und hell machte, weil es sehr dünn auseinander gebreitet, jetzt aber in die Enge zusammengetrieben und gleichsam konzentriert, darum brennt es mit Schmerzen. Wenn aber nach der Erneuerung der Erde die Sonne mit allen Sternen wieder in ihren *Æther* gegangen, wird aus diesem Licht der neue Himmel in seiner Zerteilung und Dünnmachung den neuen Himmel und neue Erde durchdringen, mit dem reinen geseelten Licht, Wasser, Feuer und Erde sich vermischen und der Himmel und neue Erde sein, dass wir der vorigen nicht mehr gedenken werden. Was aber die Natur unserer Erde anbelangt, haben die Menschen etwas weniger bessere Erkenntnis und besteht die Kraft der ganzen Natur aus den 4 Elementen, als welche 3 Reiche in sich erhält, das animalische, vegetabilische und mineralische. Diese 3 Reiche der Natur, nach ihren Kräften, Wirkungen, deren vielen Arten, Geruch, Geschmack, Leben, Regungen und Bewegungen, Zeugungen, Fortpflanzungen und Untergehung zu beschreiben, ist gleichfalls unmöglich, obwohl Salomo, welcher der Weiseste gewesen, von allen geschrieben, bis auf den Isop, der aus der Wand wächst, wo sind aber seine Bücher? Sind nicht seine meisten und besten Schriften verbrannt worden nach seinem Tode durch eines Propheten Geheiß der abgöttlichen Juden wegen? Weil er

of the fiery *sphæræ* for so long that it cannot mix with the light of the stars. Nobody has been in the astral world of air either, but because people have seen this [world] with their eyes, their spirit has not stopped researching the course of the stars and their effects; nor have they ceased to also measure their size since it is quite true that this is relevant to their effect and course. Yet measuring the size of such bodies, it seems to me, will not be possible. To the present day nobody has been able to find out what such bodies are made of, other than saying that they are lights that shine. But it is not the case that they should burn, they do not emit flames and smoke, although the sun is fire through and through, and it heats, warms, and illuminates everything. It is the spirit of the macrocosm, but it is not a combustible fire as such, but rather the fire and light which formerly illuminated the whole sky and made [the sky] clear and bright because it was spread very finely. Now, however, it has been driven into a corner and concentrated, as it were, therefore it burns with pain. Yet after the renewal of the earth, when the sun, together with all the stars, has gone back into its *æther*, this light becomes the new heaven through its dissipation and thinning. The new heaven and the new earth will be permeated by pure, ensouled light, and water, fire, and earth will mingle and become the heaven and the new earth, so that we shall no longer remember the former earth. But as far as the nature of our earth is concerned, humans have slightly better knowledge. The power of the whole of nature consists of the four elements, which contain three kingdoms, the animal, vegetabilic, and mineral. To describe these three kingdoms of nature according to their forces, effects, their many species, their smell, taste, life, impulses, movements, procreations, reproductions, and destructions is also impossible, although Solomon, who was the wisest, wrote of everything, even about the hyssop that grows out of the wall—but where are his books? Were not most of his writings, including his best works, burned after his death by a prophet's command because of the idolatrous Jews? Because he, who could see into everyone's heart,

die ganze *Magiam* und *Cabbalam* beschrieben, der allen ins Herz sehen konnte. *Item*, seine Kraft und Natur erkennen, dass er auch dem Hohen-Priester eine sonderliche Wurzel in einen Ring gemacht, welche nach ihrer rechten Influenz gegraben war, wenn der Hohe-Priester diese Hand, daran er den Ring hatte, nur einem Besessenen vor den Mund hielt, musste der Teufel alsobald ausfahren. Weil aber heutigen Tages wenig solche hoch-erleuchtete Menschen anzutreffen, so wird auch wenig oder nichts rechtes geschrieben, sondern nur schlechte und gemeine Wissenschaften, die nur den Bauch voll machen, kriechen nur um den Rand und äußersten Teil der Dinge herum, ist nur der Schulen Wissenschaft, wandelt in steter Ungewissheit, Finsternisse, zweifelhafter Dinge und tappt mit dem Stab der Mutmaßung herum auf dem Irrtums-Weg, weil fast niemand den rechten Weg der Natur kennt, absonderlich in der Medizin, da die Natur und derselben Kräfte sollen erforscht werden. Darum haben die jetzigen *Medicamenta* eine solche schlechte Kraft wider die harten und fixen Krankheiten, gegen die spagyrischen, welche die Schulen nicht lehren, weilen derselben Fundament nicht im *Centro* der Dinge, sondern nur in äußerlichen Teilen besteht, die wahre *Philosophia*, welche auch die *Gymnosophia* der Inder ist, die *Magia* der Ägypter und der Juden *Cabbala*, dringt durch bis zum Innersten hinein und lässt nichts unexaminiert, welches sie nicht vollkömmlich erforscht und desselben Kraft und Wirkung ans Licht bringt und zu dem rechten Nutzen anwendet, auch die Kraft der Natur darlegt und solches durch die Natur erweist und die verhüllte Natur entblößte und entkleidet, dass man sie erkennen kann, die doch eben so feste nicht versteckt ist, auch sich willig und gern finden lässt von denen, so ihr auf dem rechten Wege der Natur nachgehen.

2. *Was ist denn die geheime Schatz-Kammer der Natur?*

Antwort: Es ist die entblößte und entdeckte Natur in ihrer *Anatomi*, da man das innerste der Dinge, der Natur ihre Wirkung und Kraft erkennt und ist nicht genug, dass ich sage, das Feuer oder die Sonne brennt, das Wasser oder der Mond kühlt, die Luft und der Wind macht

described the whole *magiam* and *cabbalam*. Likewise, his power and nature can be seen from the fact that he also placed a special root in a ring for the high priest; the root was dug according to its proper influence, and if the high priest simply held the hand on which he wore the ring in front of the mouth of a possessed person, the devil had to leave immediately. But because few such highly enlightened people are to be found today, little to no good is actually written, but only lousy, common science, which merely fills the belly. It scuttles around the edges and outermost parts of things; it is only the science of the schools, wandering in constant uncertainty in the darkness of doubtful things, groping around with the staff of conjecture on the path of error, because almost nobody knows the correct path of nature, particularly in medicine, where nature and her powers are to be explored. This is why the current *medicamenta* have such little power against the hard, intractable illnesses compared to the spagiric ones, which the schools do not teach because the foundation of these same schools is not in the *centro* of things, but only in external parts. The true *philosophia*, which is also the *gymnosophia* of the Indians, the *magia* of the Egyptians, and the *cabbala* of the Jews, penetrates to the core, leaving nothing unexamined, fully exploring and bringing to light its power and effect and employing it for the right benefit. This expounds the power of nature and demonstrates it through nature, undressing and exposing the veiled [aspect of] nature so that one can recognize it; for it is not too tightly concealed, but willingly and gladly allows itself be found by those who follow on the right path of nature.

2. *What is nature's secret treasure-chamber?*

Answer: It is nature stripped bare and uncovered in its *anatomi*, where one recognizes the innermost aspect of things, nature's effect and power. It is not enough to say that fire or the sun burns, that water or the moon cools, that air and the wind dessicates, or that earth or the salt draws

trocken, die Erde oder das Salz zieht alles dieses in sich, sondern man muss ganz andere Erfahrung haben und aller Dinge, *Agens* und *Patiens*, untersuchen, denn durch deren Leiden und Wirken, wachsen und entstehen immer andere und andere Geburten, Kräfte und Wirkungen, in allen 3 Reichen, da muss im *Centro* der Geruch, Geschmack, Wirkung und Tugend empfunden werden; *Item*, sein Frühling, Sommer, Herbst und Winter erforscht, seine Geburt und Tod, aufsteigen und absteigen, leben, regen und bewegen. *Item*, seine rechte Zeit und Stunde, wenn es in seiner Exaltation, ob man gleiches mit gleichen oder *Contraria* anwendet durch *Contraria*, wird getötet und bekommt doch oft dadurch ein neues Leben, durch gleich und gleich, wird oftmals erhalten, wenn das erhaltende Teil stärker wird, als das so soll erhalten werden, oft erhält ein Leib den Geist, oft der Geist den Leib, ein Geist ruht auch wohl in einem Leib, aber nicht mit Bestand, wenn ihm das *Medium conjungendi* mangelt, als wie jetzt die ganze Natur und die ganze Welt steht, in drehenden Angst-Rade, bis der *Mediator* wieder ins *Centrum* eingeht, alsdann wird der Geist in den verklärten Leibe ewig wohnen, in der ganzen Natur und Creatur.

3. *Wer hat nun den Schlüssel der geheimen Schatz-Kammer der Natur?*

Antwort: Es hat ihn GOTT, welchen er gibt wem er will, und gibt ihn allein, die ihn darum bitten, es ist sein Geist und Erkenntnis und des Verstandes, des Rats und der Stärke, GOTT gibt Weisheit wem er will und erfüllt alle seine Heiligen mit Erkenntnis und Freuden, den Gottlosen aber entzieht er es. Es sind auch große Erkenntnis nicht allen Menschen nütze, weil viele solche große Gaben würden missbrauchen, so gibt und vertraut GOTT solche geheimen Schlüssel nur seinen Lieben und Getreuen, die er getreu erfunden hat, die solche großen Güter nicht verschwenden, übel anlegen oder solche Kleinodien aus der Schatz-Kammer GOTTES nicht liederlich verstreuen und den Säuen vorwerfen, wodurch GOTT mehr geschmäht und verunehrt würde, durch Missbrauch und übler Anwendung, da man würde die Natur oder derselben Kräfte höher als GOTT selber halten oder solche Kräf-

all of this into itself. One must have a completely different experience and examine the *agens* and *patiens* of all things, because through their sufferings and activities there arise other, ever-diverse births, powers, and effects in all three realms, where the smell, taste, effect, and virtue must all be felt in the *centro*. Moreover, [one must]explore its spring, summer, autumn, and winter, its birth and death, rise and fall, [how it] lives, stirs, and moves. In addition, [one must know] its proper time and hour, when it is in its exaltation, whether to use like with like or *contraria* with *contraria*; [how that which] is killed often gains new life from it, through like upon like, yet is often obtained when the sustaining part is stronger than the part that should be sustained. Often a body sustains the spirit, often the spirit [sustains] the body; indeed a spirit also rests in a body, but not with permanence if it lacks the *medium conjungendi*. As now the whole of nature and the entire world stands in the spinning wheels of fear until the *mediator* enters the *centrum* again, at which point the spirit will dwell in the transfigured body forever, in the whole of nature and creature.

3. Now who has the key to nature's secret treasure-chamber?

Answer: GOD has it, and he gives to whomever he wants. He only gives it to those who ask him for it, granting knowledge and understanding, advice and strength, through his spirit. GOD gives wisdom to whomever he wants and fills all his saints and friends with knowledge, but he withdraws it from the wicked. Even great knowledge is not useful to all people, because many would misuse such great gifts, and therefore GOD only gives and trusts such secret keys to his beloved ones whom he has found faithful, and who do not waste such great goods or use them badly, or scatter such jewels sloppily out of GOD's treasure-chamber and throw them before swine, whereby GOD would be more despised and dishonored through misuse and ill-application. For in this case one would hold nature or the same powers higher than GOD himself, or else man would despise such powers which contained

te, worin doch selbst die Kraft GOTTES ist, verlästern und sagen, es ist Teufels Werk, Hexerei, es geht nicht von rechten Dingen zu, wie es leider! heutigen Tages geht, da alles was GOTT gutes dem Menschen gegeben, dem Teufel zugelegt wird und unsern HERRN GOTT aller Ehre beraubt, als wenn er ganz ohnmächtig wäre oder die Natur und Creatur ohne alle Kräfte geschaffen, nur der Teufel muss noch alle Künste und Wissenschaften wissen, da er doch an der Laus ist zu schanden worden mit allen seinen Zaubern, dass sie haben sagen müssen: Das ist GOTTES Finger.

4. Weil unter so vielen 1000 kaum einer zu finden, der so glückselig, von GOTT so erleuchtet wird, dass er volle Erkenntnis in GOTT hat und dann dass er Natur und Creatur ihre Geheimnisse erforschen; Wie soll es nun ein anderer machen, der eben so tief nicht gründen will, sondern sich begnügen lässt, nur das mineralische Reich zu erkennen und desselben Kräfte und Nutzen?

Antwort: Er soll lieber das ganze als das halbe Teil oder nur ein Stück erwählen und nicht ablassen mit Ansuchen und Klopfen, obschon GOTT nicht gleich hört, es liegt nur an uns, dass wir nicht flugs erhört und mit seinem Licht erfüllt werden, weil wir noch so ungeschickt dazu sind, auch überschüttet uns GOTT nicht auf einmal mit seinem Segen, weil unsere Gefäße zu klein und diesen großen Überfluss des Segens dieser Wissenschaft nicht fassen können, sondern nur stückweise, das wir es erfassen und halten können, auch wenn uns GOTT in einem das Verständnis öffnet, dass wir erstlich uns lernen erkennen und hernach die Natur und natürliche Wissenschaft.

5. Man sagt nun, wer von gutem Verstande ist, soll sich befleißigen GOTT allein zu dienen und dann das hermetische Gut suchen, was ist denn das hermetische Gut?

Antwort: Wer der dreimal große Hermes Trismegistus gewesen, weiß ich nicht. Viele halten ihn für Noah, welcher die *Chymiam* in smaragdenen Tafeln beschrieben, weil er gewusst, dass die Welt in der Sintflut

GOD's strength itself, blaspheming them by claiming that they are all the devil's work, or witchcraft, or that things aren't going right. For this is unfortunately the way of the world nowadays, when everything good that GOD has given to humans is ascribed to the devil, who robs our LORD GOD of all honor, as if he were completely powerless or as if he created nature and creatures without any powers, and as though the all arts and sciences originate with the devil, although he was disgraced by the lice despite all his magic, so that they had to say: "This is the finger of GOD".[174]

4. *Because among so many thousands there is hardly one who is so blissful, so enlightened by GOD, that they have full knowledge of GOD as they investigate the mysteries of nature and creatures; how should someone else, who does not want to dig so deep, but is content with simply recognizing the mineral kingdom and its powers and uses, manage to accomplish this?*

Answer: They should rather choose the whole than half a piece or merely a single piece and not desist with requests and knocking. Even if GOD does not hear it right away, it is our own fault if we are not heard immediately and are not filled with his light, for in this case it is because we are as yet too unskilled for it. GOD does not suddenly shower us with his blessings because our vessels are too small and cannot contain this great abundance of the blessings of this science. He only disperses it piece by piece so that we can grasp and hold it. Yet if GOD opens up our understanding with respect to even a single piece, we may begin to recognize ourselves and then understand nature and natural science.

5. *It is now said that whoever is of good understanding should strive to serve GOD alone and then seek the Hermetic good, but what is the Hermetic good?*

Answer: I do not know who the thrice great Hermes Trismegistus was. Many consider him to be Noah, who wrote the *chymiam* in tablets of emerald, because he knew that the world was to perish in the deluge

sollte untergehen und dennoch Nachkommen bleiben, welche diese Wissenschaften finden sollten. Etliche sagen, es wäre Noah nicht gewesen, Noah hätte solche Tafeln nach der Sintflut gefunden, es wäre ein gelehrter Ägypter gewesen. Er mag sein gewesen wer er will, er mag auch den Namen Hermes gehabt haben oder nicht, danach haben wir eben nicht viel zu fragen, wenn wir nur verstehen, was das hermetische Gut, welches auf die smaragdene Tafel beschrieben, welches ist die rechte und wahre *Chymia*. Hermes heißt sonst in der *Cabbala* die Schlange oder ☿, die smaragdene Tafel, der ☉, es ist die Schlange, das giftige ☿al Wasser salviert dem ☉, es ist die Sonne in die smaragdene Tafel gezeichnet als der Vater, die *Luna* als die Mutter, welche ihren Sohn im Leibe getragen, die Erde als das Salz, hat ihn gezeugt, dies sind die 4 *Elementa*, die 3 Äpfel, die 3 *Principia*, worin kurz die *Chymia* beschrieben, wenn einer will das wunderbare Werk machen, worin das Oberste dem Unteren gleich und die wunderbare Verkehrung der Elemente, in welcher die irrenden Planeten auf den hohen Achsen fahren, wodurch so viele Wunder an Tag kommen und wird die rote wahre *Chymia*, die hermetische Kunst, genannt.

6. Ist doch die Chymia sonst in vielen anderen Büchern auch beschrieben?

Antwort: Sie ist viel beschrieben, wird auch wohl in Schulen gelehrt, aber es ist nicht die rechte *Chymia*, sondern alles nur Meinungen und mehrenteils falsche *Principia*, die rechte geheime *Chymia* steht nicht in Büchern klar und deutlich, sondern ist verfasst in emblematischen hieroglyphischen Figuren und in Rätseln verborgen worden.

7. Was ist das Menstruum Universale?

Antwort: Dieses wird in vielen unterschiedenen Verstande genommen, als 1. Ist dieses das *Universalissimum*, das *Acidum* der Natur, das *Nitrum Aërum*, so wir alle unsichtbarerweise in uns ziehen, davon alles lebt, was Odem hat, auch alle *Vegetabilia* dadurch wachsen, weil in der Luft eine verborgene Speise und Nahrung des Lebens und billig der Geist und Odem GOTTES heißt, denn in ihm leben, weben und sind wir,

and yet there would remain descendants who would then find these sciences. Some say it wasn't Noah, but that Noah found such tablets after the flood, and that they were written by a learned Egyptian. He may have called himself by any name he pleased, and he may also have had the name Hermes or not. This question does not really concern us as long as we understand what the Hermetic good is. This good is described on the Emerald Tablet, and it is the right and true *chymia*. Hermes is otherwise called the snake or ☿ in the *cabbala*; and the Emerald Tablet the ⊕. It is the snake, the poisonous ☿al water which dissolves the ⊕. The sun is drawn on the emerald tablet as the father, and *Luna* as the mother, *Luna* who bore her son in her womb. The earth is described as the salt, begot him, These are the four *elementa*, the three apples, the three *principia*, in which the *chymia* is briefly described. If one wants to do the wonderful work, in which what is above is like that which is below, and achieve the marvelous reversal of the elements, in which the wandering planets travel on the high axes, whereby so many miracles come to light, and the true red *chymia* is called the Hermetic art.

6. *Is not Chymia also described in many other books?*

Answer: It has been described a lot, is also taught in schools, but it is not the right *chymia*, it is all just opinions and mostly false *principia*. The correct, secret *chymia* is not clearly and unambiguously described in books, but it is written in emblematic, hieroglyphic figures and hidden in riddles.

7. *What is the menstruum universale?*

Answer: This is understood in many different ways. Firstly, it is the *universalissimum*, the *acidum* of nature, the *nitrum aërum*, which we all invisibly draw into ourselves, and from which everything lives that has breath. All *vegetabilia* also grow through it, because in the air is a hidden food and nourishment of life; and it is rightly called the spirit and breath of GOD, for in him we live, weave, and have our being, which

welches recht heißt: Er fährt auf den Fittichen des Windes. Dieser *Spiritus Universalis* oder *Nitrum Aëreum* ist immer an einem Ort häufiger, als am anderen, auch der Zeit nach, als im Frühling, wenn die Sonne im Widder tritt und im Herbst in die Waage, wiewohl es zu allen Zeiten gleich ist, aber im Sommer, wenn die Sonne im höchsten Hause des Himmels steht, in den ♌ geht, wird dieses Salz nicht so empfunden, weil es durch diese zu sehr aufgesogen und ausgetrocknet, im harten Winter aber fast nicht empfunden und durch die Sonne belebt wird, wenn aber die Sonne in der *Zona temperata*, als im Frühling und Herbst, wird es am meisten gespürt und empfunden. Die stärksten Magnete aber, so dieses anziehen, sind ♀ und ♂ oder deren Küsse, *item*, eine *Magnesia vitriolata*, *Minera martis Solaris Hassiaca*, Alaun und ⊕ Kiese, eine jegliche Erde, welche Eisen-Samen bei sich hat, wird von diesem *Sale esurino* geschwängert. Wenn hernach Wasser durch solche Gänge läuft, entstehen daher die Sauer-Brunnen und gibt ganze Berge voll solchen Salzes, welches sie aus der Luft und Regen an sich gezogen und die metallischen kleinen Leiber solches an sich gehalten; leibhaft und corporalisch gemacht, denn alle metallischen Leiber wachsen aus dem Wasser, auch die Edelgesteine und ist dieses saure *Acidum* unter das Element *aquæ* prædestiniert, welche auch in Tau und Regen-Wasser anzutreffen, die anderen *Menstrua* aber sind uneigentlich *Universal* im vegetabilischen, animalischen und mineralischen Reiche.

8. *Tun denn diese wohl, welche solches Salz durch gewisse Magneten fangen, durch Blase-Bälge, Sal Tartari, in Kolben, durch Tücher, Maien-Tau und dergleichen?*

Antwort: Auf diese jetzt erzählte Weise tun sie übel, dass sie sich vergebene Mühe machen und einen Weg von 100 Meilen zu gehen vor sich nehmen, da sie es vor der Tür haben. Man nehme die Körper, darinnen es ist, als dem selbst gewachsenen ⊕ oder die corrodierte *Minera* ⊕, welche zum Salze worden, solviere sie im Regenwasser und treibe das *Acidum* davon, so habe ja solchen auf einmal in großer Menge, was brauche mich mit solcher Weitläufigkeit zu bemühen.

rightly means: "He did fly on the wings of the wind".[175] This *spiritus universalis* or *nitrum aëreum* is always more frequent in one place than in another, and it also changes frequency in respect of the time, as in spring, when the sun enters Aries; and in autumn when the sun enters Libra, although it is the same at all times, but in summer, when the sun is in the highest house of heaven, that is to say ♌, this salt is not felt like that, because it is drawn up and dried out by it too much. Yet it is hardly felt and enlivened by the sun in the hard winter. And when the sun is in the *zona temperata*, as in spring and autumn, it is felt and sensed the most. But the strongest magnets that attract this are ♀ and ♂, or their kisses. Likewise, a *magnesia vitriolata*, *minera martis solaris Hassiaca*, alum and ⊕ pyrites, and any earth that has iron seeds within it is impregnated by this *sale esurino*.[176] When water then runs through such passages, the mineral waters are formed; and there are whole mountains full of such salt, which they have drawn from the air and rain, whereupon the small metallic bodies have held it, bodily and corporeally, because all metallic bodies grow out of water, including precious stones. This sour *acidum* is predestined under the element *aquæ*, which can also be found in dew and rain water. But the other *menstrua* are not actually *universal* in the vegetabilic, animal, and mineral kingdoms.

8. *Do they do well who catch such salt by certain magnets, by bellows, sal tartari, in flasks, by cloth, May-dew, and the like?*

Answer: In the manner now related, they resent that they waste their effort and have to walk a hundred miles to partake of what is at their doorstep. Take the bodies in which it is, like the self-grown ⊕, or the corroded *minera* ⊕, which has become salt, and dissolve them in the rainwater, driving away the *acidum*, until suddenly you find that you have a large amount of it. Why do I need trouble myself with such vastness.

9. *Dieses Acidum aber ist nicht nur universal, wenn es aus dem ⊕ kommt, sondern hat schon beigeschlafen, denn es lässt ja eine metallische Erde zurück?*

Antwort: Ist denn dies ein Beischlaf, wenn man eine Jungfrau an der Hand anrührt? Es lässt ja das corrodierte Metall zurück, das Wenige, welches davon übergeht, setzt sich nach wenigen Tagen oder Wochen als ein Pulver auf den Boden. Wenn das folgen sollte, hätte es auch bei dem Leinen-Tuch geschlafen, damit der Tau gefangen worden oder durch *Sal Tartari*, bei welchem es gar verderbt und doppelt wird, weil jenes nicht seiner Natur ist, auch liegt es nur an dem Sucher, dass er bessere Magneten legt. Zum Exempel, man nehme sehr klein gepulverte *Mineram Wismuthi*, darf aber kein Cobalt sein, setze sie alle Nacht unter freien Himmel, doch muss sie früh vor der Sonnen wieder weggetan werden, darf auch nicht darauf regnen, des Tages setzt man solches in einen frischen Keller, wenn es nicht taut, besprengt man sie mit Tau, damit verfährt man bis sie zum ⊕ Pulver wird, geht aber sehr langweilig zu und tue ich besser, ich besprenge sie gleich mit diesem *Acido*, oder digeriere diese *Mineram* in diesem *Acido*, bis es sich färbt. Es wird auch dieses *Acidum* kein Metall töten, sondern vielmehr lebendig machen, seinen ☿, ☉, ♃ nicht verbrennen, sondern zum Wachstum befördern und ist solches das rote Bad und Reinigung, darinnen die äußerlichen groben *Feces* davon geschieden werden. Dieser Essig solviert nicht mit Gewalt, dringt nur allgemach durch, bis auf den Kern, zumal wenn er von seinem fixen zurück gebliebenen Salz abgezogen wird, löst auch die metallischen Körper ohne alles Getöse, ohne Gewalt auf, ist auch kein metallischer Körper, der nicht aus diesem *Acido* gewachsen und solches noch bei sich hat und sich endlich im Innersten miteinander vereinigt.

10. *So sind die ♁, ♆, Spiritus Salis und dergleichen rechte Menstrua zu Metallen, wenn man will Arzneien oder Tincturen von ihnen machen?*

Antwort: Nein, sondern sie sind nur gemeine Solvier-Wasser, die nicht metallischer Art, auch wieder von den Metallen wegzuwaschen sind

9. *But is this acidum not only universal when it comes from the ⊕, but already has intercourse, because it leaves a metallic earth behind?*

Answer: Is it really intercourse if one only touches a virgin's hand? After all, it leaves the corroded metal behind; the little that comes over from it settles on the bottom as a powder after a few days or weeks. If that were to follow, it would also have slept by the linen cloth, so that the dew would be caught; or by *sal tartari*, in which it becomes corrupted and doubled, because that is not its nature. Furthermore, it is the responsibility of the seeker alone to ensure that he lays better magnets. For example, take very small powdered *mineram bismuthi*, (though this must not be cobalt) and put it under the open sky every night, taking care to put it away again early before the sun comes up. Moreover, it must not rain on it either. During the day you put it in a cool basement, and if it does not thaw, you sprinkle it with dew. You do this until it turns into ⊕ powder, but it is very tedious and I prefer to sprinkle it with this *acido* right away, or digest this *mineram* in this *acido* until it becomes colored. This *acidum* will not kill any metal either, but rather make it alive; and it will not burn its ☿, ☉, ♃, but will encourage them to grow and as such it is the red bath and purification, in which the external coarse feces are separated from it. This vinegar does not dissolve [metals] with force, but only penetrates gradually to the core, especially when it is removed from its remaining fixed salt. It also dissolves the metallic bodies without any noise or violence; and there is also no metallic body which was not grown out of this *acido* and that has such [*acido*] within itself; and finally each unites with the other in its innermost being.

10. *So ♅, ♆, spiritus salis, and the like are the real menstrua for metals, if you want to make medicines or tinctures from them?*

Answer: No, they are just common dissolving waters, which are not of a metallic nature, and which can also be washed away from the metals and separated from them by evaporation. They are not even *sal tartari*

und durch die Evaporation davon zu scheiden, auch nicht *Sal Tartari* oder *Spiritus Vini*, denn die Metalle wachsen aus diesen allen nicht, man findet keinen *Spiritum Vini* oder *Sal Tartari* bei den Metallen, sondern ein *Sal esurinum*, als das *Acidum* der Natur findet man allda in Bergwerken, hilft ein Kleiben und Einträncken, solviert und coaguliert sich bei den Metallen.

11. *So ist dieser saure Geist das Menstruum Universale?*

Antwort: Es ist das *Universal-Acidum*, dadurch alles wächst, sich regt, lebt und bewegt und solchen Geist unsichtbarer Weise in sich zieht. Das aber dieses Sal das *Menstruum Universale* zum metallischen Reiche sein soll, dadurch sie zu Tincturen wieder neu geboren werden, solches ist es nicht, sondern nur der Natur ihr Essig, welcher die metallische Leiber auflöst ohne Zerstörung ihres *Humidi radicalis*, weil es deren ☿ und ♄ nicht zerstört, sondern vielmehr lebendig macht, auch wenn man das rechte metallische *Menstruum* nicht finden kann, man muss solches nehmen, aus den Metallen Arzneien zu machen, auch aus den Mineralien, weil dieser *Spiritus* schon für sich Arznei genug. Wenn er aber mit Metallen vereinigt, als worin sein *Acidum* gebrochen und getötet, dass es süß und trinkbar wird, so besitzt es ungemeine Kräfte. Dies sehen wir am Sauer-Brunnen, dass dieser *Spiritus* Eisen corrodiert, welches doch nur eine gemeine Solution, und dennoch große Kräfte besitzt, geschweige denn, wenn ein anderer edler Körper, welcher voller Kräfte der Sonne und des Mondes, auch aller Planeten ist und in diesem Straußen-Magen gereinigt wird, dass er nur seinen reinsten Teil behält, alsdann destilliert und es spiritualisch wird, ist es noch herrlicher. Wenn man aber wollte Metalle in ▽, ▽ oder *Spiritus Salis* gewaltsam lassen zerfressen, wer wollte dieses Gift und Tod trinken oder sollte daraus eine Süße kommen wie Milch und Honig? Salz bleibt Salz, wenn es schon noch so oft als ein *Spiritus* übergetrieben, wenn es lange steht und alt, wird es wieder corporalisch Salz, auch bei dem ☉, wenn man solches lange stehen lässt, so lässt es das ☉ oder Metall fallen und bleibt allein, auch der ☽, der auch wieder als ☽ an-

or *spiritus vini*, because the metals do not grow from any of these, and as such *spiritum vini* or *sal tartari* cannot be found among the metals, but rather a *sal esurinum*. Nature's *acidum* is found there in mines, and it helps to glue and impregnate, dissolving and coagulating itself with the metals.

11. So this acidic spirit is the *menstruum universale*?

Answer: It is the *universal acidum*, through which everything grows, stirs, lives, and moves and invisibly draws such a spirit into itself. But this salt is not the *menstruum universale* to the metallic kingdom, through which they are reborn into tinctures, but only nature's vinegar, which dissolves the metallic bodies without destroying their *humidi radicalis*. This is because it does not destroy their ☿ and ♄, but rather brings them to life; and even if one cannot find the right metallic *menstruum*, one must take such to make medicines from the metals and from the minerals, because this *spiritus* is already a medicine enough in itself. But when it is combined with metals, in which process its *acidum* is broken and killed so that it becomes sweet and drinkable, it possesses uncommon powers. We see such a power in the mineral waters when this *spiritus* corrodes iron, which produces only a common solution though this still has great powers. Even more uncommon powers are displayed when another noble body, which is filled with the powers of the sun and the moon and of all planets, is purified in this ostrich stomach so that it retains only its purest part. It then is distilled and becomes spiritual, and it is even more wonderful. But if one wanted to have metals eaten away by force in ▽, ▽, or *spiritus salis*, who would drink this poison and death? If any dared, would they discover that sweetness like milk and honey be produced instead? Salt remains salt, no matter how often it is distilled as a *spiritus*. And if it stands for a long time and ages, it becomes corporeal salt again. This is also the case with ☉: if you leave it for a long time, the ☉ or metal falls down and it remains alone; the same with ☽, which also shoots like the ☽ and never unites with

schießt und sich nie mit den Metallen vereinigt, sondern ihnen nur von außen anhängt, rein und unrein zugleich wegfrisst.

12. *Es sind aber doch viel rare Medicamenta aus den Metallen und Mineralien zu machen, welche durch* ▽, ▼, *Spiritus Salis, und dergleichen Corrosiven solviert werden?*

Antwort: Ja, sie sind daraus zu machen, wenn das Corrosiv rein davon geschieden und durch *Spiritum Vini* corrigiert, aber alle diese *Medicamenta* gehen nur in die erste Verdauung, daher ist ihre Wirkung auch nicht groß, wer nicht bessere *Solutiones* weiß, muss mit diesen zufrieden sein und es machen wie die Freier der Penelope und so lange mit den Mägden buhlen, bis er die rechte Braut bekommt.

13. *Ist denn das rechte Menstruum Universale zum metallischen Reiche auch schwer zu machen?*

Antwort: Den Unwissenden ist es schwer, dem Wissenden aber klar und leicht, doch erfordert es viel Arbeit und muss der kluge Arbeiter der starke Herkules sein, der alles überwindet.

14. *Es hat ja Basilius Valentinus solches Werk in 12 Schlüsseln beschrieben und Ripläus in 12 Pforten, sind es denn 12 Schlüssel und 12 Pforten, wie heißen sie und was schließen sie auf?*

Antwort: Es sind 12 Schlüssel, 12 Türen und 7 Siegel. Wer aber die erste Tür mit dem rechten Schlüssel aufschließt, der kann hernach mit eben dem Schlüssel alle die anderen Türen, immer eine nach der anderen aufschließen, nicht auf einmal, sondern nach und nach, bis er zuletzt in das Cabinet der geheimen Schatz-Kammer der Natur eingehen kann.

15. *Haben denn diese beiden Männer alle Geheimnisse von Anfang bis zum Ende beschrieben und den ganzen Schatz der Natur entblößt?*

Antwort: Sie haben beschrieben den Stein zu machen, den weißen und roten, wie die *Materia* soll durch ihre dreifache Reinigung gehen

the metals, but only attaches to them from the outside, eating away pure and impure at the same time.

12. *But there are still a lot of rare medicines to be made from metals and minerals, which are dissolved by* ⩖, ⩗, *spiritus salis,*[177] *and similar corrosives?*

Answer: Yes, they can be made out of it if the corrosive is purely separated from it and corrected by *spiritum vini*, but all these *medicamenta* only go into the first digestion, so their effect is not great either. Whoever does not know better *solutiones* must be satisfied with these, and do like Penelope's suitors and court with the maids until he gets the correct bride.

13. *Is it then also difficult to make the right menstruum universale to the metallic kingdom?*

Answer: It is difficult for the ignorant, but clear and easy for the knowledgeable, though it requires a lot of work; and the clever worker must be the strong Hercules who overcomes everything.

14. *Basilius Valentinus described such a work in twelve keys and Riplæus in twelve gates. Are there twelve keys and twelve gates? What are they called, and what do they unlock?*

Answer: There are twelve keys, twelve doors, and seven seals. But if you unlock the first door with the right key, you can use the same key to unlock all the other doors, one after the other, not all at once, but gradually, until you finally get into the cabinet of nature's secret treasure chamber.

15. *Have these two men described all the mysteries from beginning to end, revealing all the treasures of nature?*

Answer: They have described how to make the stone, the white and the red, and how the *materia* should go through its triple purification and

und in die *Elementa* geschieden werden und solche wieder zusammen zusetzen gelehrt, auch wie solcher fixer Stein, wenn alle 4 *Elementa* wieder in eins verwandelt, in die Tinctur soll mit gemeinem Golde versetzt und alsdann auf die Metalle getragen werden, da denn 1 Teil etliche 1000 Teile tingirt, weiter haben sie es nicht beschrieben.

16. *Ist denn dies nicht das ganze Werk und Ende der Natur?*

Antwort: Ja, es ist das Ende der Natur des ersten Werks im mineralischen Reiche, wenn der Stein sich gesättigt und satt getrunken, dass er nichts mehr zu sich nehmen will, so wird es durch das gemeine ☉ zu Ruhe und Schlaf bracht, als wenn man den laufenden ☿ will gestehend machen, so amalgamiert man ihn mit einem Metall, so wird durch dasselbe dick und hart. Dass aber dieses soll das Ende der ganzen Natur und Kunst sein und die innerste Schatz-Kammer der Natur, das ist es nicht, sondern ist nur eine Ausruhung und eine Stillstehung, dass der Künstler kann anderen und größeren Geheimnissen nachdenken, wie er soll aus diesen Werk oder Rad der Natur eine neue Schöpfung und Generation anfangen, da das vorhergehende gering und wie nichts gegen dieses himmlische und engelhafte Geschöpf und wenn solche Bewegung Umlauf des Himmels wieder zu Ende und still steht, bringt es der Künstler abermals zur Ruhe und kann also in Ewigkeit diesen neuen Himmel und diese neue Erde wieder erhöhen und neu gebären, auch allemal in kürzerer Zeit, da er vorher etliche Monate, danach nur einen, etliche Wochen, Tage und Stunden, zuletzt etlichemal in einer Stunde, dass dieses Rad, dieser Himmel und Erde laufend wird, durch alle Farben geht und dann wieder stille steht und aufhört mit treiben, da es dann bis in Ewigkeit durch das einzige *Agens* und *Patiens* erhöht wird und an seiner Kraft unendlich zunimmt und kein Künstler bis in Ewigkeit der Natur ihr Ende ergründen kann, so wenig als GOTT und die Ewigkeit selbst. Dies ist die *Multiplication in infinitum*. Dies haben sie nicht beschrieben, ob sie es gewusst oder nicht, das weiß ich nicht. Es kann sein, dass sie sich haben begnügen lassen mit der Tinctur, so etliche 1000 Teile tingirt hat und dem Handel nicht besser nachge-

be separated into the *elementa*. They also taught how to put them back together again, as well as how such a fixed stone, when all four *elementa* are transformed into one again in the tincture, should be mixed with common gold and then applied to the metals; and since one part tinges several thousand parts, they did not describe it any further.

16. *Is this then not the whole work and end of nature?*

Answer: Yes, it is the end of nature in the first work on the mineral kingdom when the stone is saturated and has drunk enough to the point that it no longer wants to take anything more. It is brought to rest and sleep by the common ☉, as if you want to make the running ☿ stand still; thus you amalgamate it with a metal, so that it becomes thick and hard. But this is not to be thought to be the end of all nature and art and the innermost treasure-chamber of nature, for it is only a rest and a standstill, so that the artist can think about other, greater mysteries, such as how they should begin a new creation and generation from this work or wheel of nature, since the previous [work] was small and as nothing compared to this heavenly and angelic creature; and when with such movement, the orbit of heaven comes to an end and stands still, the artist brings it to rest again and can thus forever elevate and give birth again to this new heaven and this new earth, always in a shorter period of time. For before that it took several months, then only one [month], then several weeks, then mere days and hours, until finally it can be effected several times in one hour that this wheel, this heaven and earth, becomes moving, goes through all the colors, and then stands still again and stops driving, since it is then elevated through the only *agens* and *patiens* unto eternity and increases infinitely in its strength; and no artist can fathom the end of nature until eternity any more than they can fathom GOD and eternity itself. This is the *multiplication in infinitum*. They did not describe it, whether they knew it or not, I do not know. It may be that they were satisfied with the tincture, which has tinged thousands of parts and did not think

sonnen oder ob sie es gewusst und verschwiegen haben und dem Leser zu weiterem Nachdenken heimgestellt.

17. *Es sagen aber ihrer viele, dass mancher eine Tinctur, als den Stein, gemacht, die Multiplication aber haben sie nicht gekonnt und hatten daher die ganze Tinctur wieder zuschanden gemacht, da sie zuvor 1000 Teile tingiert, hatten sie danach nichts oder ein klein wenig getan und dennoch spreche man die Tinctur oder der Stein könnte nicht verderbt werden?*

Antwort: Das glaube ich wohl, wenn sie die ausgekochte Tinctur mit den unreinen bleiischen und arsenikalischen Geistern beschmeißen, ist nicht anderes, als wenn man ein Edelgestein zermalmte und unter groben Ziegelstein vermengt und alsdann zusammenschmelzte, wer will nun dies Edelgestein wieder aus dieser Erde und Kot bringen? Oder wenn sie solches mit dem lebendig-machenden Geist vereinigen, so vermengen sie solchen, wo sie sollten rot nehmen, nehmen sich weiß, nehmen anstatt des trockenen Wassers, so keine Hand netzt und Salben oder Öl genannt wird, die rohen und unreifen nassen *Menstrua*, wodurch sie der Tinctur Feuer erlöschen, das Unreife, Rohe, Ungekochte dem Zeitigen und Gekochten zusetzen, da solcher feurige Körper Feuer haben will zur Speise, geben sie ihm Jungfer-Milch, welche nur vorher zur Aufziehung dieses Kindes gebraucht worden, nun aber als ein starker Mann diese Speise nicht mehr braucht.

18. *Wenn man nun so viele Türen aufschließen muss, ehe man in die Schatz-Kammer der Natur kommt, was sieht man denn in dem ersten Zimmer?*

Antwort: Darinnen ist eben nichts Schönes zu sehen, es ist ein alter aussätziger Mann im Bade, voller Schuppen, Flecken und Aussatz, welcher durch das Bad gereinigt wird, dass er in dem Bade von seinen Schuppen und Flecken gesäubert und sein Leib rein und gesund zum Vorschein kommt.

better about the trade, or whether they knew it and kept it secret and left it up to the reader for further reflection.

17. *But many of them say that some made a tincture, i.e., the stone, but they could not multiply it and therefore ruined the whole tincture again, since they tinged a thousand parts before, and they had shortly afterwards done little or nothing; and yet they say the tincture, or the stone, could not be spoiled?*

Answer: I think so, for if they cast the impure leaden and arsenical spirits upon the boiled tincture, it is no different than if one crushed a precious stone and mixed it with coarse bricks and then melted them together. Who now wants to pull this precious stone out of the earth and dung again? Or if they unite this with the vivifying spirit, they mix them. Where they should take red, they take white. Instead of the dry water which does not wet the hands, which is called ointment or oil, they take the raw and unripe wet *menstrua*, and by doing so they extinguish the fire of the tincture. They add the unripe, the raw, and the uncooked to the mature and the cooked. Such a fiery body wants fire for food, yet they give it virgin milk, which was only previously used to rear our child. But now the child is a strong man, and it no longer needs this food.

18. *If one has to unlock so many doors before entering the treasure-chamber of nature, what does one see in the first room?*

Answer: There is nothing pleasant to see in there; there is a bath with an old leprous man full of scales, spots, and leprosy, who is purified by the bath, so that he is cleansed of his scales and spots in the bath and his body becomes clean and healthy.

19. *Ist denn dieses des Alten seine innere Reinigung und mit was für Wasser wird er so gebadet und wie lange ist er im Bade?*

Antwort: Dieses Bad wäscht nur äußerlich den Leib rein, innerlich tut dieses Wasser nichts, es hat nicht die Macht *purum ab impuro* zu scheiden und ist das Bad der Natur und ihr Essig. Die Zeit anbelanget, so werden hier viele *superflua* getan, man kann es wohl in etlichen Tagen verrichten, mancher plagt sich wohl ein halbes Jahr mit dieser Solution.

20. *Was macht man denn weiter mit diesem Alten?*

Antwort: Man drückt seine Wasser-Beulen und Schwären aus, dass sein Leib trocken und blutrot wird, das stinkende Wasser so von ihm gedunstet und aus allen seinen Gliedern geraucht, aus Händen, Füßen, Haupt und Haar, nimmt man nach seiner Rectificierung und badet abermals diesen Leib darinnen, so wird sein Geblüt gereinigt, schön, rot und sanguinischer Complexion, da er zuvor trocken, hart und dürre war, nach Art des alten, lahmen, hinkenden ♄.

21. *Ist denn dies der andere Schlüssel?*

Antwort: Er ist es wohl und kommt aus dem ersten Schlüssel, als womit man die andere Tür aufschließen muss, so findet sich, dass dieser Alte in seinem Leibe ein kleines, schönes und reines Kindlein trägt, er selber aber ist der *Vulcanus*, Schlange oder Drache, welcher nichts als Feuer und Rauch auslässt, aus Mund, Nasen, Ohren, Händen, Füßen und allen Gliedern.

22. *Ist denn dieser der ♃?*

Antwort: Nein, es ist der Stein, welchen der ♄ für seinen Sohn den ♃ gefressen und wieder ausspeien muss, das ist, den Stein flüchtig überführen, wenn der ♄ nicht einen solchen Stein verschlingt und wieder ausspeit, so ist er nicht tüchtig zum Werk.

19. *Is this the old man's inner purification, and with what kind of water is he bathed and how long is he in the bath?*

Answer: This bath only washes the body externally, the water does nothing internally, it does not have the power to separate *purum ab impuro*; it is the bath of nature and its vinegar. As far as time is concerned, a lot of *superflua* are done here, you can probably do it in a few days, some people probably struggle with this solution for half a year.

20. *What do you do with this old man?*

Answer: One squeezes out his watery boils and sores so that his body becomes dry and blood-red. The stinking water is evaporated from him and smoked out of all of his limbs: out of his hands, feet, head, and hair. After this rectification one takes [him] and bathes his body in it again, so that his blood becomes purified, beautiful, red, and sanguine of complexion, since before it was dry, hard, and arid, after the manner of the old, lame, hobling ♄.

21. *Is this the second key?*

Answer: It is, and it comes from the first key, with which one has to open the second door, so it turns out that this old man is carrying a small, beautiful, and pure child in his body; but he himself is the *Vulcanus*, the snake or dragon, which emits nothing but fire and smoke from mouth, nose, ears, hands, feet, and every limb.

22. *Is this the ♃?*

Answer: No, it is the stone which the ♄ must devour before his son, ♃, is spit out again, that is, transfer the stone in a volatile manner. If the ♄ does not devour such a stone and spit it out again, then it is not fit for the work.

23. *Wenn nun der ♄ den 🜨 Stein gefressen und wieder ausgespien, was soll man denn hernach tun?*

Antwort: Die 4 Kinder oder die 4 *Elementa* soll man reinigen, das Feuer ist das *Elementum Ignis*, aus der roten ♀ nicht aus der gemeinen, welche grün ist, ist der Pluto, der *Neptunus*, ist der Gott des Meeres oder Wassers, der ♃, Gott in der Luft, der Oberste, die Erde der *Saturnus*.

24. *Ist denn der vorige Saturnus nicht der rechte Saturnus und ist noch ein anderer?*

Antwort: Der vorige ♄ hat den rechten als ein kleines schönes Kindlein in seinem Leibe und ist erst dies der philosophische *Saturnus*, welcher nach Scheidung der 4 Elemente gezeugt wird aus der Quintessenz der 4 Elemente. Der vorige ♄ war von der groben Erde, von der Natur geschaffen und allen Artisten zur Hand, der philosophische *Saturnus* muss durch die Söhne der Kunst gezeugt werden. Dieser hat seinen Anfang von dem *Spiritus* ☿, welcher aus dem Meer aufsteigt von dem *Neptuno*. Ich habe vorhin auch schon längst gesagt, dass alle Metalle wachsen aus dem gemeinen elementischen Wasser und sind Früchte der Meteoren, die philosophischen Metalle wachsen auch aus unserem elementischen Wasser. So viel nun Metalle durch die Natur wachsen, so viel haben wir auch durch die Kunst und kann die Kunst höher als die Natur, welche nicht weiter kann, als dass sie die Metalle und Edelgesteine zeugt, denn sie hat nur die groben *Elementa*. Wir aber haben das grobe elementische Wasser imprægniert mit einem der vollkommensten Körper, welcher von *primo ente*, der ☉ und ☽, auch aller Planeten, dieses haben wir zum ersten, anderen und dritten mal gereinigt und in die 4 *Elementa* geschieden. Nun ist noch übrig die Scheidung *purum ab impuro*, welche geht über die Elemente, aus diesen Elementen zeugen wir die Metalle, diese 7 Metalle, welche daraus wachsen und weit herrlichere Körper sind als die gemeinen, aus solchen machen wir auch Edelgesteine, wenn wir sie in ihrem Wasser wieder resolvieren, sehr unterschieden von den gemeinen, denn die unse-

23. *When the ♄ has eaten the 🜨 stone and spat it out again, what should one do afterwards?*

Answer: The four children or the four *elementa* should be purified. The fire is the *elementum ignis* from the red ♀, not from the common one, which is green, is Pluto. *Neptunus* is the god of the sea or water. ♃ is the god in the air, the supreme. The earth is *Saturnus*.

24. *Is not the previous Saturnus the right Saturnus, and is there another [Saturnus]?*

Answer: The previous ♄ has the right one in his body as a beautiful little child and this alone is the philosophical *Saturnus*, which after the separation of the four elements is begotten from the quintessence of the four elements. The previous ♄ was of the coarse earth, created by nature, and available to all artists. The philosophical *Saturnus* must be begotten by the sons of art. This has its beginning from the *spiritus* ☿, which rises out of the sea, from *Neptuno*. I already mentioned a long time ago that all metals grow from the common elemental water and they are fruits of the meteors. The philosophical metals also grow from our elemental water. As much as the metals grow through nature, we grow as much through the art, and art can do better than nature, which cannot go further than the begetting of metals and jewels, for it only has the coarsest *elementa*. But we have impregnated the coarse elemental water with one of the most perfect bodies, which has the ☉ and ☽, and also all of the planets, from *primo ente*; and we have purified this for the first, second, and third time and divided it into the four *elementa*. Now there is still the separation *purum ab impuro*, which passes through the elements; and from these elements we beget these seven metals which grow from them and are far more beautiful bodies than the common ones. We also make precious stones from them when we dissolve them again in their water. They are very different from the common ones, for ours are tinctures which tinge the common metals into ☉, and the crystals or other minor rocks into bright jewels. So

re sind Tincturen, welche die gemeinen Metalle in ☉ tingieren und die Kristalle oder andere geringe Gesteine in hell leuchtende Edelgesteine. Ist also großer Unterschied unter unserem Wasser und dem gemeinen, auch unser Luft, Feuer und Erde. Alle *Meteora*, so in der großen Welt sind, kommen in unser Arbeit zu Gesicht, wenn unser Stein geboren und gezeugt wird, da gibt es Regen, Tau, Reif, Schnee, Sturm-Wind, Blitz und Leuchten, unzählige Figuren, mancherlei Farben.

25. Diese Beschreibung ist aber eine sehr langweilige Arbeit, da man doch in anderen Büchern liest, der Stein werde gemacht in 2 Nächten und 3 Tagen?

Antwort: Das dieser Stein hier so langweilig scheint beschrieben zu sein, ist die Ursache, weil kein Mensch in der Welt hat die Vorarbeit beschrieben, sondern haben alle angefangen von der Scheidung der Elemente zu schreiben und dem Sucher kein Licht vorher gegeben, wie es soll gereinigt und geschickt gemacht werden, dass es in die *Elementa* geschieden werden kann, als nur manchmal, wenn ihnen ein Wort entfahren. Weil ich nun gesehen, dass die Sache ohne dem schwer und unter 1000 Menschen nicht einem gegeben diesen *Nodum Gordium* aufzulösen, so habe aus Mitleiden und Erbarmnis dieses dem Nächsten nicht verhalten wollen. Das aber der Stein sollte gemacht werden in 2 Nächten und 3 Tagen hat diesen Verstand: Es wird zweimal schwarz und finster, die 3 Tage sind, wenn es Licht wird, weiß und leuchtend. Der andere Tag, wenn die Sonne mit schönen gelben Strahlen aufgeht. Der dritte Tag, wenn sie rot und blutig scheint.

26. Sind denn die vorigen die rechten Schlüssel?

Antwort: Es sind die rechten Schlüssel, so nur die äußere Natur aufgeschlossen, hernach die Pforten der Elemente, künftig aber folgen die geheimen Schlüssel zum Cabinet und der geheimen Schatz-Kammer der Natur.

27. Wie heißen denn die geheimen Schlüssel und wieviel sind es?

Antwort: Es sind ihrer 3, der erste heißt *Solutio*. Der andere *Conjunctio*.

there is a great difference between our water and common [water], also our air, fire, and earth. All the *meteora* that are in the macrocosm come into view in our work when our stone is born and begotten; there are rain, dew, frost, snow, storm wind, lightning, and lights of innumerable form and various colors.

25. *But this description is a very lengthy work, since one reads in other books that the stone is made in two nights and three days?*

Answer: The reason why this stone appears to be so tediously described here is because no one in the world has described the preliminary work. But everyone has begun to write about the separation of the elements and they have not given the seeker any light beforehand on how it should be purified and skillfully made so that it can be separated into the *elementa*. Only sometimes does a word escape them [on this point]. Because I have now seen that the matter was very difficult and that not one person among a thouand people was able to untie this *nodum Gordium*,[178] out of pity and mercy I did not want to withhold this from my neighbor. That the stone should be made in two nights and three days has this meaning: it becomes black and dark twice, and the three days are when it becomes light, white, and shining. The second day is when the sun rises with beautiful yellow rays. The third day is when it shines red and bloody.

26. *Are the former ones the right keys?*

Answer: These are the right keys but only those which unlock the outer nature, and thereafter the gates of the elements; but in future the secret keys to the cabinet and secret treasure-chamber of nature will follow.

27. *What are the secret keys called and how many are they?*

Answer: There are three of them. The first is called *solutio*; the second,

Der dritte *Fixatio*. Der eine schließt die innere Kammer der Erde auf und wird wieder in viele Schlüssel geteilt, als *Solutio, Coagulatio, Sublimatio, Distillatio, Conjunctio*, und wenn die Seele vom Körper durch den Geist ausgezogen wird und in der Auflösung von der Seele und Geist geschieden, daher es auch *Separatio* und *Putrefactio* genannt wird. Der andere Schlüssel, *Conjunctio*, bringt den tingierenden und färbenden Geist in das neue *Corpus*. Der dritte, *Fixatio*, figiert alles zur ewig-währenden Beständigkeit.

28. *Dieses möchte ich noch besser ausgelegt wissen?*

Antwort: Im anderen Werke wird gehandelt von der Elemente Zusammensetzung, dieselben zu figieren und zu Stande zu bringen. Das größte Geheimnis ist die Augmentation, erstlich setzen wir zusammen, lassen es verfaulen, dass ein neues Leben bekommt, das Verfaulte lösen wir auf, das Aufgelöste teilen wir, das Gereinigte vereinigen wir und figierens fix zusammen, in unserem rohen *Mercurius* wird unser rotes trockenes ☉ solviert, da es dann die Farbe eines Regenbogens bekommt, wenn sich der ♃ ins Wasser begibt, dieser Regenbogen steht im Wasser, nicht wieder im anderen Werk, in der Luft, dies wird nach 40 Tagen destilliert, in welcher die *Terra Damnata* zu Boden fällt. Im anderen Werk muss die Seele zu einem Geist, der Geist zu einer Seele und diese 2 miteinander ein Körper werden, dieser muss um 3 Zirkel gehen und ein jeder *Circul* ist eine Sonne, die erste ☉ hat ein schwarzes Gesicht, die andere ein weißes, die dritte ein rotes. Des Steins erster Teil ist die irdische ☉, ohne die irdische Sonne wird der Stein nicht gemacht, wird erstlich in 2 Teile geschieden. Der andere Teil des Steins ist der ☿, wenn er noch unbereitet und flüssig ist, so kann er die Körper auflösen, weil er sich in ihn vertieft und hart anhängt, ist mit großer Kraft begabt, sein eigenes ☉ dünne macht und wieder in sich selber in seine erste Materiam bringt, welche Kraft kein Ding in der Welt hat, drum sagt man recht, in dem ☿ ist, was die Weisen suchen und in ihm sind die Güter des ganzen Werks. Er löst sein eigenes ☉ auf, er macht es weich, er nimmt die Seele aus dem Leibe, wenn er her-

conjunctio; and the third, *fixatio*. The first [key] unlocks the inner chamber of the earth and is divided again into many keys as *solutio*, *coagulatio*, *sublimatio*, *distillatio*, and *conjunctio*; and when the soul is drawn out of the body through the spirit and separated from the soul and spirit in the dissolution, it is also therefore called *separatio* and *putrefactio*. The second key, *conjunctio*, brings the tingeing and coloring spirit into the new *corpus*. The third, *fixatio*, fixes everything into a state of everlasting permanence.

28. *I would like to know how to interpret this better.*

Answer: The second work deals with the composition of the elements, and how to fix them and to bring them to fruition. The greatest secret is the augmentation. First we put it together and let it putrefy so that it receives a new life. Then we dissolve what has been putrefied, separate what has been dissolved, unite what has been purified and fix it firmly together. In our raw mercury, our dry red ☉ is dissolved, since it then acquires the color of a rainbow; and when the ♃ goes into the water, this rainbow is [now] in the water, not again in the second work, in the air. It is distilled after forty days, during which [distillation] the *terra damnata* falls to ground. In the second work the soul must become a spirit and the spirit must become a soul. Then these two together must become a body; and this must circulate for three cycles. Each *circul* is a sun. The first ☉ has a black face, the second a white one, the third a red one. The first part of the stone is the earthly ☉. Without the earthly sun, the stone is not made. It is first divided into two parts. The second part of the stone is the ☿. When it is still raw and fluid, it can dissolve bodies because it penetrates deeply into them and adheres strongly to them. It is endowed with great power, makes its own ☉ thin, and brings it back into itself, into its first *materiam*, a power that nothing in the world has. Therefore it is said rightly that what the wise seek is in this ☿, for in it are the goods of the whole work. It dissolves its own ☉, softens it, and takes the soul out of the body. If it is sublimated with

nach mit den rechten *Aqua Vitæ* sublimiert wird, so wird es erzeugt, die Wissenschaft des Steins ist so hoch und herrlich, dass darinnen die ganze Natur, alle Dinge in der ganzen Welt, als ein heller Spiegel gesehen werden können, denn es hat die Beschaffenheit der kleinen und der großen Welt, darinnen die 4 *Elementa*, und die Quintessenz, die man den Himmel nennt, in die hat eine andere Sonne ihren Sitz gestellt, welches etliche *Philosophi* mit einem Bild der klaren Gottheit vergleichen, die weder vom Himmel noch von der Erde her sei und dieselbe haben sie die Seele der Welt genannt oder die mittlere Natur und wie GOTT allenthalben gegenwärtig ist in der ganzen Welt. Also diese Essenz überall in der philosophischen Welt, in der Phiole oder im Glas. Also reicht auch dieses Ding, so unzählbar scheint, indem es seinesgleichen erzeugt, bis an das äußerste Ende der großen Welt. GOTT hat nach seinem Sohn nichts herrlicheres erschaffen, denn dieses Ding und soll es kein Idiot, so es nicht versteht, verlästern.

29. *Ich möchte doch alle Arbeiten kürzlich nacheinander hören, absonderlich vom Solvieren?*

Antwort: Die Solution ist fünffältig, nämlich die Haupt-Solution, die Solution der Elemente, die philosophische Solution, die Solution des Ferments, die Solution des Steins und Elixirs. Die erste ist eine Reduction des unvollkommenen Körpers in einen *Liquorem* oder Chaos, die andere Solution der Elemente ist eine Separation oder Absonderung von seinem Chaos durch die Destillation in die Elemente, nämlich ☉, ♄, ☿ oder in Leib, Seele und Geist oder in Wasser, Feuer, Luft und Erde, die Luft ist das Öl; Drittens die philosophische Solution ist eine Auflösung des Mannes und Coagulierung des Weibes, zugleich in der Faulung. Viertens die Solution des Ferments ist eine Reduction oder Wiederbringung der *Luminarium* Schwefel, entweder in weißes oder rotes Öl. Fünftens, die Letzte ist eine Solution oder Auflösung des Steins und beiderlei Elixire, dieselbe augmentieren in ihrer Kraft und Wirkung. Endlich sind des ganzen Werks nur zwei *Genera* oder Geschlechte, die Solution und die Calcination, die anderen sind der

the right *aqua vitæ* afterwards, it will be engendered. The science of the stone is so high and glorious that within it the whole of nature can be seen as in a bright mirror, since it has the characteristics of the microcosm and the macrocosm, containing within itself the four *elementa* and the quintessence, which is called the sky. In this sky a different sun has placed its seat, which some *philosophi* liken to an image of the clear deity which is neither of heaven nor of earth, and which they have called the soul of the world or the middle nature. Just as GOD is everywhere present in the whole world, likewise this essence is everywhere in the philosophical world, in the vial or in the glass. So this thing, which seems to be innumerable because it creates its own kind, also reaches to the outermost limit of the macrocosm. GOD created nothing more glorious after his son than this thing; and the ignorant, who do not understand it, should not disparage it.

29. *I would like to hear about all works in brief, one after the other, especially about dissolution.*

Answer: The dissolution is fivefold, namely the main dissolution; the dissolution of the elements; the philosophical dissolution; the dissolution of the ferment; and the dissolution of the stone and elixir. The first is a reduction of the imperfect body into a *liquorem* or chaos. The second dissolution of the elements is a separation or segregation from its chaos through the distillation into the elements, namely ☉, ♃, ☿ or into body, soul, and spirit, or into water, fire, air, and earth, the air being the oil. Thirdly, the philosophical dissolution is a dissolution of the man and a coagulation of the woman, at the same time in putrefaction. Fourth, the dissolution of the ferment is a reduction or restoration of the sulphurous *luminarium*, either in white or red oil. Fifth, the last is a dissolution or resolution of the stone and both elixirs, augmenting the same in their potency and effect. Finally, there are only two *genera* or lineages of the whole work, the dissolution and the calcinations. The

Generum Species, welche erste ist ein Werk des Himmels, die Solution und *Humectatio* oder Aufrichtung: Das letzte ist ein Werk der Erde, nämlich der Calcination und Austrocknung. Auch sind die Elixire, sowohl die weißen als die roten, zweierlei, das eine, welches das *Elixir Peregrinorum* oder der Fremden genannt wird, das andere Elixir so exaltiert oder erhöht oder in ein Öl gebracht worden.

30. *Wenn aber vom ☿ geredet wird, was ist es denn eigentlich?*

Antwort: Erstlich wenn vom ☿ geredet wird, ist es nicht Quecksilber, sondern unser roher Geist. In der anderen Art ist es unser Calcinier-Wasser, so alle Körper solviert. Drittens ist es der ☿ des *Corporis*, welches er solviert hat, der Körper des weißen ♃. Der vierte ist der ☿ oder der natürliche ♃ oder ♃ *Naturæ*, erstlich ist der *Mercurius* ein stets bleibendes Wasser, ausgezogen von den Körpern durch die Destillation oder Auflösung durch das Sublimieren und durch das Subtilisieren. 2. Es sind 4 ☿, ein roher, ein sublimierter, ein ☿ *magnesia*, welches des ♃ ist und ein schmieriger, welcher ein tingierender ☿ ist. Seine *Congelatio*, man muss den ☿ congelieren mit dem Körper *Magnesiæ*, welches nicht das Quecksilber, noch die *Magnesia*, welche man anschaut, sondern durch den ☿ wird verstanden die Früchte dieser *Mixtion*, welches ist *Humidum Radicale*, das Quecksilber ist ein *Spiritus*, und wird das bleibende Wasser genannt oder der Essig, das *Menstruum*, oder das *Nutrimentum*, von ♃. Es kommt aus dem Körper und ist einer weißen Farbe, dieser ♃ muss componiert oder gemacht werden und was aus ihm kommt ist ☿ *Philosophorum*, der in einen weißen trockenen *Corpus* verkehrt worden ist. Aus diesem kommt hernach der göttliche oder himmlische und also ist es auch mit dem roten *Oleo Philosophorum*, oder dessen *Liquore*, und von den unverbrennlichen Öl, so wird das verstanden von dem Öl der Fermente oder Hefen, ehe sie ♃ sind oder wenn sie ♃ sind, von ihren Ölen oder Solution, vom roten ☿ oder Feuer des Steins, Seele und Öl, rote Tinctur, ist eines wie das andere. Der weiße ☿ ist Luft und Wasser des Steins, das Öl, sein Feuer und Seele ist noch zurück im Grunde des Geschirrs, die Erde heißt, wenn sie

others are the *generum species*, the first of which is a work of heaven, that being the dissolution and *humectatio* or erection. The second is a work of the earth, namely the calcination and desiccation. Also the elixirs, both the white and the red one, are of two kinds: one which is called the *elixir peregrinorum* or [elixir] of the strangers, and the second elixir which has been exalted or enhanced or made into an oil.

30. *But when people talk about ☿, what actually is it?*

Answer: First of all, when we talk about ☿, it is not quicksilver, but our raw spirit. In the second way it is our calcination water, which dissolves all bodies. Third, it is the ☿ of the *corporis* which it has dissolved, the body of the white ♀. The fourth is the ☿ or the natural ♀ or ♀ *naturæ*, first the *Mercurius* is an ever abiding water extracted from the bodies through the distillation or dissolution, through the sublimation and through the subtilization. 2. There are four ☿, a raw, a sublimated, a ☿ *magnesia* which is the ♀, and a greasy one which is a tinging ☿. With respect to the *congelatio*, one has to congel ☿ with the body *magnesiæ*, which are not the mercury nor the *magnesia* which one looks at, since in this case by ☿ we mean the fruits of this *mixtion*, which is *humidum radicale*. The mercury is a *spiritus*, and it becomes what is called the abiding water, or the vinegar, the *menstruum*, or the *nutrimentum*, of ♀. It comes out of the body and is white in color. This ♀ has to be composed or prepared and what comes out of it is the ☿ *philosophorum* which has been turned into a white dry *corpus*. From this comes afterwards the divine or heavenly [☿], and so it is also with the red *oleo philosophorum*, or its *liquore*, and of the incombustible oil. By this oil we mean the oil of the ferments or yeast before they are ♀, or when they are ♀, of their oils or solution, of red ☿ or fire of the stone, soul and oil, red tincture, one is like the other. The white ☿ is air and water of the stone, and the oil is its fire and soul which is still at the bottom of the retort. The earth, when it is black, is called the earth, and after

schwarz ist, die Erde, nach derselben Calcination ist sie schneeweiß, wird ihr das Salz ausgezogen, nach der anderen Elementen-Separation, wird die Erde weiß dahinten bleiben, alsdann ist es das Salz, auch wird zuweilen das Salz für den weißen ♃ genommen und der ♃ für das Salz, auch manchmal vor das *Corpus* aller fixen Salze, ist das *Corpus* und das Haus oder Wohnung und wird ausgezogen von dem *Corpore* der Metalle. Erstlich wird es Asche, danach ein Salz, aus diesem Salz ☿ *Philosophorum*, der weiße ♃ und hernach der weiße Stein. Die Kunst bedarf allezeit seiner eigenen Erde, in welcher die größte Kraft und Tugend steckt, den ☿ zu härten, dieweil dieser allezeit fließt, danach der Geist und die Seele dem Leibe wiedergegeben werden.

31. *Von der Conjunction möchte ich auch gern weitläufiger hören?*

Antwort: Ich habe ja schon viel davon geredet, sie ist dreierlei, die erste von Mann und Weib, von seinem *Corpore*, welches vor schon eine Terra oder Erde genannt werden, worunter das Salz verstanden wird, wird auch das *Matrimonium* genannt, die Verehelichung, die Putrefaction coaguliert die Erde mit dem schweren Wasser, es härtet ohne einige Hilfe oder Zutun, den lieben Sohn soll man mit seiner weißen Schwester verheiraten. Der Sohn ist die weiße Erde, die Schwester das weiße ☿al-Wasser, wenn es eine zeitlang zusammen gekocht, wird ein Ding draus und wird genannt das gute, wenn es sich sublimiert, die geblätterte Erde, ehe sich aber diese begibt, geschieht die Faulung. Aus dieser Asche wird der Phönix, welcher sich hernach durchs Feuer verbrennt, diese erste Zusammensetzung geschieht aus der Feuchte und Trockne des Wassers und der Erde und wird verrichtet ehe die Trockenheit kommt in die Faulung, die Erde ist des Wassers *Fermentum*, und das Wasser ist die Hefe der Erde, dieses wird die erste Composition genannt, ohne welche der Stein nicht zu machen ist und ohne welche das aufgelöste einmal zum Elixir gebracht wird. Die andere Conjunction ist des doppelten Leibes der geblätterten Erde mit dem Geist und die Vereinigung in Leib, Seele und Geist, wenn der weiße ♃ gewachsen ist. Die dritte ist die Vereinigung mit dem neuen roten ☿, welches eine

its calcination it is snow-white. Now the salt is extracted from it after the separation of the other elements. The earth will remain white, and it is now called the salt. Also sometimes the salt is taken from the white ♃ and the ♃ for the salt, also sometimes for the *corpus* of all fixed salts, is the *corpus* and that house or dwelling and is extracted from the *corpore* of the metals. First it becomes ashes, then a salt; and from this salt arise the ☿ *philosophorum*, the white ♃, and then the white stone. Art always needs its own earth, in which lies the greatest strength and virtue to harden the ☿, because it always flows, after which the spirit and the soul will be given back to the body.

31. *I would also like to hear more in detail about the conjunction.*

Answer: I have already talked about it a lot. It is three things, the first of man and woman, of its *corpore*, which has already been called a *terra* or earth, by which salt is meant. It is also called the *matrimonium*, the marriage. The putrefaction coagulates the earth with the heavy water, it hardens without any help or assistance. The dear son should be married off to his white sister. The son is the white earth, the sister the white ☿al water, and when they have been boiled together for a while, they become one thing and are called good. When it sublimates, it is the leafed earth, but before this happens, the putrefaction happens. From these ashes the phœnix is made, which is then burned by the fire. This first composition is made from the moisture and dryness of water and earth and is done before the dryness comes to putrefaction. The earth is the *fermentum* of the water, and the water is the yeast of the earth. This is called the first composition, without which the stone cannot be made and without which the dissolved material is made into an elixir. The second conjunction is the double body of the leafed earth with the spirit and the union in body, soul, and spirit when the white ♃ has grown. The third is the union with the new red ☿ which is

Speise, Zermalmung, *Inceratio* genannt wird. Die vierte ist des ♃ oder Steins mit seinem Ferment, oder Hefe-Öl Vereinigung, damit der Stein fermentiert werde, die 2 Quecksilber, welche die Unverständigen vom rohen ☿ verstanden, haben sich also selbst betrogen, denn die philosophischen ☿ sind die Wasser der Sonne und des Mondes, doch welche *Conjunctiones* geschehen, des Leibes mit der Erde, der Erde mit Geist und Leib und dann der Seele Zusammenkunft mit Seele, Geist und Leib, dies ist die dreifache Conjunction. Dem ersten Grad der Faulung folgt die Austrocknung durch die Luft, wenn der Stein ausgetrocknet und inceriert wird, so fängt nach demselben an der dritte Grad und wärt bis zur Perfection des Steins, wenn die Sonne in ♌ läuft, dieses ist des Himmels höchstes Haus, der Sonnen eigenes Haus, ihrer Würde und Herrlichkeit, die vorigen waren im Zwilling, welche auch machten die Farben schwarz, weiß und rot, die schwarze Farbe sieht man nach der Empfängnis, die weiße Farbe kommt im anderen Grad des Feuers, wenn die *Materia* wohl ausgetrocknet wird, die rote Farbe erscheint, wenn die *Materia* im dritten Grad des Feuers perfect und vollkommen. Es sind auch zufällige Farben in der Arbeit, so sich bei der ersten, mittleren und letzten Wärme erzeigen, zwischen der Schwärze, so das Raben-Haupt genannt wird, auch das erste Zeichen derselben ist und vor der Weiße erscheinen mancherlei Farben in dem subtilen Gestiebe, es wird gelblich, rötlich, doch keine rechte wahre Röte, es wird oftmals rötlich, oft gelblich, oft zerschmilzt es, es coaguliert sich auch oft vor der rechten Weiße, es dissolviert sich selber, es coaguliert sich selber, es rötet sich selber, es wird auch, ehe die Weiße kommt, grün, es erscheint auch vor der Weiße der Pfauen-Schwanz, alle Farben so in der ganzen Welt können erdacht werden, erscheinen vor der wahrhaftigen Weiße, welche wie Fisch-Augen leuchten, zwischen der rechten wahren Weiße und der rechten waren Röte erscheint eine Graugelbe.

32. *Ist denn kein gemeiner ♃ bei diesem Werk oder kommt keiner dazu?*

Antwort: Vor allen Dingen muss man wissen, dass unser ♃ kein gemeiner ♃ sei, sondern ein zusammengesetzter ♃ aus dem Geist und Leib

called a meal, grinding, *inceratio*. The fourth [conjunction] is of the ♃ or stone with its ferment, or its yeast-oil union, so that the stone may be fermented, the two mercuries, which the ignorant understood from the raw ☿, have thus deceived themselves, for the philosophical ☿ are the waters of the suns and of the moon, but what *conjunctiones* take place, of the body with the earth, of the earth with spirit and body, and then with the soul coming together with soul, spirit, and body, this is the threefold conjunction. The first degree of putrefaction is followed by desiccation through the air, when the stone is desiccated and incised; after this the third degree begins and lasts until the perfection of the stone, when the sun travels through ♌, which is heaven's highest house, in fact the sun's own house, the sign in which he is dignified and glorified. The former were in Gemini, which also made the colors black, white, and red. The black color is seen after conception; the white color comes in the second degree of fire when the *materia* has dried out well; and the red color appears when the *materia* is perfect and complete in the third degree of fire. There are also accidental colors in the work, such as appear with the first, middle, and the last heat; between the blackness which is called the raven's head and which is also the first sign of it, and before the whiteness, various colors appear in the subtle structures. It becomes yellowish, reddish, but not really a true red; and it often becomes reddish, often yellowish, it often melts, it often coagulates in the correct way, it dissolves itself, it coagulates itself, it reddens itself, it also turns green before the whiteness comes. The peacock's tail also appears before the whiteness; all the colors that one can imagine in the world appear before the true whiteness, which shines like fish eyes. Between the genuine, true white and the geniune, true red appears a grey-yellow.

32. *Is there not a common ♃ in this work, or is no [♃] added?*

Answer: Above all, one must know that our ♃ is not a common ♃, but a composite ♃ from the spirit and body of the *luminum*, that is to say

der *Luminum*, das ist unsers ☉ und unseres ☽ und sind zweierlei, der erste wird zu Wege bracht durch die Faulung nach derselben Solution, der andere wenn die *Lumina*, die Lichter in die Öle gebracht werden, durch das einfache ☿al Wasser oder das zusammengesetzte corrosivische ☿al-Wasser, die ♃ aber der Unvollkommenen als der Vollkommenen sind beide weiß und rot, die Unvollkommenen sind noch in Gestalt *Liquoris*. Wenn sie vereinigt werden, geben sie mancherlei Farben, zuletzt wenn die trockenen Pulver vereinigt und durch die Geister zusammengesetzt werden, wird es eine Speisung, Vereinigung, Zerreibung genannt, welche im trockenen Feuer ausgearbeitet wird, die letzte Vereinigung mit den Ölen kann billig eine Fermentation genannt werden und geben die letzten Eintränkungen mit dem weißen, gelben und roten ♃ allerhand Feuer-Farben. In der ersten Zusammensetzung heißt es *ixir*, wenn aber der ♃ zu Öl worden, heißt es Elixir.

33. *Was ist denn der Stein der Weisen?*

Antwort: Er ist die Quintessenz aus den Elementen, ein generierter und renovierter *Corpus*, der sich wie der Eisvogel verjüngt, wie der Phönix verbrennt, aus der Asche aber ein neuer lebendiger Körper wird. Ist das größte Geheimnis der ganzen Natur, dadurch wir allen Creaturen und der ganzen Natur ins Herz sehen können, der Stein der Wunder, der Spiegel, darinnen die ganze Natur entdeckt, der Stein der Gesundheit und des Reichtums, die Vereinigung des Himmels mit der Erde, die Kräfte der Obersten und Untersten, das Wunder der Natur und Kunst.

34. *Werden aber die vorigen drei geheimen Schlüssel weiter gebraucht und was schließen sie für Türen auf?*

Antwort: Der eine schließt das *Centrum* der reinen jungfräulichen Erde auf, die sonst so fest und diamantenhart verwahrt, dass sie mit keinem Schlegel zu gewinnen und alle Pfeile von allen corrosivischen Geistern vergeblich nach ihr abgeschossen werden, welche Erde auch durch nichts als diesen Schlüssel kann eröffnet werden, danach ist die

of our ☉ and our ☽. They are two things. The first is brought about by the decay after the same solution, and the second when the *lumina*, the lights, are brought into the oils by the simple ☿al water or the composite corrosive ☿al water, but both the imperfect and the perfect ♃ are both white and red, the imperfect being still in the form of a *liquoris*. When they are combined, they give various colors. Finally when the dry powders are combined and put together by the spirits, it is called a feeding, union, or trituration, which is worked out in a dry fire. The last union with the oils can simply be a called a fermentation; and the last impregnations with the white, yellow, and red ♃ give all kinds of fiery colors. In the first composition it is called *ixir*, but when the ♃ has turned into oil it is called elixir.

33. What then is the philosophers' stone?

Answer: It is the quintessence of the elements, a generated and renovated *corpus* that rejuvenates like the kingfisher, burns like the phœnix, but becomes a new living body from the ashes. It is the greatest secret of all nature, through which we can look into the heart of all creatures and all nature. It is the stone of miracles, the mirror in which all of nature is discovered, the stone of health and wealth. It is the union of heaven and earth, the powers of the highest and lowest, the wonder of nature and art.

34. But are the previous three secret keys still needed and what kind of doors do they unlock?

Answer: The first one unlocks the *centrum* of the pure virgin earths, which otherwise are kept so solid and diamond-hard that they cannot be won with a mallet; and all the arrows of all the corrosive spirits are launched at it in vain, for this earth can be opened by nothing except this key. After it is so opened the whole earth becomes transparent,

ganze Erde durchsichtig, der erste magische Spiegel, dass man alle Früchte der Erde erkennen und allen ins Herz sehen kann. Dieser tut Hilfe, dass man die Erde fruchtbar machen kann, alle Kräuter und Bäume, das Leben verlängern, den Menschen wieder in seine Jugend setzen, aus einem Alten einen Jungen machen, aus einem Kranken und Ungesunden den einen Gesunden. Weil dieser Schlüssel auch das Paradies aufschließt. Ist auch der erste Schlüssel der Solution, der rechte Schlüssel zum hesperischen Garten, denn da wird niemand eingelassen, er bringt denn ein Zeichen mit, dass er aufweisen kann, eine Phiole der Jungfer-Milch. Der andere Schlüssel *Conjunctio* bringt die geschiedenen Teile wieder zusammen und ist die Suppe, so über die kolchischen Drachen geschüttet, denn diese Drachen bewachen die güldenen Äpfel im hesperischen Garten, hernach wenn diese Drachen getötet, kann man den Garten erst recht eröffnen und die Äpfel brechen. Der dritte Schlüssel als die Fixation macht es beständig und immerwährend, dass man so oft als man will, in diesen Garten eingehen kann, doch diese beiden letzten Schlüssel noch besser zu erklären, so öffnet einer das Chaos der Luft, der andere das Feuer, dass der Luft- und Feuer-Himmel jedem seine Farben, Früchte und Wunder zeigen kann.

35. *Sind denn dieses die Schlüssel alle oder reichen sie noch mehr andere dar?*

Antwort: Sie sind es noch nicht alle, sind auch noch viel mehr andere geheime Kammern aufzuschließen, dies sind noch lange nicht die geheimen Schatz-Kammern der Natur noch derselben ihr Ende, wird auch niemand leben, der die Schätze der Natur alle erforschen oder ausgründen kann.

36. *Wie heißen die noch geheimeren Schlüssel und wer bringt solche?*

Antwort: Die allergeheimsten Schlüssel ist eins der doppelte Herolds-Stab des ☿ mit dem beiden Schlangen, welche der Adler, der Königsvogel bringt. Der andere ist das lebendige Lebens-Wasser in Seele, Geist

and it is then the first magical mirror in which one can recognize all the fruits of the earth and see into everyone's heart. It helps one to make the earth fertile, promoting the growth of all herbs and trees; and it can prolong life, restoring youth to people, making a young [person] out of an old one, or a healthy one out of a sick one. It has these powers because this key also unlocks paradise. It is also the first key of the solution, the proper key to the Hesperidian garden, for no one is admitted there, unless they bring a sign that they can show, a vial of virgin milk. The second key, the *conjunctio*, brings the separated parts back together and the broth which is poured over the Cholchic dragons, for these dragons guard the golden apples in the Hesperidian garden. Afterwards, when these dragons are killed, the garden can be opened up and the apples can be broken open. The third key, the fixation, makes it constant and everlasting so that one can enter this garden as often as one likes, but to explain these last two keys still better, we may say that one opens the chaos of the air, and the other of the fire, such that the sky of air and fire can each show its colors, fruits, and wonders.

35. *Are these all the keys then, or are there still other [keys]?*

Answer: They are not all there yet, there are still many more secret chambers to be opened up, these are by no means the secret treasure-chambers of nature nor their end, nor will anyone live who can research or explore all the treasures of nature.

36. *What are the even more secret keys called and who brings them?*

Answer: One of the most secret keys is the double herald's staff, the ☿ with the two snakes, which is brought by the eagle, the king's bird. The

und Leib, dies bringt der gekrönte Adler mit dem Stern, der gestirnte Adler. Der dritte Schlüssel ist der Schlüssel zum höllischen Kerker, das Feuer wider die Natur, der giftige Drache, das Höllen-Bad, diesen bringt der kalte Bär. Der vierte Schlüssel ist die rote blutige Sonne, welche uns der ♌ bringt.

37. *Was schließen diese Schlüssel?*

Antwort: Der Schlüssel des Feuers wider die Natur, welchen man dem kalten Bär abgenommen, der schließt uns den doppelten ☿ auf, tötet den Drachen, brennt, calciniert und figiert ihn zur weißen Tinctur. Der andere Schlüssel als das lebendige Wasser in Seele, Geist und Leib multipliciert solche, die rote blutige Sonne macht ihn rot, sie trocknet das Meer aus.

38. *Kann man denn keine größeren Wunder und Geheimnisse dadurch erfahren, als den Universal-Stein, die Tinctur zu machen?*

Antwort: Ich habe schon gesagt, dass dieser Stein nur der erste Schatz ist in der geheimen Schatz-Kammer der Natur, in dem mineralischen Reiche. Wer einmal diesen gefunden, der hat alle geheimen Schlüssel gefunden, dadurch er die ganze Natur entblößen und allem was unter dem ganzen Himmel, auf und unter der Erde und in der Erde, ins Herz sehen kann, noch mehr in die astralische Luft-Welt, in die Gestirne, ins Paradies, in den Himmel, ja er kann dadurch vor GOTTES Angesicht kommen, das ist, ihn recht lernen erkennen und sehen, in was für ein Licht er wohnt, weil ein solcher mit dem Licht der Natur erleuchtet, welches Licht auch aus GOTT kommt.

39. *Was heißt denn der doppelte ☿, wenn er mit seinem Salz erhärtet und coaguliert?*

Antwort: Es heißt der Kristall des ☿, der Wasser- und Salz-Stein oder Wasser-Stein der Weisen, der glasförmige *Azoth* des *Lullij*, und der steinerne Palast, darin der König gehen und seine Wohnung aufschlagen kann, der weiße ♁, wenn er sublimiert, ist süß von Geschmack.

second is the living water of life in soul, spirit, and body, brought by the crowned eagle with the star, called the starry eagle. The third key is the key to the infernal dungeon, the fire against nature, the poisonous dragon, the hell bath, and is brought by the cold bear. The fourth key is the red bloody sun, which the ♌ brings us.

37. *What do these keys unlock?*

Answer: The key of fire against nature, from which the cold bear was taken, unlocks the double ☿ for us, kills the dragon, and burns, calcinates, and transforms it into a white tincture. The second key, as the living water in soul, spirit, and body, multiplies this, and the red bloody sun makes it red as it dries up the sea.

38. *Is it possible to experience greater miracles and mysteries through this than by making the universal stone, the tincture?*

Answer: I have already said that this stone is only the first treasure in nature's secret treasure chamber, in the mineral kingdom. Anyone who has found this has found all the secret keys, through which one can uncover the whole of nature and see into the heart of everything under the entire sky, on the earth, under the earth, and in the earth, even more into the astral world of air, in the stars, in paradise, in heaven, indeed one can thereby come before GOD's face, that is, learn to recognize him correctly and see in what kind of light he lives, because such a person is illuminated with the light of nature, a light which also comes from GOD.

39. *What does the double ☿ mean when it hardens and coagulates with its salt?*

Answer: It is called the crystal of the ☿, the water- and salt-stone or water-stone of the wise,[179] the glass-shaped *azoth* of Lull, and the palace of stone in which the king can enter and make his dwelling. It is the white ♃ when it sublimates, at which point it becomes sweet to taste.

40. *Wie heißt denn dieser, wenn er figiert worden oder calciniert?*

Antwort: Er heißt das gläserne Meer, so mit Feuer gemengt werden kann, der natürliche Schatz, der Stein der ersten Ordnung und die erste Tinctur auf weiß, sieht etwas grünlich, sein Geschmack ist herb, welches sich doch hernach in eine Süße verkehrt.

41. *Wie heißt denn die rote Tinctur?*

Antwort: Sie heißt die erste Tinctur auf rot, erster Stein auf rot, der blutige Stein.

42. *Was ist denn der rote Mann?*

Antwort: Es ist das rote *Oleum*, das *Elementum Ignis*, aus dem philosophischen ⊕, der korallenrote Saft.

43. *Was ist denn das weiße Weib?*

Antwort: Der weiße ♄ der Natur, der nun eingetrocknet, der weiße Mond-Schein.

44. *Was ist denn der grüne ♌?*

Antwort: Es ist das allerinnerste ⊕, des ⊕ sein fixes Salz, welches der kleine grüne ♌ genannt wird, in dessen Bauchs innersten Eingeweiden die roten Gestirne ihren Sitz haben. Es ist auch die erste weiße Tinctur, welche man den natürlichen Schatz nennt und aus diesem doppelten Salz gemacht wird. Auch wird dies der grüne ♌ genannt, das erste *Oleum* ♀ aus dem ♄, so ♀ Art und Natur, so *tinea viridis girat* genannt, dadurch der ☿ præcipitiert wird, diese sind in ihrem inwendigen grün, kommen auch zur weißen Tinctur, denn durch das *Oleum* ♄ wird der ☿ fix gemacht und in das gläserne Meer verwandelt.

45. *Was ist denn der grüne und rote ♌, die miteinander in einen Freundschafts-Bund treten, als der gestirnte ♌ mit der Kronen?*

Antwort: Es ist die weiße Tinctur, welche mit den rotem ♌ vereinigt

40. *What is this called when it has been fixed or calcined?*

Answer: It is called the sea of glass, which can be mingled with fire, the natural treasure, the stone of the first order and the first tincture on white. It looks a little greenish, its taste is austere, but is later converted into sweetness.

41. *What is the name of the red tincture?*

Answer: It is called the first tincture in red, the first stone in red, and the bloody stone.

42. *What is the red man?*

Answer: It is the red *oleum*, the *elementum ignis*, from the philosophical ☉, the coral-red sap.

43. *What is the white woman?*

Answer: The white ♃ of nature that has now dried up, the white moonlight.

44. *What is the green ♌?*

Answer: It is the very innermost ☉, the fixed salt of the ☉, which is called the little green ♌, in whose belly's innermost entrails the red stars are seated. It is also the first white tincture, called the natural treasure, which is made from this double salt. This is also called the green ♌, the first *oleum* ♀ from the ♄, which has the habit and nature of ♀, which is called *tinea viridis girat*, whereby the ☿ is precipitated. These are green in their interior parts, they come also to the white tincture, because through the *oleum* ♄ the ☿ is fixed and transformed into the sea of glass.

45. *What is the green and red ♌ that enter into a bond of friendship with each other, like the starry ♌ with the crown?*

Answer: It is the white tincture which is combined with the red ♌ so

wird, dass aus beider Mund der güldene Saft ausfließt, die weiße Tinctur zur roten wird und der grüne ♌ als das aufgehende ☉ oder die Sonne, so noch hinter dem Mond steckt, in das rote ☉ verwandelt, in die rote Tinctur, und wird die weiße Tinctur, mit den Namen des gestirnten grünen ♌ belegt zum Unterschied des vorigen Salzes oder grünen ♌, welches auch *Aurum perspectible* genannt wird.

46. *Was ist denn die Sonne und der Mond?*

Antwort: Dies wird in vielen Verstande genommen, erstlich ist unser feuchter Mond, welchen wir auch Jungfer-Milch nennen, welche in unser roten Erde gesteckt, auch heißt unser weißer ♄ der Mond. *Item*, die weiße Tinctur, auch wird das unsere Sonne genannt, das ☉ Blut, ehe es noch aus der roten Erde extrahiert und hernach, wenn es zum roten durchscheinenden Gold geworden, auch ist dies, unsere Sonne, die geblätterte Erde, *item*, der rote figierte ♄ und die rote Tinctur, die rechte Sonne aber ist die zusammengesetzte und figierte *Materia* des Steins, wenn sich der ☿ in unser aufgehendes ☉ verwandelt, hernach, wenn solches ☉ gelb und rot gefärbt.

47. *Was ist denn die gekrönte Königin?*

Antwort: Es ist die weiße Tinctur.

48. *Was ist denn der gekrönte König?*

Antwort: Die rote Tinctur.

49. *Was ist denn die Jungfer-Milch?*

Antwort: Es ist das pontische Wasser, der einfache ☿ und *Circulatum minus*, wenn er mit seinen Salz sublimiert.

50. *Was ist denn der Adler?*

Antwort: Es ist der ☿ *Philosophorum*.

51. *Was ist denn die geblätterte Erde?*

that the golden juice flows out of both mouths. The white tincture becomes the red [tincture], and the green ♌ as the rising ☉ or the sun, which is still behind the moon, is transformed into the red ☉, into the red tincture, and the white tincture is given the name of the starry green ♌, to distinguish it from the previous salt or green ♌, which is also called *aurum perspectible*.

46. What is the sun and the moon?

Answer: A lot can be understood by this. First of all it is our moist moon, which we also call virgin milk, which is contained in our red earth; and our white ♃ is also called the moon. It is the white tincture, also called our sun, the ☉ blood, before it is extracted from the red earth and after its extraction, when it has become red, translucent gold, this too is our sun, the leafed earth. Likewise, the red, fixed ♃ and the red tincture, but the true sun is the composite and fixed *materia* of the stone when the ☿ transforms into our rising ☉, hereafter, when this ☉ is colored yellow and red.

47. What is the crowned queen?

Answer: It is the white tincture.

48. What then is the crowned king?

Answer: The red tincture.

49. What is virgin milk?

Answer: It is the pontic water, the simple ☿ and *circulatum minus*, when it sublimates with its salts.

50. What is the eagle?

Answer: It is the ☿ Philosophorum.

51. What then is the leafed earth?

Antwort: Es ist der ☿, welcher von seiner wässrigen Feuchtigkeit entbunden und aus dem Meer aufs trockene Land getreten.

52. Was ist denn der gestirnte Adler?

Antwort: Eben diesen doppelten ☿, der in seinem Öl aufgelöst und das große Circulat genannt wird, in Geist und Leib, das Wasser, so die Hand nicht nass macht, darinnen der König und die Königin baden.

53. Was denn der Cyllenius mit ausgebreiteten Flügeln, dessen beide Seiten Sonne und Mond beschließen?

Antwort: Eben dieses multiplicierende und vermehrende Menstruum, das vegetabilische Lebens-*Aquavit*, darinnen die Kräfte der Sonne und des Mondes.

54. Was ist denn der Spiritus ☿?

Antwort: Es ist der rohe, unreife und unzeitige mineralische Geist, der doch ein ewigwährendes *Menstruum* ist, denn er lässt alle seine Teile eins nach dem anderen auf, welche ihm auch alle abzunehmen sind, bis sie alle durch ihn geläutert und gereinigt.

55. Was ist denn der Drache mit Flügeln?

Antwort: Es ist der einfache ☿.

56. Was ist denn der Drache ohne Flügel?

Antwort: Es ist des ☿ fixer Teil und der Drache, so sich in der Erde enthält.

57. Was ist denn die Nymphe des Meers, welche mit ihren Brüsten Blut und Milch eingießt?

Antwort: Es ist des *Neptuni* Tochter, die Jungfer-Milch, wenn das Rote und Weiße zum Vorschein kommt, denn beides kommt aus einem Leibe, erst ist es Milch, wenn es erhärtet weißer ♃, der hernach mit der Gold-Seele geziert gelb und rot wird.

Answer: It is the ☿ which was delivered from its watery dampness and which stepped out of the sea onto dry land.

52. *What then the starry eagle?*

Answer: Just this double ☿ which is dissolved in its oil and is called the great *circulat*, in spirit and body, the water that does not wet the hands, in which the king and queen bathe.

53. *What then is Cyllenius*[180] *with outstretched wings, both sides of which enclose the sun and the moon?*

Answer: Simply this multiplying and increasing *menstruum*, the vegetabilic *aquavitæ* of life, in which are contained the forces of the sun and the moon.

54. *What is the spirit ☿?*

Answer: It is the raw, immature, and untimely mineral spirit, which is nevertheless an everlasting *menstruum*, for it dissolves all its parts one by one, which are also all to be taken from it, until they are all cleaned and purified by it.

55. *What is the dragon with wings?*

Answer: It is the simple ☿.

56. *What is a dragon without wings?*

Answer: It is the fixed part of ☿ and the dragon that is contained in the earth.

57. *What then is the nymph of the sea, pouring out blood and milk from her breasts?*

Answer: It is the daughter of Neptune, the virgin milk, when the red and white appear, for both come from one body. It is only milk when it hardens white ♃, which afterwards, adorned with the gold-soul, turns yellow and red.

58. *Was sind denn die 3 Brunnen des immerwährenden Wassers?*

Antwort: Es sind die 3 *Principia* ⊖, ♃, ☿, welche alle 3 aus der metallischen Schlange, so im grünen Gras liegt, entspringen, aus dem 🜨 gemacht worden, daraus Sonne und Mond und alle Gestirne ihre Kräfte bekommen und die lebendigen Wasser daraus gemacht werden, das Lebens-*Aquavit*.

59. *Was ist denn das philosophische ☉?*

Anwort: Es ist der Apollo in königlichen Kleidern, der doppelte ☿, so sich selber in Sonne gekocht und durch seinen eigenen Adler den königlichen Thron erstiegen.

60. *Was ist denn die Schlange mit 3 Hälsen?*

Antwort: Es ist die giftige Erd-Schlange, unser *Subjectum*, welche die 3 *Principia* in sich gehalten, so von ihm geschieden und wieder zusammen in einen Leib gebracht, in Seele, Geist und Leib vereinigt.

61. *Was ist denn der Pelikan, welcher seine Jungen mit Blut bespritzt?*

Antwort: Es ist unser rotes feuriges *Menstruum*, seine Kinder sind die Jungen, die 7 philosophischen Metalle, welche aus seinem Leibe kommen und aus dem weißen ☿al Wasser gewachsen, die weißen und roten ♃, welche er mit Blut bespritzt, durch die Seele lebendig macht und in seine rote Natur verwandelt.

62. *Was ist denn das grünende Erz?*

Antwort: Es ist das philosophische ☉, welches in Glase wächst.

63. *Was ist denn der Duenech?*

Antwort: Unser ♃ Öl, so den Stein figiert und flüssig macht.

64. *Was ist denn der Drache, so durch Sprossung der Sonnen pausiert?*

Antwort: Es ist der doppelte ☿ so in ☉ gekocht, welches durch ihn in

58. *What then are the three fountains of everlasting water?*

Answer: They are the three *principia* ⊖, 🜍, ☿, all three of which spring from the metallic serpent lying in the green grass, from which ⊕ was made, and from which the sun and moon and all the stars receive their powers; and the living waters are made of it, the *aqua vitæ* of life.

59. *What is the philosophical* ☉*?*

Answer: It is the Apollo in royal robes, the double ☿, which cooked itself in the sun and ascended the royal throne by its own eagle.

60. *What is the snake with the three necks?*

Answer: It is the venomous earth-snake, our *subjectum*, which held the three *principia* within itself, [which were] thus separated from it, and [then] brought together again into one body, united into soul, spirit, and body.

61. *What is the pelican that sprinkles its young with blood?*

Answer: It is our red, fiery *menstruum*. Its children are the young, the seven philosophical metals, which come out of its body and which have grown out of the white ☿al water, the white and red 🜍, which it sprinkles with blood, enlivens through the soul, transforming [it] into its red nature.

62. *What is the verdant ore?*

Answer: It is the philosophical ☉ which grows in glass.

63. *What is the Duenech?*

Answer: Our 🜍 oil, which forms the stone and makes it liquid.

64. *What then is the dragon that pauses through the sprouting of the suns?*

Answer: The double ☿ which is boiled in ☉, which has grown in his

seinem Leibe gewachsen.

65. *Ich möchte gern etlicher, sonderlich der 12 alten berühmten chymischen Vorgänger und Lehrer Sprüche erklärt wissen: als, erstlich ist der Hermes Trismegistus ein Ägypter, dessen Spruch ist gewesen dieser, nämlich: Des Ehestandes Vater ist die Sonne, der weiße Mond die Mutter, der dritte Director ist das Feuer so dabei: was soll dieses sein?*

Antwort: Hier ist angezeigt, wie das philosophische ☉ und ☽, als der weiße und rote ☿ soll zusammen gesetzt werden und durch das philosophische Feuer vereinigt, welches solches Werk dirigiert, dadurch der Sonnen-Sohn gezeugt wird, dessen Vater die Sonne und der weiße Mond die Mutter. Das Feuer ist weder Kohlen noch Flammen-Feuer, sondern das glühende rote feurige *Menstruum*, so aus dieser *Materia* kommt.

66. *Was bedeutet denn der Mariæ Hebrææ, Moses Schwester einer Palästinenserin, ihr Spruch, nämlich: Der Rauch liebt den Rauch und wird wieder von demselben geliebt. Aber des hohen Berges weißes Kraut beschließt oder fasst alles beides in sich?*

Antwort: Der Rauch ist der ☿, das rauchende Gift, wenn er seinen eigenen, vom Mutter-Leibe angeborenen Bruder, welches sein fixes Teil, solviert und solchen auch flüchtig macht, dass er mit ihm zum Rauch wird. Denn diese beiden lieben einander, steigen in die Höhe und wieder auf den Grund, ruhen einander in Armen, des hohen Berges weißes Kraut beschließt alles beides in sich, dadurch wird der doppelte ☿, die sublimierte *Terra Foliata* angezeigt, welche diese beiden in sich hält, wenn die vorigen 2 rauchenden Lösch-Brände in einem Leib gewachsen.

67. *Was meint Democritus der Grieche, mit diesem Spruch, nämlich: Damit der bewegliche Schatten des dicken Leibes beraubt werde, so müssen die feurigen Arzneien durch ein beständiges Glück es geben?*

Antwort: Der dicke Leib sind unsere 7 Metalle, so alle in einem Grad in

body through it.

65. *I would like to have an explanation of some sayings, especially from the twelve old, famous, chymical predecessors and teachers:*[181] *Firstly, Hermes Trismegistus is an Egyptian*[182] *whose saying was this: "The father of the marital union is the sun, the white moon the mother, the third that directs it all must be the fire"—what is this supposed to be?*

Answer: Here is indicated how the philosophical ☉ and ☽, i.e., the white and red ♄, are to be put together and united by the philosophical fire which directs this work, whereby the son of the sun is begotten, whose father is the sun and whose mother is the white moon. The fire is neither charcoal nor flaming fire, but the glowing, red, fiery *menstruum* that comes from this *materia*.

66. *What does Mariæ Hebræ, Moses' sister, a Palestinian,*[183] *mean by her saying: "The smoke loves the smoke and is loved again by the same. But the white herb of the high mountain closes or encloses both in itself?"*

Answer: The smoke is the ☿, which is the smoking poison when it dissolves its own brother born of the mother's womb, which is its fixed part, making it thereby volatile such that with it, it becomes smoke. Because these two love each other, they climb up and down again, resting in each other's arms. The high mountain's white herb encloses both in itself, which shows the double ☿, the sublimated *terra foliata*, which these two hold within themselves, when the previous two smoking extinguishing-fires, were grown into one body.

67. *What does Democritus the Greek*[184] *mean with this saying, namely: "In order for the moving shadow to be robbed of its thick body, the fiery medicines must give it through constant fortune"?*

Answer: The thick or fat body is our seven metals, which all emanate in

☉ ausgehen, welche durch die feurigen erleuchtenden Menstrua resolviert, dadurch erleuchtet, verklärt und durchsichtig werden, dass der dicke Leib weg kommt und zur feurigen beständigen und leuchtenden Arznei wird.

68. *Morienus, der Römer, hat diesen Spruch geführt, nämlich: Nimm, was du mit Füßen trittst, wirst du dich aber unterstehen ohne Leitern zu steigen, so versichere ich dich, dass du auf deinen Kopf herunter fallen wirst, was meint er dadurch?*

Antwort: Nimm was du mit Füßen trittst, ist die Erde, die Leiter ist das Wasser mit den vielen Sprossen, die vielen Teile und Hilfe, dadurch die Erde befeuchtet, erweicht und resolviert wird, dadurch sie endlich flüchtig und in die Höhe steigt.

69. *Avicenna, der Araber, hat diesen Spruch geschrieben, nämlich: Vereinbare die irdische Kröte mit dem fliegenden Adler, dann wirst du ein großes Meister-Stück und Geheimnis unser Kunst zu sehen haben, was meint er dadurch?*

Antwort: Durch die irdische Kröte versteht er das fixe Salz so in der Erde steckt, durch den fliegenden Adler den ☿, durch derer beider Vereinigung kommt der ☿ der Weisen an Tag, ohne welchen der Stein nicht zu machen.

70. *Albertus Magnus, ein Deutscher, hat gesagt: Dass diejenigen, welche geschrieben, dass es in einem bestehe, alle überein treffen, welcher einen in zwei Teile zerspaltenen Leib und beiderlei Glieder hat, was meint er dadurch?*

Antwort: Er meint dadurch den Hermaphrodit, den doppelten ☿, welcher beider Natur, weiblich und männlich, besteht aus Wasser und Erde, so beides aus einem *Subjecto* kommt.

71. *Arnoldus de Villa Nova oder Villa Novanus, ein Franzose, hat geschrieben: Aus dem Ehebette des Gabricus und der Beja wird unser Kleiner*

a degree in ☉, which resolves through the fiery illuminating *menstrua*, thereby becoming enlightened, transfigured, and transparent, so that the thick body disappears and becomes the fiery, permanent, and luminous medicine.

68. Morienus, the Roman,[185] used this saying: "Take what you trample under your feet, but if you try to climb without ladders, I assure you that you will fall on your head". What does he mean by that?

Answer: "Take what you trample upon" is the earth; the ladder is the water with the many rungs, the many parts and aids whereby the earth is moistened, softened, and resolved, eventually making it volatile and [able to] climb into the heights.

69. Avicenna, the Arab,[186] wrote this saying: "Unite the earthly toad with the flying eagle, then you will have a great masterpiece and secret of our art to see". What does he mean by that?

Answer: By the earthly toad he understands the fixed salt which is in the earth. By the flying eagle he means the ☿; through the union of both, the ☿ of the wise comes to light, without which the stone cannot be made.

70. Albertus Magnus, a German,[187] said: "That those who wrote, that it consists in one thing, agree with one another, which [one thing] has a body divided into two parts and both members". What does he mean by that?

Answer: By this he means the hermaphrodite, the double ☿, which is of both natures, female and male. It consists of water and earth, so both come from one *subjecto*.

71. Arnoldus de villa nova or Villa Novanus, a Frenchman,[188] wrote: "From the marriage bed of Gabricus and Beya, our little one will come out

in die Luft hervorkommen, nachdem er mit ihnen vereinigt worden, was will er dadurch anzeigen?

Antwort: Damit zeigt er an die erste Composition, das *Matrimonium, primum concubitum*. Gabricus ist der Mann, der fix; Beja das Weib so flüchtig, durch diese beiden kommt ihr Sohn, der doppelte ☿ hervor, welcher durch die Luft, das lebendige Wasser, des Königs Bad gibt, wenn er mit demselben vereinigt wird.

72. Thomas Aquinus, der Italiäner, ist genannt worden der englische Doctor, welcher geschrieben: Aus dem Quecksilber, so da mit dem eigenen Schwefel vereinbart und vermischt wird, gleich wie die Natur, also wird die Kunst alle Metalle gebären, was meint er dadurch?

Antwort: Hier beschreibt er, dass der ☿ mit seinem eigenen ♄ soll vereinigt werden, wenn er soll fruchtbar werden und Metalle gebären, der ♄ ist der Vater, der ☿ ist die Mutter, durch deren beider Vereinigung zeugen sie auch Kinder, ist der ☿ rein und schön, der ♄ auch, so wird es eine herrliche Frucht solarischer Natur, ist der ☿ aus Art ♄, so wird ☽ daraus, ist aber der ☿ ♂ und ♀ Natur, wird ☉ daraus, ist er aber solarischer Natur, wird eine Tinctur daraus: Also muss ein jeder wissen, was er vor einen ☿ præcipitiert, ob er den erhöhten auf der 7ten Stufe hat oder einen in niedrigerem Grad. Ich habe hier mit dem gemeinen metallischen ☿ nichts zu tun, sondern mit unseren 7 philosophischen Metallen, hat nun jemand das Glück, den erhöhten ☿ zu finden und præcipitiert ihn, mit seinem eigenen, vom Mutter-Leibe angeborenen ♄, so hat er gleich eine Tinctur, welche viel oder wenig tingiert, nachdem diese beiden rein oder unrein, doch kann er diesen Præcipitat in seinen eigenen Fermenten, Unguenten und Salben erhöhen, illuminieren und färben, so hoch er will, denn die erste weiße Tinctur kann er färben, durch seine eigenen Öle die Farben einbeizen, dass sie fix und beständig bleiben, zitronengelb, pomeranzengelb, Rosen-Farbe, von solcher Farbe als Granat-Körner, rot wie Spinetten, endlich Rubin-rot, dick rot, wie gestocktes dürres Blut machen, bis es genug

into the air after he has been united with them". What does he mean by that?

Answer: With this he points to the first composition, the *matrimonium, primum concubitum*. *Gabricus* is the man, the fixed; *Beya* is the woman, the volatile. Through these two comes forth their son, the double ☿, which through the air gives the king's living water bath when he is united with it.

72. *Thomas Aquinus, the Italian,*[189] *also called the angelic doctor, wrote: "From mercury, when it is blended and mixed with its own sulphur, like nature, then art will give birth to all metals"*. What does he mean by that?

Answer: Here he explains that the ☿ should be united with its own ♄ if it is to become fertile and give birth to metals. The ♄ is the father and the ☿ is the mother, and through their union they also beget children. If the ☿ is pure and beautiful, and the ♄ too, then it will be a wonderful fruit of solar nature. If the ☿ is of ♄ nature, it will become ☽, but if the ☿ is of ♂ and ♀ nature, it will become ☉; but if it is of solar nature, it will become a tincture. Thus everyone must know what he precipitates in front of a ☿, whether it is exalted to the seventh degree or if it is a lower degree. I am not dealing here at all with the common metallic ☿, but with our seven philosophical metals. If someone is lucky enough to find the elevated ☿ and precipitates it with its own innate ♄ from the mother's womb, then they have a tincture right away which tinges either a lot or else not so much, depending on whether they are both pure or impure. Yet it can increase, illuminate, and color this precipitate in its own ferments, ointments, and salves as much as one wants, for it can color the first white tincture, staining the colors with its own oils so that they remain fixed and permanent, taking on the colors of lemon, bitter-orange, rose, pomegranate seeds, spinets; and then finally ruby-red, a thick red, made dry like stagnant blood, until it is sufficiently

gesättigt und nichts mehr zu sich nehmen will. Hat man aber zu seinem Anfang einen geringen ☿ und nicht den der 7 Nächte, der 7-mal getötet und wieder lebendig worden, so muss man diesen unreifen ☿ erst durch einen güldischen ♃ überwinden und muss der Überwinder stärker sein denn der, so überwunden werden soll und hier stecken viele Verzögerungen, da etliche in einem Monat die Tinctur machen können, andere aber sich viele Monate plagen und viel Gefahr ausstehen müssen und hier sind so vielerlei Arten, Wege und Tincturen zu machen, dass man sie fast nicht zählen kann. Denn man kann durch alle metallischen und mineralischen ♃, welche güldischer Natur, den philosophischen ☿ damit præcipitieren und zum fixen tingierenden Præcipitat machen, diesen Præcipitat in seinen eigenen tingierenden Ölen und Giften wieder resolvieren, aufs neue fix machen, auch solches wiederholen so oft man will, dadurch die Tincturen immer mehr und mehr erhöht werden und viel mehr Teile als vorher tingieren. Dies hat noch kein Mensch so deutlich beschrieben, was ich jetzt tue, wer es versteht und findet, der wird erkennen lernen, was er mir für Dank zu geben schuldig ist.

73. Raymundus Lullius, ein Spanier, hat geschrieben: Das Männchen bringt des Kindes Leib in Actum, wenn ihm das Weib zugegeben wird, was meint er hierdurch?

Antwort: Er will dadurch anzeigen, wenn Mann und Weib, welche erstlich Wasser und Erde, ihr Kind, den trocken ☿ zu Wege bringen, welches hernach die andere Verehelichung wird, wenn ☿ und ♃ als Mann und Weib zusammen verehelicht, wieder ein anderes Kind zeugen, weit besser als vorher und ist die zweifache Composition. In der ersten wird der doppelte ☿ gemacht, in der anderen die Tinctur, da dieser Baum in eine andere Erde gepflanzt wird, welches der doppelte ☿ ist und zur Erde werden muss, da das philosophische ☉ in die geblätterte Erde gesät wird, das weiße Weib dem roten Mann anvertraut, dadurch sie einander ein Kind zeugen, die Tinctur.

saturated and does not want to take up anything anymore. But if one has a small ☿ at the beginning and not that of the seven nights, which has been killed and has come to life again seven times, then one must first overcome this immature ☿ with a golden ♀ and the overcomer must be stronger than the one who is to be overcome and here there are many delays, since some can prepare the tincture in a month, but others toil for many months and have to endure much danger, and here there are so many different modes, ways, and tinctures to be made that one can hardly count them. Because you can precipitate the philosophical ☿, make it a fixed tinging precipitate with all metallic and mineral ♀ which are of a golden nature, and resolve this precipitate again in its own tinging oils and poisons, fix it again, repeat this as often as you want, whereby the tinctures become more and more increased, and tinge many more parts than before. No one has yet described this so clearly as I am doing now; whoever understands and finds this will learn to recognize what they owe me out of gratitude.

73. *Raymundus Lullius, a Spaniard,*[190] *wrote: "The male brings the child's body into actum when the female is admitted to him". What does he mean by this?*

Answer: He wants to indicate that when man and woman, which are firstly water and earth, beget their child, the dry ☿, this thereafter becomes the second marriage. When ☿ and ♀ are married together as man and woman, he fathers another child again, one far better than before, and which is the double composition. In the first [composition] the double ☿ is made, in the second [composition] is made the tincture, since this tree is planted in another earth, which is the double ☿ and must become earth, since the philosophical ☉ is sown in the leafed earth, the white woman, entrusted to the red man, through which they beget each other's child, the tincture.

74. *Rogerius Bacon, ein Engländer, hat geschrieben: Wenn du gleiches Gewicht der Elemente wieder geben wirst, so wirst du mit deinen Augen angenehme Geschenke sehen, was will er dadurch anzeigen?*

Antwort: Das gleiche Gewicht der Elemente ist in vielfachem Verstande und meint er nicht, dass man soll gleiche Teile auf der Waage abwägen, sondern meint das Gewicht der Natur, *Agens* und *Patiens*, bis sich diese verglichen, alle *Elementa* und nicht mehr streiten. Denn wer will die Luft und Feuer wägen? Wenn aber die *Elementa* vereinigt und ihre gleichen Gewichte haben, sieht man die angenehmen Geschenke, welche sind die rote und die weiße Tinctur. Man bedarf aber des Gewichtes nicht eher als in der Composition, und zwar erstlich, wenn Wasser und Erde, als Mann und Weib, soll zusammengesetzt werden, so muss das Weib vielfältig sein. Denn erstlich muss es bei der Erde gerinnen, auch zur Erde werden, dann muss das Wasser die ganze Erde erweichen, auflösen und dünn machen, dass sie zu Wasser wird. Dieses Wasser muss faulen, schwarz und stinkend, zu schwarzer Erde werden, diese schwarze Erde wird hernach die bleibende Erde und wird durch den *Azoth*, ihr eigenes Wasser, weiß gewaschen und in die *Terram Foliatam* verkehrt. Zum anderen mal kommt wieder das Gewicht, wenn so viele Adler hergeführt werden, welche den ♌ zerreißen sollen, da die glücklichste Zahl von 7 bis auf 10 sein soll und wird hier nicht gemeint 7 oder 10 mal so schwer, der Adler und ein Teil des ♌, sondern es hat hier die Bedeutung der Eintränkung oder Sublimation der Adler. Wenn man alle die ersten kleinen Eintränkungen von kurzer Zeit rechnen will, so werden es 10 sein, insgemein aber rechnet man nur die letzten, welches rechte Eintränkungen sind und in siebenfacher Umdrehung des Rades bestehen, dadurch die Adler 7-mal zubereitet, das ist, sublimiert werden im Siebe der Natur. Auch ist das ein Gewicht, wenn ich diesen ✶ oder Adler in seiner Luft oder Öl auflöse, welches ungleiche Teile. Wenn nun durch diese ungleichen, widerwärtigen Dinge solche herrlichen Früchte nur durch das *Agens* und *Patiens* gezeugt und diese alle in eins zusammen bracht, dass sie nicht mehr miteinander streiten,

74. *Rogerius Bacon, an Englishman,*[191] *wrote: "If you will give equal weight of the elements, you will see pleasant gifts with your eyes". What does he mean by this?*

Answer: The equal weight of the elements can be understood differently. He does not mean that one should weigh equal parts on the scales, but rather he means the weight of nature, *agens* and *patiens*, until they are balanced, all *elementa* and no longer argue. For who will weigh the air and the fire? But when the *elementa* are united and have their equal weights, one sees the pleasant gifts, which are the red and the white tinctures. But one does not need the weight any sooner than in the composition, first of all, if water and earth, as man and woman, are to be put together, then the woman must be manifold. Because first it has to coagulate with the earth and become earth, then the water has to soften, dissolve, and dilute the whole earth so that it becomes water. The water must putrefy, [becoming] black and stinking, to become black earth. This black earth then becomes the permanent earth and is washed white by the *azoth*, its own water, and turned into the *terram foliatam*. For the second time the weight emerges, when so many eagles are brought here, which are supposed to tear apart the ♌, since the happiest number should be from seven to ten, and here this does not mean seven or ten times as heavy, the eagle and part of the ♌, but here means the impregnation or sublimation of the eagles. If one wants to count all the first small impregnations of a short time, then there will be ten, but in general one only counts the last ones, which are real impregnations and consist of turning the wheel [or: circulating] seven times, thereby preparing the eagle seven times, that is, it will be sublimated in the sieve of nature. It is also a weight when I dissolve this ✶ or eagle in its air or oil, which are unequal parts. If now through these unequal recurrent things such glorious fruits are shown only by the *agens* and *patiens*, and these are all brought together into one, in such wise that they no longer quarrel with one another, but the quarrel is settled,

sondern sich der Streit gelegt, so lass sie zusammen ruhen, *Agens* und *Patiens* sich vergleichen und keine *Turba* mehr ist, auch kein Wirken noch Leiden, denn ist die Herrlichkeit der Welt erstritten, das gleiche Gewicht den Elementen wiedergegeben, der neue Himmel und die neue Erde, rote und weiße Tinctur, in eins verkehrt und ein einziger leuchtender Himmel worden.

75. *Melchior Cibinensis, ein Ungar, hat geschrieben: Gleich wie im Anfang das zarte Kind durch schneeweiße Milch ernährt wird; Also muss dieser Stein mit reiner Milch gespeist werden, was meint er hiermit?*

Antwort: Er meint hierdurch die erste Composition, wenn er durch die Jungfer-Milch ernährt wird, dass er wächst und zunimmt und größer wird und ist der ☿, das nackende Kind, welches so lange muss gespeist werden, bis es zu seinen männlichen Jahren kommt, dann wird er durch die Luft gespeist und ist der Chamäleon, so alle Farben an sich nimmt und von der Luft lebt, zuletzt wird er durchs Feuer gespeist und ist der Salamander, so im Feuer lebt.

76. *Michael Sendivogius, ein Pole , hat geschrieben: Der ♄ befeuchtet die Erde, welche, O Sonn und Mond deine Blumen hat, was will er damit?*

Antwort: Er will zu erkennen geben, dass man die geblätterte Erde, welche die Blumen des philosophischen ☉ und Silbers hat, mit dem ♄ Öl begießen und anfeuchten soll, dadurch dieses flüchtige, feurige, lebendige ☉ und ☽, welches der ☿ *Solaris*, das weiße flüchtige ☉ ist, möge fix, zum weißen Schwefel werden, welches der ♄ mit seinem Harn waschen, das ist anfeuchten muss und hat dieses *Oleum* keinen Nutzen in der ganzen Kunst, als nur den ☿ fix zu machen, ist der jungen Kinder Blut, darin *Sol* und *Luna* zu baden pflegen.

77. *Was meinen sie denn durch die Tiere, als Widder, ♌, Ochse, Bär, Wolf, Adler, Phönix, Pfau, Rabe und Schwan, vergüldete Otter, Drache, Schlange und Kröte?*

Antwort: Durch den Widder verstehen sie ♂, weil es sein Haus am

then let them rest together, let the *agens* and *patiens* settle so that there is no longer a *turba*, neither work nor suffering, because now the glory of the world has been won. The same weights are given back to the elements, the new heaven and the new earth, the red and the white tincture, which are turned into one and become a single shining sky.

75. Melchior Cibinensis, a Hungarian,[192] wrote: "Just as in the beginning, the tender child is nourished by snow-white milk, so must this stone be fed with pure milk". What does he mean by this?

Answer: By this he means the first composition, when it is nourished by the virgin milk so that it grows and puts on weight and becomes larger; and if it is the ☿, the naked child which must be fed until it reaches its manly years, then it is fed by the air and is the chameleon that takes on all colors and lives from the air. Finally it is fed by the fire and is the salamander that lives in the fire.

76. Michael Sendivogius, a Pole,[193] wrote: "The ♄ moisturizes the earth, which, O sun and moon, has your flowers". What does he want to say by this?

Answer: He wants to make it known that the leafed earth, which has the flowers of the philosophical ☉ and silver, should be poured and moistened with the ♄ oil, thereby this volatile, fiery, living ☉ and ☽, which is the ☿ *Solaris*, the white fleeting ☉, may be fixed and become white sulphur, which the ♄ must wash with his urine, that is it must be moistened; and this *oleum* has no use in the whole art other than to make the ☿ fixed. It is the young children's blood, in which *Sol* and *Luna* used to take a bath.

77. What do they mean by beasts, such as ram, ♌, ox, bear, wolf, eagle, phœnix, peacock, raven, and swan, gilded viper, dragon, snake, and toad?[194]

Answer: Through Aries[195] they understand ♂ because it is its house in

Himmel; durch den ♌ die rote Erde, darinnen die philosophische Sonne, denn der ♌ ist der Sonnen Haus, vielen genannt, wenigen bekannt; Ochse oder Stier ist der ♀ ihr Haus, wird dadurch das *Menstruum* angedeutet, so ♀ Natur und die Sterne sind, welche Feuer und Rauch durch die Nase geblasen, der Stier hat auf dem Haupt der ☽ Zeichen, dies weist an, dass diese ♀ muss in die ☽ verkehrt werden, die ♀ lässt sich von ihr unterdrücken und lässt der ☽ das Regiment; der Bär bedeutet den alten ♄, dessen Öl, so ♀ Art, weil er so viel mit der ♀ gespielt, welches den ☿ præcipitiert zum weißen fixen ♃ macht, ist der Beischlaf des ☿ und der ♀; Der Wolf und der große Hunger in diesen ♄ Tier, daher frisst er den König auf einmal in sich; Der Adler ist der ☿, auch der Phönix, wenn er durch sein eigenes Öl gelb gefärbt, die vergüldete Otter; Dieser Phönix verbrennt sich, wird aus der Asche ein schönerer; der Pfau die bunten Farben; der Rabe die schwarze; der Schwan die weiße; wenn es durch das arsenikalische Öl angefeuchtet, dass es gelb, ist es der giftige Drache. Wenn es noch roh und giftig, Schlange oder Stein-Schlange, der ☉ ist die Schlange im grünen Grase; Kröte das Salz in der Erde.

78. *Ich möchte auch gern etliche Figuren, Sinnbilder, Texte und Sprüche M. Dan. Stolcens von Stoltzenberg , ausgelegt und erklärt wissen, als:*

Die erste Figur
Sinnbild, Text und Spruch, darinnen 4 Operationes als: Solutio, Ablutio, Conjunctio, Fixio

Der Text und Spruch ist: Siehe, hier werden durchs gleiche Los 4 Schwestern gemalt, diese erzählen wie dein Werk aussieht. Die erste befiehlt, dass man den beiseite gebrachten Leib auflösen soll. Die andere, wie man die Materie fein geschicklich waschen; Die dritte gibt dir an die Hand, die voneinander geteilten Teile wieder zusammen zu bringen. Die vierte lehrt dich, den Stein im Feuer zu figieren. *Auslegung*: Die 4 Schwestern sind die 4 Operationes in der Nacharbeit, die erste befiehlt, den beiseite gebrachten Leib aufzulösen, den doppelten

the sky; through the ♌ [they understand] the red earth, in which the philosophical sun resides, for the ♌ is the house of the sun, called by many, known only to a few. The ox or bull is the house of ♀, and thereby the *menstruum* is indicated, which has the nature of ♀ and the stars, which blew fire and smoke through the nose; the bull has on its head the sign of ☽, which shows that this ♀ must be turned into the ☽, the ♀ lets itself be suppressed by it and the ☽ has the regiment. The bear means the old ♄ whose oil has the nature of ♀, because it played so much with the ♀, which precipitates the ☿ into the white fixed ♁, this is the cohabitation of the ☿ and the ♀. Due to the wolf and the great hunger in these ♄ animals, he suddenly devours the king all at once. The eagle is the ☿, also the phœnix when colored yellow by its own oil, the gilded viper; This phœnix burns itself, becomes a more beautiful one from the ashes; the peacock [means] the bright colors; the raven the blackness; the swan the whiteness; when it is moistened by the arsenical oil so that it turns yellow, it is the poisonous dragon. If it is still raw and poisonous, serpent or stone-serpent, the ☉ is the serpent in the green grass. The toad [is] the salt in the earth.

78. *I would also like to have some figures, symbols, texts, and sayings of Magister Daniel Stolcens von Stoltzenberg*[196] *interpreted and explained, specifically:*

THE FIRST FIGURE[197]

Symbol, Text, and Saying, in which are four Operationes, such as: Solutio, Ablutio, Conjunctio, Fixio

The text and saying is: "See, here four sisters are drawn by the same lot, they tell what your work looks like. The first commands that the body that had been set aside should be dissolved. The second instructs how to wash the matter delicately skillfully. The third gives you on the hand how to bring the divided parts back together. The fourth teaches you to fix the stone in the fire". *Interpretation:* The four sisters are the four operationes in the after-work. The first sister orders the body

☿ durch Wasser und Erde zu machen, die erste Composition, wenn das Weib den Mann auflöst, der ☿ seine Erde oder fixen Teil solviert, wodurch die Schwärze erscheint. Die andere befiehlt, diese Schwärze durch die Jungfer-Milch abzuwaschen. Die dritte lehrt, die andere Composition den Stein zu machen und die geschiedenen Teile zu vereinigen. Die vierte solchen im Feuer zu figieren, fix und beständig zu machen.

2. FIGUR ODER SINNBILD
Der philosophische Stein aus Geist, Seele und Leib

Text und Spruch: Gleich wie hier die Schlange 3 Hälse hat und doch nur einen Leib, in welchem sie grausames Gift verborgen hält; Also hat auch die Fortpflanzung 3 Naturen, der Sonne und des Mondes und hat in ihrem Leibe grausames Gift; daher geben auch der Rabe, Pfau, Schwan und König die Farben, welcher du, wenn du sie mit Augen sehen wirst, versichert sein kannst. *Auslegung*: Die Schlange ist die einzige *Materia*, welche erstlich in ihrer Rohigkeit die 3 *Principia* ⊖, ♄, ☿ in sich hält und giftig ist; Aber in der Arbeit, wenn diese 3 von ihr geschieden, gereinigt und wieder zusammengesetzt, in der Fortpflanzung und Generation, da Sonne und Mond gezeigt werden, wird es noch giftiger und erstlich schwarz, hernach bunt von vielen Farben, endlich weiß, der König im Hemd und der Stein in Seele, Geist und Leib, wenn man diese sieht, ist man versichert, dass man den weißen ♄, die weiße Tinctur haben kann.

3. FIGUR ODER SINNBILD
Unser Drache

Text und Spruch: Delius erlegt durch seine warmen Pfeile den harten grimmigen Pithon, damit er sein Leben im Feuer führe. Wenn aber jemand sich unterfangen würde zu fragen, wer dieser Drache sei? Siehe so spricht ein Adler, dass es sein ♄ sei. Wenn du aber wissen willst, woher der Delius den Bogen mit den Pfeilen nehme? So wird dieses unser niederliegender ♌ sagen. *Auslegung*: Der harte grimmige Pithon,

that has been put aside to be dissolved, and she orders one prepare the double ☿ through water and earth. The first composition in which the woman dissolves the man, is when the ☿ dissolves his earth or fixed part, whereby the blackness appears. The second commands this blackness to be washed away by the virgin milk. The third teaches the other composition to make the stone and to unite the separated parts. The fourth tells us to shape it in fire, to make it fixed and permanent.

2ND FIGURE OR SYMBOL[198]
The Philosophical Stone from Spirit, Soul and Body

Text and saying: "Just as the serpent here has three necks and yet only one body in which it hides cruel poison, so reproduction also has three natures, the sun and the moon, and has cruel poison in its body; hence the raven, peacock, swan, and king give the colors which you can be assured of when you see them with your eyes". *Interpretation:* The serpent is the only *materia* that contains the three *principia* ⊖, ♄, ☿ in its rawness and which is poisonous. But in the work, when these three are separated from it, purified, and put back together again, in reproduction and generation, when sun and moon are begotten, it becomes even more poisonous and first black, then variegated with many colors, finally white; the king in the tunic, the stone in soul, spirit, and body; when one sees these, one is assured that they can have the white ♄, the white tincture.

3RD FIGURE OR SYMBOL[199]
Our Dragon

Text and saying: "Delius kills the hard, fierce Python with his warm arrows, so that he may live his life in the fire. But if someone ventured to ask who this dragon was? Behold, thus speaks an eagle, which is his ♄. But if you want to know where Delius got his bow and arrows from, so this will be told by our resting ♌". *Interpretation:*

so aus dem ☿ und Salz gewachsen, aus der Erde aus Wasser und Salz, welcher voller glänzenden Schuppen, das erschreckliche Tier, welches in seiner Höhle der Erde lauter Gift ausgetrocknet und nun der weiße ♃ und ☿ von ☿ ist, dieser ist der Drache, so muss getötet werden, ist des Alten sein ♃, hat seinen Anfang von ♄, der Erde, Delius mit seinen warmen Pfeilen erlegt ihn, ist das *Oleum* ♄, der niederliegende ♌ ist die fixe Erde, der grüne ♌.

4. FIGUR UND SINNBILD
Calcinatio, der 1. Grad der Philosophen

Text und Spruch: Hier am Tische sitzt der Cyllenius mit ausgebreiteten Flügeln, dessen beide Seiten beschließen die Sonne und der Mond. Auf dem Tisch wachsen die darauf gesetzten Kräuter mit den Blumen und der ♌ frisst die Schlange, den ☿, siehe der leichte Geist wird durch den Kalk figiert, daher freut sich derselbe, wenn er mit dem Erdreich durch das eigene Aussprossen und Grünen der Blumen vereinigt wird. *Auslegung*: Der Cyllenius, dessen beide Seiten ☉ und Mond beschließen, ist unser lebendiges Wasser, so in Seele, Geist und Leib besteht, darinnen Sonne und Mond ihre Kräfte resolviert und in diesem Wasser die herrlichen Kräuter die Tincturen wachsen, der ♌ frisst die Schlange, den ☿, macht solchen zu Kalk und calciniert ihn, weil man sich nicht auf den sublimierten sondern auf den calcinierten zu verlassen. Nun muss dieser Kalk mit den wachsenden Menstruo angefeuchtet und mit seinem eigenen Erdreiche aussprossen und grünen, dass sich eine Natur der anderen Natur erfreut und dieser leichte Geist durch den Kalk figiert wird, daher unser grünend und wachsend Erz, unser ☉.

5. FIGUR ODER SINNBILD
Der andere Grad, Solutio

Text und Spruch: Siehe der durch die Luft oder den Himmel erhitzte ♌ frisst die hell glänzende Sonne und die glänzende *Nympha* bringt die Blumen hervor. Daher wird der feurige Mann schwitzen und durch

The hard fierce Python that grew from the ☿ and salt, from the earth, from water and salt, which is full of shining scales, the terrible beast, which dried up all poison in its cave in the earth and is now the white ♃ and ☿ of ☿, this is the dragon which must be killed, the ♃ of the old man. It has its beginning from ♄ of the earth, and Delius kills it with his warm arrows. It is the *oleum* ♄, the resting ♌, the fixed earth, and the green ♌.

4TH FIGURE AND SYMBOL[200]
Calcinatio, the First Degree of the Philosophers

Text and saying: "Cyllenius is sitting here at the table with spread wings, on both sides enclosed by the sun and the moon. On the table the herbs grow with the flowers and the ♌ devours the serpent, the ☿; behold the lightweight spirit is fixed by the lime, so it is happy when it unites with the earth through its own sprouting and the greening of the flowers". *Interpretation*: Cyllenius, whose two sides are enclosed by ☉ and moon, is our living water consisting of soul, spirit and body, in which the sun and moon resolve their forces and in these waters the wonderful herbs, the tinctures, grow, the ♌ eats the serpent, the ☿, turns such into lime and calcinates it, because one does not rely on the sublimated but on the calcined [lime]. Now this lime must be moistened with the grown menstruo and must sprout and green with its own soil, so that one nature enjoys the other nature and this lightweight spirit is fixed by the lime, hence our greening and growing earth, our ☉.

5TH FIGURE OR SYMBOL[201]
Second Degree, Solutio

Text and saying: "Behold the ♌ heated by the air or the heaven devours the brightly shining sun and the shining *Nympha* brings forth the flowers. Hence the fiery man will sweat, and through the fire

das Feuer wird er seinen Leib gänzlich in die Feuchtigkeit resolvieren. Solviere nun das *Magisterium*, so durch vorgemeldeten Kalk fertig gemacht worden, damit der ☿ durch angenehme Zeichen aufstehe. *Auslegung*: Der durch die Luft oder Himmel erhitzte ♌ ist der ☿ *Solaris*, so durch die Luft oder das himmlische *Menstruum* ausgezogen, welches der Himmel der Philosophen genannt wird. Dieser rote feurige erhitzte ♌ frisst die hell glänzende Sonne, den vorigen güldischen Præcipitat, und roten ♃ und muss dieser feurige Mann, diese hitzige Sonne, schwitzen, wenn sie in ihr eigenes Haus im ♌ kommt und wird durch das feurige Blut gänzlich resolviert, nämlich das *Magisterium* so durch vorgemeldeten Kalk fertig gemacht worden, die glänzende *Nympha* wird die Blumen bringen, das ist, die Farben und der ☿ erhöht aufstehen, das ist verbessert in höhere Tinctur.

6. FIGUR ODER SINNBILD
Der dritte Gradus, Separatio

Text und Spruch: Die aufgelösten Bande sollen hinweg fallen, das Leichte soll in die Höhe steigen, das Schwere aber hernieder, damit ein jedwedes an seinem eigenen Orte sei, daher kommt die Schwere der Erden, so die Leichte von der Höhe herabstürzen wird und durch widrigen Grad die stolzen Herzen setzen. Allein die Leichtigkeit der Luft, so mit der aufgelösten Schwere vermischt, wird durch ihre Strahlen dieselbe zieren können. *Auslegung*: In allen Solutionen geschehen die *Separationes*, denn die philosophischen sind keine solche gewaltsame, wie die gemeinen, welche ohne Unterschied rein und unrein alles wegfressen, aber in dieser Solution geschieht die Separation, und auch die Conjunction, denn die leichten Geister kommen wieder nieder und führen Leib und Seele mit in Himmel in die Höhe und conjungieren es wieder und stürzen die stolze Erde, so in die Höhe gestiegen, wieder nieder, bis das Unterste das Oberste und das Oberste das Unterste wird, der Himmel die Erde und die Erde der Himmel worden und durch die himmlischen Strahlen erleuchtet, glänzend und geziert wird.

he will resolve his body entirely into moisture. Now dissolve the *magisterium*, thus completed by the aforesaid lime, so that the ☿ gets up with pleasant signs". *Interpretation*: The ♌ heated by the air or heaven is the ☿ *solaris*, thus it is drawn out by the air or the heavenly *menstruum*, which is called the heaven of the philosophers. This red, fiery, heated ♌ devours the bright-shining sun, which is the aforesaid gold-bearing precipitate and red ♀, and this fiery man must sweat this heated sun when it comes into its own house in ♌; and it will be completely resolved by the fiery blood, that is to say: the *magisterium* has been made ready by the aforesaid lime, and the shining *Nympha* will bring the flowers, that is, the colors, and the ☿ will rise increased, that is improved in the higher tincture.

6TH FIGURE OR SYMBOL[202]
The Third Gradus, Separatio

Text and saying: "The broken bonds should fall away, that which is light should rise up, but that which is heavy sinks down, so that each one is in its own place, therefrom comes the heaviness of the earth, which will throw down the light things from above and set the proud hearts through adverse degrees. Only the lightness of the air, thus mixed with the dissolved heaviness, can adorn it with its rays". *Interpretation: Separationes* happen in all solutions, because the philosophical ones are not as violent as the common ones, which eat away everything pure and impure without distinction; but in this solution the separation and the conjunction happen because the lightweight spirits come down again and carry body and soul up with them in heaven and conjoin it again. They throw down again the proud earth, which has risen on high, until the lowest becomes the highest and the highest the lowest, until heaven becomes earth and earth becomes heaven and is illuminated, resplendent, and adorned by the heavenly rays.

7. FIGUR
Der 4. Grad, Conjunctio

Gleich wie die gewünschte Sonne nach traurigen Gewölke aufsteigt, also steigt auch größere Liebe nach dem Zorn. Die Leiber, welche du voneinander getan, bringe nun wieder zusammen, damit sich der Same mit vielen Kindern erfreue. Unterdessen kann der *Neptunus* ein laulichtes Bad zurichten, damit Mann und Weib sich baden können. *Auslegung:* Wenn die Sonne ihre Finsternis ausgestanden, nach den traurigen trüben Wolken, dann hier ist keine solche Schwärze zu sehen in der Multiplication, oder vielmehr in der Zusammensetzung, sondern wird nur etwas dunkel von Farben, gar wenig schwärzlich, dunkel, himmelblau, violenblau, rötlich, bis es wieder helle Farben gibt und in seiner Kraft multipliciert. Soll es weiter multipliciert und erhöht werden, müssen die Bäder aufs neue geheizt und zugerichtet werden.

8. FIGUR
Der fünfte Grad, Putrefactio

Die Zerstörung oder Corruption, bringt der *Materia* einen grimmigen rauhen Tod, aber der Geist renoviert daher sein Leben. Daher kommt die schwarze Kugel, welche das Ebenbild eines Rabens zeichnet und daher kommt des Menschen Gerippe und der leichte Geist. Wofern nur der gesäte Same durch den gebauten Erdboden oder Grund nicht verfaulend gemacht wird, so wird deine Arbeit vergebens und unglücklich sein. *Auslegung:* Alle Faulung bringt eine neue Gebärung, ohne dieses kommt kein neuer Leib hervor, so hernach beständig ist. Alles wird in die Erde gesät, jedes hat seine eigene Erde. Also auch unser ☿, welcher in seine eigene Erde gesät, darinnen verfaulen und verherrlicht auferstehen muss, ohne diesen Weg wird nichts gezeigt, kann auch dieser nicht zum anderen und neuen Leben auferweckt werden.

7TH FIGURE[203]
The Fourth Degree, Conjunction

"As the desired sun rises after sorrowful clouds, so greater love rises after after anger. The bodies which you have separated, bring them together again so that the seed may rejoice in many children. Meanwhile, *Neptunus* can prepare a lukewarm bath so that man and woman can bathe". *Interpretation*: When the sun has endured its eclipse, after the sorrowful murky clouds, then here is no such blackness to be seen in the multiplication, or rather in the composition; but it only becomes somewhat dark of color, turning very slightly blackish, then a dark sky-blue, followed by violet-blue, and then a reddish tinge until finally there are bright colors again, when it becomes multiplied in its power. If it is to be further multiplied and increased, the baths must be reheated and made ready again.

8TH FIGURE[204]
Fifth Degree, Putrefactio

"Destruction, or corruption, brings a grim, harsh death to the *materia*, but the spirit rejuvenates its life from this. Hence there appears the black sphere, which is of the likeness of a raven, and hence comes the human skeleton and the lightweight spirit. Unless the seed sown is made to putrefy in the tilled soil or ground, your work will be in vain and unfortunate". *Interpretation*: All putrefaction brings a new birth, without this no new body comes forth which is permanent afterwards. Everything is sown in the soil, each has its own soil. So also our ☿, which is sown into its own earth, has to putrefy in it and rise up glorified, since without [doing it] this way nothing is revealed, and also it cannot be raised up to another and new life.

9. FIGUR
Der sechste Grad, Congelatio

Allhier wird der Geist der flüchtigen Luft figiert und die verborgende Feuchtigkeit ihren Wassern wieder gegeben. Viele werden zusammen getan und viel Weiche hart werden und dann so werden die krummen Glieder der Schlangen offenbar werden. Unser ☿ vereinigt sich und wirft die Flügel hinweg, trägt königliche Zepter und auch weiße Glieder zugleich. *Auslegung:* Dies ist unser doppelter ☿, welcher durch seinen eigenen fixen Leib congeliert worden, welcher durch die Luft erweicht und den Stein in Seele, Geist und Leib bedeutet. Daher dieser ☿ königliche Kronen und Zepter trägt. Denn er ist das philosophische ☉ in seinem Innersten, ob er schon äußerlich weiße Glieder hat, weil er die *Luna fixa*, das weiße ☉ ist.

10. FIGUR
Der siebente Gradus, Cibatio

Allhier wird unser Kind durch die Mutter-Milch gespeist, damit es zunehme und neue Kräfte überkomme. Die dreifältige dreifache Messung bringt den starrenden Drachen hervor, welcher die Frucht der Sonne und des Mondes bringt. Auf einer Seite weist der Triangel 3 Adler und diese kannst du den 3 Geistern übergeben. *Auslegung:* Das Kind so durch die Mutter-Milch gespeist, ist unser coagulierter ☿ oder durch die Jungfer-Milch getränkt worden, wodurch es endlich in den harten starrenden Drachen verwandelt, welcher die Frucht der Sonne und des Mondes im Leibe hat. Auf der einen Seite hat er 3 Adler, die 3 *Principia*, welches seine Geburt-Linie daraus er besteht, diese müssen den 3 Geistern übergeben werden. Der eine Geist ist das *Oleum* ♄, dadurch er fix wird. Der andere Geist ist das vegetabilische *Menstruum*, in Seele, Geist und Leib, welches diesen fixen Præcipitat solviert. Der dritte Geist ist der rote Mann, das Feuer, dadurch es zur roten Tinctur wird, wenn dieses Feuer in diesen Wasser gekocht, ist hier der kurze Weg gezeigt.

9TH FIGURE[205]
The Sixth Degree, Congelatio

"Here the spirit of the volatile air is fixed and the hidden moisture is restored to its waters. Many things will be done together, and many soft things will become hard, and then the crooked limbs of the serpents will be revealed. Our ☿ unites and throws away its wings, carries royal scepters and also white limbs at the same time". *Interpretation*: This is our double ☿, which has been congealed by its own fixed body, which is softened by the air, signifying the stone in soul, spirit, and body. Hence this ☿ bears royal crowns and scepters. This is so because it is the philosophical ☉ in its innermost being, even though it already has white limbs on the outside, for it is the *luna fixa*, the white ☉.

10TH FIGURE[206]
Seventh Gradus, Cibatio

"Here our child is fed by the mother's milk so that it gains weight and gains new strength. The threefold triple measurement brings forth the gazing dragon, which brings forth the fruit of the sun and the moon. The triangle has three eagles on one side and you can hand them over to the three spirits". *Interpretation*: The child, fed by the mother's milk, is our coagulated ☿ or has been soaked by the virgin milk, whereby it is finally transformed into the hard, gazing dragon, which has the fruit of the sun and the moon in its body. On one side it has three eagles, the three *principia*, which its birth line consists of, these must be given to the three spirits. One spirit is the *oleum* ♄, by which it becomes fixed; the second spirit is the vegetabilic *menstruum*, in soul, spirit, and body, which dissolves this fixed precipitate; The third spirit is the red man, the fire, through which it becomes the red tincture, when this fire is boiled in this water, here the short way is demonstrated.

11. Figur
Der achte Gradus, Sublimatio

Damit unser Leib in die dünne Luft verkehrt werde, so wird er aus dem niedrigen Orte wiederum in die Höhe erhoben. Der König bringt zurück den Phönix, die Königin den Schwan; und alsdann geht der Wolf aus seinen Höhlen heraus. Die wachsenden Früchte stehen auf dem apollinischen Baum, daher mäht die Zeit das reife Gras mit seiner Sense ab. *Auslegung*: Unser sublimierter Leib, der weiße Schwan, so auf dem philosophischen Meer geschwommen und zum trockenen weißen ♃ worden, durch Feuer der Calcination, durch *Oleum* ♄ verbrannt, welches der hungrige fressende Wolf, so diesen Leib verzehrt und verbessert wiedergibt, ist nun der gebratene Schwan, die Speise des Königs, weil der weiße ♃ dadurch hitziger, trockener, solarischer Natur worden, daraus der Phönix gebracht wird, der sich durchs Feuer verbrennen muss und wachsen die Früchte auf dem apollinischen Baum, auf dem fixen præcipitierten ☿, welcher der wütende Apollo genannt wird, denn seine Strahlen töten den ☿, welchen er in wahres weißes ☉ verwandelt. Die Heuernte geht an, da das grüne Gras, die weiße Tinctur, welche etwas grünlich, abgehauen wird und der rechten Ernte, der Ernte des ☉ vorgeht, der roten Tinctur, welche auch auf eben dieser Erde, dem Baum des ☿, reif wird.

12. Figur
Der neunte Gradus, Fermentatio

Der Same, welcher der gebauten Erde gegeben worden, kommt durch die Seele hervor, nachdem sie revociert worden. Nun ruft unsere Posaune und die Leiber, so dem Grabe anvertraut worden, sollen wieder aufstehen und mit neuen Körpern geziert werden. Denn ohne Ferment steigt die Sonne nicht in die Höhe und die schöne Diana erneuert ihr Leben nicht hinwiederum. *Auslegung*: Die neu erstandenen Leiber steigen durch die Exaltation stufenweise in die Höhe, durch ihre eigenen *Fermenta* und färbenden Gifte, welche diese erstandenen

11TH FIGURE[207]
Eighth Gradus, Sublimatio

"In order that our body may be turned into the thin air, it is again lifted up from the low places. The king brings back the phœnix, the queen the swan; and then the wolf goes out of his cave. The growing fruits stand on the Apollonian tree, therefore time mows the ripe grass with its scythe". *Interpretation*: Our sublimated body, the white swan, which has swum on the philosophical sea and become dry, white ♁ by the fire of calcination, burned by the *oleum* ♄, which is the hungry devouring wolf, which consumes and improves this body, and is now the roasted swan, the food of the king, because the white ♁ has become hotter, drier, solar in nature, from which the phœnix is brought, which has to be burned by the fire and the fruits grow on the Apollonian tree, on the fixed precipitated ☿ which is called the fierce Apollo, because its rays kill the ☿, which it transforms into true white ☉. The hay harvest begins, since the green grass, the white tincture, which is somewhat greenish, is cut down and it comes before the real harvest, the harvest of the ☉, the red tincture, which also ripens on this very earth, the tree of ☿.

12TH FIGURE[208]
The Ninth Gradus, Fermentatio

"The seed given to the tilled soil comes forth through the soul after it has been revoked. Now our trumpet calls, and the bodies that have been entrusted to the grave shall rise again and be adorned with new bodies. For without ferment, the sun does not rise, and the beautiful Diana does not renew her life in return". *Interpretation*: The new created bodies gradually rise up through exaltation, through their own *fermenta* and coloring poisons, which cannot kill these created bodies,

Leiber nicht töten können, denn sie sterben bei diesen Leibern und werden zu Tincturen und sind 2 *Fermenta*, eins der Sonne und das andere des Mondes, welches nicht das gemeine ☉ und ☽, auch die fermentiert werden sollen, sind gleichfalls nicht das gemeine ☉ und ☽, sondern unsere ausgekochten figierten ☿, weiß und rot, welche weit edler als gemeines ☉ und Silber, haben auch andere Eltern, das gemeine ☉ und ☽ kann kein Ferment sein, sie haben schwere fixe compacte Leiber, wie können sie denn diese roten und weißen ☿ erheben, als ein Teig gärend machen und flüchtig in den Himmel des Glases erheben und wieder zurück auf die Erde bringen? Darinnen es schmilzt, sich aufs neue gradiert, den Leib mit in die Höhe führt, veredelt und höher steigt und solchen himmlisch macht durchs verbessernde Ferment, welche die *Unguenta*, Öle und Balsam sind, dadurch sie sich *in infinitum* multiplicieren, doch hat es hier auch seine gewissen Absätze und Himmel, wenn es durch jedere Sphären ausgeholfen, wenn *Agens* und *Patiens* in gleichem Grad der Waage, die Natur gesättigt, keine Speise noch Trank mehr zu sich nehmen wollte, der unverständige Arbeiter aber wollte nicht aufhören mit seinen Fermenten, immer mehr und mehr zusetzen, so überfüllt er den Leib die Tinctur, dass sie erstlich gar zerschmilzt und in lauter Öl verkehrt, welches zuletzt so flüchtig würde, dass es alle Gefäße durchdringe und gar verschwinde und also sein ganzer Schatz in einen Geist als sein Ferment verwandelt werde. Wenn er nun diesen Leib erfüllt, so ist es hohe Zeit ihm güldene Klammern anzuwerfen, dadurch er erhalten, denn er bekommt dadurch ein *Patiens*, da er als *Agens* darein wirken kann, welches er zerstört und zerschmelzt und durch diesen fixen Leib wieder erhalten wird, da dieser Geist einen Leib annimmt, in dem er ruhen kann, sonst wäre er unwiederbringlich verloren, müsste wieder in sein Chaos verschwinden.

13. FIGUR
Der zehnte Gradus, Exaltatio

Der König und die Königin werden auf einen glänzenden Thron gesetzt, in der Mitte bringt ein schöner Baum die Früchte. Auf des

for they die with these bodies and become tinctures. Moreover, there are two *fermenta*, one of the sun and the other of the moon, which are not the common ☉ and ☽, also the [☉ and ☽] which are to be fermented, are likewise not the common ☉ and ☽, but are our boiled fixed ☿, white and red, which are far nobler than common ☉ and silver. They also have other parents. The common ☉ and ☽ cannot be a ferment, they have heavy, fixed, compact bodies, how can they raise these red and white ☿, make it fermenting like dough and raise it fleetingly in the heaven of the glass and bring it back to earth? There it melts, ferments itself anew, leads the body up, ennobles and rises higher and makes it heavenly through the improving ferment, which are the *unguenta*, oils, and balms, whereby they multiply *ad infinitum*, but it also has here its certain landings and heavens when it is helped out by each sphere, when *agens* and *patiens* are in the same degree of the scale, nature is saturated, no longer wants to take food or drink, but the ignorant worker does not want to stop with their ferments, but adds more and more until they overfill the body, the tincture, so that it first melts and then turns into sheer oil, which finally becomes so volatile that it penetrates all vessels and even disappears and thus transforms his entire treasure, as his ferment, into a spirit. If he now fills this body, then it is high time to throw golden brackets on him, through which he will be stopped, because through this he gets a *patiens*, where it can act with the *agens*. This it destroys and melts, and it is preserved again through this fixed body, since this spirit assumes a body in which it can rest, otherwise it would be irrevocably lost, would have to disappear again into its chaos.

13TH FIGURE[209]
Tenth Gradus, Exaltatio

"The king and queen are seated on a resplendent throne, in the midst of which a beautiful tree bears fruit. On the steps of the throne there

Thrones Stufen stehen 2 mal 7 Löwen, zwei mit den Schädeln zusammen gefügten Flügeln. Also wird unser Stein in der ganzen Welt erhöht, also freut er sich durch seine eigene Ehre und also glänzt er auch. *Auslegung*: Die Exaltation geschieht durch die siebenfachen magischen Zahlen, die dreimal Umdrehung des Rades, durch den ersten Himmel, durch das geseelte himmlische Wasser, läuft der ☿ durch alle 7 Planeten, bis er den Krebs erlangt, das Haus des ☽, den weißen ♃, durch die anderen 7 Stufen der Exaltation, durch den astralischen Luft- und Sternen-Himmel, läuft der ☿ bis er alle unsere Metalle in leuchtende Sterne verwandelt, in Tincturen. Im dritten als im Feuer-Himmel werden die 2 Tincturen, rot und weiß, zusammen vereinigt. Diese sind danach die 2 ♌, der grüne und der rote in einem Leibe gefügt, welche in ewiger Freundschaft beisammen wohnen, sind auch durch ihren Adel, Kraft und Natur vollkommen worden, ohne andere Hilfe, herrschen über die 3 Himmel, in welchen sie Götter sind, haben sich selbst gezeugt, sich selbst geboren, sich selbst erzogen, sich selbst erhöht, können auch durch sich selbst sich unendlich renovieren und neu gebären, immer noch mehr und mehr erhöhen, neue Himmel und Erde schaffen, denn es hat alles bei sich, sich selbst vollkommen zu machen, braucht keines anderen Hilfe. Dies kann ja mit recht ein Bildnis der klaren Gottheit genannt werden, jeder Himmel hat seine sonderliche Sonne und Licht und dieses also in Ewigkeit und diese Sonne oder Licht ist in der *Materia* selbst enthalten. Also ist auch GOTT als das göttliche Licht in allen Himmeln, in der Erde und unter der Erde und die Finsternis hat es nicht begriffen. Man kann keinen Zirkel reißen und sollte er auch nur wie ein Punct sein, darinnen das göttliche Licht nicht wäre und in ihm leben, weben und sind wir. Darum sieht GOTT alles, hört alles und richtet alles.

14. FIGUR
Der elfte Gradus, Multiplicatio

Die glänzende Königin wird auf einen großmütigen ♌ getragen und hält, o Pelikan, deine Jungen. Aber der ♌ ernährt durch sein eigenes

are two-times seven lions, two wings joined together with the skulls. So our stone is exalted throughout the world, so it rejoices in its own glory, and so it shines too". *Interpretation*: The exaltation happens through the sevenfold magic numbers, the threefold rotation of the wheel, through the first heaven, through the ensouled heavenly water, the ☿ runs through all seven planets until it reaches Cancer, the house of the ☽, and the white ♀. The ☿ runs through the second seven stages of exaltation, through the astral heaven of air and stars, until it transforms all our metals into shining stars, into tinctures. In the third, as in the fire-heaven, the two tinctures, red and white, are united together. These are then the two ♌, the green and the red, joined in one body, who live together in eternal friendship, and which have also become perfect through their nobility, strength, and nature. Without any help they rule over the three heavens, in which they are gods, having begotten themselves, born of their own power, and then having raised themselves. They can also infinitely renovate themselves and give birth again through themselves, always increasing more and more, creating a new heaven and earth, because it has everything to make itself perfect and does not need anyone else's help. This can rightly be called an image of the clear divinity, for every heaven has its particular sun and light and this is so for eternity; and this sun or light is contained in the *materia* itself. So GOD is also the divine light in all heavens, in the earth, and under the earth, and the darkness did not comprehend it. One cannot draw a circle, even if it should only be like a point, in which the divine light would not be, and in which we live, weave, and exist. That is why GOD sees everything, hears everything, and judges everything.

14TH FIGURE[210]
The Eleventh Gradus, Multiplicatio

"The shining queen is borne upon a magnanimous ♌ and holds, O pelican, your young. But the ♌, by his own flesh, nourishes countless

Fleisch unzählig viele Junge, welche mit ihrem Vater scherzen: Also kann er auch viele Junge des Steins hervor bringen und kann sie ohne Ende multiplicieren. *Auslegung*: Die glänzende Königin ist die weiße Tinctur, der grüne ♌, durch diesen ist sie in das gläserne Meer verwandelt, welches nun mit dem Feuer, dem roten ♌, soll vereinigt werden, dadurch er sich in viele Teile multipliciert und auf solche Weise kann sie multipliciert werden.

15. Figur
Aus den 4 Elementen besteht alles

Aus den ersten 4 Elementen besteht alles, auch dasjenige, welches du für reine Elemente hältst. Denn es wird nichts Reines gefunden, der Schöpfer der Welt hat mit seiner künstlichen Hand alle *Elementa* vermischt, daher steht unsere Sonne auf und nichts minder auch unser Mond und daher empfängt auch unser Mägdelein ihren Ursprung. *Auslegung*: Obschon die *Elementa* rein geschieden sein, so hat doch noch ein jedwedes alle *Elementa* in sich, sowohl in der großen als in unser kleinen Welt, muss auch also sein durch göttliche weisliche Vorsehung, denn die simplen *Elementa* wären ohne Kraft, solches sieht man in der Natur, wie alles aus dem Wasser wächst, die Metalle, Steine, Salze, Bäume und Kräuter und ist die Erde der Mann, das Wasser das Weib, so auch in unseren Elementen, aus unserer Erde und Wasser wachsen auch Metalle und Steine, so viel derselben durch die Natur kommen, so viel hat man derselben durch die Kunst, nur dass diese edler sind. Aber durch unsere Luft oder Himmel fahren unsere Planeten auf den hohen Achsen, unsere Metalle werden die Gestirne und Planeten, durch das Feuer aber werden sie unzerstörlich, denn das Feuer ist die größte Reinigung, dadurch erscheint der neue Himmel und die neue Erde leuchtend. In Wasser und Erde aber fault alles und wächst darinnen, steht auch zu einem herrlichen Leben wieder auf, es könnte aber nicht faulen, wenn nicht Luft und Feuer rein davon geschieden worden, welches beides Geist und Seele ist, so lange die in

young, who frolic with their father: So he can also bring forth many young from the stone, and can multiply them endlessly". *Interpretation*: The shining queen is the white tincture, the green ♌, through this she is transformed into the sea of glass, which is now to be united with the fire, the red ♌, whereby it multiplies into many parts and in this way it can be multiplied.

15TH FIGURE[211]
Everything Consists of the Four Elements

"Everything consists of the first four elements, including what you consider to be pure elements. Because nothing pure is found, the creator of the world has mixed all the *elementa* with his artificial hand, that is why our sun rises and no less our moon and that is also where our little maiden comes from". *Interpretation*: Although the *elementa* are separated purely, each one still has all the *elementa* in itself, both in the macrocosm and in our microcosm, it must also be like that through the wise, divine providence, because the simple *elementa* would be without power, you can see that in nature, how everything grows out of water, the metals, stones, salts, trees, and herbs and if the earth is man, the water is woman, so also in our elements, from our earth and water also metals and stones grow, as much of it comes from nature, so much of it comes from art, except that these are nobler. But our planets move on the high axes through our air or heaven, and our metals become the stars and planets—but through fire they become indestructible, because fire is the greatest purification through which the new heaven and new earth appear radiant. But in water and earth everything putrefies and grows therein, also rises again to a glorious life, but it could not putrefy if air and fire had not been separated cleanly from it, which is both spirit and soul, as long as these two dwell in one body,

einem Körper wohnen, ist kein Tod, sobald aber die Seele und Geist den Leib verlassen, ist der Leib tot und fault in seiner eigenen Erde.

16. FIGUR
Sieben Metalle

Allhier siehst du das inwendig entdeckte der irdischen Welt und die hohen Gestirne auf den bergigen Orten. Nämlich das Erdreich selbst bekommt die eigenen Planeten, welchen die *Elementa* ihre Kräfte darreichen. Wenn du zweifelst, welche es sind, so siehe nur mit wachsamen Sinnen die Metalle an, also wird dir die Spitze des Himmels oder die Höhe bekannt werden. *Auslegung*: Unsere reine metallische Erde oder Salz bekommt durch Einträkung des metallischen Wassers unsere Planeten, in der ersten Einträkung gewinnt es, das Salz zieht das Wasser mit seiner strengen Herbe und Trockne in sich und weil es keine Geister hat, so gesteht und friert es, wird also der irdische ♄, der den Himmel zum Vater hat, unser Jungfer ☿, welcher nun der Zeuge-Vater aller Götter, da die güldene *Genealogia* herstammt, wenn dieser ♄ wieder durch das metallische Wasser, das Weib, als die *Beja*, eingetränkt, wird dieses Wasser wieder zur Erde und der ♃, welches schon *Meteora* gibt, Wind, Wolken, Farben in der *Materia*, denn der Leib ist noch nicht flüchtig, sondern ruht im Stuhl, ♃ ist schon ein größerer Herr und hat ein hohes Gestühle, welches die Spitzen dieser Berge zeigen, die dritte Einträkung weist den ☿, diese Erde wird nun erweicht und amalgamiert sich und fangen die Dünste an etwas weniges vom Leib mit in die Höhe zu nehmen, als doppelte Geister, bei der vierten Einträkung, wenn das Salz oder Erde alles aufgelöst, wird es grün und fängt an mit in die Höhe zu steigen, lässt sich auch der ☉ Spiegel sehen und zeigt die glückselige Grüne, dass dieser Stein wächst und sich vermehrt. Nach der Solution geschieht die *Coagulatio*, wird oben hart als Eisschollen, wenn sie schwer werden, brechen sie, fallen auf den Boden, gefriert aber und erhärtet bald wieder, welches abermals zu Boden fällt, bis alles Wasser zu gelbem Sand worden, welches auf der Erde schmilzt, wodurch es Blasen bekommt, woraus schwarze,

there is no death, but as soon as soul and spirit leave the body, the body is dead and putrefies in its own earth.

16TH FIGURE[212]
Seven Metals

"Here you see what has been discovered internally from the earthly world and the high stars on the mountainous places. That is to say, the earth itself gets its own planets to which the *elementa* give their powers. If you doubt what they are, just look at the metals with an alert mind, and the peak of heaven or the height will be known to you". *Interpretation*: Our pure metallic earth or salt receives our planets by impregnation with the metallic water, it is obtained in the first impregnation, the salt draws the water into itself with its severe astringency and dryness; and because it has no spirits, it becomes solid and freezes, so it becomes the earthly ♄, which has heaven for its father, our virgin ☿, which is now the begetter of all gods, from whence comes the golden *genealogia*. When this ♄ is impregnated again by the metallic water—the woman, the *Beya*—then this water turns again into earth, and the ♃ which already gives *meteora*, wind, clouds, and colors in the *materia*, because the body is not yet volatile, but rests in the throne, ♃ is already a greater lord and has high thrones, which the peaks of these mountains show. The third impregnation indicates the ☿. This earth is now softened and amalgamated, and the vapors begin to take a little of the body with them into the cave, like double spirits, at the fourth impregnation, when the salt or earth dissolves everything. It turns green and begins to rise with it, and the ☉ mirror can also be seen; and the blissful green shows that this stone grows and multiplies. After the solution the *coagulatio* happens, and so it gets hard on top just as when ice floes become heavy they break and fall to the ground; but it freezes and soon hardens again, then falls to the ground once more until all the water has turned to yellow sand, which melts on the earth, causing it to bubble. From this black, grey-colored cloud vapors

grau-farbige Wolken-Dünste aufsteigen, welche wieder zurück auf die *Materia* schlagen und wie schwarze dicke Pfeffer-Brühe sieht, endlich wird es schwarz und dick, wie schwarzes Pech, schwarze stinkende Erde, wird genannt Kröten-Drachen-Otter- und Schlangen-Pulver, daraus hernach der Theriak gemacht wird, so wider alles Gift dient, ist die Schlange, so sich selber gebissen und verwundet, sich selber getötet, wird sich auch selber lebendig machen. Die Farben sind rot und gelb, wenn sie in ihrem Blut und Gift schwimmen, von Geschmack bitter, dieser schwarze *Laton* wird weiß gewaschen durch sein eigenes Wasser, bis die Metalle alle auf und absteigen und nur die 2 stärksten blieben, als ☉ und ☽, welche aus den Kräften der Elemente gewachsen und am höchsten Ort des Himmels stehen bleiben, wenn der Himmel des ♄ verschwunden und ausgelassen.

17. Figur
Der Ehe-Stand

In unserem Himmel glänzen zugleich 2 Lichter, welche die großen Lichter des Himmels wiederbringen. Die 2 füge zusammen, gleich wie das Weib mit dem Manne zusammen gegeben wird, auf dass dir daher ein rechtmäßiges Ehe-Bett komm, daher muss auch die Verkehrung der Elemente metrisch werden, auf dass die runde Form ihre Kräfte hervor bringe. *Auslegung:* Die 2 Lichter des Himmels sind unsere Sonne und Mond, der Mond ist unser fixer weißer ♀, die Sonne unser roter ♀, diese 2 als Mann und Weib werden verheiratet, damit die *Elementa* verkehrt werden, der Mann in das Weib, und das Weib in Mann, der ☿ in ♀ und der ♀ in ☿, Feuer im Wasser gekocht, so ist die ganze Natur des Wassers ins Feuer verkehrt, welches nicht eher geschehen konnte, das Wasser musste denn ausgetrocknet, in Erde verkehrt und die Erde in Luft, alsdann konnte das Feuer in der Luft und Wasser brennen, weil es durch die Feuchte des Wassers nicht ausgelöscht werden konnte.

rise which fall back onto the *materia* and look like a thick, black pepper broth. Finally it becomes black and thick, like black pitch, black stinking earth, and it is called toad, dragon, otter, and serpent powder, from which afterwards the theriac is made, which serves against all poisons, being the serpent which has bitten and wounded itself, killed itself, and which will also revivify itself. The colors are red and yellow as they swim in their blood and poison, bitter in taste. This black loam is washed white by its own water until the all metals rise and fall and only the two strongest have remained, which are ☉ and ☽, which have grown out of the forces of the elements and remain in the highest place in heaven when the heaven of ♄ has disappeared and been removed.

17TH FIGURE[213]
The Marital Status

"Two lights shine in our sky at the same time, which bring back the great lights of the sky. Put the two together, just like the woman is put together with the man, so that you get a lawful marriage bed, therefore the reversal of the elements must also become metric, so that the round form produces its powers. *Interpretation*: The two lights of heaven are our sun and moon. The moon is our fixed white ♃, and the sun is our red ♃; and these two are married like man and woman so that the *elementa* are reversed, the man into woman and the woman into man, the ☿ into ♃ and the ♃ into ☿, the fire boiled in water, and the whole nature of water is turned into fire. This could not happen before, for the water had to be dried up and turned into earth and the earth turned into air, then the fire could burn in the air and water because they cannot be extinguished by the moisture of the water.

18. Figur
Der Mercurius

In dem ☿ ist, was die Weisen suchen und in ihm sind die Güter der ganzen Welt. Diesen schadet die hohe Flamme des feuerspeienden Vaters nicht, denn er flieht alsobald von dem warmen Herde. Wo du aber denselben bei den brennenden Altären behalten wirst, so wirst du durch sein zartes Kind glückselig sein. *Auslegung*: In dem ☿ der Weisen ist, was die Weisen suchen, nämlich in dem ☿ die Körper, welche doch nicht die 7 Metalle sind, sondern von seinen eigenen Körpern, welche alle aus dem Körper der *Magnesia* kommen, denn der unsere in seiner äußerlichen Reinigung erscheint in Gestalt eines klaren Wassers, wie Brunnen-Wasser, wenn aber solche Schlange seine eigene rote Erde solviert und dieses Wasser schleimig und weiß, dick und schwer, heißt es das pontische Wasser. Wenn nur dieses seine eigene centralische Erde oder fixen Körper, den kleinen grünen ♌ solviert, mit selbigem als mit seinem eigenen fixen Teil in die Höhe gestiegen, dann ist es erst der ☿, der Körper, in diesem ist was die Weisen suchen, denn dieser, die geblätterte flüchtige Erde, fürchtet sein eigenes Feuer nicht, so von seinem Vater kommt, es ist sein angeborener ♃, haben als Bruder und Schwester in einem Leibe gelegen und ob er gleich, wenn ihn dieses Feuer berührt, flieht, so flieht dieses Feuer diese Flamme mit ihm, weil sie aber beide nicht entlaufen können, so erhält eine Natur die andere und erfreuen den Arbeiter durch sein Kind die Tinctur, so sie gezeugt, durch welches er glückselig wird.

19. Figur
Tinctura

Das grausame Tier verfolgt die zarte Jungfer durch unwegsame Felder und knirscht durch erschrecklichen Klang, ☿ schreit: Eile hierher, du schöne Nymphe, mit geschwinden Füßen und fürchte die grausame Gefahr deines Grabes nicht. Wer mit mir stirbt, der wird auch mit mir, nach traurigem Unglück, fröhlich zu den erwünschten Gütern aufer-

18TH FIGURE[214]
The Mercurius

"In the ☿ is what the wise seek, and in it are the goods of the whole world. The high flame of the fire-breathing father does not harm them, because it immediately flees from the warm stove. But if you keep him by the burning altars, you will be blessed by his tender child". *Interpretation*: The ☿ of the wise contains what the wise seek, that is to say, in the ☿ [are] the bodies, which are not the seven metals, but [are] from its own bodies, which all come from the body of *magnesia*; for our [☿] appears in the form of clear water in its external purification, like well water, but when such a serpent dissolves its own red earth and this water is slimy and white, thick and heavy, it is called the pontic water. Only when it dissolves its own central earth or fixed body, the little green ♌, and ascends on high with it, as with its own fixed part, only then is it the ☿, the body. What the wise seek is within it, for it is the leafed volatile earth which does not fear its own fire, and which comes from its father. It is its inborn ♃. They have lain as brother and sister in one body and although it flees when this fire touches it, this fire flees this flame with it; but because neither of them can escape, one nature preserves the other and enjoys the worker by its child, the tincture, if he has conceived it, through which he becomes blissful.

19TH FIGURE[215]
Tinctura

"The cruel beast pursues the tender virgin through impassable fields, stomping with a terrifying crushing sound. ☿ cries: "Hasten here you, beautiful nymph, with swift feet, and do not fear the cruel danger of your grave. Whoever dies with me will also rise with me joyfully, after

stehen. *Auslegung*: Das grausame Tier ist der Drache, welcher auch das Höllen-Bad genannt wird. Die zarte Jungfer ist die geblätterte Erde, weil solches doppelter, männlicher und weiblicher Natur, wird es bald für den Mann bald für das Weib genommen, der ☿ so ihr nachschreit, ist das *Menstruum*, so aus dem ☿ bereitet und in der Seele, Geist und Leib besteht, welches unser lebendiges Wasser ist, in selbigem nun muss der weiße fixe ♃, die weiße Tinctur, ersinken, zerschmelzen und darinnen neu geboren werden, sich multiplicieren und zu Reichtums-Gütern auferstehen, hier wird angezeigt die weiße Tinctur, den weißen Stein der ersten Ordnung zu figieren, zu augmentieren und multiplicieren, worein kein Feuer kommt, sondern die weißen Geister.

20. Figur
Das Wasser der Weisen

Bisweilen mischt das Weib die mancherlei Farben des Wassers und daher wäscht sie geschwinde die Leinwand oder das Tuch. Aber das Wasser geht heraus und exhaliert in die dünne Luft, die gefärbte Leinwand aber bleibt in der gesuchten Farbe: Also geht auch das Wasser der Weisen durch die Glieder der Metalle und macht durch seine geschwinde Flucht die gefärbten Leiber. *Auslegung*: Das Wasser der Weisen ist nur ein Wasser, der *Nilus*, so aus dem Paradies entspringt, ergießt sich in 4 Ströme, die Elemente. Aus diesen entspringen die Wasser der Gesundheit und des Reichtums, das elementische ist das bleibende Wasser bei der Erde, es schwärzt und weißt auch die Erde, aus diesem entspringt der Baum des lebendigen Wassers, so aus dem Heiligtum fließt, das Wasser der Gesundheit und des Reichtums, die anderen Wasser sind die färbenden Gifte, wenn es heißt, färbt das Blatt mit Gift, ist *Venenum Tingens*. Mit dem feurigen Wasser aber wird diese Leinwand rot gefärbt, welches beides beständige Farben sind, die erste weiße Farbe ist das bleibende Wasser und unsere weiße ausgetrocknete Erde oder Stärke, darauf, wenn sie dünne ausgebreitet, zum dünnen Blatt sublimiert, zum *Folio* worden, so stärken wir dies *Folium*, welches hungrig und durstig, kühlen es mit dem Öl ab, dadurch es sanft, gelinde und

sorrowful misfortune, to the desired goods". *Interpretation*: The cruel beast is the dragon, which is also called the hell bath. The tender virgin is the leafed earth, because such things are of a double nature, male and female. It is taken now for the man, now for the woman, the ☿ which cries after her; and it is the *menstruum* which was prepared from ☿ and consists of soul, spirit, and body. This *menstruum* is our living water, and in it must sink the white fixed ♃, the white tincture, which must melt and be reborn in it, whereupon it will multiply and rise to become rich goods. Herewith it is indicated to augment and multiply the white tincture, the white stone of the first order, into which no fire comes, but the white spirits.

20TH FIGURE[216]
The Water of the Wise

"Sometimes the woman mixes the various colors of the water, and so she quickly washes the linen or cloth. But the water goes out and exhales into the thin air, but the colored canvas remains in the color sought: so also does the water of the wise go through the limbs of the metals and makes the colored bodies by its swift flight". *Interpretation*: The water of the wise is only one water, the *Nilus*, which springs from paradise and pours into four streams, i.e., the elements. From these [streams] spring the waters of health and wealth. The elemental [water] is the enduring water on earth. It also blackens and whitens the earth, whence springs the tree of living water, so that from the sanctuary flows the water of health and wealth. The other waters are the coloring poisons, and when these color the leaf with poisons, they are known as the *venenum tingens*. With the fiery water, however, this canvas is colored red, both of which are permanent colors. The first white color is the abiding water, and when our white, dried-up earth or starch is spread thin with this water it is sublimated into a thin sheet, becoming a *folio*. We strengthen this *folium*, which is hungry and thirsty, cooling it with the oil, whereby it becomes gentle, mild, and soft. We soak the colors with the oil so that they remain permanent and cannot be washed

weich wird, tränken durch das Öl die Farben ein, dass sie beständig bleibt und nicht wieder abzuwaschen ist, durch das Feuer-Bad wird es vollends geläutert, gereinigt und blutrot gefärbt, der weiße Zündel in ein gelbes Tüchlein gewunden und hernach im Feuer rot gebrannt, welche rote Farbe die letzte beständige Farbe ist und alles aus einem Felsen entsprungen und wieder zu einem harten Stein coaguliert, welcher doch als Wachs, flüssig, das vorher der harte Fels war, daraus GOTT klares Wasser geben und einen harten Stein.

21. FIGUR
Zweierlei Schwefel

Siehe da kommen 2 Löwen mit zusammen gefügten Hälsen und treten miteinander in einen starken Freundschafts-Bund. Du, der du das Ferment suchst, füge die 2 ♄ zusammen, auf dass du deine Last multiplicieren kannst, der eine sei beständig, der andere steige in die Höhe, aber wenn sie miteinander vereinigt, müssen sie in einen einmütigen Grad stehen und bleiben. *Auslegung*: Die 2 Löwen sind hier die 2 Schwefel, der rote und der weiße, welche zusammen in einen ♄ müssen gebracht werden, welches nicht eher geschehen kann, er ist denn trokken und fix, sonst hat der rote keinen Ingres darein und schwimmt stets oben auf, durch das rechte Mittel aber, das *Medium conjungendi*, gehen sie ineinander, der fixe hält den flüchtigen und werden eins.

22. FIGUR
Elixir

Wer das Elixir, die Vögel, durch begierigen Sinn erkennen will, dass daher die Metalle geschickt gemacht werden zu deinem Werk, so musst du dir ein solches *Medicament* zuvörderst suchen, welches vor der geschwinden Flucht des ☿ herfließe, muss endlich diese Königin mit dem ♄ des Königs vereinbaren, also wird der davonfliegende Vogel durch den Vogel betrogen werden. *Auslegung:* Hier ist die Figierung des weißen ♄ beschrieben, weil er der flüchtige Adler ist, so muss ihn der andere Vogel, welcher anderswo der fliegende ♌, auch der fliegen-

away again. Then with the fire bath it is completely refined, purified, and dyed blood-red. The white tinder is wound in a yellow cloth and then burned red in the fire, and this redness is the last permanent color. And everything sprung from a rock and coagulated again into a hard stone, which nevertheless is as fluid as wax, which was formerly the hard rock, from which G o d gave clear water and a hard stone.

21ST FIGURE[217]
Two Kinds of Sulphur

"Behold, there come two lions with joined necks and they enter into a strong bond of friendship with each other. You who seek the ferment, put together the two ♄, so that you can multiply your burden, one stays constant, the other rises up, but when they are united together, they must stand and remain in a unanimous degree". *Interpretation*: The two lions here are the two sulphurs, the red and the white, which must be brought together into one ♄. This cannot happen before it is dry and fixed, otherwise the red [♄] has no ingress into it and always swims up; but through the right means, the *medium conjungendi*, they merge, the fixed one keeps the volatile and they become one.

22ND FIGURE[218]
Elixir

"If you want to recognize the elixir, the birds, through an eager sense, so that the metals are made suitable for your work, you must first seek such a *medicament* that will flow before the swift flight of the ☿. You must finally reconcile this queen with the ♄ of the king, so the bird that flies away will be deceived by the bird". *Interpretation*: Here the fixation of the white ♄ is described, because it is the volatile eagle, and therefore the other bird, which elsewhere is called the flying ♌ also the flying ♄,

de ♄, genannt wird, zugesetzt werden, auch der rote Mann in das weiße Weib, unser flüchtiges ☉ und ☽, wenn diese beiden recht zusammen vereinigt, dem Gewicht der Natur nach, so wird daraus der natürliche Schatz, das weiße Elixir, dieses ist etwas hart zur Schmelzung und der Wasser-Stein der Weisen genannt, wo aber das Gewicht nicht getroffen, steigt nur einer von diesen Vögeln in die Höhe und geht durch, es hilft kein Siegel, weil das hermetische Siegel nicht recht angewendet, entweder das Gefäß tut einen Schlag oder der weiße ♃, der ☿ raucht aus und wird von seinem widerwärtigen nicht gehalten. Wer aber dieses trifft, der hat alles gewonnen und werden diese beiden in die Höhe steigen und wieder auf den Grund gehen, bis sie beide ein schwarzes Pech worden. Dies währt eine gute Zeit und kann der Künstler bei dieser anderen Schwärze nicht helfen, wie bei der Ersten, die er durch das Wasser von dem schwarzen *Laton* abwaschen konnte, sondern er muss Geduld haben, bis aus der *Materia* ein Geist aufsteht, sich von Leib und Seele scheidet, circuliert, auf den toten Körper fällt, ihm ein neues Leben einbläst, sich mit Leib und Seele wieder vereinigt, die Schwärze in die Luft verschwindet und das Wunder-Tier herausgeht und sich beschauen lässt. Es wird oft fließen und hart werden und wieder fließen, bis es endlich die weiße Tinctur, die erste auf weiß wird.

23. FIGUR

Conjunctio

Die ertappten Vögel schließe in 2 Gläser ein, deren Mund-Löcher du wohl zusammenfügen musst, der eine wird davon fliehen, der andere wird den Lauf aufhalten und wird sich nicht mehr in die Höhe begeben. Habe nur Geduld, diese Arbeit wird dich nicht betrügen durch vergebene Hoffnung, sondern der schöne Baum wird dir reife Frucht geben. *Auslegung*: Die ertappten Vögel sind die vorhergehenden ♃, welche zur weißen Tinctur figiert. Diese werden mit dem weißen Ferment eingeschlossen, mit dem Öl der Luft, dem lebendigen *Aquavit*, wodurch die weiße Tinctur *in infinitum* vermehrt wird, dadurch die Tinctur güssig

must be added to it. Also the red man in the white woman, our volatile ☉ and ☽, and when these two properly unite together, according to nature's weight, it becomes the natural treasure, the white elixir. This is somewhat hard to melt and is called the water-stone of the wise, but where the weight was not met, only one of these birds climbs up and passes through. No seal is of assistance, because [if] the Hermetic seal is not applied properly, either the vessel or the white ♃ makes a break, the ☿ wafts out and is not contained by its repelling [principle]. But whoever encounters this has obtained everything and both of them will rise up and descend to the bottom again until they both have become black pitch. This lasts for a good time and the artist cannot help with this second blackness as with the first, which he was able to wash away from the black loam with water; but he must have patience until a spirit arises from the *materia*, separating itself from body and soul. This spirit circulates and falls on the dead body, breathing new life into it, thus reuniting body and soul. The blackness disappears into the air and the miraculous beast comes out and can be viewed. It will often flow and harden and flow again until at last it becomes the white tincture, the first [tincture] on white.

23RD FIGURE[219]
Conjunctio[220]

"Lock the captured birds in two glasses, whose mouth-openings you have to put together, one [bird] will flee, the other will stop the run and will not go up any more. Have patience, this work will not deceive you with lost hope, but the beautiful tree will give you ripe fruit". *Interpretation*: The captured birds are the previous ♃, which are fixed to the white tincture. These are impregnated with the white ferment, with the oil of the air, with the living *aquavit*, whereby the white tincture is multiplied *ad infinitum*, making the tincture pourable and fluid when

und flüssig gemacht, wenn sie mit ihrem Öl conjugiert wird und kann man die reifen Früchte von dem Silber-Baum abbrechen.

24. Figure
Augmentatio

Hier hast du 3 Gläser zu sehen, desgleichen des ♌ starke Mäuler, auch den ☿, welchen der geschwinde Flügel leicht macht. Nimm 3 Teile der Arznei, tue sie mit dem fliegenden zusammen, so wird er seine Flügel in dem heißen Feuer bald verderben. Daher wird unser ♌ sein, welchen du mit eben der Sorge nach deinem Gefallen in den Ofen wieder multiplicieren kannst. *Auslegung*: Die 3 Gläser sind der Kristall des ☿ oder glasförmige *Azoth* des *Lullij*, welcher das gläserne Meer wird, denn der ☿ muss von des ♌ starken Mäulern gefressen werden, vom grünen ♌ und vom roten ♌, der erste verwandelt ihn in die weiße Tinctur, der andere macht ihn rot, in die rote Tinctur, das rote Glas, diese 3 Steine sind alle zur Arznei und können vermehrt werden, die geschwinden Flügel des ☿ sind die beide vermehrende und multiplicierende *Menstrua*, sowohl der weißen als der roten Tinctur, wird auch hier etlicher Maßen das Gewicht angezeigt, dass des fixen Teils 3 oder 4 mal schwerer sein muss, als des flüchtigen, sonst könnte es solches nicht coagulieren und fix machen.

25. Figur
Cibatio

Das zarte Kind wird durch die Mutter-Milch ernährt, aber der Mann nimmt mit dem Munde Weizen-Speise. Einem jedweden Magen, so es begehrt, muss man Speise geben, so ihm gut ist und sich zum Magen schickt, doch es vermehrt sowohl der Mann, als der Knabe seine Kräfte. Also musst du auch unserem jungen Knaben Speise geben, so ihm gut ist und sich zum Magen schickt, damit derselbe durch seine gewünschten Kräfte wachse. *Auslegung*: Das Kind ist unser nackender unbekleideter ☿, so erstlich durch die Jungfer-Milch, als ein kleines Kind erzogen wird, bis es zu seinen männlichen Jahren kommt, her-

conjoined with its oil, and one can see the ripe fruits breaking from the silver tree.

24TH FIGURE[221]
Augmentatio[222]

"Here you can see three glasses, also the strong mouths of the ♌, also the ☿, which is made light by the swift wing. Take three parts of the medicine, put them together with the flying one, so it will soon spoil its wings in the hot fire. Therefore, our ♌ will be there, which you can multiply again with the same concern for your pleasure in the oven". *Interpretation*: The three glasses are the crystal of ☿ or the glassy *azoth* of Lullus, which becomes the sea of glass, for the ☿ must be eaten by the ♌'s strong mouths, by the green ♌ and by the red ♌. The first transforms it into the white tincture, and the second makes it into the red tincture, the red glass. These three stones are all for medicine and can be multiplied. The swift wings of the ☿ are both increasing and multiplying *menstrua*, both the white and the red tincture. We are here also given some indication as to the weight required: the fixed part must be three or four times heavier than the volatile part, otherwise it could not coagulate and fix it.

25TH FIGURE[223]
Cibatio[224]

"The tender child is nourished by the mother's milk, but the man takes wheat food with his mouth. Any stomach that desires it, must be given food that is good for it and suits the stomach, but it increases the strength of both man and boy. So you must also give our young boys food that is good for them and suits their stomach, so that they can grow as much as they want". *Interpretation*: The child is our naked, unclothed ☿; at first it is brought up through the virgin milk, like a small child, until

nach bekommt er den Straußen-Magen, der Eisen verdauen kann, bis er auch *Sol* und ☽ verschluckt, die Wasser der Sonne und des Mondes, danach er gespeist und getränkt wird, danach bekommt er auch einen Leib, bald ist er das trockene, bald das nasse *Menstruum*, bald gar der Acker, daher besteht alles in ihm, an seiner Speise und Trank und Auferziehung.

26. Figur
Die Wunder unseres Steins

4 Schwestern teilen den Schatz gar artig unter sich und nimmt jedwede ihren Teil davon. Der Erste kann alle menschlichen Krankheiten heilen. Der andere nimmt die kranken Glieder der Metalle hinweg. Der Dritte verändert die gemeinen Steine in Edelgesteine. Und der vierte macht, dass man durch seine Hilfe das Glas ziehend machen kann. *Auslegung*: Die 4 Schwestern sind die 4 General-Tugenden des Steins, es ist aber der Stein des ☿ schon Arznei genug, die weiße Tinctur noch mehr, die rote über diese beiden und diese 3 Steine sind die Steine des *Universals*, und müssen zusammen in einen einzigen Stein kommen, denn der Kristall-Stein des ☿ wird verwandelt in die weiße Tinctur, solche in die rote, welche beiden letzten die Metalle tingieren, in ☉ und ☽ verwandeln. Ehe aber die weiße Tinctur ganz rot wird, können die Kristalle und andere geringe Steine in Edelgesteine dadurch verwandelt werden, wenn sie aber sehr rot und flüssig, so macht solche auch das Glas ziehend und biegend, wenn sie in ziemlicher Quantität darauf geschmelzt wird, auch machen die eigenen Öle ihre eigenen Geister, welche in fixe Gläser verwandelt, fließend und biegend, wenn solche oft mit den Ölen imbibiert werden.

27. Figur
Aqua Vitæ

Allhier entspringt lebendiges Wasser aus lebendigen Brunnen, welches dir viel Nutzen deines Leben geben kann. Allhier erweitern die resolvierten Gestirne ihre Kräfte, daher der Mond und Sonne ihre Ge-

it comes to its manly years, and afterwards it gets the ostrich stomach, which can digest iron, until it also swallows *sol* and ☽, the waters of the sun and the moon. After that it is fed and watered, and then after that it also gets a body, sometimes it is the dry, sometimes the wet *menstruum*, sometimes even the field. Therefore everything consists in it, in its food and potion and upbringing.

26TH FIGURE[225]
The Wonders[226] of Our Stone

"Four sisters politely share the treasure among themselves and each takes their share of it. The first can cure all human diseases. The second takes away the diseased limbs of the metals. The third changes the common stones into precious stones, and the fourth makes it so that with its help one can draw the glass". *Interpretation:* The four sisters are the four general virtues of the stone, but the stone of the ☿ is already medicine enough, the white tincture even more so, and the red is beyond these two; and these three stones are the stones of the *universal* and must be combined into one single stone, because the crystal stone of the ☿ is transformed into the white tincture, which [is transformed] into the red [tincture], which both tinge the metals into ☉ and ☽. But before the white tincture turns completely red, the crystals and other lesser stones can be transformed into precious stones, but if they are very red and liquid, the glass also makes them stretch and bend if they are melted on it in large quantities, also their own oils make their own spirits, which are transformed into fixed glasses, flowing and bending, when such are often imbibed with the oils.

27TH FIGURE[227]
Aqua Vitæ[228]

"Living water springs up here from living fountains, which can give you much benefit in your life. Here the revolving stars expand their powers, therefore the moon and sun wash their faces, so hasten here and

sichter waschen, drum eile du geschwinde hierher und lösche deinen Durst bei heißer Sonne. *Auslegung*: Hier werden die 2 multiplicierenden *Menstrua* beschrieben, eins der Sonne, das andere des Mondes, welche beide unsere flüssige Sonne und Mond sind, die rote Sonne und ☉ und das *Aqua Benedicta*, so die Hand nicht nass macht, das mineralische Bad und philosophische *Aqua Regis*, so die Metalle wieder neu gebiert, die resolvierten Gestirne erweitern darinnen ihre Kräfte, welche sind unser weißer und roter ♃, wenn sie darinnen resolviert werden, kommen sie immer zu höheren und höheren Tincturen. Auch kann das gemeine ☉ und ☽ darinnen gewaschen und wieder renoviert werden und zu Tincturen gebracht, ob sie schon nicht so viel tingieren, als das große Werk der Natur. Wer jenes nicht finden kann, wird seinen Durst schon hier löschen. Wer nicht den himmlischen Nectar zu trinken bekommen kann, der muss mit Wein vorlieb nehmen, er stärkt auch.

28. FIGUR
Das philosophische Erz

Unser grünendes Erz ist das philosophische ☉, denn es reinigt durch seine Strahlen unsere Leiber. Wenn du aber dieses mit unserem *Duenech* nicht weißt zusammen zu fügen, so wird es kein gewünschtes Ehebett des Ehestandes sein. Wenn aber 2 Löwen, zwei starke Herzen, zusammen kommen, so wird aus ihrem Munde der güldene Saft hervorfließen. *Auslegung*: Das philosophische ☉ ist dreierlei, denn unsere 3 Himmel haben jeder seine eigene Sonne, erstlich das astralische, darinnen unsere Gestirne aufgehen, welches ist das güldene Büchlein, so mit den *Duenech*, in der anderen Composition des grünenden Erzes der Smaragd geworden, *Aurum Perspectibile*, da die Sonne noch hinter dem Mond verborgen, bis sie zuletzt hoch sanguinisch rot aufgeht. Das Dritte ist, wenn die 2 Löwen zusammen vereinigt werden, der grüne und der rote ♌, die rote Tinctur.

quench your thirst in the hot sun". *Interpretation*: Here the two multiplying *menstrua* are described, one of the sun, the other of the moon, both of which are our liquid sun and moon. The red sun and ☉ and the *aqua benedicta*, which does not wet the hands, the mineral bath and philosophical *aqua regia*, in which the metals are reborn again, the revolving stars expand their powers there, which are our white and red ♃, when they are resolved there; and they always come to higher and higher tinctures. The common ☉ and ☽ can also be washed in it and be renovated again and made into tinctures, although they do not tinge as much as the great work of nature. Those who cannot find this will quench their thirst here. Anyone who cannot get to drink the heavenly nectar must make use of wine, it also strengthens.

28TH FIGURE[229]
The Philosophical Ore

"Our verdant ore is the philosophical ☉, because it purifies our bodies with its rays. But if you do not know how to combine this with our *Duenech*, then it will not be a desired wedding bed for the matrimony. But when two lions, two strong hearts, come together, the golden juice will flow out of their mouths". *Interpretation*: The philosophical ☉ is of three kinds, because our three heavens each have their own sun, first the astral, in which our stars rise, which is the little golden book, so [that] with the *Duenech*, in the second composition of the greening ore, it has become the emerald, *aurum perspectibile*, since the sun is still hidden behind the moon until it finally rises in a high sanguine red. The third is when the two lions are united together, the green and the red ♌, the red tincture.

29. FIGUR
Das philosophische Feuer

Dieses Feuer ist lieblich, warm und zugleich feucht, welches alles erhält und nichts umkommen lässt. Nämlich es ist gleichförmig und geschickt, die Sachen zu generieren und temperiert mit seinem *Temperie* jedweders. Von diesem nun flieht die Sonne mit ihrem eigenen Eheweibe keines Weges, sondern geht mit ihr liebreizend, gleich wie in die angenehmen Bäder. *Auslegung*: Das philosophische Feuer verbrennt die Hände nicht, sondern zu schmilzt die Leiber, reinigt sie in Seele, Geist und Leib, renoviert sie, unser Feuer macht aus unserer Sonne und Mond Tincturen. Diese, wenn sie aufs neue gewaschen und darinnen gebadet werden, sind sie dadurch aufs neue erhöht in *Quantitate* und *Qualitate*.

30. FIGUR
Das philosophische Bad

Du musst dergleichen Wasser haben, dadurch nicht ein einziges Glied nass wird, in diesem nun verschaffe, dass sich die Sonne mit dem Mond bade. Welches, wenn du es zu Werke gerichtet haben wirst, so wird sich der leichte Geist mit ihnen vereinbaren und du wirst mit deinen 2 Lilien sehen, daher wird jeder Baum seine eigenen Früchte bringen und dann kannst du die Äpfel brechen. *Auslegung*: Das philosophische Bad ist das *Menstruum* in Seele, Geist und Leib, ist das Wasser, so die Hand nicht nass macht, löst den Mann und Weib, *Sol* und *Luna*, auf. Wenn diese beiden darinnen gebadet, wird sich das giftige Wasser mit ihnen vereinigen und die 2 Lilien zu sehen sein, die rote und weiße Tinctur, oder vielmehr der rote und weiße ☿, dann kann man die Äpfel aus dem hesperischen Garten brechen und ihre Früchte genießen.

31. FIGUR
Conjunctio

Allhier liegt *Gabricus* mit seiner geliebten *Beja* zusammen gefügt, aber

29TH FIGURE[230]
The Philosophical Fire

"This fire is lovely, warm, and moist at the same time, it sustains everything and lets nothing perish. Namely, it is uniform and skillful in generating things and tempering everything with its *temperie*. From these the sun does not flee with its own wife, but goes with it in a charming way, just like in the pleasant baths". *Interpretation*: The philosophical fire does not burn the hands, but melts the bodies, purifies them in soul, spirit, and body, and repairs them. Our fire turns our sun and moon into tinctures. These, when they are washed anew and are bathed in them, are thereby increased anew, in *quantitate* and *qualitate*.

30TH FIGURE[231]
The Philosophical Bath

"You must have water of this kind, so that not a single limb gets wet, in which you now cause the sun to bathe with the moon. Which, when you will have put it to work, the light spirit will reconcile with them and you will see with your two lilies, therefore each tree will bear its own fruit and then you can pluck the apples". *Interpretation*: The philosophical bath is the *menstruum* in soul, spirit, and body, is the water which does not wet the hands, it dissolves man and woman, *Sol* and *Luna*. When these two bathe in it, the poisonous water will unite with them and the two lilies will be seen, the red and white tincture, or rather the red and white ♀. Then you can pick the apples from the Hesperidian garden and enjoy their fruits.

31ST FIGURE[232]
Conjunctio

"Here *Gabricus* lies together with his beloved *Beya*, but she encloses her

sie schließt mit ihrem Bauche den Liebhaber; Also wachsen diejenigen, welche 2 gewesen, zusammen in einen Leib, auf dass sie dir die angenehme Geschenke ihres angenehmen Ehebettes geben mögen. Gleich wie nun die Henne eines Hahns bedürftig, also bedarf auch der Mond der Sonne, daher er anders verlangt, sein Geschlecht zu multiplicieren. *Auslegung*: Wer *Gabricus* und *Beja* ist, das habe ich schon ausgelegt, es ist Wasser und Erde, ☿ und ☉; Das Salz ist das fixe Teil, so ihn coaguliert, dieser einfache ☿ schlägt alle Vermählung aus, ohne seinen eigenen Ehegatten, von dem nimmt es den Beischlaf an, darum heißt es, diese Schlange hat sich selber gebissen, diese beiden sind die Drachen, deren einer Flügel, der andere keine Flügel hat, welche das güldene Vlies bewachen, die Äpfel im hesperischen Garten. Diese *Beja*, der Jungfer ☿, fasst dieses Salz in seinem Leib, bis sie ihren Mann flüchtig, auch zum ☿ macht, dann heißt er der ☿, der Körper, sollten sie sich nun vermehren, muss aufs neue eine Verehelichung geschehen, damit sie ihr Geschlecht multiplicieren.

32. Figur
Animæ extractio sive Imprægnatio

Die Leiber, welche in einem verschlossenen Grabe verfault wären, empfangen nunmehr wiederum glänzende Gaben einer neuen Seele. Mit diesen werden sie nun vereinbart und werden zugleich von denselben behalten und also wird eins durch des anderen Hilfe erleuchtet. Habe du nur Geduld und sei sowohl mit der Hand als scharfsinnigem Nachdenken hurtig, damit der Geist nicht von dem warmen Herde hinweg fliege. *Auslegung*: Die Leiber sind unsere Sonne und Mond, wenn sie in der Faulung liegen, Seele und Geist von ihnen geschieden, aber nach einer gewissen Zeit fängt sich an, aus der *Materia* ein Geist zu circulieren, welcher wieder zurück in die Erde geht und den Leib mit einer neuen Seele begabt, welche sich mit diesen Leibern vereinigt, Seele mit Seele und Geist mit Leib und Seele, dadurch sie ewig nicht zu scheiden, und Sonne und Mond erleuchtet zu einem einzigen durchsichtigen Leib wird, wenn die Sonne im Wasser bei der ☽ schläft,

lover with her belly, so that those who have been two grow together into one body upon which they may give you the pleasant gifts of their pleasant marriage bed. Just as the hen needs a rooster, so also the moon needs the sun, therefore it demands otherwise to multiply its kind". *Interpretation*: I have already laid out who *Gabricus* and *Beya* are, it is water and earth, ☿ and ☉. The salt is the fixed part which coagulates it, this common ☿ rejects all marriages, without its own spouse, from whom it accepts intercourse. Therefore it is said that this snake bit itself. These two are the dragons, one of which has wings, while the other has no wings. They guard the golden fleece, and the apples in the Hesperidian garden. This *Beya*, the virgin ☿, takes this salt into its body until it makes its husband flee, and also [makes it] to the ☿. Then it is called the ☿ of the bodies, and should they now multiply, a new marriage must take place so that they can have their kind multiplied.

32ND FIGURE[233]
Animæ extractio sive Imprægnatio[234]

"The bodies that would have rotted away in a locked grave now receive the brilliant gifts of a new soul. They are now united with these souls and at the same time are kept by them and thus one is enlightened by the help of the other. Have patience, and be quick both with your hands and with keen reflection, so that the spirit does not fly away from the warm stove". *Interpretation*: The bodies are our sun and moon when they lie in decay, soul and spirit separated from them; but after a certain time a spirit begins to circulate out of the *materia*, which goes back into the earth and the body is endowed with a new soul, which unites with these bodies, soul with soul and spirit with body and soul, whereby they become eternally inseparable and illuminate sun and moon. They become one transparent body when the sun sleeps in the

gebiert die ☽ in Wolken ihr Kind.

33. FIGUR
Ablutio vel mundificatio

Ein ganz stiller Regen fällt von dem Himmel herab und wäscht mit seinen Tropfen unsere Leiber ab. Daher verschwindet jene schwarze Farbe in die dünne Luft und alsobald haben die Glieder die liebliche Weiße. Aus was Ursachen werden solche also rein? Darum, weil der reine Geist mit ihnen nicht vereinbart wird, denn sie sind ohnedem schon rein. *Auslegung*: Wenn die Leiber ihre Zeit im Grabe gelegen und der Geist sich anfängt zu circulieren, steigt er in die Höhe, macht schwitzend, dann fällt ein Regen nieder, die Tropfen waschen die Schwärze weg, wenn die weißen Strömlein der *Luna* kommen, so vereinigt sich der reine Geist mit dem Leib und wird zusammen ein trockener Geist.

34. FIGUR
Animæ Jubilatio, sive sublimatio

Nun ist es dereinsten Zeit von dem Ehe-Bette aufzustehen und das traurige Grab zu verlassen, nun klingt das süße liebliche Jauchzen der Seele, nämlich sie pflegt erfreut zu werden, wenn der Leib derselben sowohl gehorsam leistet, als auch Dienstleistung; Also erlöst ein Vogel den anderen, welcher in den klaren hellen Wassern eingetaucht wird, von der Sorge der Ersaufung und Untertauchung. *Auslegung*: Wenn der Geist zurück in den Leib geht, macht er Leib und Seele lebendig, erfrischt und erfreut solche und werden diese beiden dem Geist gehorsam, lassen sich von ihm in die Höhe führen und folgen dem Geist, gehen mit ihm auf und ab und nimmt der flüchtige Vogel, welcher in hellen Wassern eingetaucht, mit ihn in die Höhe, da sich der Wasser-Vogel in der Luft abtrocknet und wieder mit ihm untertauchen kann, bis es ein einziger Vogel wird.

water by the ☾, the ☾ gives birth to her child in clouds.

33RD FIGURE[235]
Ablutio vel mundificatio[236]

"A very still rain falls from the sky and washes our bodies with its drops. Hence that black color vanishes into the thin air, and immediately the limbs have the lovely whiteness. From what causes, then, do such become pure? This is because the pure spirit is not united with them, for they are already pure anyway". *Interpretation*: When the bodies have lain their time in the grave and the spirit begins to circulate, it rises up, making you sweat. Then a rain falls and the drops wash away the blackness. When the white straws of *Luna* come, then the rain unites the pure spirit with the body and together they become a dry spirit.

34TH FIGURE[237]
Animæ Jubilatio, sive sublimatio[238]

"Now it is the time to get up from the marriage bed and to leave the rough grave, now the sweet, lovely rejoicing of the soul sounds, that is, they want to be pleased when the body renders obedience as well as service. So one bird relieves the other, which is submerged in the clear, bright waters, from the worry of drowning and submersion". *Interpretation*: When the spirit goes back into the body, it revitalizes body and soul, refreshing and delighting them. Both become obedient to the spirit, letting themselves be led up. They then follow the spirit, walking up and down with it and taking the fugitive bird which is immersed in clear waters up with it, because the water-bird dries itself in the air and can submerge with it again, until they become a single bird.

35. Figur

Germinato

Siehe ein zweiköpfiger Leib, unter welchem Monds-Hörner sind, in den Händen 3 Schlangen und einen Vogel tragend, wobei auch ein Baum mit Ästen und herrlichen Früchten geziert, von welchem du nun unzählig viele Güter abzuzählen, wofern du aber die rechte Wurzel nicht recht wirst erkannt haben, so wird dieser Baum mit seinen Früchten wohl sicher vor dir bleiben. *Auslegung*: Der zweiköpfige Leib ist Wasser und Erde, ☿ und Salz, unter welchen die Monds-Hörner, welches der weiße ♄, in Händen 3 Schlangen, dieser ♄ besteht schon in Seele, Geist und Leib, obgleich kein sichtbarer Geist damit einverleibt, so hat doch Wasser und Erde alle beide Geist in sich, denn die simplen *Elementa* wirken nicht, wenn die Erde keinen *Spiritum* oder Geist hätte, wäre sie tot und wenn das Wasser keinen Geist aus der Luft bei sich hätte, könnte nichts aus ihm wachsen, wenn aber das Wasser die Erde als das Salz aufschließt, so wird der Geist, welcher im Salz gebunden, los von seinen Banden und er greift im Wasser, als in seinem Weiblein, derselben ihren Geist und vereinigen sich miteinander und also kommt der doppelte Leib hervor, in welchem der Dritte, der Geist auch ist, denn es ist ein lebendiger Leib, so aus der Ersterbung und der Ertötung das neue Leben bekommen, einen Leib, so besteht in Seele, Geist und Leib, den Vogel so er dabei hat, ist das lebendig-machende *Menstruum*, welches flüchtig ist und in Seele, Geist und Leib besteht, in selbigem müssen die 3 Schlangen, welches die Schlange mit 3 Hälsen ist, resolviert werden, so wird daraus wieder ein anderer neuer Leib hervorwachsen, in weit besserer Gestalt als vorher und können durch öftere wiederholte Arbeit unzählig viele Früchte abgezählt werden, wer aber die Wurzel des Baums, den ☿ der Weisen nicht recht erkannt, noch zubereitet, der wird weder den Baum noch die Früchte erkennen und genießen, sondern wird alles vor ihm bleiben.

35TH FIGURE[239]
Germinato[240]

"See, a two-headed body, under which are the horns of the moon, holding in its hands three serpents and a bird, whereby a tree is also adorned with branches and splendid fruits, from which you can count innumerable goods, but if you do not recognize the right root, this tree with its fruits will remain safe from you". *Interpretation*: The two-headed body is water and earth, ☿ and salt, under which are the horns of the moon, which is the white ♄, in the hands three serpents. This ♄ already consists of soul, spirit, and body, although no visible spirit is incorporated with it, yet water and earth both have spirits in themselves, for the simple *elementa* would not function if the earth had no *spiritum* or spirit, since in that case it would be dead. Nor would they function if the water had no spirit from the air with it, since then nothing could grow from it; but if the water opens up the earth as a salt, so the spirit, which is bound in the salt, is loosed from its bonds and it grasps in the water, as in its wife, its spirit, each uniting with the another. Thus the double body comes forward, in which the third, the spirit, also subsists; for it is a living body which received new life from dying and death, a body, consists of soul, spirit, and body, the bird which it has with it is the life-giving *menstruum*, which is volatile and consists of soul, spirit, and body, in which the three serpents, which is the serpent with three necks, must be dissolved. Then another new body will grow out of it again, one that will be in far better form than the one before; and countless fruits can be counted through more often repeated work, but whoever does not properly recognize or prepare the root of the tree, the ☿ of the wise, will neither recognize nor enjoy the tree nor the fruit, but everything will stay away from him.

36. Figur
Fermentatio

Tue das wahre Ferment mit dem eigenen ♄ zusammen, auf dass du feine leichte Vögel zeugen kannst. Diese 2 musst du hernach fröhlich in unser Meer werfen, daher wird die lebendige Flamme seine Kräfte heraus tun. Also kannst du den Samen der geblätterten Erde geben, welche ganz artig zum warmen Herde geschickt sein wird. *Auslegung*: Die Fermentation ist etlichemal im philosophischen Werke, erstlich ist des Wassers Ferment die Erde und die Erde ist das Wasser ihre Hefe, dadurch beides ein Teig wird und gärt, danach ist eine Fermentation des roten Mannes und des weißen Weibes, auch ist das eine Fermentation zu nennen, wenn der ☿ figiert wird, da ein flüchtiger Vogel den anderen Vogel halten muss, wenn diesen Falken die rechten Klauen angeworfen werden. Die rechte Fermentation ist, wenn der weiße figierte ♄ in seinem weißen Öl solviert und der Vogel in den klaren hellen Wassern eingetaucht wird, welches das Wasser ist, so die Hand nicht nass macht. Das andere *Fermentum* ist das rote Öl der Sonnen ♄, das Feuer des Steins, hier aber wird angezeigt wie der Gold-Same soll in die geblätterte Erde gesät werden, sobald diese beiden einander berühren, fliehen sie voneinander, werden glühend-heiß und fängt sich der Stein an, *Agens* und *Patiens*, hier ist das andere Mal das Gewicht wohl in acht zu nehmen, trifft man dieses nicht, dass diese Feinde nicht beide fliehen, so vereinigen sie sich auch nicht und wird weder die weiße noch die rote Tinctur zu machen sein. Wo man aber das Gewicht trifft, gehen sie ineinander und geht der Streit an, steigen in die Höhe, der Wind aber schlägt sie wieder zurück, nach vielem Sturm kommt Regen, endlich schmelzt diese geblätterte Erde und wird das schwarze Toten-Grab, darinnen der königliche Leib tot liegt, bis er von der Schwärze gewaschen, nach dem Tod aufersteht, in dem weißen verklärten Leibe, dies wird genannt des Königs Hemd.

36TH FIGURE[241]
Fermentatio[242]

"Put the real ferment together with your own ♃ so that you can beget fine, lightweight birds. Afterwards you must happily throw these two into our sea, so the living flame will exert its power. So you can give the seed to the leafed earth, which will be quite nicely sent to the warm stove". *Interpretation*: Fermentation occurs several times in the philosophical work. First the ferment of the water is the earth, and the earth is the yeast of the water, whereby both become a dough and ferment. After that there is a fermentation of the red man and the white woman. We also call it fermentation when the ☿ is fixed, since a fleeing bird must hold the other bird when the right claws are thrown at that falcon. The right fermentation is when the white, fixed ♃ is dissolved in its white oil and the bird is immersed in the clear, bright waters, which is the water which does not wet the hands. The other *fermentum* is the red oil of the sun ♃, the fire of the stone, but here it is shown how the gold seed is to be sown in the leafed earth, as soon as these two touch each other they flee from each other, becoming red-hot; and if the stone starts, *agens* and *patiens*, then here is the second time you have to be careful with the weight. If you do not get it right, these two enemies do not flee, and yet they do not unite either and there is neither the white nor the red tincture to be made. But where one hits the [right] weight, they go into each other and the fight begins. They rise up, but the wind blows them back again. After many storms comes rain, and finally this leafed earth melts and becomes the black grave where the royal body lies dead, until the blackness is washed from it. It rises again after death in a white transfigured body, this is called the king's tunic.

37. Figur
Illuminatio

Siehst du wie allhier die Sonne mit ausgespannten Flügeln leuchtet und wie sie hurtig die Höhlen des hohen Brunnens verlässt, aber der Mond durchschießt sie durch einen reissenden Schuss des Pfeils und also fällt sie geschwind wieder in ihren Brunnen, daher empfängt der ganze Brunnen goldfarbige Haar-Farbe und glänzt gleich wie der schöne Apollo mit seinen Strahlen. *Auslegung*: Wenn Sonne und Mond zusammengesetzt, so flieht die Sonne vor des Mondes Kälte, steigt als der hitzige und feurige Geist in die Höhe, die kalten Dünste aber der Luna erhaschen sie durch deren Kälte und wird durch sie niedergedrückt und fällt wieder in den Brunnen des Mondes, daher wird der ganze Brunnen gefärbt und der Mond in die Sonne verwandelt, dass sie auch das schöne glänzende Gold wird.

38. Figur
Nutrimentum

Da ist ein schöner lustiger Garten, allwo unser Stamm und Geschlecht sich niedergelegt. Hermes steht und ist fertig mit geschwinden Füßen sich hinweg zu machen, drum allerschönster König ergreife die gegenwärtigen Leute, es wird wohl keine Geschickter zu deinem Magen sein, gebrauch dich der Ertappten, damit die jungen Glieder nicht welk werden, wenn sie durch den reißenden Hunger geschwächt werden. *Auslegung*: Der schöne lustige Garten Eden, darinnen unser Adam und Eva in der Lust und Freuden gelebt, ist der doppelte ☿, so in einem Leib gewachsen und sich selber zu einem hungrigen und weißen Kalk reverberiert und unser königliches Geschlecht ist. Hermes, welcher in seinem Wappen die Schlange führt, welche auch *Aphrodita*, die ♀ genannt und die rote Schlange ist oder vielmehr die Schlangen-Zähne, die müssen in diese Erde gesät werden und muss solches geschwind geschehen, ehe die jungen Glieder welk werden. Die weiße doppelte Blume männlicher und weiblicher Natur, der Hermaphrodit sich sel-

37TH FIGURE[243]
Illuminatio[244]

"Here you see how the sun shines with outstretched wings and how it quickly leaves the cave of the high well, but the moon shoots through it with a ripping arrow and so it quickly falls back into its well, so that the whole spring receives gold-colored hair and shines like the beautiful Apollo with his rays". *Interpretation*: When sun and moon come together, the sun flees from the cold of the moon, rising up as the hot and fiery spirit; but the cold vapors of *Luna* catch it through its coldness and [the sun is] pressed down by this and falls again into the well of the moon. Hence the whole fountain is colored and the moon is turned into the sun so that it also becomes the beautiful shining gold.

38TH FIGURE[245]
Nutrimentum[246]

"There is a beautiful, lovely garden where our tribe and family have settled. Hermes is standing there and is ready to depart with quick feet. Seize therefore, most beautiful king, the present people; there will probably be no one else appropriate for your stomach. Use what is snared, so that the young limbs do not wither when they are weakened by ravaging hunger". *Interpretation*: The beautiful, lovely Garden of Eden in which our Adam and Eve lived in lust and joy is the double ☿ grown in one body which has turned itself into a hungry and white lime, and this is our royal family. Hermes, who has the serpent in his coat of arms, which is also called *Aphrodita* the ♀, is the red serpent, or rather the serpent's teeth, which must be sown in this earth and this must be done quickly before the young limbs wither. The white double flower of male and female nature is the hermaphrodite which eats and

ber frisst, verzehrt und gar verschwindet, darum muss es durch das hermetische Siegel geschwind verschlossen werden, damit dem ☿ die Augen geblendet, welche funkeln und leuchten, dass er nicht entfliehen kann und seine rechte Speise für seinen Hunger findet, so wird das Blatt mit Gift gefärbt und wird kein *Spolium* sondern das *Folium*.

39. FIGUR
Fixatio

Allhier in diesem Grabe liegt unser ehrwürdiger Leib, bei diesem steht der Geist, aber das Gemüt, Vernunft und Sinn kommt zu dem Himmel wieder, derhalben verschaffe, dass dieses sich wieder nach dem Himmel begebe und davon wiederum auf das Allerunterste fliege! Also wird es sich mit den angenehmen Kräften vereinbaren und durch seine Hilfe den Leib wieder lebendig machen. *Auslegung*: Dies ist die Figierung des weißen ♄ und nun wahrhaftig in Seele, Geist und Leib zum vollkommenen Manne gewachsen, durch die viele Auf- und Absteigung hat es sich erstlich in Seele, Geist und Leib geschieden, nachdem aber sich wieder vereinigt, denn wenn diese Dünste in die Höhe steigen, ergreift Seele, Geist und Leib einander und weil keine wässrige Feuchtigkeit hier mehr vorhanden, sondern alles glutinosisch ist, so ergreift im Aufsteigen eins das andere und figiert es, dass es gleich fix zu Boden fällt, was aber noch vom Geist übrig und noch keinen Leib ergriffen, kehrt wieder zurück, ergreift den Leib, trägt ihn in die Höhe, damit er von der Seele gefasst und gehalten wird und ist der Geist der Mediator, so Leib und Seele vereinigt und das rechte Band, Leib und Seele zu binden.

40. FIGUR
Multiplicatio

Allhier fällt ein großer Regen von den flüssigen hellen Wolken und befeuchtet das zarte Kind mit der Mutter. Also muss unser Wasser anfänglich in die leichte Luft solviert werden, bald wird es von den hohen Örtern sich in die Tiefe begeben, daher wasche man den zarten *Lato-*

consumes itself and finally completely disappears. Therefore it must be closed quickly with the Hermetic seal, so that the ☿'s eyes, which sparkle and shine, are blinded so that it cannot escape and find the right food for its hunger. Thus the leaf is colored with poison and becomes not *spolium* but *folium*.

39TH FIGURE[247]
Fixatio[248]

"Here in this grave lies our venerable body, near to it the spirit stands, but heart, reason, and sense return to heaven, so make sure that it returns to heaven and that from here it flies again to the very bottom! So it will unite with the congenial forces and with their help bring the body back to life". *Interpretation*: This is the fixation of the white ♃ and has now truly grown into a perfect human being in soul, spirit, and body. Through many ascents and descents it has first separated into soul, spirit, and body, but then reunited, because when these vapors rise up, soul, spirit, and body seize each other and because there is no longer any watery moisture here, but everything is glutinous, one seizes the other as it rises and fixes it so that it immediately falls to the ground. But that which still has spirit but has not yet seized a body, returns again, seizes the body, carries it up so that it can be seized and held by the soul. And the spirit is the mediator which unites body and soul and [is] the proper bond to bind body and soul.

40TH FIGURE[249]
Multiplicatio[250]

"Here a great rain falls from the bright, liquid clouds and moistens the tender child with the mother. Hence, our water must be initially dissolved in the lightweight air, for soon it will descend from the high places into the depths, so wash off the delicate loam with the liquid

nem durch die flüssige Feuchtigkeit ab, damit derselbe eine Vermehrung seiner Glieder nehme. *Auslegung*: Hier wird die Abwaschung der schwarzen Erde gelehrt, welche hier die Mutter genannt wird, weil die Erde aller Dinge Mutter ist, das zarte Kind ist unser ☿, welcher noch unerzogen, liegt gleichsam in seinem Schweiß, Kot und Urin, nass, faul, stinkend und schmutzig, das muss nun als Mutter und Kind von seiner Schwärze weiß gewaschen und gereinigt werden und bedeutet dies auch die Schlange Hydra, wenn ihr ein Kopf abgehauen, wuchsen ihr allzeit mehr, je öfter solches gewaschen, je mehr der Erde wird, denn diese Erde figiert das Wasser, dadurch sie sich multipliciert.

41. Figur
Revificatio

Allhier steht der glänzende Mond mit seinen hellem Bruder aus dem reinen Brunnen mit frischem Gesicht wieder auf, daher besprizt der Pelikan seine Jungen, seine allerliebsten Kinder mit Blut, aus seiner eigenen Brust, davon sie hernachmals neue Kräfte in ihre Flügel und Herzen bekommen und die lebendig-machende Kunst erwärmt die zarten Glieder. *Auslegung*: Diese *Revificatio* hat den Verstand, dass der figierte ☿ muss wieder flüchtig gemacht werden durch das wiederbringende *Menstruum*, so in Geist, Seele und Leib besteht und eben aus dem ☿ *Philosophorum* gemacht ist, dies muss den ☿ wieder zurückrufen, dass er aufgeschlossen und in die Luft hinein gehen kann, hier heißt es: Gib dem ☿ den ☿, so wird der fixe ☿ von diesem ☿ zu Öl gemacht und vereinbart sich mit diesen himmlischen Kräften, daher er auch himmlisch wird.

42. Figur
Perfectio

Siehe, der König trägt in seiner linken Hand eine starke Kette, welche er um des ♌ Hals gewickelt, in der rechten hält er eine Rute und zeigt die ungeheure Schlange. Wobei der stolze Baum mit deinen Früchten, O Sonne, ist, willst du sie abbrechen, so suche erst die Wurzel. Es wird

moisture, so that it increases its limbs". *Interpretation*: Here the washing off of the black earth is taught, which is called the mother here, because the earth is the mother of all things. The tender child is our ☿ which is still uneducated, lying as it were in its sweat, feces, and urine, being wet, lazy, stinking, and dirty, and which must now be washed white and purified of its blackness like mother and child. When the head of the serpent Hydra was severed it always grew back more [heads]; here this means that the more often such a thing was washed, the more of the earth is created, for this earth fixes the water, thereby multiplying it.

41ST FIGURE[251]
Revificatio[252]

"Here the gleaming moon rises again with its bright brother from the pure well, with a fresh face, therefore the pelican sprinkles its young, its dearest children, with blood from its own breast, from which they afterwards get new strength in their wings and hearts and the life-giving art warms the delicate limbs". *Interpretation*: The meaning of this *revificatio* is that the fixed ☿ must be made volatile again, through the re-establishing *menstruum*, which consists of spirit, soul, and body and is made of the ☿ *philosophorum*. This must call back again the ☿, so that it is opened and can go into the air. Here it says: give the ☿ to the ☿, and then the fixed ☿ will be turned into oil by this ☿, uniting with these heavenly forces, so it also becomes heavenly.

42ND FIGURE[253]
Perfectio[254]

"Behold, the king carries in his left hand a strong chain, which he wraps around the neck of the ♌, in his right [hand] he holds a staff and shows the abominable serpent. There stands the proud tree with your fruits, O sun, if you want to break it off, first look for the root. There will be

keine Frucht als nur von den eigenen Samen sein. *Auslegung*: Der König ist allhier die rote Tinctur oder der rote figierte ♃, die starke Kette, so er um des ♌ Hals gewickelt, wenn er von diesem ♌ weggeführt oder aufgelöst, er solchen ♌ fest an sich hält und gleichsam mit einer Kette fest mit ihm verbunden, dadurch zur roten perfecten Tinctur wird, die Rute damit er die Schlange zeigt, ist die natural Berg-Rute, aus welcher der König selbst seinen Anfang hat und nun alles zu einer einzigen Schlange geworden, wobei die reifen güldenen Früchte abzubrechen. Wer sie aber brechen will, muss erst die Wurzel dieses Sonnen-Baums suchen, welches ist ☿ *Philosophorum* aus der Schlange gemacht, dies ist der Baum, so die ☉ und ☽ Früchte trägt.

43. Figur
♌ *Viridis*

Was bedeutet der gestirnte ♌? Was das Ebenbild der glänzenden Sonne, welche er hat? Das sage mir doch, du liebe Muse? Das ist unser Held, der grüne ♌, in dessen Bauchs innersten Eingeweiden die roten Gestirne verborgen liegen, daher gibt jener faule apollinische Farben von sich und ist im Meer der niederschlagende Glanz. *Auslegung*: Der grüne ♌ ist unser ☉, in die weiße Tinctur verwandelt, das Weib so von der Sonne schwanger und den Mond unter ihren Füßen, wird nun bald die aufgehende Sonne. Es ist das gläserne Meer, so mit Feuer gemengt werden kann, warum es aber grünlich sieht, ist die Ursache, das rote *Oleum veneris* so in ♄ gespielt und eben die ♀ ist, die bei dem ☿ geschlafen und in die ☽ verwandelt, weil ihre roten Wangen weiß gefärbt worden, hat ihre Grüne in das weiße Glas gegeben, welche man doch in etwas schimmern sieht, ist aber nun in inwendigen ☉, daher gibt er der ☿, wenn er in der Fäulung steht und die ♀ in sich verschlungen hält, apollinische, das ist, güldische Farben von sich, dass es auf dem Meer einen güldischen Glanz macht, den Sonnen-Spiegel.

no fruit but of its own seed". *Interpretation:* The king here is the red tincture or the red fixed ♃. The strong chain is that which he wraps around the ♌'s neck when he is led away from this ♌, or is dissolved, at which point he then holds such ♌ firmly to itself with this chain, thereby becoming the red, perfect tincture. The rod with which he shows the serpent is the natural mountain rod, from which the king himself has his beginning, and now all has become one serpent, whereby the ripe golden fruits are to be plucked. But whoever wants to pluck them must first seek the root of this sun tree, which is ☿ *philosophorum*, made from the serpent. This is the tree that bears the ☉ and ☽ fruits.

43RD FIGURE[255]
♌ *Viridis*[256]

"What does the starry ♌ mean? What is the likeness of the shining sun which it has? Tell me that, dear Muse? That is our hero the green ♌ in whose belly the red stars lie hidden, therefore that lazy [♌] emanates Apollonian colors and is the precipitating shine in the sea". *Interpretation:* The green ♌ is our ☉ transformed into the white tincture, the woman who is pregnant with the sun and the moon under her feet. This will soon become the rising sun. It is the sea of glass mingled with fire, but it looks greenish due to the red *oleum veneris*, which played in ♄ and is precisely the ♀, which slept with the ☿ and was transformed into the ☽, because its red cheeks were colored white; and it put its greenness in the white glass, which you can still see somewhat shimmering. It is now in the inner ☉, so it gives the ☿ when it stands in the putrefaction and holds the ♀ entwined within itself, that is, the Apollonian, gold-bearing colors that makes a gold-like shine on the sea, the mirror of the sun.

44. Figur
Aurum nostrum

Allhier sitzt auf dem Thron der Apollo in königlichen Kleidern und bei ihm sitzt der schöne Mond mit ausgebreiteten Haaren, der Sohn in der Mitte, welchen Sie zugleich miteinander krönen, auf dass er bereit sei ihre glückseligen Reiche zu regieren. Was bedeutet nun dieses? Ohne die Sonne und den Ehegatten derselben, wird dieser unser Kleiner zum Reiche ungeschickt sein. *Auslegung*: Der Apollo in königlichen Kleidern ist unser roter ☿, die figierte *Materia* des Steins, der König im güldenen Stück, der Mond mit ausgebreiteten Haaren, die weiße Tinctur, welche beide sich aufs neue vermählen und den Sonnen-Sohn zeugen, welchen sie auch beide miteinander krönen. Denn eines kann nicht generieren und ohne diese beiden wäre auch der Dritte nicht.

45. Figur
Resuscitatio Regis

Nun ist unserem König die höchste Gewalt gegeben und solche große Gaben trägt er von dem väterlichen Geschenk. Ob er nun wohl alles überwindet, so wird er doch von einem Knecht überwunden: Dieser löst jene Glieder durch seine Hervorsprossung auf, jedoch so, sammelt die Mutter die aufgelösten Glieder, daher hat sowohl die Mutter als der Knecht große Ehre davon. *Auslegung*: Unsere multiplicierte rote Tinctur, als der Sonnen-Sohn, hat nun die größte Gewalt im metallischen Reiche, kann allen güldene Kronen aufsetzen, will er sich aber weiter multiplicieren, muss er sich von einem Knecht töten lassen, weil ihm niemand gleich, so sind seine vorigen Öle und Geister sehr gering gegen seine jetzige Hoheit und Adel, alldieweil sie aber aus seinem vorigen Geblüt entsprossen, von seinen Groß-Eltern, welche nur arme Bauern und Untertanen, so lösen sie seine Glieder auf. Weil aber der König so großer Tugend und in seinem vorigen Blut und Schweiß-Bade, so wird das geringe Blut durch das königliche gebessert. Der fixe ☿ erhält den Flüchtigen und der Flüchtige solviert und wird aber-

44TH FIGURE[257]
Aurum Nostrum[258]

"Here sits Apollo enthroned in royal robes, and with him sits the fair moon with spread-out hair, the son in the midst, whom they crown together, that he may be ready to rule their blissful kingdoms. Now what does this mean? Without the sun and the wife of the same, this little one of ours will not be able to rule the kingdom". *Interpretation*: Apollo in royal clothes is our red ♀, the fixed *materia* of the stone, the king in the golden piece; the moon with her hair spread out is the white tincture; both marry again and beget the son of the sun, whom they both crown together. For one alone cannot create, and without these two the third would not exist either.

45TH FIGURE[259]
Resuscitatio Regis[260]

"Now our king has been given supreme authority and he bears such great gifts from his father's gift. Though he overcomes everything, yet he is overcome by a servant: He [the servant] dissolves these sprouting limbs, but in such a way that the mother collects the dissolved limbs, hence both the mother and the servant have great honor from it". *Interpretation*: Our multiplied red tincture, the son of the sun, now has the greatest power in the metallic kingdom. He can put golden crowns on everything, but if he wants to multiply further, he has to let a servant kill him, because nobody is like him. Thus his previous oils and spirits are very small compared to his present highness and nobility, but since they sprang from his former ancestors, from his grandparents, which were only poor peasants and subjects, they dissolve his limbs. But because the king is of such great virtue and is bathed in blood and sweat in his former life, the poor blood is improved by the royal [blood]. The fixed ♀ preserves the fugitive [♀] and the fugitive [♀]

mals der König durch diesen Streit hoch und herrlich, erweitert seine Macht und Herrschaft, sein Land, die Erde, wird vermehrt, tingiert noch viel weiter als vorher.

46. FIGUR
Tres fontes perennis aquæ

Wir 3 wässrige Brunnen werden in einer Höhle eines verborgenen himmlischen Felsens gesehen und werden mit wunderlichen Wassern geziert. Und gibt ☉ und Mond ihre Kräfte und alle Gestirne sind uns durch ihre Strahlen günstig. Wer von uns trinken wird, der wird nach gebührender renovierter Jugend leben und seine gewünschte Weise continuieren. *Auslegung*: Die 3 wässrigen Brunnen in den verborgenen Höhlen des himmlischen Felsens, ist das weiße Wasser, die Jungfer-Milch; das Andere, das Blut und das Dritte das Öl, so alles dreies der harte Fels gegeben, welche 3 Brunnen die Gewichte des philosophischen Goldes und Silbers vermehren können. Auch wird ☉ und ☽ daraus wachsen und ihre Früchte darein geben, wenn sie die Wasser der Gesundheit geworden. Wer davon trinkt, wird in steter Jugend leben, das ist in steter Gesundheit.

47. FIGUR
Die Nymphe unseres Meers

Ich, der ich vor Zeiten von einem flüssigen Vater erzeugt und geboren, welcher so oftmals den ganzen Erdboden umläuft, ich gieße dir mit unseren Brüsten Milch mit Blut ein, welche 2 gekochte angenehme Gewichte des Goldes geben werden: Also wird dem Besitzer der angenehme Lohn kommen und seine Arbeit wird nicht vergebens sein. *Auslegung*: Die Nymphe ist unsere ☽, gezeugt von dem einfachen flüssigen ☿ der Weisen, sie gibt den weißen und roten ♃, Milch und Blut, die angenehmen Gewichte des Goldes, wenn rot und weiß zusammen vereinigt werden und beide rot gefärbt.

dissolves; and through this quarrel, the king becomes high and glorious again, expanding his power and dominion, multiplying his land on the earth, tingeing much further than before.

46TH FIGURE[261]
Tres Fontes Perennis Aquæ[262]

"We, the three watery fountains, are seen in a cave of a hidden, heavenly rock and are graced with wondrous waters. And ☉ and moon give their powers and all the stars are favorable to us with their rays. Anyone who drinks from us will live according to a duly renovated youth and continue [to live] in their desired manner". *Interpretation*: The three watery fountains in the hidden caves of the heavenly rock are [firstly] the white water or the virgin's milk, the second is the blood, and the third is the oil. All three of these were given by the hard rock, and these three wells can increase the weights of the philosophical gold and silver. Also ☉ and ☽ will grow from it and put their fruits in it when they become the waters of health. Whoever drinks of it will live in perpetual youth—that is to say, in perpetual health.

47TH FIGURE[263]
The Nymph of Our Sea

"I, who was long ago begotten and born of a liquid father, who so often circled the face of the earth, I will pour you milk with blood from our breasts, which will give two boiled, congenial weights of gold: so will the possessor's agreeable reward come and their work will not be in vain". *Interpretation*: The nymph is our ☽, begotten of the simple liquid ☿ of the wise. It gives milk and blood to the white and red ♃, the pleasing weights of gold when red and white are united together and both are colored red.

48. Figur
Unser ♃

Verschaffe, dass der Adler mit dem ♌ durch brennende Liebe vereinbart werde und vergieße geschwächt seine Tränen, danach vereinbare die Tränen mit dem ♌ Blut, so wirst du fröhlich die Güter der ganzen Erden haben. Denn wenn sie zugleich vereinbart nach Art des Salamanders durchs Feuer, so werden sie Kräfte durch stößige und mutwillige Flammen verachten. *Auslegung*: Hier wird beschrieben, unseren flüchtigen weißen ♃ zu figieren, da der Adler und ♌ den Kopf halten müssen, sobald der ♌ die Adler zerreißt, gibt es Tropfen und Dünste. Diese fallen auf die Erde, auf die Adler und ♌ bis eins das andere verschluckt und in weiße trockene Tinctur verwandelt. Diese trockenen Tränen werden durchs Feuer, dem roten ♌, in die rote Tinctur gebracht, so ist es der Salamander, so im Feuer lebt.

49. Figur
Das philosophische Ei

Allhier ist unser Drache anwesend, welcher durch die Sprossung der Sonnen paust und welchen ein jedweder durch ein geringes Gold sich schaffen kann, worüber dieser liegt, eben dieses ruht auch zugleich in ihm, daher wird dein Ei die tiefe *Turba* genannt. Was bedeuten aber so viele glänzende Kronen? Er bereitet seinen Brüdern güldene Geschenke. *Auslegung*: Unser Drache ist der ☿ der Weisen, welcher Sonne und Mond in sich verschlungen hält, denn er ist selber das ☉ und ☽, so er verschluckt und diesen Fresser in eine andere Gestalt verwandelt, worüber er liegt, dies ruht auch in ihm, das Ei, das rote und das weiße, in welcher Vereinigung die *Turba* zu sehen, dieser Drache kostet wenig Geld und wenn er all sein Gift in sich gefressen, kann er die metallischen Leiber in ☉ verwandeln.

48TH FIGURE[264]
Our ♃

"Make the eagle unite with the ♌ through burning love and shed its tears weakly, then unite the tears with the ♌ blood, and you will happily have the goods of the whole earth. For if they are united at the same time, like the salamander through fire, they will despise powers of violent and wanton flames". *Interpretation*: Here it is described how to fix our volatile, white ♃, since the eagle and ♌ have to hold their heads, and as soon as the ♌ tears the eagles apart there are drops and vapors. These fall to the earth, onto the eagles and ♌ until one swallows the other and turns into the white, dry tincture. These dry tears are brought into the red tincture by the fire, the red ♌, so it is the salamander that lives in the fire.

49TH FIGURE[265]
The Philosophical Egg

"Here our dragon is present, which pauses through the sprouting of the suns and which anyone can create for themselves with a small piece of gold, upon which it lies, and which also simultaneously rests within it. Therefore your egg is called the deep *turba*. But what do so many brilliant crowns mean? It prepares gifts of gold for its brothers". *Interpretation*: Our dragon is the ☿ of the wise, which keeps sun and moon swallowed up within itself, because it itself is the ☉ and ☽. If it swallows and transforms this devourer into another shape, over which it lies, it also lies within it, the egg, the red and the white, in which union the *turba* can be seen. This dragon costs little money and when it has devoured all its poison, it can turn the metallic bodies into ☉.

50. FIGUR
Generatio

Siehe, da ruht ein Alter in einer scheußlichen Höhle eines Felsens, welchem ein Rabe durch ein Verbündnis zugefallen ist, dessen Geist und zugleich der Sinn die Glieder verlassen, auf dass er von der Strafe der bösen Tat erlöst und befreit sein, indem sie nun zusammen kommen und mit eben demselben wieder vereinbart werden, so wird unser Apollo aus diesen dreien geboren. *Auslegung*: Der Alte ist unser ♄, so in Seele, Geist und Leib muss geschieden werden, denn der Geist und Seele müssen den Leib verlassen, sonst kann er nicht sterben, der Geist und Seele ruht wohl darinnen, aber nicht mit Bestande. Wenn aber der Leib verfault und mit dem Geist und Seele nach seiner verklärten Auferstehung wieder erleuchtet und belebt, so ist es der Apollo, unser königlicher Leib, der keiner Corruption mehr unterworfen, sondern sie wohnen hernach darinnen mit Bestande. Dies ist die neue Geburt.

51. FIGUR
Conjunctio

Sonne und Mond tragen in der Hand die schönsten Lilien und mitten an dem Orte wird unser Apollo gesehen, daher vereinbaren sie alle beide die eigenen Kräfte und daher trägt er mit starker Hand die königlichen Zepter, sein Glanz wird in der ganzen Welt gerühmt und seine Ehre und sein Lob fließt in die herumschweifende Rotte. *Auslegung*: Die schönen Blumen oder Lilien, so Sonne und Mond tragen, sind die beiden ☿, der rote und der weiße, aus diesen wächst die Tinctur gar sehr, wenn sie ihre Kräfte vereinbaren, daraus hernach der großmächtige starke König wird, der in dem ganzen metallischen und mineralischen Reiche herrscht, der Sonnen-Sohn.

52. FIGUR
Mortificatio

Nun wird die Ehre auf unserem glänzenden König neidisch und die-

50TH FIGURE²⁶⁶
*Generatio*²⁶⁷

"See, there rests an old man in a desolate cave of a rock, to whom a raven has fallen through alliance, his spirit and at the same time his mind forsake the limbs, so that he may be redeemed and freed from the punishment of the evil deed, by coming together now to be reconciled with the same, our Apollo is born of these three". *Interpretation*: The old man is our ♄, which must be separated in soul, spirit, and body, for spirit and soul must leave the body, otherwise it cannot die, and spirit and soul rest well within, but not with lasting effect. But when the body putrefies and is enlightened and enlivened again after its transfigured resurrection with spirit and soul, then it is the Apollo, our royal body, which is no longer subject to corruption, but they live in it afterwards with permanence. This is the new birth.

51ST FIGURE²⁶⁸
*Conjunctio*²⁶⁹

"The sun and the moon carry in their hands the most beautiful lilies, and in the middle of the place our Apollo is seen, therefore they both unite their own powers, and therefore he carries the royal scepters with a strong hand, his splendor is praised throughout the world and his honor and his praise flows to the roving herd". *Interpretation*: The beautiful flowers or lilies that the sun and moon carry are the two ♃, the red and the white, from these the tincture grows profusely when they unite their powers, from which the mighty and powerful king emerges, who reigns in the whole metallic and mineral kingdoms, the son of the sun.

52ND FIGURE²⁷⁰
*Mortificatio*²⁷¹

"Now honor is jealous of our shining king and ten young men from the

sen töten 10 Jünglinge von dem gemeinen Pöbel, alles wird verwirrt, Sonne und Mond geben viele Zeichen ihrer Traurigkeit nach entstandenen Finsternissen an Tag. Ein Regenbogen mit vielen Farben gemalt erscheint, welcher dem Volk angenehme Post des Friedens bringt. *Auslegung*: Der König so getötet, ist der rote ♁, die Tinctur, so in seinen vorigen Wassern wieder resolviert, darinnen fault die Sonne und der Mond, wenn sie im Drachen-Haupt verfinstert. Endlich kommt das Gnaden-Zeichen, der Regenbogen, mit vielen Farben, wenn sich der Streit gelegt und der Himmel des ♄ ausgelaufen, so ist die Tinctur vermehrt, die 10 Jünglinge sind 10 Teile des *Menstrui*.

53. Figur
Putrefactio

Die Stadt wird von unzähligen Feinden und Feuer verwüstet, daher geht auch der tolle Pöbel mit seinem König unten. Die schwarzen Raben verzehren die niedergeschlagenen Körper, jedoch so bleibt der geistliche Teil unversehrt. Welcher, wenn er wieder mit dem starken Körper, mit dem König vereinbart werden wird, so wird allen ein Ursprung des neuen Lebens auferstehen. *Auslegung*: Die Stadt, so gestürmt von Feinden und Feuer verwüstet, ist die reiche Stadt, so von allen Göttern erbaut, darinnen unser König seine Residenz hat, der Pöbel ist unter des ♄ Influenz, dessen rote Geister den ☿ figieren, den königlichen Leib tötet, wenn dieser verfault und wieder mit den starken doppelten Geistern vereinbart wird, so wird dieser König zum neuen Leben wieder verbessert auferstehen.

54. Figur
Albificatio

Siehe, eine wieder lebendig gewordende Schwester geht ihrem geliebten Bruder vor und trägt auf kristallische Art weiße Glieder, sie klagt die Brüder an, dass sie nach Verachtung des himmlischen Geschenks nichts als Lasten der Erden geliebt, vermahnt die Gleichen des Bruders, dass sie dem Gestirne möchten wieder gegeben werden und ihren

common mob kill him, everything is confused, sun and moon give many signs of their sorrow after darknesses have occurred during the day. A rainbow painted with many colors appears, bringing pleasant messages of peace to the people". *Interpretation*: The king who was slain is the red ♃, the tincture which is resolving in its former waters, in which the sun and moon putrefy as it darkens in the dragon's head. Finally the sign of grace appears, i.e., the rainbow with many colors, when the quarrel subsides and the sky of the ♄ has run out, the tincture is increased, and the ten youths are the ten parts of the *menstrui*.

53RD FIGURE[272]
Putrefactio[273]

"The city is devastated by countless enemies and fire, so the mad rabble goes down with its king. Black ravens consume the struck-down corpses, but the spiritual part remains intact. When this is reconciled with a strong body, with the king, then a source of new life shall arise for all". *Interpretation*: The city that is being stormed by enemies and devastated by fire is the rich city built by all the gods, in which our king has his residence; and the rabble is under the influence of ♄, whose red spirits fix ☿, kill the royal body, whereupon it putrefies and unites with the strong double spirits, such that the king will rise again rectified with new life.

54TH FIGURE[274]
Albificatio[275]

"Behold, a sister who has been brought back to life precedes her beloved brother and, wearing white limbs of a crystalline nature, accuses the brethren of having despised the heavenly gift: they have loved nothing but the burdens of the earth. She admonishes her brother's peers so that they may be returned to the stars and their heads adorned with

Kopf einen neue Krone Zierde. *Auslegung*: Hier ist erstlich beschrieben die weiße Tinctur, welche der roten vorher geht, welche durch den grünen ♌ zubereitet werden, wenn solcher wohl rectificiert und der grüne *Duenech* von seinem 🜍 höchst rein geschieden, sieht sie wie Diamanten, wenn sie multipliciert worden, wenn es aber der Stein der ersten Ordnung ist, als der natürliche Schatz, sieht er etwas grün, denn wenn der lautere ☿ialische Himmel sein süßes Wasser aufschließt, die ♀ als der grüne ♌, so in dem ♄ gespielt, darinnen himmlisch wird, ihre Grüne und Bittere in das Süße gibt, sieht es also schimmernd, zumal wenn das Salz auch nicht recht rein vom roten 🜍 geschieden, denn 3 Dinge das Werk verrichten, als ♌ *viridis*, *Aqua fœtida* und *Fumus albus*, der weiße Rauch ist der doppelte ☿, das stinkende Wasser das 🜍 Öl, der grüne ♌, welcher auch der Engel ♄ genannt wird oder *Saturiel*, ist des ♄ seine Sonne, welche in seinem Kerker verschlossen gewesen oder vielmehr dessen rotes Öl, so ♀ Art, weil es inwendig grün ist.

55. FIGUR
Rubificatio

Der König, so von dem Tode wiederkommt, bringt allen neue Freude und ergötzt die traurigen Herzen durch seinen Geschenke. Ihre Mitgenossen bekrönt er mit glänzendem Gold, daher bedeckt er auch die zarten Gliedmaßen mit schönem Purpur. Das ist der wahre *Azoth*, das ist die rechte Arznei der Weisen, welche durch ihre Strahlen die kränklichen Glieder wieder repariert. *Auslegung*: Der König in Purpur ist die rote vollkommene Tinctur. Erst war es der König im Hemde, darauf wurde ihm das güldene Stück angelegt, seine zarten und glänzenden Glieder sind mit dem Purpur bedeckt, von satt roter Farbe, welche die anderen alle bedeckt. Hier ist die Vereinigung aller widerwärtigen Dinge, weil aus dem wässrigen *Azoth* nun der rote und trockene *Azoth* worden, die wahre Medizin für Menschen und Metalle. In seinem Anfang war es das Natur Salz, welches feucht, schmierig, auch unvollkommen, jetzt ist es das trockene blutrote Salz, welches färbt und beständige Farbe ist, so nicht abgeht, die allen Metallen kann güldene Kronen aufsetzen.

a new crown". *Interpretation*: First, the white tincture, which precedes the red one, is described, which is prepared from the green ♌, when such is well-rectified and the green *Duenech* is extremely pure, having been separated from its ♃, it looks like diamonds when they have been multiplied. When, however, it is the stone of the first order, the natural treasure, it looks somewhat green, for when the pure ☿ial heaven opens up its sweet waters, so that ♀, as the green ♌, which played in the ♄, becomes heavenly within it, gives its greenness and turns its bitterness into sweetness, and thus appears shimmering, especially since the salt is not properly separated from red ♃. For three things accomplish the work: ♌ *viridis*, *aqua fœtida*, and *fumus albus*. The white smoke is the double ☿; the stinking water is the ♃ oil; the green ♌, which is also called the angel ♄, or *Saturiel*, is the sun of ♄, which was shut up in its dungeon, or rather its red oil of ♀ nature, because it is green within.

55TH FIGURE[276]
Rubificatio[277]

"The king who returns from the dead brings new joy to all, and delights sorrowful hearts with his gifts. He crowns their fellows with glittering gold, therefore he also covers their delicate limbs with beautiful crimson. This is the true *azoth*, this is the right medicine of the wise, which repairs sick limbs with its rays". *Interpretation*: The king in purple is the perfect red tincture. First it was the king in the tunic, then the golden piece was put on him, his delicate and brilliant limbs are covered with purple of a rich red color, which covers all others. Here is the union of all repellant things, for the watery *azoth* has now become the red and dry *azoth*, the true medicine for humans and metals. In the beginning it was the salt of nature, which is moist, greasy, and also imperfect, whereas now it is the dry, blood-red salt, which colors and is of permanent color, so it does not come off, and all metals can wear golden crowns.

56. FIGUR
Der Traum oder das Gesichte.

Ein klarer Mann ist da, um die stille Nacht-Zeit und die geheimsten Glieder erscheinen meinen Augen. Seine Gliedmaßen werden mit einem grünen Kleide bedeckt und sein Haupt ziert eine schöne Krone, über ihm ist ein heller Stern, so die dicken Finsternisse erleuchtet und indem ich mich verwundere, wer er sei, so befiehlt er mir, dass ich ihm folgen soll. *Auslegung*: Den philosophischen ☿ zu figieren und die Composition des Steins zu machen, ist nicht einerlei Weg, obschon einerlei Materien genommen werden, denn viele *Philosophi* haben ihn in 7 Monaten gemacht, etliche in einem Monat, hier ist eben nicht der kürzeste Weg beschrieben, sondern wie der ☿ soll mit dem grünen ♌ figiert werden, wenn der ☿ mit der ♀ Eigenschaft præcipitiert wird und grünlich sieht, wenn er aber durch seinen eigenen ☿ wiederum zurück gerufen und revificiert in dem Lebens-*Aquavit*, so wird er hell und weiß, denn der ☿ muss ihm folgen und erleuchten, die Grüne in die Weiße verkehren, wenn ein *Corpus* soll verwandelt werden, muss es durch einen Geist geschehen und muss der, so ihn überwinden will, stärker sein, wenn man nun dem ☿ den ☿ gibt, mit dem wahren weißen Ferments vereinigt, so ist das Weiße mehr als die Grüne, diese ♀ Tinctur überwunden und in die weiße ☾ Tinctur transmutiert.

57. FIGUR
Das Kennzeichen oder Wahlspruch unser Jungfer

Unsere Jungfer erzählt dir mit wunderlichen Gesichte Wunder, auf das du alles mit aufmerksamen Gemüt merken sollst. Von dem Nord-Wind mag dir der Bär warme Feuer geben und daher wird der Süd-Wind löwische Wasser bringen. Diese miteinander vermischten zwei werden fliegen, nachdem du sie einem heißen Ofen anvertraut und werden von den hohen Orten wiederum nach dem Untersten zurückgehen. *Auslegung*: Unsere Jungfer ist der Jungfer ☿, dem folgen die anderen *Menstrua*, als der Bär vom kalten Nord-Wind, des alten ♄

56TH FIGURE[278]
A Dream or Vision

"A clear man is there in the stillness of night, and the most secret limbs appear to my eyes. His limbs are covered with a green robe, and his head is adorned with a beautiful crown, above him is a bright star illuminating the thick darknesses, and while I wonder who he is, he commands me to follow him". *Interpretation*: It is not the same way to fix the philosophical ☿ and to make the composition of the stone, although the same materials are used, because many *philosophi* have done it in seven months, and some in one month. Here the shortest way is not being described, but rather how the ☿ is to be fixed with the green ♌, when the ☿ is precipitated with the property of ♀ and looks greenish. Yet when it is called back by its own ☿ and revived in the living *aqua vitæ*, it becomes bright white, for the ☿ must follow it and enlighten it, turning the green into the white. If a *corpus* is to be transformed, it must be done by a spirit, and the one who wants to overcome it must be stronger. If one now gives the ☿ the ☿, with the true white ferment, the white is more than the green, this ♀ tincture is overcome and transmuted into the white ☽ tincture.

57TH FIGURE[279]
The Sign or Motto of Our Virgin

"Our virgin tells you miracles with a strange face, you should note everything with an attentive mind. From the north wind the bear may give you warm fires and hence the south wind will bring leonine waters. Once mixed, theses two will fly, and then you are to commit them to a hot furnace so that they will return from the high places to the lowest in turn". *Interpretation*: Our Virgin is the virgin ☿, followed by the other *menstrua* like the bear from the cold north wind, the sharp seal of the

seine scharfen Siegel, sein Öl oder Feuer, der Süd-Wind, das weiße ♃ Öl, bringt die roten löwischen Wasser, diese beiden Feuer miteinander vermischt sind *Contraria*, fliegen voneinander, eins wird das glühende flammende Schwert ♄ genannt, seine Sense, damit er dem ☿ die Flügel behaut, das andere ist das glühende rote Feuer der ☿ der Sonnen, das Feuer der Sonne und werden alle durch den jungfräulichen ☿ gewiesen, welcher allen vorgeht, die anderen mit Fingern zeigt und durch ihn an den Tag gebracht werden müssen.

58. Figur
Der neue Wahlspruch oder Kennzeichen

Ich bin die schöne Göttin, allzusehr in dem Grunde des Meeres entsprossen, welches durch den Lauf die ganze Erde besieht und um solche herum geht, unsere Brüste ergießen die 2 Bächlein, nämlich des Bluts und der Milch, welche du wohl kennen kannst, diese 2, wenn sie vermischt sind, überlasse sie dem leichten *Vulcano* zuzubereiten und dann wird der Mond und Apollo deinem Wunsch zustatten kommen. *Auslegung*: Die Göttin, so in dem Grunde des Meeres entsprossen, welches durch den Lauf die ganze Erde besteht, ist unser Wasser, wenn es bei der Erde coaguliert wird, die Brüste von Blut und Milch, so sie gibt, sind der rote und weiße ♃, diese 2, wenn sie vermischt durch das philosophische Feuer, so werden beide Tincturen unseren Wunsch erfüllen.

59. Figur
Ein anderes Kennzeichen oder Wahlspruch

Indem jene Tränen vergießen wird, wirst du den Adler martern, alsdann wird der schwache ♌ durch einen erschrecklichen Tod hinfallen. Dessen Blut ist der größte Schatz in der ganzen Welt. Diesen sollst du mit Tränen mit dem Adler vereinbaren und dann so wirst du reich werden. Denn sie baden zugleich und bald werden sie durch Liebe verzehrt und sind deiner Natur, Salamandern gleich. *Auslegung*: Sobald als die 7 Adler und der ♌ zusammengesetzt, verwundert der ♌ die Adler, dass sie Tränen durch sein Blut lassen, daher fängt es an zu

old ♄, its oil or fire, the south wind, the white ♃ oil, brings the red leonine waters. These two fires mixed together are *contraria*, and they fly away from each other accordingly. One is called the glowing, flaming sword or scythe of ♄ by which is cut the wings of ☿, while the other is the glowing, red fire of the ☿ of the sun. These are all pointed out by the virgin ☿, which precedes all, to which others point fingers and must be brought to light by it.

58TH FIGURE[280]
The New Motto or Sign

"I am the beautiful goddess sprung from the uttermost depths of the sea, which in its course comprises the whole earth and surrounds it; our breasts pour out two little streams, one of blood and one of milk, which you can certainly know. When these two are mingled, leave them to light *Vulcano* to prepare, and then the moon and Apollo will grant your wish". *Interpretation*: The goddess who sprang from the bottom of the sea, which, due to its course, consists of the whole earth, is our water when it is coagulated by the earth. The breasts of blood and milk that she gives are the red and white ♃, and when mingled by the philosophical fire both tinctures will fulfill our desire.

59TH FIGURE[281]
Another Sign or Motto

"With the shedding of those tears, you will martyr the eagle, then the weak ♌ will fall with a terrible death. Its blood is the greatest treasure in all the world. You should unite this with tears, with the eagle, and then you will become rich. For they bathe at the same time and soon they are consumed by love and are [of] your nature, like salamanders" *Interpretation*: As soon as the seven eagles and the ♌ are put together, the ♌ surprises the eagles so that they release tears through his blood. Thus

schwitzen, bis es Tropfen gibt, welche etwas blaulich scheinen, weil aber die Adler viel und stets auf ihn zufliegen, verwunden sie den ♌ auch und werden von seinem Blute beschmutzt, welcher Streit und Kampf währt, bis sie alle tot dahin gefallen, der ♌ und Adler, dieses grünen ♌ Blut wird mit des Adlers Tränen vermischt, mit dem vermehrenden *Menstruo*, welches in Seele, Geist und Leib besteht und das Ferment des ☽ ist, so wird sich die weiße Tinctur multiplicieren, denn der ♌ und Adler baden zugleich in diesem Brunnen und sind 10-fältig vermehrt, sind des Feuers Natur, dem Salamander, der roten Tinctur gleich, weil sie nun wie der Salamander im Feuer leben kann, denn der Mond oder die weiße Tinctur ist hitziger Natur worden.

60. FIGUR
Die erste Materia

Ich bin der Drache und spritze tödliches Gift von dem Leibe, der rote ♌ nebst dem grünen ♌ liebt mich. Ich werde von vielen das angenehme Ei der Natur genannt und fliehe davon, wenn du mich nicht wohl wirst anbinden. Ich bin die Art Farbe, welche du willst, ich bin auch der Mann und Weib, ich ergötze den menschlichen Leib und auch zugleich alle Metalle. *Auslegung*: Die erste *Materia* ist giftig und tödlich, denn wenn die beiden Drachen, der flüchtige und fixe zusammengesetzt und einander töten sollen, so spritzen sie Gift aus, doch wurden sie das angenehme Ei der Natur genannt. Wenn diese beiden Drachen zu einem einzigen geworden, aus diesen Ei kommt das Rote und Weiße, der weiße und rote ♀, der grüne ♌ und der rote ♌ lieben ihn, der grüne ♌ figiert ihn in die weiße Tinctur, der rote ♌ bringt ihn zur roten Tinctur, er ist die Farbe, welche man will, er lässt sich schwarz, weiß, gelb und rot machen.

61. FIGUR
Die erste Operation

Allhier liege ich, gleich wäre ich durch groben Sand begraben und unser Geist flieht mit der Seele in die Höhe. Sie werden aber nicht entlau-

it begins to sweat until there are drops, which appear a little bluish; but because the eagles fly much and always towards it, they also wound the ♌ and are stained by its blood. This strife and struggle lasts until they all fall dead, the ♌ and eagles together. The blood of the green ♌ is mixed with the eagle's tears, with the increasing *menstruo*, which consists of soul, spirit, and body and is the ferment of the ☽. Then the white tincture will multiply, because the ♌ and eagle bathe in this fountain at the same time and are multiplied tenfold; and they are of fire's nature, equal to the salamander, the red tincture, because it can now live in the fire like the salamander, for the moon or the white tincture has become more ardent in nature.

60TH FIGURE[282]
The First Materia

"I am the dragon and I spit deadly poison from my body, the red ♌ along with the green ♌ loves me. I am called by many the congenial egg of nature; flee from me if you will not bind me properly. I am the kind of color you want, I am also the man and the woman; I delight the human body and also all metals at the same time". *Interpretation*: The first *materia* is poisonous and deadly, for when the two dragons, the volatile and the fixed, are put together and intend to kill each other, they sprout poison, yet they have been called the convivial egg of nature. When these two dragons become one, from this egg comes the red and the white, the white and red ♃, which the green ♌ and the red ♌ love, the green ♌ fixes it into the white tincture, the red ♌ brings it to the red tincture. It is whatever color you want, and it can be made black, white, yellow, and red.

61ST FIGURE[283]
The First Operation

"Here I lie as if I were buried by coarse sand and our spirit flees with the soul on high. But they will not escape because the tomb will be closed

fen, weil das Grab durch Kunst verschlossen wird, sie werden aber gezwungen, mein Leben mir wiederzugeben. Ich bin dem Raben gleich, bis dass 14. Licht zurückläuft, alsdann erneuere ich die glänzenden Zepter meines Geschlechtes. *Auslegung*: Hier wird die Begrabung des Leichnams oder unser ☿ beschrieben, welcher in dem groben Sand, so sich auf den Boden des Gefäßes setzt und endlich das ganze Grab als grober Sand wird, der Leib, Grab und Sand ist alles eins, denn der königliche Leichnam ist in allen dreien enthalten. Wenn dieser Leib nun alles zu Sand worden und darein verscharrt, steigt in seinem Tod der Geist mit der Seele in die Höhe. Weil aber das Gefäß verschlossen, werden sie gezwungen dem Leib ein neues Leben wiederzugeben, den schwarzen Raben in die weiße Taube oder vielmehr den weißen Schwan zu verwandeln, welches auch der König im Hemd kann genannt werden, weil sie als wie die Geister weiß sind oder die Leiber, so vom Tod auferstanden, so ist dieser glänzende König, alsdann kann er erneuert werden und sein Zepter noch glänzender werden, das ist, gelb, in ☉ verwandelt, dass die gelbe Farbe auf die weiße getragen wird, wie die Maler auf Weiß Gelb und auf Gelbe Rot malen, die gelbe Farbe heißt dem König im Hemd das güldene Stück anlegen, wodurch er sein Geschlecht fortpflanzt.

62. FIGUR
Die andere Operation

Wenn viele und mancherlei Farben erscheinen werden, so werde ich scheinbar mit einem roten Leib hervor gehen. Niemand wird mir gleich in der ganzen Welt gefunden und die Sonne und Mond werden meinen Händen unterworfen. Überdies wird der Kräuter Macht uns unterworfen, durch welche die Krankheiten der Menschen von dem menschlichen Leibe hinweg getrieben werden. *Auslegung*: Hier wird beschrieben, wenn dem Könige über sein güldenes Stück der Purpur-Mantel umgelegt wird, so erscheinen die vielen bunten Farben, anstatt der Schwärze, als rot, violenbraun, gelbrot, blitzblau und andere bunte hohe Farben, je mehr und höher dieser König in Purpur eingeklei-

by art, but they will be forced to give my life back to me. I am like the raven, until the fourteenth light runs back, then I renew the shining scepter of my lineage". *Interpretation:* Here the burial of the corpse or our ☿ is described, which settles in the coarse sand at the bottom of the vessel and finally the whole grave becomes coarse sand. The body, grave, and sand are all one, for the royal corpse is included in all three. When this body has now all turned to sand and is buried in it, the spirit in its death rises up with the soul. But because the vessel was closed, they were forced to give the body new life again, to transform the black raven into the white dove, or rather the white swan, which can also be called the king in the tunic. Because they are as white as the spirits are, or the bodies which have risen from death, then this is the shining king, and now he can be renewed and his scepter can become even more brilliant—that is to say more yellow—transformed into ☉, so that the yellow color is applied to the white, like the painters' pigment: yellow upon white, red upon yellow. The yellow color means to apply the golden piece to the king in the tunic, through which he propagates his lineage.

62ND FIGURE[284]
The Second Operation

"When many and varied colors appear, so will I seem to emerge with a red body. No one will be found like me in all the world, and the sun and moon will be subjected to my hands. Moreover, the power of herbs is subjected to us, by which the diseases of humans are driven out of the human body". *Interpretation:* Here is a description of the placing of the purple cloak upon the king, over his golden piece, at which time, instead of blackness, many bright colors appear such as red, violet-brown, yellow-red, lightning-blue, and other bright colors. The more this king is clothed in purple, the bolder and thicker the col-

det, je fetter und dicker die Farben sich sehen lassen. Die Sonne und Mond sind dieser Tinctur unterworfen, sie müssen sich verkriechen und diese Tinctur über sich herrschen lassen, weil jene nur als nichts gegen ihr sind und muss der Schwache dem Starken weichen, weil hier ein größerer König den Thron eingenommen, er kuriert alle Kräuter, denn er selbst ist das größte *Vegetabile*, so alle Krankheiten circuliert, weil er aller Dinge Quintessenz in sich hat.

63. FIGUR
Die dritte Operation

Siehe 2 mal 5 Männer bringen mich ums Leben, jedoch werden sie auch gezwungen, mit mir durch den Tod umzukommen und unterzugehen. Die Sonne hat mit mir Mitleiden und der Mond mit schwarzem Gesicht und der Regenbogen, ein Zeichen vom Himmel ist gegenwärtig. Derhalben, so wasche ich alle meine Feinde von dem Unflat ab und bringe geschicklicher Weise den triumphierenden Klang. *Auslegung*: Wenn der König in 10 Teile seines Wassers des männlichen *Menstrui* solviert, weil wir 2 Wasser machen, eins zum solarischen Körper, das andere zum lunarischen und der König in seinem Wasser fault, so stirbt das *Menstruum*, auch der königliche Leib und faulen miteinander. Weil aber der König edler Natur, sich sein Geist mit jenen Geistern vereinigt, so gehen sie beide wieder zurück in Körper, waschen und reinigen ihn. Endlich wird er vom Tod auferweckt. Wenn der Regenbogen erscheint, ist das neue und verbesserte Leben gefunden, der König größer und vermehrt, das ist multipliciert.

64. FIGUR
Die vierte Operation

Damit man aber erkenne, was für große Tugenden ich habe, so vergelte ich bald darauf mit meinen Gütern den mir zugefügten Schaden. Denn wenn gleich die Raben die feindlichen Körper verzehren, so bleibt doch derselben Geist und Seele beständig zurück. Diese vereinbare ich mit mir und gebe das volatilische fix wieder, damit jedweder

ours that allow themselves to be seen. The sun and moon are subject to this tincture, and they must retreat and let this tincture rule over them, because they are nothing against it. The weak must give way to the strong, because here a greater king has taken the throne. He cures all herbs, for he himself is the greatest *vegetabile*, which circulates all diseases, because they contain the quintessence of all things.

63RD FIGURE[285]
The Third Operation

"Behold, two times five men kill me, but they are also forced to die and perish with me. The sun has pity on me and the moon with a black face and the rainbow, a sign from heaven, is present. Therefore, I wash all my enemies from the filth and skillfully bring the triumphant sound". *Interpretation*: When the king is dissolved in ten parts of his water, the male *menstrui* and the king putrefy in this solution, because we prepare two waters, one for the solar body, the other for the lunar. The *menstruum* also dies, as does the royal body, and they putrefy together. But because the king is of noble nature and unites his spirit with those spirits, they both go back into the body, washing and purifying it. Finally it is raised from the dead. When the rainbow appears, the new and improved life is found, and the king is greater and magnified—that is to say, multiplied.

64TH FIGURE[286]
The Fourth Operation

"But so that people can see what great virtues I have, I soon repay the damage done to me with my goods. For even though the ravens consume the enemy's bodies, their spirit and soul always remain behind. I unite these with myself and make the volatile fixed so that everyone may

die Geschenke meiner rechten Hand sehen möge. *Auslegung*: Dies ist eben die vorige Operation, welche etwas besser hier ausgeführt wird, dass der fixe Geist den Leib wieder annimmt, solchen erhöht und vermehrt, *in Quantitate & in Qualitate*, damit er seinen Brüdern, den Metallen, die Geschenke austeilen möge und ihre unreinen Leiber in fixe und beständige Leiber verwandeln.

65. FIGUR
Fünfte Operation

Ich stehe in Gestalt einer Königin von der Last des Grabes auf und bringe meinen Brüdern fröhliche Botschaft. Der keusche Bräutigam hat mich die Braut heißen zu euch legen, auf dass ich durch meine Geschenke eure traurigen Herzen wieder erneuere. So nehmt denn mit angenehmen Sinn die weißen Kronen, bis der Apollo von seinem Grab wieder aufstehe. *Auslegung*: Dies ist die weiße Tinctur, die weiße Königin, welche der roten vorher geht und alle Metalle in weißes ☉ verwandelt, wenn sie mit gemeinem ☉ fermentiert wird, denn die weiße Tinctur ist das Weib mit der Sonne bekleidet und ist selber das weiße ☉, ist auch durch das Wasser, so in Seele, Geist und Leib besteht, im königlichen Wasser aufgelöst und darinnen vermehrt worden in der Quantität und Qualität.

66. FIGUR
Sechste Operation

Siehe, ich habe alle meine Feinde tapfer überwunden und habe würdige Sieges-Zeichen meines Geschlechtes davon getragen. Derohalben ihr Brüder, ihr lieben Herzen (wofern noch Liebe bei euch ist) macht euch mit geschwindem Fuße hierher und nehmt hin die glänzende Krone, welche ich euch gebe und multipliciert euch bis auf das zehnte Geschlecht. *Auslegung*: Hier werden die philosophischen Metalle in der weißen Tinctur mit der güldenen Krone geziert und mit dem gelben Öl angefeuchtet, welches die aufgehende Sonne mit gelben Strahlen genannt wird, wenn der weiße Brunnen gelbe Farbe bekommt. Hier

see the gifts of my right hand". *Interpretation*: This is simply the previous operation, which is carried out a little better here, so that the fixed spirit accepts the body again, increasing it and multiplying it, *in quantitate & in qualitate*, so that it may distribute the gifts to its brothers (who are the metals) thereby transforming their impure bodies into fixed and permanent bodies.

65TH FIGURE[287]
Fifth Operation

"I rise from the burden of the grave in the form of a queen and bring glad tidings to my brethren. The chaste bridegroom has bidden me lay the bride with you, so that I may renew your sorrowful hearts with my gifts. So take the white crowns in a congenial sense, until Apollo rises from his grave again". *Interpretation*: This is the white tincture, the white queen, which precedes the red [queen] and turns all metals into white ☉ when fermented with common ☉, for the white tincture is the woman clothed with the sun, and is itself the white ☉ too, being also dissolved by the water consisting of soul, spirit, and body, and dissolved in the royal water and multiplied therein in quantity and quality.

66TH FIGURE[288]
Sixth Operation

"Behold, I have valiantly conquered all my enemies, and have borne therefrom worthy tokens of victory of my lineage. Therefore, you brothers, you dear hearts (if there is still love within you) make yourselves here with hasty feet and accept the shining crown which I give to you and multiply yourselves to the tenth generation". *Interpretation*: Here the philosophical metals are adorned in the white tincture with the golden crown and moistened with the yellow oil, which is called the rising

vermehrt sich der Stein 10-fältig in diesem lebendig-machenden und färbenden Gift, welches das güldene Stück genannt wird, so dem König über sein weißes Hemd angelegt worden.

67. FIGUR
Das ganze philosophische Ei

Alles dasjenige was vorher in so vielen Figuren eingeschlossen, siehst du nun alles unter einem einzigen *Circul* verschlossen. Unser Alter zeigt dir den Anfang und den Schlüssel und das Werk gibt der ♃ und der ☿ mit dem Salz: Dafern du nun allhier nichts siehst, warum suchst du ein mehreres? Denn also wirst du mitten im Lichte blind sein. *Auslegung*: Wenn der Sonnen-Sohn erschienen und alles, was aus dem einzigen kommen, wieder in eins gebracht, die 7 Gestalten der Natur vorbei, jeder Sonne ihren Himmel oder Kreis durch den ordentlichen Lauf absolviert und durchlaufen, so ist es wieder eins, welches vorher auch eins gewesen; aber nun in viel 1000-fältiger Verbesserung. Der alte ♄ ist der Anfang, welcher alles dies gezeigt und in seinem Leibe verborgen gehabt, dieser ♄ hat den Schlüssel, sie sind alle in seinem Kerker verwahrt und gefangen gelegen, das Werk aber wird aus den ☿, ♃, ☉ gemacht, der ☿ steigt vor sich selbst durch wunderbare Stufen und Treppen in die Höhe, bis er den königlichen Sitz einnimmt, denn er ist in seinem Anfang ein nackendes elendes Kind, voller Blöße und Feuchtigkeit, wenn aber sein Wasser durch sein eigenes Salz eingetrocknet, dann ist er das Kraut, welches wider den Tod dient, er vertritt Mannes und Weibes statt, wegen seines doppelten Leibes und wird durch seinen eigenen Gift und Zorn getötet, das ist, figiert und hernach in seinen eigenen Ölen weiß, gelb und rot gefärbt, dass er flüssig, das unverbrennliche Öl wird.

68. FIGUR
Das ♄ Kennzeichen oder Wahlspruch

Afrika gibt warme ♌ mit roten Herzen, welche alle unseres Drachens Geschenke haben. Der kalte *Septentrio* aber gibt die starrenden Bären,

sun with yellow rays when the white fountain obtains a yellow color. Here the stone multiplies tenfold in these life-giving and coloring poisons, which is called the piece of gold placed upon the king's white tunic.

67TH FIGURE[289]
The Whole Philosophical Egg

"Everything that was previously enclosed in so many figures you now see enclosed under a single *circul*. Our old man shows you the beginning and the key, and the ♄ and the ☿ together with the salt provide the work: if you do not see anything here, why do you seek anything further? Because then you will be blind in the midst of the light". *Interpretation*: When the son of the sun appeared and everything that came from unity was brought back into one, the seven forms of nature passed by, each sun completing and traversing its heaven or circle through its proper course. For that which had been one before is also one again; but now it has improved a thousandfold. The old ♄ is the beginning. He begot all of this for he had it hidden in its body. This ♄ holds the key, and they are all kept imprisoned in his dungeon. But the work is made of the ☿, ♄, and ☉. The ☿ rises and lifts itself up by marvelous steps and stages until it takes the royal seat, for in its beginning it is a naked, miserable child, completely nude and full of dampness, but when its water has dried up by its own salt (for this is the herb which serves against death), it represents man and woman because of its double body, and it is killed by its own poison and anger—that is to say, it is fixed and then colored white, yellow, and red in its own oils so that it will become liquid, the incombustible oil.

68TH FIGURE[290]
The mark or motto of ♄

"Africa gives warm ♌ with red hearts, which have all our dragon's gifts. But the cold *Septentrio* gives the gazing bear, which brings your eagle

welche dir du schöner Apollo deinen Adler mitbringen, daher wird der feuchte Saft mit der Flamme in die Höhe steigen, diese 2 befiehlt dir, die kluge Jungfer zu nehmen. *Auslegung:* Der warme ♌ mit roten Herzen, ist das Feuer des Steins, sein Blut und Seele, welche alle des Drachens Geschenke, denn von ihm kommen sie, der starrende Bär ist das Feuer ♄, welcher der schöne Apollo dem Adler den ☿ *Philosophorum* bringt, daher der feuchte Saft des Adlers mit der Flamme des Feuers, des *Olei* ♄, in die Höhe steigen, diese befiehlt dir die kluge Jungfer zu nehmen, die *Sophia*.

69. FIGUR
Das Wappen unseres Helds

Unser Held, nachdem er im Krieg alle seine Feinde überwunden, hat nun als ein Überwinder die größte Belohnung seiner Tugend. Denn er stellt die Sonne dar, mit schönen Strahlen umgeben und unter dieser sind die Monds-Hörner, in der Mitte zieren die Sonne 3 runde Äpfel und diese Zeichen werden dir den Ursprung mit den Taten geben. *Auslegung:* Der Held ist unser König, die rote Tinctur, der Sonnen-Sohn, er hat alle seine Feinde zu Wasser und Land nieder gelegt, er ist die ☉ und der Mond, in welchen beiden er steht mit Glanz umgeben, denn die Sonne und Mond waren seine Eltern, unter ihm die Monds-Hörner, der abnehmende Mond, den man nun vor der Sonnen Glanz nicht sehen kann, die 3 runden Äpfel sind die 3 *Principia*, diese geben den Ursprung und dem Werk die Kräfte und das ganze Vermögen.

70. FIGUR
Materia des philosophischen Steins

Es ist zwar nur ein Ding, es sind aber ihrer auch zwei, ja auch drei, wer klug ist wird wohl erfahren, dass ich ihm dieses nicht für die Langeweile gesagt habe. Denn es ist nur ein Drache, so in schlechtem Wert und alt von Jahren, welcher der Königs Brunnen und richtige Vergeltung hat. Es sind 2 ☿, davon du den Beständigen oder Fixen mit dem Fliegenden zu vereinbaren hast. Es ist eine Seele, ein Leib und ein leichter

to you, beautiful Apollo, so the moist sap will rise up with the flame, the clever virgin orders you to take these two". *Interpretation*: The warm ♌ with red heart is the fire of the stone, its blood and soul, which are all the dragon's gifts, for they come from the dragon. The gazing bear is the fire of ♄, which brings the beautiful Apollo, the ☿ *philosophorum*, to the eagle, hence when the moist sap of the eagle with the flame of fire of the *olei* ♄ is soaring up, it commands you to take the clever virgin, Sophia.

69TH FIGURE[291]
The Coat of Arms of Our Hero

"Our hero, having conquered all his enemies in war, being a conqueror, now has the greatest reward for his virtue. For he represents the sun surrounded by beautiful rays, beneath which are the horns of the moon. The middle of the sun is adorned with three round apples and these signs will give you the origin and the actions". *Interpretation*: The hero is our king, the red tincture, the son of the sun. He laid low all his enemies on water and land. He is the ☉ and the moon. He stands in both, surrounded with splendor, for the sun and moon were his parents. And beneath him are the horns of the waning moon which now cannot be seen against the sun's brilliance. The three round apples are the three *principia*, and these give full capability to the origin and the powers of the work.

70TH FIGURE[292]
Materia of the Philosophical Stone

"It is only one thing, but there are also two of them, even three; whoever is clever will probably find out that I did not tell him this a long time ago. For it is only a dragon, so inferior in worth and old in years, that has the king's fountain and right vengeance. There are two ☿, of which you have to unite the permanent or fixed one with the volatile. It is

Geist. *Auslegung*: Unsere *Materia* ist nur eine einzige *Materia*, worin unser Anfang und Ende. Es sind ihrer auch zwei, nämlich Mann und Weib, *Agens* und *Patiens*, es sind ihrer auch 3, die 3 *Principia* ☉, ♁, ☿ und ist doch nur ein ein einziger Drache, der alte giftige und rauchende ♄, der hat den Königsbrunnen, den reichen Schatz in seiner Verwahrung. Es sind 2 ☿, der fixe und der fliegende, Bruder und Schwester, welche beide zu einem Leibe werden müssen; Es ist eine Seele, das rote Blut, ein Leib, der doppelte ☿, ein leichter Geist, das Öl oder Luft-*Spiritus* ☿ genannt.

So weit M. Stolcens Figuren, Sinnbilder, Texte und Sprüche ausgelegt. Folgt nun weiter die Continuation der angefangenen Fragen.

79. *Alle diese Texte sind schwer und dunkel, auch die Auslegung, können sie denn nicht leichter gegeben werden?*

Antwort: Sie sind nicht schwer noch dunkel, sondern alles klar und leicht, es ist nur verdeckt denen, welchen es GOTT nicht geben will, die es nicht haben sollen, die es auch übel anwenden würden, denn GOTT sieht das Herz an, erforschet Herz und Nieren. Dass es aber deutlicher beschrieben werden kann ist nicht möglich, denn Dinge, so auf Erden nicht gefunden worden, sondern durch die Söhne der Kunst gezeugt, diese haben keinen Namen, keine Eltern, sind neue Geburten und Gestalten der Natur aus der ersten, anderen und dritten Reinigung, aus so vielen *Compositionibus*; Aus einem kommen die *Elementa*, diese zeugen die 3 *Principia*, die 3 *Principia* die Metalle, diese werden in hell leuchtende Sterne verwandelt und in so vielerlei Tincturen, dazu muss man so vielerlei *Menstrua*, bald nasse, bald trockene Geister, auch Salz, Wasser, Öl und Feuer haben. Wie heißen nun die alle und wie soll man sie nennen? Und diese alle kommen doch aus einem einzigen Ding, gemahnt mich als einen Mann, welcher mit seinem Weibe viele Kinder soll zeugen, die Kinder nun sind in seinen Lenden, wer sieht sie aber oder wer will sie zählen wie viel ihrer werden sollen oder wie sollen sie aussehen? Sie sehen wohl alle aus wie Menschen, aber keiner von Gesicht wie das Andere, eins dem Vater, das Andere

one soul, one body, and one lightweighted spirit". *Interpretation*: Our *materia* is only one *materia*. Within it is our beginning and our end. There are also two of them, namely man and woman, *agens* and *patiens*; there are also three of them, the three *principia*, ☉, ♃, ☿, and yet there is only one single dragon, the old poisonous and smoking ♄, which has the king's fountain, the rich treasure,[293] in its safekeeping. There are two ☿, the fixed and the volatile, brother and sister, both of which must become one body. It is one soul, the red blood, one body, the double ☿, one lightweighted spirit, called oil or air-*spiritus* ☿.

So far we have interpreted M. Stolcen's figures, symbols, texts, and sayings. Now we turn our attention back to answering the questions that we began examining above.

79. *All these texts are difficult and obscure to interpret. Can they not be given more easily?*

Answer: They are neither difficult nor obscure. Everything is clear and easy, and their sense is only hidden from those whom GOD does not want to give understanding to, for they should not have it as they would abuse this knowledge. For GOD looks at the heart, examines the heart and kidneys. However it is not possible for these things to be described more clearly, for such things as are not found on earth, but instead are only begotten by the sons of art, and have no name and no parents. These are new births and forms of nature, from the first, second, and third purification, and from as many *compositionibus*. From one come the *elementa*, and then these beget the three *principia*; the three *principia* beget the metals, and then these are transformed into brightly shining stars and into so many different tinctures that you have to have so many different *menstrua*, sometimes wet and sometimes dry spirits, as well as salt, water, oil, and fire. What are all of these called and what should one call them? And these all come from one thing. This reminds me of a man who will father many children with his wife—for even now the children are in his loins. But who sees them, or who wants to count them? How many are to be sired? What

der Mutter gleich, eins ist männlicher, das andere weiblicher Natur, kommen doch von einem Vater und einer Mutter. Also hier auch in unserem Werk, wo ♄, ♃, ☿ die Oberhand haben, zeugen sie einen weiblichen *Corpus*, so *Luna* genannt wird, wenn aber ♂ und ♀, als der ♀ die Oberhand hat, zeugen sie ein männlichen *Corpus*, so *Sol* genannt wird und ist der ☿ beiden Theilen behilflich, denn er ist der Mediator, so alles beides vereinigt und in sich fasst, er ist der Mann auch das Weib, wie man will. Wenn sie nun die Nachfolgenden gern durch Schriften unterrichten wollen, haben sie müssen etwas aussinnen, dass sich mit solchem vergleichen ließe und diesen Kindern Namen geben ihrer Natur nach. Wenn es noch unreif, sauer, tödlich und giftig, muss es heißen Schlange, Drache, Basiliske, Gift, Drachen-Blut, Kröten, Ottern. Wenn es von der Schärfe und medizinalisch, heißt es auch Milch, Wein und Blut. Wenn es aber himmlisch worden, ist es die Speise der Götter, Nectar und Ambrosia. Ist es tingiert oder metallischer Art, hat es die Namen der Planeten, wie sie nach ihren Graden wachsen. *Summa*, es hat ihm ein jeder Namen gegeben, seiner Art nach in was für Lieberei er ihnen gefunden, doch allemal seiner Natur gleich, den nassen einfachen ☿ haben sie genannt Jungfer-Milch, Milch der Vögel, den doppelten haben sie ✶ genannt, das doppelte Salz, den Hermaphrodit, den Mond, die weiße Sonne, Sonne und Mond in einem Leibe, den weißen ♀ die geblätterte Erde, den gestirnten Adler, den fliegenden Drachen, ☿ *Sublimatum*, den ☿ von ☿, ist er fix gewesen, haben sie ihn genannt den gezähmten Adler, den erlegten Drachen, den figierten weißen ♀, die erste Tinctur auf weiß, ☿ *Præcipitatum*, das grünende Erz, das philosophische ☉, König in Hemd, der grüne ♌. Ist solcher gelb gefärbt worden, ist es die aufgehende ☉ *horicontale*, der gekrönte König im güldenen Stück, ist er rot, so ist es der blutige Stein, die rote Tinctur, der König in Purpur, der Sonnen-Sohn und großer *Universal-Stein der Weisen*.

might their faces and appearances be like? They all look like human beings, but none of their faces are alike. One is like the father, another is like the mother; one is male, another is female in nature, and yet they come from a father and a mother. So too here in our work: where ♄, ♃, ☿ have the upper hand, they beget a female *corpus*, which is called *Luna*, but when ♂, ♀, and also ♄ have the upper hand, they beget a male *corpus*, which is called *Sol*; and ☿ is helpful to both aspects, because it is the mediator which unites both and contains everything. It is the man and the woman, if you like. If they now want to teach those who come after them through writings, they must think of something that can be compared with such things, and give these children names according to their nature. When it is still immature, sour, deadly, and poisonous, it must be called a serpent, dragon, basilisk, poison, dragon-blood, toads, or a viper. When it is pungent and medicinal, it is also called milk, wine, and blood; but when it has become heavenly, it is the food of the gods, their nectar and ambrosia. If it is tinged or metallic, it has the names of the planets as they increase in degree. All in all, everyone gave it a name according to its nature and after their own preference, but always after consideration of its nature. They called the wet simple ☿ virgin's milk, milk of the birds, the double [☿]. They called ✶ the double salt, the hermaphrodite, the moon, the white sun, sun and moon in one body. They called the white ♄ the leafed earth, the starry eagle, the flying dragon, ☿ *sublimatum*, the ☿ of ☿. If it was fixed, they called it the tamed eagle, the slain dragon, the fixed white ♄, the first tincture in white, ☿ *præcipitatum*, the verdant ore, the philosophical ☉, king in a tunic, the green ♌. If such has been colored yellow, it is the rising ☉ *horicontale*, the crowned king in the golden piece; and if it is red, then it is the bloody stone, the red tincture, the king in purple, the son of the sun and the great *universal* stone of the wise.

80. *Es haben es aber andere ganz leicht beschrieben und ausdrücklich befohlen, man soll das Gold in seinem eigenen Wasser aufschließen, als dem Menstruo Universale und es darinnen zum Stein kochen?*

Antwort: Ein jeder hat ihn in der Lieberei beschrieben, wie er ihn gefunden, Basilius und andere haben ihn erst aus dem gemeinen Golde gemacht und als sie die Größe seiner Wirkung nicht gefunden, gingen sie wieder zu der Natur und forschten nach, wie dieser doppelte ☿ mit Geist, Seele und Leib vereinigt, das ☉ veredeln und verbessern könnte und sein roter Bruder solches sehr hoch röten und ihm einen Tinctur geben, so dachten sie, weil er der Same des Goldes, ob er nicht auch der Acker und selber Gold werden könnte, wenn sie ihn præcipitierten, dass er sein eigener Acker wäre, dass sein eigener ♃ oder ☉ darein gesät würde? Und da sie solches versuchten und befanden, dass es sich tun ließe, weil es eine Tinctur gab, die sich viel weiter in der Projection hervortat, als die vom gemeinen ☉ auch in viel kürzerer Zeit fertig wurde, so sonnen sie nach, ob sie es nicht noch weiter vermehren könnten durch die wiederholte Arbeit der Eintränkung? Sie befunden aber, dass dieses, wenn es sich einmal ersättigt durch sein eigenes Fleisch und Blut, solches nicht wieder annahm oder sich damit coagulierte, darauf dachten sie wieder: Wenn sie die Tinctur mit gemeinem ☉ schmelzten, damit sie von der Spiritualität zur Corporalität gebracht würde, ob es denn sich tun ließe? Als sie es nun zu Werke richteten, so sahen sie mit Verwundern, wie dieses feurige *Menstruum* sich damit vermischte und darinnen agierte, denn es fand an dem gemeinen Golde ein *Patiens*, dabei es sich erhitzen konnte, dadurch haben sie gefunden die Tinctur *in infinitum* zu vermehren und zu multiplicieren. Also hat den Stein ein jeder beschrieben wie er ihn gefunden oder woraus er seinen gemacht. Es kann auch ein tingierender Stein gemacht werden von dem ♃, ♂ und ♀. *Item*, aus dem ☿ und anderen, so tingierende Geister haben, aber nicht anders als durch den Brunnen des *Universals*. In dieser gewürzten Brühe werden alle diese Speisen gut und wohlschmeckend gemacht, zu tingierenden Steinen, doch ist keines der

80. *But others have described it as very easy and expressly ordered that one should open the gold in its own water, i.e., the menstruo universale, and to boil it in this until it becomes the stone?*

Answer: Everyone has described it after his own preference depending upon how he discovered it. Basilius and others first made it out of common gold and when they did not find the magnitude of its effect, they went back to nature and investigated how this double ☿ united with spirit, soul and body, could ennoble and improve ☉ and its red brother would redden such a thing very brightly and give it a tincture. They thought that because it is the seed of gold, could it not also become the field [of gold] and become gold itself, if they precipitated it so that it became its own field, that its own ♃ or ☉ was sown into it? And since they tried this and found that it could be done, because there was a tincture that was much more prominent in the projection than that of common ☉, and was also finished in a much shorter time, they pondered whether they could not increase it further by the repeated work of impregnation. They found, however, that once it had been saturated through its own flesh and blood, it did not accept it again, or coagulate with it, and therefore they thought further: if we melted the tincture with common ☉, so that it would be turned from spirituality to corporeality, could it be done?' When they set it to work, they saw with astonishment how this fiery *menstruum* mingled with it and acted within it, for it found a *patiens* in the common gold whereby it could heat up, and so they found how to increase and multiply the tincture *ad infinitum*. Thus everyone described the stone, how they found it, or what they made theirs out of. A tingeing stone can also be made from the ♃, ♂, and ♀. Likewise this may be done from ☾ and from others which have such tingeing spirits, but not otherwise than through the wellspring of the *universal*. In this seasoned broth, all these dishes are made good and tasty, into tingeing stones, but none into the *universal*

Universal-Stein, sondern nur Zweiglein vom großen Baum gebrochen, ist auch nicht allen gegeben, die vielen Geheimnisse des großen Steins zu erforschen, so müssen sie zufrieden sein mit dem, was ihnen GOTT gibt und wie weit sich ihr Verstand erstreckt. Daher kommt die vielerlei Beschreibung. Es hat noch kein Mensch in der Welt das *Universal* ohne Ruhm so deutlich beschrieben, als ich in meinen vorigen 2 Traktätlein dem *Mineralischen Glutine* und dem *Philosophischen Perl-Baum* und auch hier getan, da ich alle Vor- und Nach-Arbeiten in rechter Ordnung, ohne Hinterhalt und alle Geheimnisse in der Multiplication, weitläufig ausgeführt und nackend und alles bloß dargestellt, damit ein jeder seine bisherigen Irrtümer sehen und erkennen lernt, dass er seine Fehler verbessern kann.

81. *Sie haben ja die Vorarbeit beschrieben unter dem Weinstein oder wie das Sal Tartari soll flüchtig gemacht und mit dem Spiritu Vini vereinigt werden, wie räumt sich dies hierzu?*

Antwort: Sie mussten ein *Subjectum* suchen, dass diesem gleichstimmig oder gleichförmig wäre, so sehen sie das einzige, den Wein, welcher in sich hatte den *Spiritum Vini*, welches sein ♁, denn er brennt wenn er angezündet wird, danach bleibt der Weinstein, das erste Salz, aus solchem macht man den *Spiritum Tartari*, das ist sein rauchender ☿ und stinkendes Öl, zurück bleibt das andere Salz, als das *Sal Tartari*. Will man nun dieses fixe Salz flüchtig machen und doppelt, muss es durch seinen eigenen und rauchenden ☿ geschehen, den *Spiritum tartari*, diese beiden setzt man zusammen, bis der Flüchtige allen beiden Fixen bleibt und sich zur *Terra Foliata* sublimiert, alsdann kann sich der alcoholisierte *Spiritus* mit dem Leibe, Seele und Geist vereinigen und ein einziger Geist werden, den hernach sein eigenes Öl solviert und wird dieser Stein der *Universal*-Stein genannt, aus dem vegetabilischen Reiche; Und hierinnen ist die wahrhaftige Beschreibung des ganzen *Menstrui Universalis*, besser haben sie es nicht können beschreiben. Wer nun die Composition des Steins aus diesen nicht machen kann, der muss in diesem Wasser Fleisch und Fische kochen, wie er will und

stone, only small twigs broken from the large tree, and not everyone is given the opportunity to explore the many mysteries of the great stone, as they must be content with what GOD gives them and how far their understanding stretches. Hence the multiple descriptions. I do not mean to boast, but no one in the world has yet described the *universal* as clearly as I have done in my previous two little treatises on the *Mineral Gluten* and the *Philosophical Pearl Tree*; and also here, since I have dealt with all the preliminary and after-work in the correct order, without ambush, and with all the mysteries of the multiplication, amply performed, all laid bare and simply presented, so that everyone may learn to see their previous errors and to recognize that they can correct their mistakes.

81. *They have described the preparatory work under the tartar or how the sal tartari is to be made volatile and combined with the spiritu vini. How does this come about?*

Answer: They had to look for a *subjectum* that would be equal or similar to it, so they saw only one, the wine, which had the *spiritum vini* in itself, which is its ♄ because it burns when it is lit. After that the tartar remains, this being the first salt. From this, the *spiritum tartari* is made, that is to say its smoking ☿ and stinking oil, and what remains is the other salt or *sal tartari*. If one now wants to make this fixed salt volatile and double it, it must be done with its own, smoking ☿, the *spiritum tartari*. These two are put together until the volatile remains in both fixed and sublimates into *terra foliata*, and then the alcoholized *spiritus* can unite with body, soul, and spirit and become one sole spirit which afterwards dissolves its own oil; and this stone is called the *universal* stone, from the vegetabilic kingdom. And this is the true description of the whole *menstrui universalis*, nor could they have described it any better. Anyone who cannot make the composition of the stone out of these must boil meat and fish in this water as they are

kann, wird nichts darinnen verderben, sondern alles verbessern und neu gebären und weil es ein metallisches *Menstruum*, bleibt es ewig bei den Metallen. Denn wer kann Wasser von Wasser scheiden, Öl von Öl? Kriecht wegen seiner Fettigkeit in die Metalle als Öl in Lösch-Papier, hängt allem an, geht in ihr Herz, Seele, Geist und Leib und kann sich kein Stäublein verbergen im innersten Kern, daher es in allerhand feuerbeständigen Metalle aushält, solche auch beständig macht, je mehr sie Feuer leiden, je fester es darein kriecht, so vor dem Feuer beschirmt und die Metalle mit ihm, wird auch seiner Fettigkeit wegen, des Goldes erstes Wesen genannt, weil dies so fett ist, die anderen aber alle hart und spröde. Und überdies hat es lauter blutroten ♃ bei sich, auch den gelben Riemen um den Leib, das gelbe Öl und den weißen unverbrennlichen ♃, dieses hat kein einziger metallischer *Corpus* in der ganzen Natur und deswegen ist es der Metalle erstes Wesen, ihre *Prima Materia*, sonderlich des Goldes und Silbers, welche in der Erde aus solchen Geistern gewachsen und ist ihr rechtes *Menstruum*, solche wieder neu zu gebären, welches kein Salz-Geist tun kann, kein ▽ noch ▽, denn diese solvieren nur die Metalle, zerfressen sie von auswendig und sind wieder davon zu scheiden, weil sie nicht metallischer Natur und auch der *Spiritus* ⊕ nicht, obschon in demselben viel vom *Spiritu Universali* enthalten, so ist doch dieses *Acidum Materia remota* und würde unser Leben zu kurz sein, wenn wir wollten daraus einen tingierenden Stein machen. Wir haben in der Natur viel eine nähere, welche wir in diesem *Acido* reinigen ohne Zerstörung der Grund-Feuchtigkeit, doch geschieht dadurch keine innerliche Reinigung, sondern nur der äußerliche grobe Leib wird nur dadurch von dem Berg geschieden, dass die äußerlichen und groben Schlacken davon fallen, damit es hernach durch sein eigenes Wasser und Essig kann gereinigt werden, denn das ganze *Corpus* dient zu diesem Werk nicht, erst nimmt man das Erz durch den Essig vom Berg, dies wird nicht gerechnet, noch eine Solution genannt, sondern unsere erste Solution ist, wenn von dieser *Materia* sein selbst eigener Essig abdestilliert und sein eigener *Corpus*, der rote *Laton*, darinnen solviert und gereinigt wird, das ist seine erste

willing and able; nothing in it will spoil, but it improves everything and gives birth again, and because it is a metallic *menstruum*, it stays forever with the metals. For who can separate water from water, or oil from oil? Because of its greasiness it creeps into the metals like oil into blotting paper. It clings to everything, going into their heart, soul, spirit, and body and since not even a speck of dust can lay concealed in the innermost core, it withstands all kinds of fire-resistant metals accordingly. It also makes them resistant, and the more fire they suffer, the harder it creeps into them. Thus it is shielded from fire and the metals with it, and because of its fatness it is also called the first essence or being of gold, because it is so fat, but the others are all hard and brittle. And moreover, it has nothing but blood-red ♄ within it, also the yellow belt around its body, the yellow oil, and the white incombustible ♄. This does not have a single metallic body in all of nature, and that is why it is the first being of metals, their *prima materia*. This is especially true of gold and silver, which grew in the earth from such spirits, and it is their proper *menstruum* through which they give birth to it again. This is something which no salt spirit can do, and no ▽ nor ▽, because these only dissolve the metals, eating them up from the outside—and then they have to be separated from them again because they are not of a metallic nature and neither is the *spiritus* ☉, although it contains a lot of the *spiritu universali*. Yet this *acidum* is *materia remota*, and our lives would be too short if we wanted to make a tingeing stone from it. In nature we have a much closer one which we purify in this *acido* without destroying the fundamental moisture. Yet this does not result in any inner purification, but only the outer, coarse body is separated from the mountain solely by the fact that the outer and coarse slags fall from it, so that it can afterwards be purified with its own water and vinegar. This is so because the whole body is not used for this work. First the ore is taken from the mountain with the vinegar, and this is not counted, nor is it called a solution; but our first solution is when its own vinegar is distilled from this *materia* and its own *corpus*, the

Reinigung, dadurch es bequem gemacht wird in die *Elementa* zu scheiden und muss das *Confusum Chaos* in seinem eigenen Wasser faulen, hernach kommt die Reinigung der Elemente, wonach folgt die Scheidung *purum ab impuro*, welche geht über die Elemente, und dann die rechte Composition.

82. *Wie können aber aus dieser bleiischen Materia oder Minera so viele Dinge kommen, die doch unrein, giftig, flüchtig, gering und verachtet sind?*

Antwort: Die *Materia* ist die Blume der Erde, ja aller Metalle, hat viel arsenikalisches Gift bei sich, doch ist er darinnen nicht so giftig als in Cobalt, Wasser-Kies oder rotgülden Erz, weil dieser noch ein schrecklich *Acidum* bei sich hat, daher jener Vehement wirkt. Wo ihn ein animalischer Körper in den Magen bekommt, so macht er darinnen schwarze Brandflecke, wo er sich anlegt und tötet auch, wenn man ihn auf ♀ wirft, macht es solches weiß, aber es läuft gleich an der Luft an und wird schwarz. Hergegen unser Arsenik aus dem Blei ist etwas milder und süßer, kann in Brust-Krankheiten, so von zähen Flüssen herrühren, in kleiner Dosis eingenommen werden, da es alsbald eine Fermentation macht und solchen durch ein Vomitiv hebt, am meisten aber unten ausführt, da sonst nichts als der Tod zu hoffen wäre und augenblickliche Erstreckung und nur zu halben bis ganzen Grauen und durch *Spiritum Vini* wohl corrigiert. Auch macht solcher das ♀ nicht schwarz, wenn es aber zu Öl worden, ist es der Basiliske und weit giftiger als vorher, ist *Venenum Tingens*. Sein ☿ aber ist ohne Gift und das Wunder der Natur, sein rotes Öl ist das Blut der Metalle, Drachen-Blut, ♌ Blut; Aber alle diese Dinge sind noch nicht die hohe Arznei, wiewohl sie schon mehr vermögen als alle Krämer-Pillen, in kleinen und großen Krankheiten, welche schon fix und eingewurzelt sind, aber noch herrlicher, wenn solche süß, himmlisch, wohlriechend, leuchtend und brennend, die Blumen der Sonne und des Mondes, das rohe unreife und unzeitige ☉, ja alle 7 Metalle, alle *Vegetabilia*, welche in ihrer schönen Blüte abgebrochen werden, haben ihre größte Kraft, denn die Blüte ist ihr Sommer, ihre reife Frucht oder Samen ist ihr

red loam, is dissolved and purified within. That is its first purification, whereby it is made convenient to separate into the *elementa* and the *confusum chaos* must putrefy in its own water. Then after that comes the purification of the elements, followed by the separation *purum ab impuro*, which goes over the elements and then the right composition.

82. *But how can so many things come from this leaden materia or minera, although it is impure, poisonous, volatile, mean, and despised?*

Answer: The *materia* is the flower of the earth, indeed of all metals; and it has a lot of arsenical poison within it, but it is not as poisonous as it is in cobalt, water pyrite, or red-gold ore, because these still have a terrible *acidum* within them, hence they act vehemently. If an animal body takes them into the stomach, they make black burn marks inside, where they attach themselves and they also kill. If you throw them on ♀, it makes such things white, but it immediately starts to tarnish in the air and turn black. On the other hand, our arsenic from lead is somewhat milder and sweeter. It can be taken in small doses in chest diseases resulting from tenacious fluxes, since it immediately starts a fermentation and raises it upwards with a vomitive, but mostly discharges it downwards. Otherwise nothing but death could be hoped for, and momentary extension, and only half to full horror, surely corrected by *spiritum vini*. Nor does such a thing make the ♀ black, but when it has turned to oil, it is the basilisk and far more poisonous than before, it is *venenum tingens*. But its ☿ is without poison and the wonder of nature, its red oil in the blood of metals, dragon blood or ♌ blood. But all these things are not yet the high medicine, although they already have more power than all the grocer's pills in both minor and major diseases that are already fixed and rooted. But they are even more wonderful when they are sweet, heavenly, fragrant, luminous, and burning. They are the flowers of the sun and the moon, the raw immature and untimely ☉. Indeed all seven metals, all *vegetabilia*, which are broken off in their beautiful blossom, have their greatest power, because the blossom is their summer, their ripe fruit

Herbst, hernach verdorrt es, dass ist sein Winter. Nun haben ja die Blumen mehr Geruch und Kraft als ihr Samen, hier in dieser Blume oder königlichem Kraut auch, welches billig die weiße Lilie und rote Rose, auch die gelbe Sonnenblume möchte genannt werden, diese ist conträrer Natur gegen andere Kräuter, wächst im Winter, in der Kälte wird sie constringiert, in der schwarzen Erden fault sie und bekommt eine schwarze Wurzel, welche sich in alle Farben verkleidet, dann fängt die weiße Lilie an zu blühen, das königliche regalische Kraut, das wider den Tod dient, dann kommt die Sonnenblume und die rote Rose, welche in Blut getunkt wird, durch welche viele Arbeit es an seiner Kraft unendlich zunimmt, nicht dass dieses Vermögen vorher gehabt, wiewohl es auch ohne Wirkung und viele Tugenden hat, aber durch die Arbeit vielmehr bekommt, denn wenn der rote mineralische Geist seinem Körper zugefügt wird, so verbrennen diese 2 widerwärtigen Fechter ineinander, als das trockene, hitzige mineralische Salz und der kalte ♄ und wird aus ihnen beiden ein vollkommenes Metall, dieses beides hat schon andere Art als vorher, wenn solches wieder in sein erstes *Menstruum* aufgelöst. Ist es aber ein anderer Leib und einen Grad höher an Kraft und Tugend, wird damit fortgefahren bis es den höchsten Grad der Natur erreicht, so ist es der Lebens ☿, so wider den Tod dient. Weil aber die Kunst über die Natur, so steigt sie auch mit diesem noch einen Grad höher. Nun dieser erhöhte und vollkommene ☿ läuft durch den astralischen Luft-Himmel, conjungiert sich mit allen Sternen, nimmt aller Natur an sich und wird davon erhöht, wenn er die Tincturen bekommt, absonderlich seine eigene, dadurch er sich am allerliebsten solvieren und gradieren lässt, sich dadurch veredelt, resolviert und neu gebiert, da die Natur noch lange nicht so hoch kann, sondern ihr höchstes Grad ist das ☉, die Kunst aber macht die ☉ zur weißen Tinctur. Diese wird, als der volle Mond, so lange durch die färbende Luft oder Öl abgetrocknet und eingedrückt, bis er gelb und die rote Rose wird, die aufgehende Sonne, dahinter sich der Mond versteckt, dass er nicht mehr gesehen werden kann. Endlich wird diese Rose ins Blut getaucht und bekommt die Purpur-Farbe.

or seed is their autumn, afterwards it withers and this is their winter. Well, the flowers have more scent and strength than their seeds, so too here in this flower or royal herb, which might be called the white lily and red rose, or the yellow sunflower. This is of a nature contrary to other herbs, for it grows in winter; it is constricted in the cold; and it putrefies in the black earth while developing a black root which disguises itself in all colors. Then the white lily begins to bloom. This is the royal regalian herb, which again serves death, for then come the sunflower and the red rose which is dipped in blood, and through much work it increases its strength infinitely. It did not possess this ability before, although it also has potency and many virtues without [this work], but rather it acquires this ability through work. For when the red mineral spirit is added to its body, these two repelling swordsmen burn into each other, much like the dry, hot, mineral salt and the cold ♄ both become a perfect metal. When they are dissolved again in their first *menstruum*, they both have different nature than before; but if it is a different body and a degree higher in strength and virtue, it is to be continued with it until it reaches the highest degree of nature, then it is the ☿ of life, which serves against death. But because art is above nature, it also climbs a degree higher than nature. Now this exalted and perfect ☿ runs through the astral air-heaven, conjoining with all the stars, absorbing all their nature. It becomes exalted by this when it receives the tinctures, especially its own, which makes it quite easy to dissolve and gradate it. It is thereby ennobled, resolved, and reborn, since nature is far from capable of bringing it that high, its highest degree is ☉, but art turns the ☉ into a white tincture. Like the full moon, this is dried and pressed in by the coloring air or oil until it turns yellow and becomes the red rose, the rising sun, behind which the moon hides so that it can no longer be seen. Finally, this rose is dipped in blood and becomes purple in color. Now one

Nun kann man leicht denken, was für himmlische Kraft, Wirkung und Vermögen in einem solchen erhöhten, so vielmals neu generierten spiritualischen, geistlichen Leibe ist, der nur durch seinen Odem und Anhauchen alles in ☉ verwandelt, durch sein himmlisches Licht und Feuer alles erleuchtet, alle Krankheiten vertreibt, den Lebens-Geist anzündet, stärkt, vermehrt, doppelt, ja vielfach wiedergibt, so verloren waren, das *Humidum Radicale* erhält bis ins hohe Alter und vor allen Krankheiten bewahrt. Durch die Kunst wird es so hoch gebracht, von der Natur hat es so viel Stärke nicht.

83. *Warum sagt denn Hermes: Die Sonne ist sein Vater, der weiße Mond seine Mutter, die Luft, der Wind hat ihn in Bauch getragen, die Erde hat ihn gesäugt?*

Antwort: Die Sonne ist sein Vater, dass ist unser Feuer, Sonnen ♄, die Mutter, der weiße Mond unser *Menstruum* ☽, welches wir auch Jungfer-Milch nennen, der Wind hat ihn im Bauche getragen, weil er im runden Gefäß auf den Fittichen des Windes, als ein subtiles Gestiebe getragen worden, die Erde hat ihn gesäugt, diese hat ihn erzogen, er ist in ihr gewachsen und groß geworden, dieses sind unsere *Elementa*.

84. *Ist denn nicht mehr als ein Weg diesen rohen Körper von dem groben Berg zu scheiden oder sind ihrer mehr?*

Antwort: Es sind mehr Wege und werden in dieser Materie viel *Superflua* getan, doch muss es gereinigt werden, dass ihn sein *Humidum Radicale* nicht verbrennt, welches ist sein wachsendes Würzlein, denn wenn man seinen ☿ verbrennt, so kann er hernach seinen eigenen *Corpus* nicht solvieren und wenn sein ♄ zur toten Erde verbrannt, so kann diese tote Erde den ☿ nicht gradieren, noch erweichen, güssig, flüssig, zum unverbrennlichen Öl machen.

85. *Wie kann aber sein ☿ solche wunderliche Solution machen und wie geht das zu?*

Antwort: Solches sehen wir an dem gemeinen ☿, wie er als ein trockener

can easily imagine what heavenly power, effect, and capacity there is in such an increased spiritual body, newly generated so many times, which transforms everything into ☉ simply through its breath and exhalation, illuminating everything through its heavenly light and fire, driving away all illnesses, kindling the spirit of life, strengthening, multiplying, doubling, indeed reproducing many times over what was lost, preserving the *humidum radicale* to old age, and protecting from all illnesses. It is brought to such a high level by art, but it does not have so much strength by nature.

83. *Why does Hermes say: "The sun is its father, the white moon its mother, the air and the wind it carried in its belly, the earth suckled it?"*[294]

Answer: "The sun is its father," this means our fire, ♄ of the sun, whereas the mother is the white moon, our *menstruum* ☽, which we also call virgin milk. "The wind carried it in its belly" because it was in the round vessel on the wings of the wind, carried like a subtle gesture; the earth nursed it, brought it up, and it grew in her and became large, these are our *elementa*.

84. *Is there not more than one way to separate this raw body from the rough mountain, or there are even more?*

Answer: There are more ways and many *superflua* are done in this matter, but it must be purified so that its *humidum radicale* does not burn, which is its growing root, because if one burns its ☿, it cannot dissolve its own *corpus* afterwards. And if its ♄ is incinerated to dead earth, this dead earth cannot gradate the ☿; it can not soften it, make it pourable, make it liquid, or turn it into an incombustible oil.

85. *But how can its ☿ make such a strange solution and how does it work?*

Answer: We see this in the common ☿, how it dissolves the metals as

Körper die Metalle solviert, wenn er aber durch *Salia* sublimiert, noch mehr. Aber wenn der corrosivische sublimiert durch *Spiritum Salis* solviert und endlich als ein Öl übersteigt, solviert er noch besser. Also auch unser ☿, wenn er die Jungfer-Milch ist, solviert er die ☽, bleibt bei ihr und wird zu Silber, dies tut der gemeine ☿ nicht, ob er sich schon bei ihr præcipitiert, wenn er aber auf die Capelle kommt, geht es wieder fort und hat man nicht mehr *Luna* als vorher, warum? Er hängt den ☽ nur von außen an, hat keinen ☽ Schwefel bei sich, als der unsere, weil er die *Lunam* erweicht, in alle Adern, so zu reden, kriecht, sich damit veredelt und weil er eines Wesens oder ihr erstes Wasser, bleibt er bei ihr und wird dadurch zum beständigen *Corpus* ☽, aber unser doppelter ☿ ist weit mehr erhöhter und in weit größerer Kraft, wenn er sublimiert zur *Terra Foliata* worden, so ist es der volle Mond, der weiße ♃, der alle Metalle *radicaliter* solviert, das wahre *Menstruum Universale*, ist der ✶, so in die Metalle einkriecht, das Herz sucht, solche flüchtig und fix macht, danach man seine Arbeit anstellt, außen weiß, inwendig blutrot, mit allen Farben geziert, nimmt die metallischen Leiber an und wird eins mit ihnen. Wenn aber dieser ✶ in seinem eigenen Öl solviert und zum Öl worden, so ist es das *Aqua permanens, Aqua benedicta*, das Lebens-Wasser und unverbrennliche Öl, da der ☿ in Öl gebracht worden, er ist in diesem Wasser revificiert, wiedergebracht in ☿, in Geist, Seele und Leib und den *Spiritum* ☿. Was mit diesem Wasser für Wunder auszurichten, wird kein Mensch glauben, dieses Wasser solviert das Feuer und kocht solches, dabei wird es blutrot und hart als ein gestocktes Blut, fließt in gelinder Wärme wie Wachs, macht den ☿ und das Gold zum roten Glas, dieses auf die unreifen Metalle getragen, verwandelt sie in ☉.

86. *Wie kann aber das philosophische ☉ und ☽ andere Eltern haben als das gemeine ☉, da es doch dem gemeinen gleich, welches die Sonne der großen Welt zeigt?*

Antwort: Das philosophische ☉ ist besser als das gemeine ☉ und bringt es in eine Tinctur. Es solviert das gemeine ☉, denn es hat seinen

a dry body, but even more so when it was sublimated through *salia*. Yet when the corrosive [☿] sublimated by *spiritum salis* dissolves and finally distills over as an oil, it dissolves even better. Thus also our ☿, if it is the virgin milk, dissolves the ☽, staying with it and turning it into silver. The common ☿ does not do that, even though it is already precipitating itself with it; but when it comes to the cupel, it goes away again and one does not have more *Luna* than before. Why? It only clings to the ☽ from the outside; it has no ☽ sulphur with it, like our [☿], because it softens the *Lunam*, creeping into all its veins, so to speak, ennobling itself with it. And because it is one being or its first water it is with it and thereby becomes the permanent *corpus* ☽; but our double ☿ is far more heightened and in far greater power when it has been sublimated into *terra foliata*, so it is the full moon, the white ♁, which dissolves all metals *radicaliter*, that real *menstruum universale*. It is the ✳ which creeps into the metals, searches the heart, makes them volatile and fixed, and then one does one's work, white on the outside, blood-red on the inside, adorned with all colors, accepting the metallic bodies and becoming one with them. But when this ✳ dissolves in its own oil and becomes oil, it is the *aqua permanens*, the *aqua benedicta*, the water of life and the incombustible oil, since the ☿ has been brought into oil. It is revived in this water, brought into ☿ again, in spirit, soul, and body and the *spiritum* ☿. No one will believe what miracles can be achieved with this water, this water dissolves the fire and boils it, it becomes blood-red and hard like stagnant blood, flowing with gentle warmth like wax, turning ☿ and gold into red glass. Applying this to the immature metals transforms them into ☉.

86. *But how can the philosophical ☉ and ☽ have other parents than the common ☉, since it is equal to the common [☉] that shows the sun to the macrocosm?*

Answer: The philosophical ☉ is better than the common ☉ and brings it into a tincture. It dissolves the common ☉, because it still has its

Lebens-Geist noch bei sich, das gemeine ☉ ist tot, das philosophische lebendig, dieses unser edles ☉ hat seinen Anfang von der Sonne und dem Geist der großen Welt, sowohl als das gemeine ☉ aber die Natur hat in unser *Subjectum* alle ihre Kräfte gesteckt und die Sonne hat ihr eigenes Wesen, diese kleine Sonne, dem ♄ zu verwahren gegeben, der sie gleichsam gefangen im Kerker verschlossen und mit großer Mühe und Arbeit zu erlösen ist. Bei dieser Sonne ist auch das Weib ☽ zu finden und hat die Natur und Kunst gewiesen, wie der Mond soll in Blut verwandelt werden, das ist, zur Sonne werden, alsdann die Sonne mit der Sonne fermentieren, mit unserer roten flüssigen Sonne, alsdann ist es der Sonnen-Sohn und hat dieses freilich edlere Eltern, als das gemeine ☉, welches zwar eben aus den arsenikalischen Geistern gewachsen und aus den Elementen der großen Welt, die Sonne aber am Himmel tut wenig bei Wachsung des Goldes, weil die Sonne in die Gebirge nicht scheint, sondern die Metalle wachsen aus Wasser und Erde, wie unser ☉ auch. Weil wir aber in unseren Gefäßen die allerlauterste und reine Erde haben, auch das lautere reine himmlische Paradies-Wasser, so muss freilich ein edler Leib daraus wachsen, weil hier keine unreine Erde zuschlagen kann, oder unreiner ♁ sich damit vermischt, noch unrechtes Wasser aus unrechten Brunnen, unser Wasser schöpft der *Neptunus* aus dem Meer.

87. *Sind denn die Elemente, welche der Anfang des Steins sind, anders als die Elemente der großen Welt?*

Ant. Sie sind freilich weit herrlicher, denn die *Elementa* der großen Welt. Denn der Geist der Natur hat aller Kräfte Quintessenz in diesen mineralischen Körper gesteckt, welcher die ganze Natur und ihre Wirkung in sich hat, auch die Geister der Planeten. Er ist ein Auszug der großen und kleinen Welt, die Blumen der Sonne und des Mondes. Diesen Körper reinigen wir von seinem groben Berg durch ein Bad, so seiner Natur nicht zuwider, hernach solvieren wir es in seinem eigenen Wasser, scheiden das *Confusum Chaos* in die *Elementa*, dem vorigen Körper haben wir nur sein Blut und Seele, das Herz, sein Innerstes

spirit of life with it, for while the common ☉ is dead, the philosophical [☉] is alive. This, our noble ☉, has its beginning from the sun and the spirit of the macrocosm, as does the common ☉, but nature has put all its powers into our *subjectum* and the sun has given it its own being, this little sun, to ♄ in order to keep it locked in a dungeon, and it is redeemed with great effort and work. The woman, ☾, can also be found with this sun and nature and art have shown how the moon should be transformed into blood, that is, to become the sun, then ferment the sun with the sun, with our red, liquid sun, for then it is the son of the sun, and this certainly has nobler parents than the common ☉, which has indeed grown from the arsenical spirits and from the elements of the macrocosm, but the sun in the sky does little in the growth of gold, because the sun does not shine into the mountains, rather, the metals grow from water and earth, like our ☉ too. But because we have the very clearest and purest earth in our vessels, and also the clear, pure, heavenly water of paradise, a noble body must of course grow from it, because no impure earth can touch it here, no impure ♁ can mix with it, nor can any false water from false wellsprings [mix with it], for *Neptunus* draws our water from the sea.

87. *Are the elements that are the beginning of the stone different from the elements of the macrocosm?*

Answer: They are certainly far more glorious than the *elementa* of the macrocosm. This is because the spirit of nature has put the quintessence of all its forces into this mineral body, which contains all of nature and its effects, including the spirits of the planets. It is an excerpt of the macro- and microcosm, the flowers of the sun and moon. We purify this body of its coarse mountain with a bath which is not contrary to its nature. Afterwards we dissolve it in its own waters, and then we separate the *confusum chaos* into the *elementa*. We have only taken the blood and soul, the heart, the innermost part, of the previous body,

genommen, solches in die *Elementa* geschieden, die sind nun freilich gegen die Elemente der großen Welt ganz anderer Gestalt und himmlisch, denn jene sind nun Mütter dieser *Minera*, die *Minera*, die Frucht oder das Kind, so sie gezeugt. Aus dieser Frucht kommen unsere Elemente, weil wir es wieder zurück bringen müssen, aber nicht in ein solches Wasser, wie Brunnen- oder Fluss-Wasser, sondern unser Wasser ist weiß, dick und schwer, wie Leim, hat lauter weißen ☿ialischen Geist in sich, welcher von Natur den Samen des Goldes oder Silbers hat, ist voll von schmutziger Feuchte, darum ist es der Leim des Adlers, der Geist ist sehr rein davon geschieden, deswegen gerinnt es auch bei seiner Erde und wird ein schönes Perllein in seiner *Coagulatio*, weil es kein einfältiges simples Wesen ist, sondern das Paradies-Wasser, *Lac Virginis*. Wenn ein Tropfen in einen Löffel *Spiritus Vini* getan wird, wird es weiß wie Milch, die Erde aber ist das Salz und köstlicher als Gold und ☉ und ♂ verwandt, auch der ♀, welches die gelben, roten und grünen Farben weisen. Wenn dieses Salz solviert, wieder coaguliert und die Arbeit so oft wiederholt wird, dass ihm die Bitterkeit vergeht und als eine Wolke übersteigt, so ist es der Schlüssel zum königlichen Gemach und das doppelte Wasser. Die Luft ist ein lauteres Öl, voll von 🜂, ☽, welches arsenikalisch und hat den gelben ☉ Schwefel auch in sich, ist ein rechtes Gradier-Öl, färbt alle weißen Leiber in gelb, erweicht, macht güssig, flüssig, reinigt, erneuert und erhöht, trägt die Farben auf, beißt solche als ein Öl ein, dass sie nicht weggewaschen werden können, das Feuer aber ist sehr brennend rot, der Sonne ihr Wesen, ein blutroter 🜂, ganz purpurfarben, dergleichen in der Natur nicht zu finden, färbt sehr hoch rot. Hier sieht man, was für Unterschied zwischen den Elementen der großen Welt und zwischen unsern Elementen ist.

88. *Es braucht aber sehr großer Mühe, Gefahr und lange Zeit, dieses Werk zu verfertigen?*

Antwort: Ja, es erfordert eine ziemliche Zeit und großen Verstand die *Compositiones* vorzunehmen, zumal in der ersten, wenn man den doppelten ☿ machen will, dass man das Salz auflöst, aus selbigem seinen

separated such into the *elementa*, which are now of course completely different in shape and heavenly in comparison to the elements of the macrocosm, because they are now the mothers of this *minera*, the *minera*, the fruit or the child, which they have begotten. Our elements come from this fruit, because we have to bring it back again, but not in water such as well or river water, rather, our water is white, thick, and heavy, like glue; and it has lots of white ☿al spirit in it, which by nature has the seed of gold or silver, being full of dirty moisture, and so it is the glue of the eagle. The spirit is very pure once separated from it, and that is why it also coagulates with its earth and becomes a beautiful little pearl in its *coagulatio*, because it is not a gullible, simple being, but the water of paradise, *lac virginis*. If a drop is put in a spoonful of *spiritus vini*, it becomes white as milk; but the earth is the salt and more precious than gold, and ☉ and ♂ are related, as well as the ♀, which show the yellow, red, and green colors. When this salt is dissolved, then coagulated again, and the work is repeated so many times that its bitterness passes away and it rises as a cloud, it is the key to the royal chamber and the double water. The air is a pure oil, full of ♃, ☽, which is arsenical and also has the yellow ☉ sulphur in it. This is a very gradating oil, and it colors all white bodies yellow. It softens, makes it pourable and liquid, purifies it, renews and increases it, applies the colors, bites like oil [penetrating] so that they cannot be washed away. But the fire is an intense burning red, the essence of the sun. It is blood-red ♃, completely purple, the likes of which cannot be found in nature, and it stains very bright red. Here you can see what a difference there is between the elements of the macrocosm and our elements.

88. *But it takes great effort, danger, and a long time to complete this work?*

Answer: Yes, it requires quite a bit of time and great understanding to undertake the *compositiones*, especially since in the first [composition] if you want to prepare the double ☿, you must dissolve the salt and

Geist bringt. Der Leib vermag nichts, denn nach seiner Solution und Flüchtigmachung, wenn es zum ✳ worden, ist es der geheime Schlüssel zu des Königs Schatz-Kammer. Es muss aber das Salz selbst durch seinen eigenen Schlüssel aufgeschlossen werden, welcher ist das Wasser, wodurch es gereinigt; Erstlich aber gerinnt es bei diesem Körper so lange, bis es ihn in seinem Innersten erweicht und dissolviert, da es dann sehr bitter und entzündet ist, gibt farbige Dünste, der bittere Geschmack ist solarischer und ♀ Natur, etwas süßes, welches von ♂ das diese Bittere nicht so greulich zu empfinden, sonst wäre es noch bitterer, die bleiischen, jovialischen Leiber sind in der ersten Extraction davon geschieden worden und nur die Farben der anderen Planeten genommen, diese sind übergestiegen, haben den solarischen, martialischen und venerischen Leibern ihr Blut und Seele übergeführt, welches der Dieb, der ☿ getan, der hat sie alle beraubt, dass ihre Leiber nackend und bloß zurück blieben, dennoch ist in ihrem Salz der bittere Geschmack, welcher muss süß gemacht werden. Die größte Gefahr ist, diese dürre Erde mit dem Wasser seiner Art anzufeuchten und zu rechter Zeit zu tränken. Es will einen rechten Termin haben, zu rechter Zeit und Stunde und muss ein solcher den Geburtsort des Steins wohl wissen, denn diese Erde bricht zuletzt für Durst, wenn sie den Raben, die Schwärze hat von ihr fliegen lassen, so lässt sie die weiße Taube aus, diese muss das Öl-Blatt im Munde bringen. Dieser Stein schießt von sich selbst und gibt Mehl, als wie der ungelöschte Kalk an der Luft zerfällt, wer hier nicht wohl bewandert ist, wird alles verlieren: Wer nun diese geblätterte güldische Erde erheben will, der muss wissen, wenn der Bruch geschehen will, wann die Geister und Drachen, so solche bewahren, unruhig werden, ist solches ein Zeichen, dass sich der Schatz entblößen will, der schuppige, funkelnde Drache, welcher so lange in einer Höhle gewohnt, sich von der Erden und deren Schlamm genährt, aus dem Drachen ohne Flügel und dem Drachen mit Flügeln ein einziger Drache worden mit großen schrecklichen Flügeln, funkelnden Augen und Schuppen, welche schimmern und leuchten und dieser Drache stets in seinem Gefängnis Gift ausgespien, den fixen Drachen

extract its spirit. The body can do nothing, because after its dissolution and volatilization, when it has become a ✱, it is the secret key to the king's treasure chamber. But the salt itself must be unlocked by its own key, which is the water by which it is purified. First, however, it coagulates in this body until it softens and dissolves it in its innermost being, since it is very bitter and inflamed and gives colored vapors; the bitter taste is more solar and ♀ in nature, something sweet, which [being] of ♂, does not feel the bitter taste so terribly. Otherwise it would be even more bitter. The leaden, Jovial bodies were separated from it in the first extraction and only the colors of the other planets were used. These were then distilled, and the solar, martial, and venusian bodies have their blood and soul distilled. What the thief, ☿, did resulted in them all being robbed, so that their bodies were left naked and bare, yet there is a bitter taste in their salt, which must be made sweet. The greatest danger is moistening this dry earth with the water of its nature and watering it in due time. It has to happen at the right date, at the right time and hour; and such a person must know the birthplace of the stone well, because this earth breaks at last for thirst when it has let the raven, the blackness, fly from it. When it lets the white dove out, it must bring the olive leaf in its mouth. This stone shoots from itself and gives forth flour like quicklime disintegrating in the air. Whoever is not well versed here will lose everything. Now whoever wants to lift up this leafed, golden earth must know when the break is about to happen, when the spirits and dragons who protect it become restless, for this is a sign that the treasure wants to uncover itself. The scaly, sparkling dragon, which dwelt in a cave for so long, fed on the earth and its mud, which became one dragon from the wingless dragon and the winged dragon, with great terrible wings, sparkling eyes, and scales that shimmer and shine. And this dragon always spat out poison in his prison, then killed the fixed dragon and devoured itself, fattened itself

getötet und in sich verschlungen, sich davon gleichsam gemästet und nun mit ihm davon fliegen will, wenn man dieses Drachens ansichtig wird, muss man ihn nicht vor Furcht entfliehen lassen und wohl Achtung geben, wenn er dem Felsen zerreißen und fliehen will, so muss man ihn im Moment, in seinem Flug erhaschen und die zugerichtete Suppe über diesen kolchischen Drachen gießen, dem Drachen seinen eigenen Zorn und Gift trinken lassen, Phöbus schießt seine Pfeile in Pithon, den harten grimmigen Drachen, sobald er die Suppe anfängt in sich zu saufen, fängt sein ganzer Leib an zu rauchen und zu dampfen, die giftige Suppe aber fährt mit in die Höhe tötet dem Drachen seinen Geist in der Luft und bringt ihn wieder zu seinen Schuppen und Gebeinen, bis es zuletzt ein schwarzes Grab wird, dies ist die andere Schwärze und wärt lange, dass der Arbeiter denkt, er hat verloren. Aber es steigt ein Geist, ein Licht von dieser *Materia* in die Höhe und geht wieder herab, circuliert sich, bis es den Leib abwäscht, fließt und coaguliert sich oft, ehe es ganz fix, rein und weiß wird, eben als wenn ein Metall auf dem Test abgeht und oft stehen bleibt, wenn es heißgrädige Unart bei sich hat, endlich von sich selber wieder anhebt zu treiben, bis dieser gebratene Schwan in weißen Stein verwandelt und die Speise des Königs wird.

89. *Der harte grimmige Drache Pithon ist ja der doppelte* ☿ *was ist aber des Phöbus sein Pfeil?*

Antwort: Ja, es ist der schuppige Drache, die geblätterte Erde, welche mit Fisch-Augen leuchtet und wie Fisch-Augen schimmert, Phöbus sein Pfeil, welcher auch Delius genannt wird, ist das giftig-rauchende Öl ♄, die Sonne, so dieser Alte in seinem Leibe gehabt, welches auch der jungen Kinder Blut genannt, darinnen Sonne und Mond zu baden pflegen, des ♄ seine Sense, mit welcher er dem ☿ die Flügel abhaut, das ist, figuriert, hier muss das Glas nicht aufgemacht werden, weder davon noch dazu getan, bis der Stein der ersten Ordnung fertig, denn hier ist keine Speisung noch Einträkung, als wie vorher in Zubereitung

as it were and now wants to fly away with it. When you see this dragon, you must not let it flee out of fear. Be careful, for if it wants to tear apart the rock and flee; you must catch it in the moment of its flight and pour the prepared broth over this Colchian dragon, and then let the dragon drink its own anger and poison. Phœbus shoots his arrows into Python, the hard fierce dragon; and as soon as it starts to drink the broth its whole body starts to smoke and steam, but the poisonous broth rises up and kills the spirit of the dragon in the air, and brings it back to its scales and bones, till at last it becomes a black grave. This is the second blackness, and it lasts so long that the laborer thinks he has lost. But a spirit, a light, rises up from this *materia* and descends again, circulating until it washes off the body, flowing and coagulating often before becoming quite fixed, pure, and white, just like a metal which comes out of the cupel; and it often stands still when it is in a state of extreme vice, finally starting to be driven off on its own accord again until this roasted swan is turned into the white stone and becomes the food of the king.

89. *The hard, ferocious dragon Python is the double ☿, but what is the arrow of Phœbus?*

Answer: Yes, it is the scaly dragon, the leafed earth, which shines and shimmers with fish-eyes, the arrow of Phœbus, who is also called Delius. It is the poisonous smoking oil of ♄, the sun which this old man had in his body, which is also called the blood of the young children, in which the sun and moon bathe, the scythe of ♄, with which it cuts off the ☿'s wings, that is, fixes. Here the glass must not be opened, neither should anything be taken away from it or anything added to it until the stone of the first order is ready, for here there is no feeding

dieses ☿, sondern seine Figierung und Calcinierung, darum heißt es das hermetische Siegel.

90. *Sind denn keine Abkürzungen in diesen Werke?*

Antwort: Ja, es sind Abkürzungen, dass man den roten ☿ reif und güldisch macht, durch güldische ♁, dieses Wasser ein Kot und zur Erde macht, doch muss dieser nasse ☿ sein eigenes Salz erst auflösen und von seiner Terrestrität scheiden, solchen fixen Præcipitat hernach in seinem eigenen Öl auflösen und wieder fix machen, dann mit der Luft erweichen, mit dem roten Öl übergießen und zu figieren einsetzen. Die lange Zeit, so auf dieses Werk geht, ist nur den ☿ in ☉ Grad zu setzen.

91. *Ich glaube endlich, dass der Stein noch wohl zu machen ist, aber seine Multiplication ist schwer zu finden, wenn er immer und ewig sollte vermehrt werden?*

Antwort: Es ist auch nicht schwer, es wird der weiße Stein in dem weißen Öl oder *Menstruo*, so in Seele, Geist und Leib besteht und der rote Stein im roten Öl aufgelöst, welche beide genannt werden das Wasser der Sonne und das Wasser des Mondes, doch sind noch andere Arten der Fermente, dass man ihn aufs neue Faulen lässt, alsdann erst illuminiert und multipliciert, welcher Weg langwieriger aber auch von größerer Kraft.

92. *Wer nun das große und geheime Werk der Natur nicht finden und den Stein machen kann, hat aber die Scheidung der Elemente gelernt und weiß in der Composition nicht fortzukommen, will gleichwohl einen tingierenden Stein verfertigen, wie fängt man solches an, dass man Nutzen davon hat?*

Antwort: Dieser Wege sind unzählig viele Arten. Wer kann sie alle erzählen? Was in diesen Wassern gekocht wird, verbessert sich. Wer die *Elementa* zu scheiden gelernt, hat das Wasser der Sonne und des Mondes, in welchen beiden das gemeine ☉ und Silber kann aufgelöst und

nor impregnation, as previously in the preparation of this ☿, but only its fixation and calcination, hence it is called the Hermetic seal.

90. *Are there no shortcuts in this work?*

Answer: Yes, there are shortcuts that one can take in order to make the red ☿ ripe and gold-bearing. With gold-bearing ♃, this water turns into feces and into the earth; but this wet ☿ must first dissolve its own salt and separate it from its terrestrial nature, dissolve such fixed precipitate afterwards in its own oil and fix it again, then soften it with air, pour the red oil over it and use it for fixing. The great time that goes into this work is only necessary to transfer the ☿ into the ☉ degree.

91. *Finally I believe that the stone can still be made, but its multiplication is hard to find if it is to be multiplied forever and ever?*

Answer: It is also not difficult, the white stone is dissolved in the white oil or *menstruo*, which consists in soul, spirit, and body, and the red stone is dissolved in red oil, which are both called the water of the sun and the water of the moon; but there are still other kinds of ferments which allow one to putrefy it anew before illuminating and multiplying it, which is a more tedious way but also of greater power.

92. *Anyone who cannot find the great and secret work of nature and cannot make the stone, but has learned the separation of the elements, and yet does not know how to get ahead in the composition, but still wants to make a tingeing stone—how does one start in such a way that one benefits from it?*

Answer: These ways are of innumerable kinds. Who can explain them all? What is boiled in these waters improves. Whoever has learned to separate the *elementa* has the water of the sun and the moon, in both of which the common ☉ and silver can be dissolved and brought into

durch wiederholte Arbeit in Tincturen gebracht werden, auch kann in dem weißen Wasser die *Luna* solviert und im roten Wasser ☉, diese beiden zusammen vereinigt, wird die ☽ ganz in ☉ verkehrt sein.

93. *Auf was vor Art solviert es denn die Metalle?*

Antwort: Sie werden dadurch ohne Gewalt aufgelöst, ohne Zerstörung ihrer Grund-Feuchtigkeit, durch dieses fette, wiederbringende und vermehrende *Solvens*, und werden die metallischen Leiber darinnen zerschmolzen, ohne alles Getöse, als wie Eis in warmen Wasser, bleiben als ein schleimiges fettes Wesen, weil ihm dieses *Menstruum* sehr fest anhängt und sich nicht davon scheiden lässt. Ob man wohl nach befundener Not etwas davon überziehen will, bleibt doch der übrige Körper als ein Schleim zurück, worauf man unser Öl, als die Luft, welches auch der *Spiritus* ☿ genannt wird, gießen kann, so wird es den ♃ vom Metall in sich ziehen und in Gestalt eines Öls überführen, die ♃ des ☉ und ☽ conjungiert man, ihre ☿ auch, digeriert sie beide, alsdann vereinigt man sie. Wer aber, wenn diese beiden Körper vereinigt, solche mit einem anderen güldischen ♃ vereinigen kann, er sei aus den ♄, ♀, ♂ oder wo er sonst her will, rot färben und hernach, wenn sie trokken, mit ihren eigenen ♃ in *Spiritu* ☿ auflöst, mit dem roten Ferment vereinigt und fix macht, solcher hat eine weit ausstreckende Tinctur, welcher Weg kurz und dies hat noch kein Mensch gelehrt, was ich jetzt dem Nächsten zum Besten schreibe.

94. *Braucht man aber in diesen Arbeiten Feuer und auch stark?*

Antwort: In den Solutionen nicht, es müsste denn nur so gelinde sein als der Sonnen Wärme, wegen Flüchtigkeit dieser Geister, absonderlich des Öls oder Luft, welches sehr fest muss vermacht werden, auch ist dieser Geruch des Öls gefährlich und ein rechtes Gift, welches färbt, in der Coagulation aber, wenn es componiert, muss man es in einem Gradier-Feuer halten, bis es fix. Auch hat man bei diesem Werk nicht solche Gefahr als bei dem *Universal*, da man gewisse Zeiten, Termine, Tage und Stunden haben muss, wenn man es eintränken will. Wo

tinctures through repeated work, the *Luna* can also be dissolved in the white water, and in the red water ☉ [can be dissolved]. If these two are united together, the ☽ will be completely transmuted into ☉.

93. *In what way does it dissolve the metals?*

Answer: They are dissolved thereby without violence, without destroying their basic moisture, by this fat, restoring, and increasing *solvens*; and the metallic bodies are melted in it, without any noise, like ice in warm water. They remain as a slimy, fat being because this *menstruum* is very attached to them and cannot be separated from them. Whether one wants to distill some of it over, if one finds the need, the rest of the body remains as a slime, upon which you can pour our oil. As the air, which is also called the *spiritus* ☿, it will draw the ♃ of the metal into itself and transfer in the form of an oil. The ♃ of the ☉ and ☽ are conjoined, their ☿ too; digest them both, then they are united. But when these two bodies are united, whoever can unite them with another gold-bearing ♃—whether from ♄, ♀, ♂, or wherever one wishes—colors them red. And afterwards, when they are dry, they dissolve them with their own ♃ in *spiritu* ☿ and then combine them with the red ferment and fix them, producing a tincture of extensive power. The path to this is short and as no one has yet taught this, I am now doing my best to write this down for my followers.

94. *But does one need a strong fire in these works?*

Answer: Not in the solutions, as it would then only have to be as mild as the heat of the sun because of the volatility of these spirits, especially of the oil or air, which must be bequeathed very firmly. This smell of the oil is also dangerous and a real poison that stains in the coagulation. However, when it is to be composed, one must keep it in a gradating fire until it is fixed. In this work there is not such a danger as with the *universal* [work], in which one must have certain times, dates, days, and hours if one wants to impregnate it. But if one does not know [these times] and makes mistakes, the stone dies. It no longer drives

man aber solche nicht weiß und fehlt, so stirbt der Stein, treibt nicht mehr, denn er ist kalt worden. Hier aber bei den *Particular*-Werken mögen die Eintränkungen und *Conjunctiones* geschehen, wenn man denkt es sei fix genug, hat keine Gefahr, hergegen bei dem *Universal* ist das Feuer, der Ofen und wie er seinen eigenen Gefäße soll zugefügt und geheizt werden, das größte Geheimnis, darinnen ein Autor über 200-mal geirrt, ob er schon in der rechten *Materia* gerarbeitet. Denn ein Arbeiter muss wohl verstehen, was *vas, fornax, ignis, pondus, cum tempore latent* bedeutet.

95. *Den rechten ♄ möchte noch besser erklärt wissen?*

Antwort: ♄ ist ein Sohn des Himmels und der Erde, [Vater] der Göttin Vestæ, und ein Ehemann seiner Schwester Opis, die helfende und erhaltende Kraft der Dinge, der alte Demogorgon, welcher seine Kinder verschluckt und wiedergibt, wird gemalt mit weiß-grauen Haaren, in der Hand eine Sichel oder Sense und eine Schlange, deren Kopf in den Schwanz beißt, zeigt an seine sich selbst helfende und wieder gebärende Natur, mit der Sense haut er dem ☿ die Flügel ab und hat um seinen Leib einen gelben Riemen, in seinem Herz und inwendig ist es ☉, hat rote Stiefel an den Füßen, sein fixer ♃, ist unsere rote ♀, die rote Erde, Ton oder Letten, unser *Polus* und rotes *Atramentum*.

96. *Ist denn nun die Quintessenz dieses Bleies das wahre und allgemeine Solvens, und wie wird solches gemacht?*

Antwort: Es ist freilich der Anfang unser Arbeit und muss dieser Alte in sein Bad der Reinigung geführt werden, damit der grobe unreine Berg zu Boden fällt und nur das reine Teil zu unserem Werk genommen wird. Wenn er nun in Essig digeriert und wohl gefärbt, dann gießt man ihn ab und anderen darauf, bis sich keiner mehr färben will. Den Essig evaporiert man bis zur Honig-Dicke, destilliert solches *Gradatim*, bis die Retorte glüht und keine Dünste mehr gibt. Den übergangenen *Spiritum* rectificiert man, darinnen ist enthalten ein *Spiritus*, welcher wie *Spiritus Vini* brennt, ein weißes *Phlegma* und zuletzt ein rotes Öl,

off, for it has become cold. Here, however, in the case of the *particular* works, the impregnations and *conjunctiones* may happen. If one thinks it is fixed enough, there is no danger, whereas in the case of the *universal*, the fire, the furnace, and how it should be added to its own vessels and heated, this is the biggest secret, about which an author has been wrong more than two hundred times, whether he has already worked with the right *materia*. For a worker must understand what *vas, fornax, ignis, pondus, cum tempore latent*[295] means.

95. *I would like to know more about the right ♄?*

Answer: ♄ is a son of heaven and earth, [father][296] of the goddess Vestæ, and a husband of her sister Opis, the helping and sustaining power of things, the old Demogorgon, which swallows his children and gives them back, and who is painted with white-grey hair, holding a sickle or scythe in his hand, as well as a snake, whose head bites into its tail, showing thus its self-helping and reproductive nature. He cuts off the wings of the ☿ with his scythe and has a yellow strap around his body; and on his heart and inside of it is ☉. He has red boots on his feet, his fixed ♃ is our red ♀, the red earth, clay or loam, our *polus* and red *atramentum*.

96. *Is the quintessence of this lead the true and general solvent, and how is this prepared?*

Answer: It is of course the beginning of our work and this old man must be led into his bath of purification so that the coarse, unclean mountain falls to the ground and only the clean part is used for our work. If it is now digested in vinegar and well colored, then you pour it off and other [vinegar] on it until nothing wants to color anymore. The vinegar is evaporated to the thickness of honey, and distilled *gradatim* until the retort is glowing and no more vapors are emitted anymore. The distilled *spiritum* is rectified. It contains a *spiritus* that burns like *spiritus vini*, a white *phlegma*, and finally a red oil, which must be

welches in starkem Feuer muss übergetrieben werden und vor der Luft verwahrt, sonst verraucht es alles. Alles diese Dinge sind rohe, giftig, noch nicht medizinalisch und muss man sich davor hüten. Diesen ♄ flüchtigen rohen Geist nenne ich und andere *Spiritum* ☿, welcher unser erster geheimer Schlüssel.

97. *Nun möchte ich auch gern vom ⊕ seinem roten Öl hören?*

Antwort: Dies beschreibt Paracelsus ganz klar, wenn er sagt: Nimm unser ⊕ *veneris*, zum höchsten auf spagyrisch zubereitet, darauf das Element *Aquæ* and *Aëris* zuvor behalten und solvier es, &c. So weit Paracelsus: Unser roter ⊕ *veneris*, so mit dem ♄ gespielt, die solviere in unserem *Spiritu* ☿, welches unser destillierter Essig oder unser *Aquafort*, auch die giftige Schlange und Drache ist. Hier heißt es der Drache tötet das Weib und diese denselben und werden zugleich mit Blute durchgossen. Es ist eine überaus schöne Solution, es fängt aber bald in dieser Solution an zu faulen und zu stinken, hat einen Geruch der toten Körper und wird sich die schöne Farbe verlieren und wie Schusters Schwärze werden, nach welcher Erscheinung man es im *Balneo* abdürstet, bis es als ein dickes schwarzes Pech stehen bleibt, den übergestiegenen *Spiritum* rectificiert man, und gießt den flüchtigen Teil, welcher als Branntwein brennt und krumme Äderlein macht, auf dieses schwarze Pech, ohne alle Wärme, rüttelt und schüttelt dies vielmals, damit der rote fixe ♃ sich in seinem eigenen Öl oder *Spiritu* ☿ auflöse und das gebenedeite Öl zur Quintessenz gebracht werde. Es muss lange stehen und oft geschüttelt werden. Wenn es anfängt wunderbar schön an Farben und Geruch zu werden, so ist es Zeit zu destillieren, es wird dieses philosophische ☉ der trockne rote ♌, wenn er anfängt sich in sein eigenes Wesen zu resolvieren, am Rande des Glases alle Farben eines Regenbogens sehen lassen, wunderschön, dann destilliert man es, so steigt der ☿ialische Teil als Tränen über, rund als Tropfen, welche sich im Glase anhängen, der *Spiritus* ☿ aber als ein Öl und färbt den Helm gelb, zuletzt kommt das gebenedeite rote Öl, färbt den Helm rot, mit wunderschönen Farben. Alle diese Teile werden allein gefangen,

distilled over in a strong fire and kept out of the air, otherwise everything will smoke up. All these things are crude, poisonous, not yet medicinal, and one must guard oneself against them. This ♄ I call the volatile, raw spirit, others call it the *spiritum* ☿, which is our first secret key.

97. *Now I would also like to hear about the red oil of ⊕?*

Answer: Paracelsus describes this very clearly when he says: "Take our ⊕ *veneris*, prepared spagyric to the highest, add the element *aquæ* and *aëris* previously retained and dissolve, etc." this is what Paracelsus wrote: Our red ⊕ *veneris*, which had played with the ♄, dissolve in our *spiritu* ☿, which is our distilled vinegar or our *aqua fort*, and also the poisonous snake and dragon. Here it is said that the dragon kills the woman, who kills the dragon at the same time; and they are suffused with blood.[297] It is an exceedingly beautiful solution, but it soon begins to putrefy and stink in this solution. It has the smell of a dead body and will lose its beautiful color and become like black shoe polish, after which appearance one evaporates it in *balneo* until it becomes like a thick black pitch; the spirit that has distilled over is rectified and the volatile part, which burns like brandy and makes crooked veins, is poured onto this black pitch without any heat. Then rattle and shake it many times so that the red fixed ♃ dissolves in its own oil or *spiritu* ☿, and the blessed oil will be brought to the quintessence. It has to stand for a long time and be shaken often. When it starts to become wonderfully beautiful in color and smell, it is time to distill it. This philosophical ☉ will be the dry red ♌, when it begins to resolve itself into its own being, showing all the colors of a rainbow at the edge of the glass. It is beautiful, because if you distill it, the ☿al part rises like tears, round as drops, which cling to the glass; but the *spiritus* ☿ [is] like an oil and colors the helm yellow, Finally, the blessed red oil comes and colors the helm red, with beautiful colors. All of these parts will be caught alone, and

auch jedes allein rectificiert, das *Phlegma* mit schneeweißen Steinen und runden Tropfen wie Perlen, ist unser viskosischer ☿, wenn weißes Salz aufsteigt, wird es zu den weißen Tropfen getan, die Luft oder das Öl muss rein davon, auch von der Luft das Feuer geschieden werden und zum Feuer getan, das Feuer muss auch rectificiert werden, damit die Luft und Erde so es bei sich hat, davon gebracht wird, ist nun der rote ♌ rein von der Erden geschieden, so bleibt die Erde dieses 🜨 rein und weiß zurück, glänzend, weiß, das Silber sieht dagegen wie Blei, diese Erde hat nun in ihren einen *Spiritum* und Geist, der gelb und rot und die wahren Blumen der Sonne und des Mondes trägt, diese wird in ihre *Phlegmate* gereinigt und ist das *Matrimonium* und die Mutter des Glücks, unser metallisches Salz, daraus der ☿ *Philosophorum* gemacht wird, denn wer kein metallisches Salz hat, der kann auch nicht ☿ *Philosophorum* machen, dass dieses Salz flüchtig zur geblätterten Erden wird, dieser flüchtigen Erden gießt man ihr eigenes Öl zu, dadurch sie gebunden und fix wird, durch die Luft wird sie gelb und flüssig, durch das Feuer aber rot. Es hätten diese Dinge wohl etwas kürzer gegeben werden können, weil ich aber nicht aller *Adeptorum* Fluch auf mich laden wollen, dass ich diese Dinge zu gemein gemacht, als habe ich etliche Regeln in acht genommen, dass nur die Würdigen, so sich das Studieren in Büchern und die *Praxis* lassen angelegen sein, solches finden, die anderen Faulen aber, als Unwürdige davon ausgeschlossen bleiben. Ist auch keine Wissenschaft für die Bauern, wie sie den Wagen schmieren sollen, sondern eine göttliche Kunst und Wissenschaft für die Söhne der Weisheit, welche GOTT und die Natur lieben, durch fleißiges Lesen und Arbeiten diese Geheimnisse finden und GOTT zu Ehren, dem notleidenden Nächsten zum besten anwenden, da denn die wilden Menschen und gottlosen Mast-Säue des Teufels, welche nur denken ihren Gold-Kasten zu füllen und vor Geiz nimmermehr satt werden können, weit davon ausgeschlossen bleiben. Und ich glaube gewiss, dass es viele Menschen haben sollen, ausgenommen die geizigen, solches sind die allerlasterhaftigsten Menschen, weil der Geiz eine Wurzel allen Übels, diese auch den Gold-Klumpen für ihren GOTT

each will also be rectified alone. The *phlegma* with snow-white stones and round drops like pearls is our viscous ☿. When white salt rises, it is added to the white droplet. The air or the oil has to be purely separated from this, also from the air and fire, and added to the fire. The fire must also be rectified so that the air and earth that it has within it can be removed from it. At this point it is now the red ♌ separated entirely from the earth, so that the earth of this ☉ remains pure, white, and shining, so white that it makes silver looks like lead. This earth now has in it a *spiritum* and a spirit which bears yellow, red, and the true flowers of the sun and the moon. These are purified in their *phlegmate* and [it] is the *matrimonium* and the mother of fortune, our metallic salt, from which the ☿ *philosophorum* is made, because whoever lacks metallic salt cannot make ☿ *philosophorum* either, so that this volatile salt becomes a leafed earth. Their oil is then poured into this volatile earth, whereby it becomes bound and fixed. Through the air it becomes yellow and liquid, but through the fire it becomes red. These things could probably have been described a little more concisely, but because I do not want to take all the curse of all *adeptorum* upon myself for publicizing these things, I have therefore observed a number of rules so that only the worthy student who does both theoretical and practical study will be worthy enough to find such a thing. As for the other, lazy ones, they will be excluded from this as unworthy. Nor is it a science for the peasants, like lubricating a wagon, but a divine art and science for the sons of wisdom who love GOD and nature, through diligent reading and work, to find these mysteries and to honor GOD, to the best of the neighbor in need, since the wild people and the godless, fattening sows of the devil, who only think about filling their gold chest, which can never be full due to avarice, remain far from it. And I certainly believe that many people should have it, except for the stingy. Misers are the most vicious people, because stinginess is the root of all evil, and they also regard the nugget of gold as their god, and they

achten, eher GOTT und Himmelreich als die Liebe zum Geld lassen. Diese sage ich, werden es nicht bekommen, GOTT wird es ihnen auch nicht geben, denn sie viel zu unwürdig dazu sind.

98. *Was macht man nun mit dieser Erden oder metallischen Salze?*

Antwort: Man bringt ihn zu dem Becher der Liebe, dem *Universal-Geist*, so wird diese jungfräuliche Erde geschwängert, bringt Früchte ihrer Art, weil sie eine reine Jungfrau, die Königin *Alma*, wenn sie mit dem *Spiritu* und *Menstruo Universali* geschwängert, wachsen aus ihrem Schoß metallische Fünklein, endlich die allerreinsten und vollkommensten Metalle. Die Erde. wenn sie mit den Spiritu angefeuchtet, ist sehr trocken und dürr, daher sie dieses Wasser als ein Schwamm in sich säuft, wird bei der ersten Eintränkung wegen des wenigen Wassers nicht rauchen oder doch so wenig, dass man es nicht empfindet. Wenn das Wasser bei ihr geronnen, feuchtet man sie abermals an, dann fängt sie an zu brodeln und Dünste zu geben, aber noch keine Nässe, sondern Rauch, Dampf und Nebel. Wenn dieses abermals coaguliert und erhärtet, sieht es wie spitzige Lanzen und wird genannt das Feld *Martis*, auch der saubere Jungfern ☉, wenn es noch mehr eingetränkt, wird endlich die Erde erweicht und amalgamiert sich der ☿ mit dem Salz oder Erde, seinem fixen Teil, wird als eine dicke Butter, dazu bringt man nach Gelegenheit so viel, bis die ganze Erde der ☿ in sich gefasst und als ein glänzender ☿ erscheint, bei mehreren Eintränkungen wird die Erde solviert. Erst musste das Wasser notwendig coaguliert werden, weil solches viel, der Erde aber wenig, so musste dieses Wasser zur Erde werden. Nun aber muss es verkehrt gearbeitet werden. Was coaguliert, muss wieder solviert und zu Wasser werden, sobald diese Erde solviert, wird sie grün, welches die glückselige und wachsende Grüne ist, das Gefäß wird sich färben wie ein grünes Nussbaumblatt, mit einem güldischen Glanz, dem Gold-Spiegel und vielen schönen Farben, absonderlich gelb und rot, riecht trocken und bitter, auch ganz eklig. Die Farben beweisen dass diese Erde flüchtig wird und der Geist, so in ihr gebunden lag, los wird von seinen Banden, sie muss

would rather renounce GOD and the kingdom of heaven than their love of money. These, I say, will not obtain it, GOD will not give it to them either, because they are much too unworthy for it.

98. *What do you do now with this earth or metallic salt?*

Answer: One brings it to the cup of love, the *universal* spirit, so that this virgin earth is impregnated, bearing fruits of its kind because it is a pure virgin who becomes the Queen *Alma*, after being impregnated with the *spiritu* and *menstruo universali*. At this point she grows metallic sparks from her womb, and then finally the purest and most perfect metals. The earth, when moistened with the *spiritu*, is very dry and arid, so it soaks up this water like a sponge; and because of the low amount of water, it will not smoke during the first impregnation, or at least so little that you will not recognize it. When the water has congealed with it, it is moistened again, then it begins to seethe and give off vapors, not yet dampness, but smoke, steam, and mist. When this coagulates and hardens again, it looks like pointed lances and is called the field *Martis*. It is also the clean virgin ☉; and when it soaks even more, the earth is finally softened and the ☿ amalgamates with the salt or earth, its fixed part becoming like thick butter. Add as much as you can until the whole earth of the ☿ is in it and appears as a shining ☿, with several impregnations the earth is dissolved. First the water must have been coagulated, because there was so much [water] but little earth, so this water had to become earth. But now it must be worked in reverse. What coagulates must be dissolved again and become water, once this earth dissolves it will turn green, which is the blissful and growing greenness. The vessel will tint like a green nut tree leaf with a goldish glow, the gold mirror and many beautiful colors, and peculiarly, yellow and red. It smells dry and bitter, and is also quite disgusting. The colors prove that this earth is volatile and that the spirit that was thus bound in it has been released from its bonds. It must

ganz aufgelöst sein, gibt in den Solutionen und den Eintränkungen, Dünste und Rauch, welche schwarz, gelb, himmelblau, rot und dergleichen durcheinander gesehen werden, es wird als gelber Sand sich auf den Boden setzen, bis es alles wieder dick wird und erhärtet, darauf fängt es an zu fließen und muss vorher der Körper in in diesem Wasser sieden und prudeln, wo dies nicht geschieht, ist die Arbeit vergebens: Denn es ist weder *Agens* noch *Patiens*, wird auch keinen Sturm-Wind und Regen geben. Wenn aber dieser Sand oder Letten schmilzt, fließt es als Pech, gibt Blasen, solche reißen auf und fährt schwarzer rußiger Rauch heraus, welcher doch voller gelber, roter und blauer Dünste. Wenn die Feuchte vergeht, sind die Farben eines Regenbogens nach der Schwärze zu sehen, aber nicht in der *Materia*, sondern im Glase, welche wie subtiles Geflitter herumstiebt und dieser Stein oder Erde wird ganz in die Höhe getragen, geht wieder zurück, erfasst den Leib, trägt ihn wieder in die Höhe, sublimiert sich augenblicklich und schmilzt wieder, bis dieser männliche und weibliche Körper in einem Leib gewachsen, der Mann in das Weib und das Weib im Mann verkehrt und der Hermaphrodit worden, der Manns als Weibes statt vertreten kann, ist durch das Sieb der Natur so oft gesiebt worden, so subtil, dass sie als ein Blatt auf dem Boden des Gefäßes bleibt, in trockener, schwarzer Gestalt, als ein Achat oder schwarzer beblätterter Kalk, doch dichter, auf Pech-Art, garstigen Geruchs, welcher nicht mehr bitter, sondern faul nach Leichen-Art riecht, dieser schwarze *Laton* ist das *Æs Hermetis*, die geblätterte Erde, dieses muss von der Schwärze und Schmutz gereinigt werden und wird diesen schwarzen Raben das Haupt weggenommen, damit die weiße Taube an seine statt kommt, des Salzes war sehr wenig, als es in seinen *Phlegmate* gereinigt, da es weiß worden, wurde es vermehrt, nun aber wenn dem Tier so in diesem Gehölz, die Schwärze weggenommen wird, vermehrt sich durch den weißen Geist, mit welchem diese Schwärze weggewaschen wird. Die Schwärze kommt nicht auf einmal, sondern wird erstlich oben auf rußig, als Kohlen-Staub oder schwarzen Pfeffer aufgestreut, bis es immer schwärzer und schwärzer wird, je dicker solcher wird, je

be completely dissolved, and there are vapors and smoke in the solutions and the impregnations, which are black, yellow, sky-blue, red, and the like appear mixed with one another. It will settle on the ground as yellow sand until it all thickens again and hardens. Then it begins to flow, and the body must first boil and bubble in this water. If this should not take place, the work is in vain, for it is neither *agens* nor *patiens*, nor will there be storm winds and rain. But when this sand or clay melts, it flows like pitch, giving rise to bubbles, which rupture and emit black sooty smoke, which is full of yellow, red, and blue fumes. When the moisture disappears, the colors of a rainbow can be seen after the blackness, although not in the *materia*, but rather in the glass, which swirls around like subtle flint; and this stone or earth is carried all the way up. It then goes back again, seizes the body, and lifts it up again, and then it instantly sublimates and melts again, doing so until these male and female bodies have grown into one body, with the man becoming a woman and the woman a man. At this point it has become the hermaphrodite which can represent man as woman, and which has been sifted through the sieve of nature so many times, so subtly that it remains as a leaf at the bottom of the vessel, in dry, blacker form, like an agate or black-leafed hide, only denser, pitch-wise, with a nasty odor, which no longer smells bitter, but smells rotten like corpses. This black loam is the *æs Hermetis*, the leafed earth. This must be purified of the blackness and dirt, and it will take away the head of the black raven so that the white dove can take its place. There was very little of the salt when it was purified in its *phlegmate*, though after it became white it was multiplied; but now when the blackness is taken away from the animal in this wood, it is multiplied by the white spirit by which this blackness is washed away. The blackness does not come all at once, but first becomes sooty on top, sprinkled like coal dust or black pepper, until it becomes blacker and blacker, the thicker it becomes, the blacker it becomes, but before it turns black, it turns as yellow as

schwärzer, aber ehe es schwarz wird, wird es gelb als Achatstein, auch rot, vor die Schwärze hat man nicht zu sorgen, sie kommt gewiss, wenn es aber durch die weiße Milch abgewaschen, stellt sich die Weiße ein und darf der Rabe nicht wieder eingelassen werden, aber die weiße Taube darf wohl wiederkommen und das Öl-Blatt im Munde bringen. Es wird nicht aufgehört mit weißen, weil man hier nicht zuviel tun kann, denn die ☽ kann nicht zu weiß werden, auch muss man wohl Achtung geben, wenn dieser Vogel fliegen will, wenn man will warten, bis es ihm selbst gefällt, so geht er fort und lässt einem das ledige Nest. Ich habe es aber in diesem Buche schon gedacht, was für Zeichen vorgehen, wenn dieser Fels zerreißen wird und dieser Stein sich zermalmt, darum wartet man nicht so lange, nimmt etwas davon, wirft es auf ein glühendes Blech, geht es in die Höhe, lässt wenig oder nichts auf dem Blech, so sublimiert man es in die *Terram Foliatam*, teilt solches in 2 Teile, eines zur Vermehrung, das andere zum Stein.

99. *Wie macht man das vermehrende Ferment?*

Antwort: Der ☿ *Sublimatum* der 7 Nächte wird in seinem eigenen ♐︎, Öl oder Luft solviert und von diesem ✷ abgezogen, dies ist das rechte philosophische ♒︎ und mineralische Bad, wenn der ☿ zum Öl gemacht worden, zum Wasser, so die Hand nicht nass macht und ist das rechte Gradier-Öl, das Bad, darinnen Mann und Weib sich baden können, denn es löst auf und conjungiert sie.

100. *Was tut man denn hernach?*

Antwort: Man macht die andere Composition, nämlich des Steins.

101. *Wie macht man denselben?*

Antwort: Der ☿ muss fix gemacht werden, so ist es unser weißer ♃.

102. *Ist denn dies der Stein, wenn sie sagen: Sät eure ☉ in die geblätterte Erde?*

Antwort: Ja, er ist es, wenn man das Gold in seine eigene Erde sät,

an agate stone. It also turns red. You don't have to worry about the blackness, for it will certainly come; but when it is washed off with the white milk, the whiteness will appear and the raven may not be let in again, but the white dove may well come back bringing the olive leaf in its mouth. We do not stop with whitening, because one cannot do too much here, because the ☽ cannot become too white. One must also be careful when this bird wants to fly, if one wants to wait until it likes it, then it goes away and leaves you with the empty nest. But in this book I have already considered what the signs are like when this rock is torn apart and this stone crushes itself. So do not wait too long. Take some of it and throw it on a glowing sheet of metal. If it goes up, leaving little or nothing on the sheet, it is sublimated into the *terram foliatam*. Divide it into two parts, one for multiplication, the other for the stone.

99. *How to make the multiplicating ferment?*

Answer: The ☿ *sublimatum* of the seven nights is dissolved in its own ▽, oil, or air, and distilled from this ✶. This is the right philosophical ▽ and mineral bath. When the ☿ has been made into oil, into a water that does not wet the hands and which is the right gradating oil, this is then the bath in which men and women can bathe in, because it dissolves and conjuncts them.

100. *What do you do afterwards?*

Answer: One makes the other composition, namely of the stone.

101. *How does one make this same [stone]?*

Answer: The ☿ has to be fixed, so it is our white ♄.

102. *Is this then the stone, when they say: sow your ☉ in the leafed earth?*

Answer: Yes it is, when one sows the gold in its own soil, previously we

vorher sind wir durch eine sehr geheime Tür der ersten Conjunction unvermerkt gegangen, welche Hochzeit wurde gehalten in dem Hause der Natur, da der Himmel und Erde sich zusammen schloss und auf dem Thron der Freundschaft stiegen, welche Hochzeit ziemlich lange gehalten wurde, da indes der ☿ diesen Himmel 7 mal umlaufen musste, nebst der Sonne, Mond und anderen Planeten, ☿ aber hat das meiste zu tun, denn er war ihrer aller Bote und hatte den Schlüssel zu diesem Hochzeitshause, welchen er vom ♄ empfangen und war nun der ☿ ihrer aller Freund, denn er war ihnen allen behilflich, wenn sie sich durch zu viel Bewegen staubig und schmutzig gemacht, sich erhitzt hatten, so badet er sie, wusch sie ab, tränkte sie, schloss ein geheimes Zimmer nach dem anderen auf, dass die Götter und Göttinnen konnten zusammen kommen und miteinander buhlen, weil sie alle in diesen Hochzeitshause adulterieren, auch Bruder und Schwester einander beschlafen, er selber aber, der ☿, wird durch viele Arbeit, Mühe und Laufen ermüdet, dass er zur Erde fällt und Erde wird zur Strafe der Sünden, er wird in Erde gesät und wird selber zur schwarzen Erde, diese ist schwarz, aber gar lieblich, die Sonne hat sie so verbrannt, bis er von der Erbsünde abgewaschen, weiß und rein gebadet, sich wieder abgetrocknet, seinen Lauf im Himmel angetreten, weil er durch diesen Tod das neue Leben erlangt und der gekrönte Adler worden, so in die Sonne fliegen kann und doch hat man diesen geheimen Schlüssel nicht in acht genommen, weil ich kein Wesen davon gemacht. Der ☿ wurde in die Erde gesät, nun aber, wenn er das fliegende Erdreich worden, so wird er der Acker, in welchem das ☉ oder die Schlangen-Zähne gesät werden.

103. *Wie soll man aber den Drachen töten?*

Antwort: Man soll ihn in *Aqua Stygia* ertränken, in diesem giftigen Wasser muss er sterben, denn solches Wasser ihn ganz und gar verzehrt.

passed unnoticed through a very secret door of the first conjunction, marriage was held in the house of nature where heaven and earth are close together and ascend upon the throne of friendship. This marriage was held for quite a long time, since meanwhile the ☿ had to go around this heaven seven times, together with the sun, moon, and other planets; but ☿ has the most to do, because it was their messenger and had the key to this wedding house. It received the key from ♄; and now ☿ was friend of all of them, because it helped them all, for when they got dusty and dirty from too much movement, becoming hot, it bathed them, washed them, soaked them, and unlocked one secret room after the other, so that the gods and goddesses could come together and court with one another, because they all adulterated in this wedding house. Brother and sister also slept with each other, but the ☿ itself is tired by a lot of work, toil, and running, so that it falls to the earth and becomes earth as punishment for the sins. It is sown in the earth and itself becomes the black earth, this is black, but very lovely. The sun burned them until it was washed from original sin, bathed white and pure, dried itself again, started its course in heaven, because through this death it gained new life and became the crowned eagle that can fly into the sun. And yet no one took heed of this secret key, because I made no fuss about it. The ☿ was sown in the earth, but now when it has become the flying soil, it becomes the field in which the ☉ or the serpent's teeth are sown.

103. *But how should one kill the dragon?*

Answer: It should be drowned in *Aqua Stygia*.[298] It must die in this poisonous water, for such water completely consumes it.

104. *Ich möchte doch dieses Geheimnis besser erklärt wissen, den Kampf zwischen dem ♌ und Adler?*

Antwort: Der gestirnte Adler, der Königs Vogel, so sich durch seine Flügel hoch geschwungen, dass er in die Sonne gesehen, wird unvermutet aus seinen Hause in ein anderes getrieben, der alte Bär bringt löwische Wasser, das tödlich stinkende Wasser ist inwendig grün, daher es auch der grüne ♌ genannt wird, voller bitterer Galle, welches durch den Kampf des Adlers, wenn sie beide miteinander getötet und miteinander verfault, soll zu einer Süße werden, wie aus Simsons Aas im ♌ der Honig, sobald der ♌ den Adler ergreift, fallen ihm gleich Federn weg, der Adler flieht, der ♌ erhascht ihn, aber doch frisst der Adler auch von des ♌ Fleisch und Blut in sich, dieses geheime Zimmer, darinnen dieser Kampf gehalten, verwahrt der fliegende Cherub mit dem flammenden Schwert das hermetische Siegel, wenn es hier wohl angebracht, bleiben diese beiden in Kampf und Streit, bis die Adler zerrissen und man sehr wenig Federn noch von ihnen auf dem Boden liegen sieht, durch den Rauch, Dampf, brodelndes Blut und Schweiß dieser beiden, sieht man nichts als ein grünes Meer, welches oben hart wird und gefriert, bis es durchaus zum Eis worden, wenn diese Schollen auf dem Boden erharten. In der Luft gibt es Tropfen, welche sich anhängen, gegen das Licht etwas bläulich sehen, wenn sie anheben zu sterben und beide faul werden, scheiden sich Geist und Seele vom Leibe, verfolgen einander als Feinde, auch in der Luft und vereinigen sich und kommen wieder zum Leibe, hier muss der Arbeiter Geduld haben, nicht zu verzweifeln, wenn er anstatt des Adlers den Raben sieht, diese Schwärze kann er nicht scheiden als wie im ersten Werk, sondern muss Geduld haben, bis nach der von der Natur bestimmten Zeit der Geist aufersteht, welcher sich circulieren, den Leib mit sich in die Höhe nehmen und an der Luft abtrocknen wird. Es wird vielmals fließen und wieder gestehen ehe es ganz hart wird, wird auch der Regenbogen, das Gnaden-Zeichen, gewiss erscheinen und zuletzt erwartet man das Ende, wenn es nicht mehr zum Fluss kommen kann, so gesteht und

104. *I would like this secret better explained, the fight between the ♌ and the eagle.*

Answer: The starry eagle, the king's bird, which swept itself up upon its wings so that it could see into the sun, is unexpectedly driven from its house to another. The old bear brings lion's water. The deadly stinking water is green on the inside, hence it also called the green ♌, full of bitter bile, which through the fight of the eagle, when they both were killed and putrefied together, shall become sweet, like honey from Samson's carrion in the ♌. As soon as the ♌ seizes the eagle, its feathers fall away. The eagle flees, and the ♌ catches it, but the eagle also feeds on the ♌'s flesh and blood. The secret room, where this battle is held, is kept by the flying cherub with the flaming sword. If the Hermetic seal is well applied here, these two remain in battle and strife, till the eagles are torn apart and very few of its feathers are seen lying on the ground; and through the smoke, steam, seething blood, and sweat of these two, one sees nothing but a green sea which hardens on top and freezes until it has become ice throughout, when these clods harden on the ground. In the air there are drops which cling, look a little bluish against the light, when they begin to die, and both become lazy; spirit and soul separate from the body, pursue each other like enemies, also in the air, and unite and come back again to the body. Here the worker must have patience not to despair when he sees the raven instead of the eagle. He cannot separate this blackness as in the first work, but must have patience until, after the time determined by nature, the spirit rises, which will circulate, take the body up with it, and dry it in the air. It will flow many times and coagulate again before it gets really hard. The rainbow, the sign of mercy, will certainly appear, and finally one expects the end, when it can no longer flow, so

erhärtet es in helle Diamanten, so etwas grünlich, doch danach der grüne ♌ stark oder schwach gewesen, dieses ist die erste weiße Tinctur.

105. *Was tut man weiter mit diesem ♌?*

Antwort: Der ♌ ist der Sonne ihr Haus und muss durch die Sonne erhitzt werden, dass die Sonne dieses Zeichen in ihrer Exaltation durchläuft, so wird er von der Sonnen-Natur rot, gebrannt und feuriger Natur werden, aber erst muss dieser ♌ durch die Luft oder Himmel erhitzt werden, dass er die helle Sonne fressen kann, das ist, in sein Haus nehmen.

106. *Warum heißt man aber diese beiden feurigen Menstrua ♌?*

Antwort: Darum weil sie wegen ihrer grimmigen Natur alle metallische Leiber fressen und ihren Hunger damit stillen, aber sie geben sie alle verbessert wieder, der grüne ♌ frisst den Adler, der rote ♌ die glänzende Sonne.

107. *Was ist denn die hell glänzende Sonne?*

Antwort: Es ist die figierte und beständig gemachte Materia des Steins, unser rot figierter ♃ und erste Tinctur auf rot, unser gekrönter König, die Sonne, die durch des grünen ♌ Haus laufen müssen und in ihrer Morgenröte hoch glänzend aufgeht.

108. *Vorher wurde vom ☿ geredet, dass er der Schlüssel wäre, was schließt er auf?*

Antwort: Er ist der allergeheimste Schlüssel und gleichsam der Capital-Schlüssel, ohne diesen Schlüssel kann die verschlossene Tür zu des Königs verschlossenen Palast nicht eröffnet werden, er muss aber erst vorher das *Centrum* der Erde aufschließen, welches Stahl, Diamant, fest und hart, mit keiner Gewalt, ja mit der größten nicht zu zwingen und alle Pfeile, welche danach abgeschossen, zurückprallen. Die Göttin Vesta hat ihr Schloss in diesem Felsen, der ☿ aber, als ein listiger geschwinder Geist, der durch den Himmel geflogen, bringt himmlisches

it coagulates and hardens into bright diamonds, which are somewhat greenish, but depending on how strong or weak the green ♌ has been, this is the first white tincture.

105. *What do you do next with this ♌?*

Answer: The ♌ is the house of the sun and must be heated by the sun, which sun passes through this sign in its exaltation. It will be burned red by the nature of the sun' and become fiery in nature. But first this ♌ must be heated by the air or sky so that it may devour the bright sun, that is, take it into its house.

106. *But why are these two fiery menstrua called ♌?*

Answer: Because of their fierce nature, they devour all metallic bodies and satiate their hunger with them, but they reproduce all of them improved. The green ♌ eats the eagle, the red ♌ the shining sun.

107. *What is the bright, shining sun?*

Answer: It is the fixed *materia* of the stone made permanent, our fixed, red ♃ and first tincture in red, our crowned king, the sun that must pass through the house of the green ♌ and which in its dawn rises high and brilliant.

108. *Before, it was said of ☿ that it was the key, what does it unlock?*

Answer: It is the most secret key and, as it were, the capital key, for without this key the locked door to the king's closed palace cannot be opened. First it has to unlock the center of the earth, which is solid and hard as steel or diamond. It is so strong that no force can conquer it, indeed, not even the greatest [force], and all arrows shot towards it bounce back. The goddess Vesta has her closed palace in this rock, but the ☿, as a swift, cunning spirit who has flown through the sky, brings

Wasser, damit erweicht er diese Erde und harten Felsen nach und nach bis er die Spring-Wurzel, so dieses Schloss aufsprengt, tief eingegraben, denn es hat die Art, dass es erst von innen muss eröffnet werden. Wenn nun dieses himmlische Wasser, welches auch der schärfste Essig genannt wird, das *Centrum* dieser harten Erde erweicht, dann schließt der ☿ sie gar auf, insinuiert sich bei dieser keuschen Königin, die sich sonst keinem Mann unterwerfen will, weil aber dieser ☿ Cælus, oder ein himmlischer Sohn genannt wird und einer von den Göttern ist, so lässt sie sich seine Liebe gefallen, weil er aber himmlischer Natur und sie irdischer, so sein sie widerwärtige Eheleute, so oft, als er sie umfangen will, verwundet er seinen Leib an ihr, wegen ihres harten und dürren Leibes, dass er allezeit unleidig wird, flieht auf und will von ihr, sie aber erhält ihn feste aus Liebe bis sie von ihm, durch so viele Umfangung, schwanger, wird sie ganz linde und weich, zeugt den neidischen ♄, dieser die anderen Götter, wegen großer Liebe, so sie zu ihrem Mann dem ☿ hat, gibt sie ihm den Schlüssel zu allen ihren Schatz-Kästen, dass er daraus darf nehmen, was er will und wieviel er will, weil die Königin Vesta wunderschönen Schmuck, schöne Kleidung, der köstlichsten Perlen und Edelgesteine von allerhand Farben, aber weil sie sich gänzlich aller Mittel entblößt und ihrem Mann Cælus alle ihre Schätze gegeben und gleichsam ein stillschweigendes Testament gemacht. Denn nach Überreichung des Schlüssels, wenn sie ihrem Mann alle ihre Schätze gegeben, alle ihre Schönheiten gezeigt und sich aller Mittel entblößt, stirbt sie dahin und vergeht gar. Weil aber ihr Mann Cælus himmlischer Natur, so nimmt er ihren Geist, Leib und Seele mit in den Himmel, denn eben von ihr sollte ein himmlisches Geschlecht aufkommen, zeigt solches allen Göttern und Göttinnen, welche dieses güldene Vlies weit höher als alle Kleinodien lieben und es in ihren größten Ehren-Tagen, als das herrlichste Geschmeide an ihrem Hals hängen, wenn sie in ihrer hochzeitlichen Freude und Wohlleben sind.

the heavenly water, with which he softens this earth and hard rock little by little until he has planted the spring root deeply, so that this locked palace will burst open, because it is of such a nature that it must first be opened from the inside. Now when this heavenly water, which is also called the hottest vinegar, softens the center of this hard earth, then the ☿ competely unlocks it, insinuates itself with this chaste queen, which does not want to submit to any other man, but because this ☿ is called Cælus[299] or a heavenly son and is one of the gods, she accepts his love. Yet because he is of a heavenly nature and she is of an earthly nature, they are miserable as a married couple. For whenever he tries to embrace her, he wounds his body because of her hard and dry body; he always becomes displeased, flees, and wants to leave her; but she keeps him firm out of love, until she becomes pregnant by him, and through so many embraces she becomes very mild and soft. She begets the envious ♄, and this the other gods, because of her great love, which she has for her husband, the ☿, she gives him the key to all her treasure chests so that he can take from them whatever he wants in whatever quantity he wishes; for Queen Vesta has beautiful jewelry, beautiful clothes, the most precious pearls, and jewels of all kinds and colors, but she stripped herself entirely of all means and gave to her husband Cælus all her treasures and made a tacit testament, as it were. Since, after the handing over of the key and all her treasures to her husband, and after having shown all her beauties and having stripped herself of all means, she dies and perishes altogether. But because her husband Cælus is of a heavenly nature, he takes her body, soul, and spirit with him to heaven; for from her a heavenly generation shall arise, begetting all the gods and goddesses who love this golden fleece far more than all jewels, and who hang them from their necks in their greatest days of glory, when they are in their nuptial joy and well-being.

109. *Was schließt denn nun der ☿ mit dem Schlüssel, so er von der Vesta bekommen?*

Antwort: Er schließt den Leib der Planeten auf, aller Götter ihre Schatz-Kästen, welche er spoliert, der ☿ ist ein Gott der Kaufleute und der Diebe, er stiehlt selber gern, denn er ist mit seinem Stand nicht zufrieden, will gern der oberste Gott sein, fängt an den armen lahmen ♄ zu bestehlen, stiehlt ihm seine Stelzen, dass dieser gar nicht mehr gehen kann, sondern muss auf der Erde liegen bleiben. Er stiehlt dem ♃ seine Krone, sein Scharlachen-Wams mit ☉ bordiert, dem ♂, weil er streitbar ist, verwundert er, dass er Blut gibt, dieses Blut vermischt er mit seinem Schnee und färbt seine weißen Wangen damit, bei die ☽ kommt er mit schmeicheln, stiehlt ihren Schleier und beschläft sie, die ♀ ist seine liebste Göttin, die hat er ganz zu seinem Willen, beschläft sie oft in Gestalt eines Drachens, eines Schwans und Adlers, bis er sie ganz ihrer Zierde beraubt und ihre Schätze gestohlen und sich damit bekleidet; Ja er schont der Sonne, des großen Planeten, nicht, er conjungiert sich mit ihr, stiehlt ihr Feuer, zündet seinen Leib damit an und will, wie die Sonne, auf 4 Rädern fahren, weil er nun so hitzig und trocken wie die Sonne, hört er nicht auf, bring so oft Post und Liebes-Briefe, bis er die ☽ und ☉ in Liebe entzündet, dass sie sich vermählen und den Sonnen-Sohn zeugen, davon er abermals den besten Teil davon trägt und durch seinen Flug durch den obersten Feuer- und Luft-Himmel bekommt er einen Schlüssel zu des Königs verschlossenen Gemach, welches er aufschließen kann.

110. *Was ist denn für ein Schatz verwahrt in des Königs verschlossenem Gemach?*

Antwort: Es sind lauter Juwelen und Kronen darinnen, von lauter Edelgesteinen und aus einen Stück, nicht hier und da zusammen geheftet und von solchem hohen Wert, dass sie kein Künstler bezahlen kann, die güldenen Kronen sind hier als Blei geachtet, eine ist ein kristallenmagischer-Spiegel, dadurch man das Innerste der Erde und alle ihre

109. *What does ☿ unlock with the key it got from Vesta?*

Answer: He unlocks the bodies of the planets, the treasure chests of all the gods, which he polishes. The ☿ is a god of merchants and thieves. Because he is not satisfied with his status he likes to steal. He would like to be the supreme god, and he begins to steal from the poor, lame man ♄, pilfering his stilts so that he can no longer walk at all and has to lie on the ground. He steals from ♃ his crown, his scarlet doublet, bordered with ☉. Because he is choleric, the ♂ is surprised that he gives blood. He mixes this blood with his snow and dyes his white cheeks with it. He comes to ☽ with flattery, steals her veil and sleeps with her. The ♀ is his dearest goddess, he has her just as he wants, and he sleeps with her often in the form of a dragon, a swan, and an eagle, until he robs her entirely of her ornaments, steals her treasures, and clothes himself with them. Indeed, he does not spare even the sun, the great planet, but he conjoins himself with it, steals fire from it, sets his body on fire with it and wants to ride on four wheels like the sun, because he is now as hot and dry as the sun. He will not cease bringing mail and love letters so frequently that he kindles the ☽ and ☉ in love until they marry and produce the son of the sun, from which he again takes away the best part, and through his flight through the highest heavens of fire and air, he receives a key to the king's locked chamber, which he can unlock.

110. *What treasure is kept in the king's locked chamber?*

Answer: There are nothing but jewels and crowns inside, nothing but precious stones of one [whole] piece (not fastened together here and there) and of such high value that no artist can afford them. The golden crowns are treated like lead here. One is a magical, crystal mirror through which one can see the innermost parts of the earth and all

Schätze sehen kann, wer solche Kronen zubereiten kann, ist ein *Magus*, kann allen Dingen ins Herz sehen, der ganzen Natur, ins Paradies. Wer solche besitzt, kann lachen und fröhlich sein bis in seinen Tod. Die andere Krone ist von einer ganzen Perle, wegen ihrer Kostbarkeit nicht zu schätzen, auch nicht zu kaufen, ob es schon der Krösus wäre, auch wird das güldene Vlies hier verwahrt; Die dritte Krone ist von Opal, weil alle Farben darinnen durch die vielen *Uniones*, diese Krone trägt die weiße Königin; Danach findet man Kronen von Jaspis, von Rubin, eine von einem ganzen Karfunkel und noch eine von himmlischem Glanz, davor der Karfunkel dunkel scheint, der allergrößte Schatz ist hier das verwahrte Feuer, welches dieser allergeheimste Schlüssel aufschließt, ob es schon in einer brennenden roten Gestalt hier verwahrt ist, welches leuchtet, so kann es doch seine Wunder nicht zeigen, wenn es nicht resolviert und aufgelöst und sich im ganzen Himmel ausbreitet, dies ist der rechte Schatz-Kasten des Königs, der Schlüssel ist himmlisch, es hat ihn ein Sohn der Götter, dieser schließt das Feuer mit dem himmlischen Wasser auf, dass dieses Sonnen-Feuer sein Licht und Glanz im ganzen Himmel ausbreitet und der ganze Himmel lauter Licht und Glanz wird und weil vorhin durch diesen Schlüssel die verklärte Erde aufgeschlossen und himmlisch worden, so geht dieses Licht und Feuer der himmlischen Sonne wieder in seinen verklärten Himmel und Erde und breitet sich dieses concentrierte Licht in dem ganzen Luft-Himmel aus, erleuchtet die verklärte Erde und Himmel, denn dieser allergeheimste Schlüssel conjungiert das Feuer mit dem neuen Himmel und der neuen Erde, weil das Vorige vergangen und alles neu worden und weil die Sonne sich nun wieder zerteilt und den ganzen Himmel eingenommen, so brennt sie nicht mehr, geht auch nicht mehr unter, wird auch nicht mehr verfinstert, denn dieser Himmel leuchtet selber und ist der Mediator wieder ins *Centrum* getreten, hat den *Polum arcticum* zu dem *Polo ant arctico*, und also in einen gebracht und diesen mit dem astralischen Luft-Himmel vereinigt, mit dem Feuer conjungiert und zwischen den Widerwärtigen Friede gestiftet, dass die alte Widerwärtigkeit beigelegt, alles zur Ruhe und Stille

its treasures. Whoever can prepare such crowns is a *magus* who can see into the heart of everything, into of all nature, and into paradise. Whoever possesses this can laugh and be joyous until their death. The second crown is of a whole pearl, not to be valued because of its preciousness, nor to be bought, even if it were Crœsus who offered to purchase it.[300] The golden fleece is also kept here. The third crown is of opal, because all colors are in it through the many unions. The white queen wears this crown. After this [crown] one finds crowns of jasper, of ruby, one of whole carbuncle, and yet another one of heavenly splendor, before which even the carbuncle appears dim. The greatest treasure here is the guarded fire which is unlocked by this most secret key. Even if it is already preserved in a burning red form which shines but cannot show its wonders unless it is resolved and dissolved and spreads throughout heaven, this is the right treasure chest of the king. The key is heavenly, and it is in the possession of a son of the gods. This one unlocks the fire with the heavenly water, so that this solar fire spreads its light and splendor throughout the whole sky and the whole sky becomes pure light and splendor; and because the faded earth was opened up by this key and became heavenly, so this light and fire of the heavenly sun goes back into its transfigured heaven and earth, and this concentrated light spreads into the whole air-heaven and illuminates the transfigured earth and heaven. This most secret key conjoins the fire with the new heaven and the new earth, because the former has passed away and everything has become new. Because the sun now divides again and occupies the whole sky, it no longer burns, no longer goes down, and is also no longer darkened. For this sky shines in and of itself, and the mediator has stepped back into the center, having united the *polum arcticum* with the *polo antarctico*, thereupon joining them with the astral air-heaven. It then conjoined it with fire and made peace between their hostile natures, so that the old repugnance was settled.

gediehen, die Erde zum Himmel worden und der Himmel zur Erden gebracht und alles mit dem himmlischen Licht erleuchtet, dies ist die Kunst, Feuer im Wasser zu kochen.

111. *Vorhin wurde des Aquæ Stygiæ gedacht, was ist selbiges und wie braucht man es zu dem Drachen und wer ist der Drache?*

Antwort: Der immer-wachende Drache ist der ☿ *Universalis*, welcher Feuer und Rauch aus der Nasen bläst und sind ihr erst 2 einer ohne Flügel, der andere mit Flügeln, welche ohne Schlafen die güldenen Äpfel in hesperischen Garten bewachen. Diese muss Jason auf seiner Reise und Schifffahrt nach der Insel Kolchis, wenn er das güldene Vlies holen will, bestreiten und wenn er seine *Labores* zu Ende gebracht, diesen einzigen Drachen umbringen, welcher den anderen in seinen Leibe verschlungen, so muss er ihn in dem höllischen Wasser ersticken, welches das Feuer wider die Natur genannt wird. Wenn diese zugerichtete Suppe aufgegossen, der Drache darinnen solviert und figiert, so hat Jason gewonnen, kann das güldene Vlies wegnehmen und seinen alten Vater wieder gesund machen. NB. Der feuerspeiende Drache ist kein ☉, die Medea zeigt noch die Fermentation mit der *Pasta* dieses Drachens, nach welcher letzten Arbeit er gesehen, was er vor einen doppelten Schatz besitze, des Reichtums und der Gesundheit, Cadmus bringt den Drachen um und muss derselbe, Jason, sein Gelehrter sein, ohne diesen hätte Jason nichts ausgerichtet, Cadmus ist selbst die Schlange.

112. *Was ist denn Acetum Philosophorum?*

Antwort: Es ist *Lac Virginis*.

113. *Was ist denn des Adlers Gluten-Zubereitung?*

Antwort: Es ist die Machung unsers schleimigen Wassers, wenn solches coaguliert worden zu einem Kristall.

Everything grew to peace and stillness, the earth became heaven and heaven was brought to earth and everything was illuminated with the heavenly light. This is the art of cooking fire in water.

111. *Earlier, the Aquæ Stygiæ was considered. What is this, and how is it needed for the dragon, and who is the dragon?*

Answer: The ever-guarding dragon is the ☿ *universalis*, which blows fire and smoke from its nose. There are only two of them, one without wings, and the other with wings; and these two guard the golden apples in the Hesperidian garden without sleeping. Jason must fight these on his voyage to the island of Colchos if he wants to fetch the Golden Fleece; and when he has finished his *labores*, he must kill this singular dragon which has swallowed the other one into its body. He must choke it in the infernal waters called the fire against nature. When this prepared broth is poured out, the dragon within it is dissolved and fixed. Then Jason has won and can take away the Golden Fleece and heal his old father again. NB. The fire-breathing dragon is not ☉, Medea still shows the fermentation with the *pasta* of this dragon, after this last work it saw what sort of a double treasure he possessed, a treasure of both wealth and health. Cadmus kills the dragon and the same Jason must be its scholar; without it Jason would have done nothing. Cadmus itself is the serpent.

112. *What is acetum philosophorum?*

Answer: It is *lac virginis*.

113. *What is the eagle's gluten preparation?*

Answer: It is the making of our slimy water when such has been coagulated into a crystal.

114. *Was ist denn Antimonium oder Plumbum Philosophorum?*

Antwort: Es ist der rote ♌ und das *Lili* der Alten, das ☿ ist zweierlei, eins das gemeine Schwarze, das andere ist das Weibchen, das weiße, heißt *Marcasitta plumbea*.

115. *Wie haben sie die Arbeit beschrieben, so damit soll vorgenommen werden?*

Antwort: Sie haben es beschrieben, man soll es in destilliertem Essig solvieren, welches aber kein Essig ist, den die Weiber hinter dem Ofen machen, heißt philosophischer Essig.

116. *Wie macht man denn des ♌ Blut?*

Antwort: Man muss seinen eigenen philosophischen Essig auf das rot calcinierte ☿ gießen, so färbt er sich hoch rot, dies destilliert man gelinde herüber, so kommt das gebenedeite Öl dieser *Minera*, in blutroten Tröpflein als rechtes Blut, denn der ganzen Welt-Schatz nichts zu vergleichen.

117. *Wie solviert man denn Silber nach philosophischer Weise?*

Antwort: Es wird in 6 Teilen unseres Calcinier-Wassers, der Jungfer-Milch solviert und gelinde abdestilliert, bis es als Öl erscheint, das grüne Wasser, hernach wieder solviert und abdestilliert, in dreimal ist die ☽ zu Öl, darauf wird unser *Spiritus Vini solventis* größer, so nimmt er den ☿ der ☽ in sich, welchen man scheiden und wieder vereinigen kann.

118. *Wie solviert man denn ☉?*

Antwort: Man solviert es in der roten Jungfer-Milch und cohobiert es, bis es in Öl gebracht oder scheidet auch durch *Vinum solvens* den ☿ davon und conjungiert sie wieder oder man solviert sie beide durch unser Lebens-*Aquavit*, so in Geist, Seele und Leib besteht.

114. *What is antimonium or plumbum philosophorum?*

Answer: It is the red ♌ and the lily of the old, the ☿ is of two kinds, one is the common black, the other is the female, the white, called *marcasita plumbea*.

115. *How did they describe the work to be done with it?*

Answer: They have described it as dissolving it in distilled vinegar, which, however, is not the vinegar that women make behind the stove, but it is called philosophical vinegar.

116. *How do you make the ♌ blood?*

Answer: You have to pour your own philosophical vinegar on the red calcined ☿, so it turns bright red. You distill this gently over, then the blessed oil of this *minera* comes in blood-red droplets like real blood, for there is nothing to compare with the treasure of whole world.

117. *How does one dissolve silver in a philosophical way?*

Answer: It is dissolved in six parts of our calcinating water of the virgin milk and gently distilled until it appears like an oil, the green water, which is hereafter dissolved and distilled off again; after three times, the ☽ becomes an oil, then our *spiritus vini solventis* becomes larger, so it takes the ♃ of the ☽ in itself, which one can separate and reunite.

118. *How does one dissolve the ☉?*

Answer: One dissolves it in the red virgin milk and cohobates it, until it is turned into oil or separates the ♃ from it through *vinum solvens* and conjoins them again, or else one dissolves them both through our *aqua vitæ* of life, which exists in spirit, soul, and body.

119. *Was ist denn das Aqua ardens oder Aqua Vitæ?*

Antwort: Es ist kein Branntwein noch *Spiritus Vini*, sondern unser Lebens-Wasser, das lebendig macht, welches der *Spiritus* ☿ und seinen Anfang von ♄ hat.

120. *Was ist denn das philosophische Feuer, ist es einerlei oder wird es in vielen Verstande genommen?*

Antwort: Erstlich ist es das glühende weiße Feuer, welches auch *Balneum Mariæ* genannt wird, auch unser Roß-Mist und geheimer Ofen, welcher seinem eigenen Gefäß zugefügt werden muss, welches Gefäß *Vas Viride* genannt wird, in welchem Feuer Pontanus über 200 mal geirrt, ob er schon in der rechten *Materia* gearbeitet. 2. Haben wir das Feuer wider die Natur, so heftiger brennt als das höllische Feuer, das *Aqua Stygia*. 3. Haben wir ein Feuer, welches ein gelbes 🜍 Öl. 4. Haben wir ein Feuer, unser Sonnen-Feuer, das glühende rote.

121. *Was ist der 🜍 der Natur?*

Antwort: Es ist unser weißer 🜍.

122. *Was macht man damit?*

Antwort: Er muss in Öl solviert werden, dass er das unverbrennliche Öl wird.

123. *Was tut man mehr damit, macht man ihn alles zu Öl oder nur einen Teil?*

Antwort: Man macht einen Teil fix, den anderen solviert man in Öl.

124. *Womit macht man ihn fix?*

Antwort: Mit seinem eigenen 🜍 oder Feuer, dass die Hände nicht verbrennt.

125. *Was tingiert der Stein?*

119. *What is the aqua ardens or aqua vitæ?*[301]

Answer: It is not brandy nor *spiritus vini*, but our life-water, which vivifies, which is *spiritus* ☿, and has its beginning from ♄.

120. *What is the philosophical fire, is it one thing or can it be understood in different ways?*

Answer: First it is the glowing white fire, which is also called *Balneum Mariæ*. It is also our horse-dung and secret furnace, which must be added to its own vessel. This vessel is called the *vas viride*, and in this fire Pontanus erred over two-hundred times, although he already worked in the right *materia*. Second, we have the fire against nature that burns more fiercely than the hellish fire, the *aqua stygia*. Third, we have a fire which is a yellow ♃ oil. Fourth, we have a fire, the fire of our sun, the glowing red [fire].

121. *What is the ♃ of nature?*

Answer: It is our white ♃.

122. *What does one do with it?*

Answer: It must be dissolved in oil so that it becomes the incombustible oil.

123. *What else does one do with it, does one turn it all into oil or just a part of it?*

Answer: One fixes one part while the other one is dissolved in oil.

124. *How does one fix it?*

Answer: With its own ♃, or the fire that does not burn the hands.

125. *What does the stone tinge?*

Antwort: Er vermag die ganze Welt zu tingieren.

126. *Warum soll und kann man denn nicht das große Werk aus dem ☉ und Silber machen, da doch die meisten ihren Stein aus ☉ und Silber gemacht?*

Antwort: Darum, dass es nicht möglich ist, aus den reifen Früchten wieder Blüten zu machen oder aus einem Mann ein Kind, man kann ihn nicht wieder in seiner Mutter Leib bringen, dass er aufs neue geboren wird; Aber der Kunst ist zugelassen, das rohe, unreife zum Wachstum zu befördern, solches zu befördern, solches zu veredeln und auf den höchsten Staffel zu führen, was die Natur nicht kann; dass aber viele ihre Steine von gemeinem ☉ und ☽ gemacht, weil sie das große Werk als die philosophische ☉ und ☽ nicht finden konnten, so mussten sie die *Menstrua*, färbende nasse und trockene Geister zu dem gemeinen ☉ und Silber brauchen, solche darinnen färben und zu einer Tinctur bringen, welche viel oder wenig tingierte, danach sie oft darinnen aufgelöst und in Öl gebracht worden.

127. *Kann man auch aus den Mineralien tingierende Steine machen?*

Antwort: Ja, sonderlich aus dem ☿ nach dem nassen und trockenen Weg, wenn die Jungfer-Milch ihn solviert oder unser trockener ☿ der 7 Nächte damit vereinigt wird, so solvieren sie beide einander, dass sie eins werden und aufsteigen, wie die schöne Morgenröte. Wenn man nun dieses ferner reinigt und scheidet und den raubenden flüchtigen ☿ die Flügel bindet, dass er wieder zurück in sein *Centrum* geht, so wird man erfahren, dass ihm seine diebische und räuberische Art genommen und er der ☽ ein güldenes Stück anziehen kann.

128. *Wie kann aber dieses Wasser solche großen Dinge tun, dass es auch die rohen und unreifen Mineralien kann veredeln und zu herrlichen Tincturen verwandeln?*

Antwort: Das einfache simple Wasser kann es nicht, sondern unser erhöhtes vollkommenes Wasser, wenn es himmlisch worden und sei-

Answer: It is able to tinge the entire world.

126. *Why shouldn't one and why can't one make the great work out of ☉ and silver, since most of them have made their stones out of ☉ and silver?*

Answer: Because it is not possible to make blossoms again out of ripe fruits, or make a child out of a man; he cannot be brought back into his mother's womb to be born again. But for the sake of the art it is permitted to promote the growth of the raw and immature, and to promote such things or to ennoble such things and to lead them to the highest level, which nature cannot do. Yet many made their stones from the common ☉ and ☽ because they could not find the great work, i.e., the philosophical ☉ and ☽, so they had to use *menstrua*, coloring wet and dry spirits for the common ☉ and silver, coloring such in it, and bringing them into a tincture, which tinges much or little. After this they were often dissolved in it, and made into an oil.

127. *Can one also make tingeing stones from the minerals?*

Answer: Yes, especially from the ☿ after the wet and dry way; when the virgin milk dissolves it, or our dry ☿, which is united with it for seven nights, they both dissolve each other so that they become one and rise like the beautiful dawn. If one now further purifies and separates this and ties the wings of the predatory fleeing ☿ so that it goes back to its *centrum*, one will experience what its thieving and rapacious nature has taken from it, and it can clothe the ☽ with a golden piece.

128. *But how can this water do such great things that it can ennoble even the raw and unripe minerals and transform them into glorious tinctures?*

Answer: The plain simple water cannot do it, but our exalted perfect water when it has become heavenly, and its coloring poisons, which

ne färbende Gifte, diese haben solche giftige Tugend, müssen sie sich doch selbst veredeln und neu gebären, wenn das große Werk aus ihnen gemacht wird. Denn da nehmen wir nichts fremdes dazu als die Elemente, fügen die widerwärtigen Dinge zusammen und erwarten was uns selbige für Früchte geben, denn aus den Elementen besteht alles, sowohl in der großen als in der kleinen Welt. Wachsen nicht aus Erde und Wasser alle Metalle, alle *Mineralia*, alle Edelgesteine, auch wachsen aus dem Tau Perlen, so viel haben wir auch durch die Kunst. Denn unser Wasser wächst so lange, bis es der ☿ialische Baum wird, welcher die Früchte der Sonne und des Mondes trägt. Die erste Frucht, die Lilie, dem weißen ♃ bringen wir in sein Öl, als in sein Ferment, einen Teil figieren wir, das der ☿, der weiße ♃, fix wird, ist die wahre Erde der Sonne und des Mondes, lösen ihn in seinem Ferment auf, bis er gelb wird und röten ihn, den roten ♃ lösen wir auf in rotem Öl, den Weißen in weißem Öl, conjungieren sie und färben sie hernach aufs neue auf so vielerlei Art wir wollen, treiben einen scharfen Pfeil nach dem anderen darein, erhöhen die Geister, wie es uns gut dünkt, sonderlich in der Multiplication und Illumination, da denn nicht nur ein Weg sondern viele sind. Wir malen auf Schwarz Weiß, auf Weiß Gelb, auf Gelb Rot und gar Purpur-Farbe. Der rote Purpur-Mantel deckt die anderen Farben alle zu, denn wenn die Sonne rot und hitzig scheint, sieht man vor der Sonnen-Glanz die anderen Sterne nicht mehr am Himmel: Also haben sie hier auch die anderen Planeten nebst dem Mond unter der Sonnen Glanz versteckt, dass man sie nicht mehr sieht.

129. *Was ist doch der Demogorgon?*

Antwort: Der Demogorgon ist unser Alter, welcher die ganze Natur in sich hält, er ist auch unser Drache, hat ein centralisches Feuer und Dampf in sich, nämlich den *Spiritum Mundi* oder ☿, denselben gibt er aus seinem Herzen heraus, *item*, eine reine Erde, welche künstlich zubereitet werden muss.

130. *Was ist denn das Feuer wider die Natur?*

have such poisonous virtue, these must ennoble themselves and give birth again when the great work is made of them. For then we take nothing extraneous to it but only the *elementa*. We put together the recurring things and expect the fruits they will give us, because everything consists of the elements, both in the macrocosm and in the microcosm. Do not all metals, all *mineralia*, all precious stones grow from earth and water, and do not also pearls grow from dew, since we also have so much through art? Yes, for our water grows until it becomes the ☿al tree, which bears the fruits of the sun and the moon. The first fruit, the lily, the white ♄, we bring into its oil, i.e., into its ferment. We fix one part, so that the ☿, the white ♄, becomes fixed. It is the true earth of the sun and the moon, and we dissolve it in its ferment until it turns yellow. Then we redden it, and the red ♄ we dissolve in the red oil, and the white [♄] in white oil, joining them and then coloring them again in as many ways as we like, shooting one sharp arrow after the other into it, elevating the spirits as we see fit, especially in multiplication and illumination, since there is not just one way but many. We paint white on black, yellow on white, red and even purple on yellow. The purple cloak covers all the other colors, because when the sun shines red and hot, you can no longer see the other stars in the sky in front of the splendor of the sun. So here they have also hidden the other planets along with the moon under the splendor of the sun, so that you can no longer see them.

129. *What is the Demogorgon?*

Answer: The Demogorgon is our old man, who holds the whole of nature within. It is also our dragon. It has a central fire and vapor within it, i.e., the *spiritum mundi* or ☿, which it issues from its heart. Likewise, it has a pure earth which must be prepared artificially.

130. *What then is fire against nature?*

Antwort: Ich habe es schon gesagt, es ist das corrosivische stinkende Wasser, das *Aqua Stygia*, darinnen der Drache getötet, der ☿ fix gemacht, des fliegenden ♄ seinen Sense, damit er dem ☿ die Flügel abhaut, das höllische Feuer so calciniert und in einen fixen Leib verwandelt, den ☿ *Philosophorum* in einen fixen Præcipitat.

131. *Warum soll aber dieses Feuer den ☿ figieren, da doch vorher gemeldet worden, dass er soll in der Erde gehärtet werden und darinnen coaguliert?*

Antwort: Vorhin musste der schlackenreiche und wässrige ☿ durch das eiserne Salz seine eigene Erde oder Salz gehärtet werden, denn eine jede Erde hat Macht ihren eigenen ☿ zu härten und wenn er solche dürre Erde oder Salz in sich fast, ist seine Feuchtigkeit vertrocknet, dass er nicht mehr laufend ist, sondern erstarrt, erhärtet und coaguliert, denn er muss sich mit seinem eigenen fixen Körper vereinigen, darum heißt er ein ☿, der Körper, *item*, ☿ *duplicatus*, wenn er solchen Körper flüchtig gemacht hat, denn ohne das Salz kann der ☿ *Philosophorum* nicht gemacht werden, viel weniger der Stein. Wenn er nun unser ☿ *sublimatus* ist, machen wir 2 Teile davon, einen zum Öl, durch seinen eigenen *Spiritum Vini*, zu dem lebendigen Wasser in Seele, Geist und Leib, den anderen Teil præcipitieren wir durch seinen eigenen 🜔, welches der tödlich stinkende Becher ist und lassen ihn in dem höllischen Pfuhl ersticken, dadurch er mortificiert und zum herrlichen Leben durch den Tod aufersteht, dies ist der höllische Schlangen-Stich und der Cherub mit dem feurigen Schwert vor das Paradies gelagert, zu bewahren den Weg zu dem Baum des Lebens; durch dieses Feuer werden die Körper gereinigt und geläutert, wenn sie in die Zahl der Götter aufgenommen werden, dass sie hernach unsterblich sein.

132. *Ist denn dieses Sal Martis des Cosmopolites, seine Minera derselben Insel des Goldes und Stahls?*

Antwort: Ja, denn dieses Salz ist martialischer und solarischer Natur, so heißt es der Cosmopolit seinen *Mineren*, derselben Insel des Goldes und Stahls, welche ihm *Neptunus* gezeigt, weil diese *Mineren* die *Insul*

Answer: I have already said it, it is the corrosive, stinking water, the *aqua stygia*, in which the dragon is killed, [in which] the ☿ is fixed, the scythe of the flying ♄, with which it cuts off the wings of ☿, the infernal fire which calcinates and transforms into a fixed body, the ☿ *philosophorum* in a fixed precipitate.

131. *But why should this fire fix the ☿, since it was previously reported that it should be hardened in the earth and coagulated therein?*

Answer: Formerly the slag-like and watery ☿ had to be hardened by the iron salt, which is its own earth or salt, for each earth has power to harden its own ☿; and when it has almost taken up this dry earth or salt, its moisture has dried, so that it is no longer flowing, but solidifies, hardens, and coagulates. It does so because it has to unite with its own fixed body, and that is why it is called a ☿ body. Likewise, it is called ☿ *duplicatus* when it has made such a body volatile, because without the salt the ☿ *philosophorum* cannot be made, much less the stone. If it is now our ☿ *sublimatus*, we make two parts from it: one for the oil through its own *spiritum vini*, for the living water in soul, spirit, and body; and the other part we precipitate through its own ♆, which is the deadly stinking chalice, and we let it suffocate in the mud hole of hell, whereby it is mortified and rises to glorious life through death. This is the serpent's sting from hell and the cherub with the fiery sword placed before paradise to guard the way to the tree of life. By this fire the bodies are purified and cleaned as they are received among the ranks of the gods, so that afterwards they are immortal.

132. *Is this sal martis of the Cosmopolitan,[302] his minera, the same island of gold and steel?*

Answer: Yes, because this salt is martial and solar in nature, and so the Cosmopolitan calls it his *minera*, the same island of gold and steel that *Neptunus* showed him, because these *minera*s are the islands

selber sind, welche im philosophischen Meer schwimmen und eine *Insul* mitten auf diesem Meer gesehen wird, welche täglich abnimmt, denn die *Minera* zerschmelzt darinnen und obwohl, schreibt er, alle Dinge nach Wunsch allda zu haben, so hat doch das Wasser gemangelt, hat auch niemand über 10 Teile schöpfen können und als er die Weiße des Wassers betrachtet, hat er sich sehr verwundert und ob sich schon viele unterstanden von allen Orten Wasser dahin zu leiten, so hat er doch nichts gedacht, sondern hierzu unrein und vergiftet, dies aber wird aus den Strahlen der Sonne und des Mondes geschöpft, welches der *Neptunus* bei hellem Mond-Schein schöpft.

133. *Was meinen die Adepti, wenn sie sagen, man soll das Werk in der 10 Zahl beschließen, wie versteht man solches?*

Antwort: Es ist also zu verstehen, aus einem einzigen, unserem *Subjecto*, werden 2 gemacht, der Himmel und die Erde, aus diesen die 4 *Elementa*, und die 3 *Principia*, diese alle conjungiere, setze sie zusammen, beschließe das Werk in der 10 Zahl, so ist es vollkommen: 1. ein Ding, 2. Himmel und Erde, 3. die 3 *Principia*, 4. Die 4 *Elementa*, diese Zahlen zusammen: 1 + 2 + 3 + 4 sind 10.

134. *Ich möchte doch eine Repetition hören wegen etlicher geheimer Schlüssel in Machung und Verfertigung des Steins der Weisen und zwar vom Allerersten bis auf den Letzten?*

Antwort: Der erste Schlüssel, wenn unser roter *Laton* solviert wird, wenn sich die Farben als im Regenbogen im philosophischem Essig geben, wenn die rote Nuss soll durchscheinend gemacht werden, die Zubereitung und Geschicktmachung des *Confusum Chaos* durch seinen eigenen Geist, der auf dem Wasser schwebt, wenn es finster ist und solches teilt und scheidet, dass es kann in die *Elementa* geschieden werden, dieses heißt die aufgesperrte Pforte der 4 Elemente, durch welche man gehen muss, nach der Pforte der himmlischen Conjunction. Der erste Schlüssel solcher himmlischen Conjunction ist himmlisch, er schließt die Erde auf, vereinigt den Himmel mit der Erde, macht

themselves, swimming in the philosophical sea. And an island in the midst of the sea is seen, an island which shrinks daily due to the *minera* which melt in it; and although he writes that all things desired can be obtained there, the water was still lacking, and nobody was able to draw more than ten parts. And when he looked at the whiteness of the water, he was very astonished; and although many dared to direct water there from all places, he thought nothing of it, but that it was impure and poisoned. Yet this is drawn from the rays of the sun and the moon, which is drawn by *Neptunus* in bright moonlight.

133. *What do the adepti mean when they say that the work should end with the number ten, how do you understand this?*

Answer: From this it is to be understood that from one single thing, i.e., our *subjecto*, two [things] are made: the heaven and the earth. From these come the four *elementa* and the three *principia*, and when we put them all together, we complete the work with the number ten, and this is how it is it perfect: (1) One thing; (2) Heaven and earth; (3) The three *principia*; (4) The four *elementa*. Thus 1 + 2 + 3 + 4 = 10.

134. *I would like to hear a repetition, after all, about a number of secret keys in the making and manufacture of the philosophers' stone, from the very first to the last?*

Answer: The first key—when our red loam is dissolved, when the colors in the philosophical vinegar appear as in the rainbow, when the red nut is to be made translucent, the preparation and making of the *confusum chaos* by its own spirit floating upon the water, when it is dark and divides and separates such that it can be separated into the *elementa*—this means that the gate of the four elements is open, through which one must proceed to the gate of the heavenly conjunction. The first key of this heavenly conjunction is heavenly, it unlocks the earth and unites heaven with earth, rendering the latter heavenly

solchen himmlisch, dass sie die Kräfte der Oberen und der Unteren bekommt und himmlische paradiesische Früchte gebiert, vereinigt die Erde mit dem Himmel, dass solche auch in Himmel fliegt, dieser Schlüssel hat viel andere geheime Schlüssel, als da ist, *Solutio, Putrefactio, Coagulatio, Resolutio, Generatio, Sublimatio, Distillatio, Separatio, Fixatio* und alle diese Schlüssel schließen doch nur ein Gemach, welches mit 7 Siegeln versiegelt und heißt die Natur und die 7 Grade der Natur, wenn man da hinein kommt, ist es mit lauter weißen Silberstükken aufgeputzt, die Natur sieht man auf einem erhabenen Thron in himmlischer Gestalt sitzen, welcher Thron von allen Göttern erbaut, den ersten Grund-Stein hat gelegt ♄, danach alle Götter, *Sol* aber den Stuhl oder die letzte Stufe darauf die Natur in Gestalt einer Königin sitzt mit ausgebreiteten silbernen Haaren, welche schimmern und leuchten, ihr Kleid ist silbernes Stück, welches gelb schimmert und leuchtet, ihr ganzer Leib ist lauter Sternen-Glanz, ihre Brüste, eine ist die Sonne, die andere der Mond, auf ihrem Haupt hat sie eine Krone von Opal, in ihrer Hand ein Büchlein vom *Electro minerali in maturo arctificiale*, um ihren Hals eine Schnur von den allerteuersten Perlen, ihre Wangen Milch mit Blut gemengt, ihre Zähne Elfenbein, ihr Kleid mit lauter weißen Lilien bestreut und obwohl ihr Leib mit einem güldenen Stück bedeckt, ist es doch inwendig der Sonnen Glanz und ihre ganze Kraft und Wesen, sie reicht ihr Buch, wer darinnen lesen kann, der versteht die ganze Natur, besitzt die Schätze der Natur und herrscht über die Natur, denn er kann die Natur noch vollkommener machen, da sie gleichsam in vielen Stücken als wie geschwächt und ohnmächtig, auch kraftlos worden, kann er ihr zu Hilfe kommen und sie erhöhen, weil er der Natur lieber Sohn und die geheimen Schlüssel hat, kann er sie aus diesem Hause in ein anderes führen, darinnen sie immer mehr und mehr verherrlicht wird, dass sie den Zirkel der Sonne und des Mondes durchlaufen kann.

so that it receives the powers of those above and below, whereupon the earth bestows heavenly paradisiacal fruits. It unites earth with heaven such that [earth] also flies into heaven. This key has many other secret keys such as *solutio, putrefactio, coagulatio, resolutio, generatio, sublimatio, distillatio, separatio,* and *fixatio*; and all these keys only unlock one chamber, which is sealed with seven seals and is called nature and the seven grades of nature. When one gets inside, it is decorated with nothing but pieces of white silver, and you see nature on an exalted throne, sitting in heavenly form. This throne was built by all the gods, and the first foundation stone was laid by ♄. After this stone, all the gods contributed, but *Sol* built the seat or the last step upon which nature sits in the form of a queen, with her silver hair extended, shimmering and shining. Her dress, which shimmers and shines yellow, is of silver; and her whole body is the pure brilliance of stars. One of her breasts is the sun and the other is the moon, while on her head she has a crown of opal and in her hand a small book on the *Electro minerali in maturo artificiale*. Around her neck is a string of the most costly pearls. Her cheeks are the color of milk mingled with blood. Her teeth are ivory, and her dress is strewn with nothing but white lilies; and although her body is covered with a piece of gold, internally it is the sun's splendor and all its power and being. She presents her book, and whoever can read it understands the whole of nature, possesses the treasures of nature, and rules over nature, because they can make nature even more perfect, since in its ordinary state it is in many ways weakened and powerless. Yet whoever can come to her aid and elevate her, because they are nature's dearest son and have the secret key, can get her out of this house and lead her to another in which she is more and more glorified, so that she can traverse the circle of the sun and the moon.

135. *Wie heißen denn die Häuser darinnen die Natur soll verherrlicht werden oder wie heißt dieser Ort, darinnen die Natur mit ihrem Büchlein gesehen wird?*

Antwort: Der Ort, wo die Natur mit ihrem offenen Büchlein gesehen wird oder wo sie ihren Thron und Sitz hat, ist das Resident-Haus der Sonne, in diesen Häusern wird sie aufs neue verkleidet und ihr ganzes Wesen veredelt, denn jede *Sphæra*, da sie durchgeführt wird, dessen Himmel und Influenz unterwirft sie sich, weil sie nicht störrig noch widerspenstig, lässt sich leiten und führen und gehorcht ihrem Diener so auf sie wartet, mit ganzem Fleiß, lässt sich von ihm bloß sehen und sich entkleiden, weil ihrer Keuschheit dadurch kein Abbruch geschieht, sondern bleibt allezeit die reine und unbefleckte Natur.

136. *Wer hat denn den Schlüssel zu diesem Residenz-Haus der ☽?*

Antwort: Solcher Schlüssel ist das renovierende, vermehrende und multiplicierende *Menstruum*, dadurch die weiße Tinctur *in infinitum* vermehrt wird, welches in Seele, Geist und Leib besteht, dieser himmlische Schlüssel schließt den *Circul* der *Lunæ* auf, durch welches Haus sie in der Sonne Haus gelangt, wenn ihr der solarische Schlüssel gereicht wird, da sie dann von der Sonnen Hitze im Hause des ♌ ganz rot gebrannt wird, doch muss die Natur ihr Büchlein von sich geben und solches zerreißen lassen, eher kann sie nicht ins Haus des Mondes und ins Haus der Sonne gelangen.

137. *Wer zerreißt denn ihr Buch des Electrum artificiale?*

Antwort: Es zerreißt solches der giftige Drache und höllische Cerberus, welcher solches frisst, davon der Drache stirbt und bersten muss und wird dieser Fresser selber in eine andere Gestalt verwandelt und beide nicht mehr zu sehen, sondern in das güldene Vlies verwandelt, steht nun da als eine weiße Königin, welche künftig selber über den Mond herrschen wird, wenn sie diesen Zirkel durchlaufen.

135. *What are the names of the houses in which nature is to be glorified, or what is the name of this place in which nature is seen with her little book?*

Answer: The place where nature is seen with her open booklet, or where she has her throne and seat, is the house of the sun's residence. In these houses she is disguised anew and her whole being is ennobled, for every sphere she is led through sees her submitting to its heavenly influence, for she is neither stubborn nor rebellious, but allows herself to be guided and led, and obeys her servant, allowing herself to be undressed by him and seen naked, because this does not affect her chastity, which remains pure and undefiled at all times.

136. *Who has the key to this house of the ☽'s residence?*

Answer: This key is the renovating, augmenting, and multiplying *menstruum*, whereby the white tincture is augmented *ad infinitum*, which consists of soul, spirit, and body. This heavenly key unlocks the *circul* of the *Lunæ*, through which house she enters the house of the sun, when the solar key is handed to her, since she is then burned red from the sun's heat in the house of the ♌. But nature must give up her little book and have it torn up before she can enter the house of the moon and get into the house of the sun.

137. *Who tears up her book of the electrum artificiale?*

Answer: It is torn to pieces by the poisonous dragon and the hellish Cerberus, who devour the book. From this the dragon must die and burst, and the devourer itself is transformed into another form and both are no longer to be seen, but are transformed into the golden fleece, now standing there as a white queen, who will henceforth rule over the moon itself when she goes through this circle.

138. *Wie heißt denn der Schlüssel, so das Residenz-Haus der Sonne aufschließen muss und wer hat ihn?*

Antwort: Es hat ihn ein roter feuriger Mann, welchen er einem roten ♌ vom Halse genommen, heißt ☿ *Solaris*, auch Jungfer-Milch der Sonne.

139. *Wer kann so viele Schlüssel fordern oder alle nennen oder kennen, ist es doch nicht möglich?*

Antwort: Wer diese Schlüssel nicht alle weiß und kennt, der wird die Geheimnisse der Natur nicht sehen, auch kein einziges Gemach aufschließen, noch die nackende Diana im Bade sehen, viel weniger küssen, man hat sich die Schlüssel eben so schwer nicht einzubilden, man sehe nur, dass man mit dem alten ♄, dem Türhüter wohl daran ist, der wird einem schon einen gewissen Schlüssel reichen, wenn der erste Schlüssel nur wohl angebracht ist und das erste Schloss wohl eröffnet, so findet sich eine Schatz-Truhe, in welcher viele magische Schlüssel liegen.

140. *Wie macht man den fixen ☿ Philosophorum, oder diesen Præcipitat oder Stein zu Öl?*

Antwort: Man solviert ihn in dem lebendigen *Aquavit*, welches in Seele, Geist und Leib besteht, wenn dieses 3-mal geschieht, ist die weiße Tinctur in Öl gebracht oder der weiße ♃.

141. *Wie aber mit dem Roten?*

Antwort: Diesen solviert man in seinem roten Öl, wiederholt es bis der rote ♃ als Öl erscheint.

142. *Wie macht man denn die Composition des Steins?*

Antwort: Man calciniert den flüchtigen ☿ durch *Ignis fortissimus*, zu einem fixen Glase, welches etwas grünlich aussieht, weil er mit der ♀ Eigenschaft oder deren Geblüt præcipitiert, dieses ist das *Aurum horicontale*, wird durch sein erhöhendes und ernährendes Lebens-Wasser

138. *What is the name of the key that unlocks the house of the sun's residence and who has it?*

Answer: A red fiery man has it, who took it from the neck of a red ♌. It is called ☿ *Solaris*, also the virgin milk of the sun.

139. *Who can ask for so many keys or name them all, or know them all; is this really possible?*

Answer: Anyone who does not recognize and know all of these keys will not see the secrets of nature, will not unlock a single chamber, nor see the naked Diana in the bath, much less kiss her. It is not so difficult to imagine the keys, just see that one is well disposed with the old ♄, the door keeper, who will give you a special key; if the first key is properly fitted and the first lock is properly opened, then there is a treasure chest in which many magical keys lie.

140. *How does one turn the fixed ☿ philosophorum, or this precipitate or stone, into oil?*

Answer: You dissolve it in the *aqua vitæ* of life, which consists of soul, spirit, and body. If this is done three times, the white tincture is brought into oil or the white ♃.

141. *But how about the red one?*

Answer: You dissolve this one in its red oil, repeat it until the red ♃ appears like oil.

142. *How do you prepare the composition of the stone?*

Answer: The volatile ☿ is calcined by the *ignis fortissimus* into a fixed glass, which looks somewhat greenish, because it precipitates with the ♀ property or its blood. This is the *aurum horicontale*, and it is brought

hierher gebracht, bis das Glas güssig und flüssig wird, kann auch gar in Öl-Gestalt gebracht werden.

143. *Dies ist aber nur der weiße Stein, wie wird der rote gemacht?*

Antwort: Wenn dieser weiße fixe ♃, welcher in Gestalt eines hellen sehr schweren Glases, welches etwas grünlich und helle Brüche hat, in dem weißen Ferment aufgelöst, so wird dieser weißen Königin der rote König im Purpur-Gewand zugeführt, dass sie mit ihm Beilager hält und den Sonnen-Sohn von ihn empfängt.

144. *Ist denn nun der rote Stein hier fertig?*

Antwort: Ja er ist fertig, und ist ein kurzer Weg, aber dass es der hochrote blutige Stein sein sollte, das ist es nicht, es ist die rote Tinctur, der Stein der ersten Ordnung auf rot, und kann solcher viel höher und wieder zur Multiplication gebracht werden.

145. *Aber auf welche Weise wird er höher gebracht?*

Antwort: Durch seine fruchtbringende Gemahlin, welcher er oft beiliegen muss, denn diese Tinctur muss mehrmals in dem roten Öl aufgelöst werden, oder damit eingetränkt, welches seine Fermentation ist.

146. *Sind aber nicht noch mehr andere Wege und Abkürzungen?*

Antwort: Deren sind sehr viel, wer will sie alle beschreiben? Will ihrer aber 2 setzen, einen Weg zum Weißen, den anderen zum Roten. 1. Auf weiß, so solviert man den weißen fixen ♃ in seinem weißen Öl, als in seinem Ferment, bis er auch in Öl gebracht ist, tut ☿ *Currentem*, gemeinen ☿, wie man ihn im Kram kauft, in ein Gefäß und dieses Öl darauf, hält es im Feuer bis es fix ist, dieser Præcipitat wird einen sehr großen Teil anderes Erz, es sei Blei oder ♀ in Silber verwandeln und also solviert man auch 2. den rot figirten ♃ durch sein rotes Ferment, bis er in Öl-Gestalt erscheint und tut dieses Öl auf ☿, welcher vorher ins Gefäß muss getan werden, hält es gleichfalls im Feuer, bis er fix ist, so

here by its elevating and nourishing water of life until the glass becomes viscous and liquid. It can even be brought into the form of oil.

143. *But this is only the white stone, how is the red one prepared?*

Answer: When this fixed, white ♃, which has the form of a bright, very heavy glass, which is somewhat greenish and has light-colored cracks, is dissolved in the white ferment, then the red king in purple robes is brought to this white queen, so that she lies down with him in matrimony and receives from him the son of the sun.

144. *Is the red stone finished here?*

Answer: Yes, it is finished, and it is a short way, but it should be the crimson, bloody stone, which it is not; it is the red tincture, the stone of the first order in red, and this can be brought much higher and multiplied again.

145. *But how is it taken higher?*

Answer: Through its fruit-bearing wife, with which it must often sleep, for this tincture must be dissolved or soaked several times in the red oil, which is its fermentation.

146. *But are there not still other ways and shortcuts?*

Answer: There are very many of them, but who can describe them all? Yet I want to name two [ways]: one way to the white [stone], the other to the red. (1) For the white, you dissolve the white, fixed ♃ in its white oil, i.e., in its ferment, until it is also brought to an oil. You then put ☿ *currentem*, common ☿, as one buys it in the general store, in a vessel and pour this oil on it. Then keep it in the fire until it is solid. This precipitate will transmute a very large part of other ores, be it lead or ♀, into silver and so one also dissolves: (2) the red, fixed ♃ through its red ferment until it appears in the form of an oil. Pour this oil onto ☿, which must be placed into the vessel beforehand, and also keep it in the fire until it is fixed.

wird dieser Præcipitat einen sehr großen Teil anderer Metalle in ☉ verwandeln.

147. *Sind dies die besten Wege der Abkürzung oder kann man noch bessere haben?*

Antwort: Es sind noch bessere Wege, welche ich auch 2 setzen will und sind doch noch nicht die besten. Wenn der fixe ♃ in sein Öl gebracht, dass er alle zu Öl und einer dünnen Salben worden, so nimmt man einen reinen Silber-Kalk, nimmt man aber hierzu ☉ Kalk, so bekommt man in der Projection weißes ☉, welches alle ☉ Proben aushält, diesen Kalk solviert man mit diesen Öl bis es alles ein fixer Stein zusammen worden, so hat man bald eine Tinctur auf den ☿ zu tragen.

148. *Wie wird es aber auf rot gemacht?*

Antwort: Man bringt den roten fixen ♃ durch sein rotes Öl auch zu Öl, vermengt einen ☉ Kalk damit, bis es ein roter fixer Stein worden, so hat man eine Tinctur, welche alles in ☉ verwandelt.

149. *Weil aber noch bessere Wege sind, so möchte solche auch gern wissen?*

Antwort: Wer will sie alle beschreiben? Die besten Wege sind, wenn man das gemeine ☉ und Silber durch die beiden *Fermenta*, als das rote und weiße, solviert, hernach die weiße und rote Tinctur gleichfalls in den roten und weißen Ferment, solche zusammensetzt, conjungiert und hernach figiert.

150. *Die Philosophi schreiben, als Adfar, die Materia wäre saurer als das Saure selber; Item Flamellus, man soll sich hüten, dass einem der Essig mit seiner Schärfe nicht in die Augen schlage, item, Johannes de Padua, wenn er schreibt von Scheidung der Elemente, dass man soll eine Messer-Klinge vorhalten, so lange als die Tropfen auf der Messer-Klinge nicht sieden und schwarz werden, soll man es nicht nehmen, denn es nicht zum Werke dient,*

This precipitate will then transmute a very large proportion of other metals into ☉.

147. *Are these ways the best short cuts, or are there even better ones?*

Answer: There are still better ways, of which I will also name two, and yet they are not the best. If the fixed ♃ is put into its oil so that it has all become oil and a thin ointment, then you take a pure silver lime, but if you take ☉ lime for this, you get white ☉ in the projection, which withstands all ☉ tests; this lime is dissolved with this oil until it has all become a solid stone, so one soon has a tincture to apply to ☿.

148. *But how is it done on red?*

Answer: You also bring the red, fixed ♃ to an oil with its red oil, mix a ☉ lime with it until it has become a red, fixed stone, so you have a tincture that turns everything into ☉.

149. *But because there are even better ways, I would also like to know them.*

Answer: Who wants to describe them all? The best ways are when you dissolve the common ☉ and silver through the two *fermenta*, like the red and the white [ones]. Then dissolve the white and red tincture likewise in the red and white ferment. Place these together, conjoin them, and then fix them.

150. *The Philosophi write, for example Adfar,[303] that the materia is more acidic than the acid itself. Likewise Flamellus[304] states that one should be careful not to let the sharpness of the vinegar hit one's eyes. So also does Johannes de Padua,[305] when he writes about the separation of the elements and says that one should hold up a knife's blade. For as long as the drops on the knife-blade do not seethe and turn black, one should not take it, because it is not useful for the work, but only for rinsing out the vessels. How can such a*

sondern nur die Gefäße auszuspülen, wie kann nun solcher scharfer Essig die Wasser der Gesundheit sein, sie fressen ja den Leib entzwei, Magen und Därme?

Ant. Wer will die Elemente, Luft und Feuer fressen? Freilich ist es hier Gift und aus diesem Gift muss doch die Arznei kommen, durch die Composition, da ein Drache und ein Gift den anderen muss ertöten, dadurch ein neues Leben erwachsen, dass aber die Elemente doch nicht ohne Kraft und Wirkung sein, ausgenommen die Luft oder das Öl, dies ist ein sehr großes Gift, wer solches gebraucht, wäre ärger als esse er Basilisken-Augen, das *Phlegma* aber, der ☿, so weißlich, auch in runden Tröpflein aufsteigt, ist nicht giftig. Wenn etliche Tropfen in *Spiritus Vini* getan werden, so ist es eine Jungfer-Milch, deren Dosis ein halber oder ganzer Löffel voll, macht ein gelindes Erbrechen, wegen seiner Rohigkeit oder vielmehr wegen seiner ☿ialischen Natur, deren Tugend aufwärts wirkt, weil sie feucht und luftig ist, ist weit besser als ein Perlen-Trank, kann auch der Perlen-Trank dadurch gemacht werden. Es ist gut in der Wassersucht, Podagra, Gicht, Zipperlein, in der Rosse und Entzündung des Geblüts, kühlt und löscht das microcosmische Feuer, öffnet alle *Obstructiones*, vermehrt die natürliche Feuchtigkeit, kühlt die entzündete und erhitzte Leber, ist das rechte *Aqua* ♄ und Sauer-Brunnen, macht Herz und Lunge frisch, ist gut wider alle Schwulste von innen und außen, wider den kalten Brand, wenn schon die Kronen angelaufen und schwarz sind, wird die Fässlein auf die schwarzen Kronen gelegt, werden solche wieder weiß und heil, wer es machen kann, wird es auch gebrauchen wissen wider alle Schäden am Menschenleibe, welche groß sind und keine Heilung annehmen, wird doch diese Arznei noch größer sein. Wenn er aber der doppelte ☿ worden, sein seine Kräfte und Vermögen ungemein groß. Der ☿ *Vitæ* und Baum des Lebens, das Feuer aber und blutrote Öl ist auch kein Gift, wenn es mit *Spiritu Vini* corrigiert und gelinde überzogen, ist es eine himmlische Arznei. Etliche Tröpflein 10 bis 12 Tage nacheinander gebracht, wird dem Menschen seinen ganzen Leib erneuern von allen

sharp vinegar be the water of health when it eats the body away, including the stomach and intestines?

Answer: Who wants to eat the elements air and fire? Admittedly, it is poison here, and the medicine must come from this poison through the composition, since one dragon and one poison must kill the other. A new life grows from this, but since the elements are not without power and effect, this is a very great poison except for the air or oil, and whoever uses it would be worse off than eating basilisk eyes; but the *phlegma*, the ☿ which is whitish, which also rises in round droplets, is not poisonous. If a few drops are put in *spiritus vini*, it is virgin milk, the correct dose of which is a half spoonful to a spoonful. This induces mild vomiting because of its rawness, or rather because of its ☿al nature, whose virtue works upwards because it is moist and airy. It is far better than a pearl potion, though the pearl potion can also be made through it. It is good for dropsy, podagra, gout, minor pains, shingles, and inflammation of the blood. It cools and quenches the microcosmic fire, opens all *obstructiones*, increases the natural humidity, and cools the inflamed and heated liver. It is the right *aqua* ♄ and mineral water. It invigorates the heart and lungs, and it is good against all internal and external swelling. It is also effective against cold burns. When the crowns are already tarnished and black, the little cask is placed on the black crowns, and they become white again and heal. Whoever can do this will also know how to use it against all severe damage to the human body which does not admit to healing, in which case this medicine will be even greater. But when it becomes the double ☿, its powers and abilities are immense. [It is] The ☿ *vitæ* and tree of life, but the fire and blood-red oil is also no poison. If it is corrected with *spiritu vini* and gently distilled over, it is a heavenly medicine. A few drops taken ten-to-twelve days in a row will regenerate the whole body, freeing it

Schmerzen und Krankheiten befreien, seine Seele und Geist stärken, dass er denkt, er sei im Paradies oder gar im Himmel, leicht von Gemüt und Geist, auch alle dunklen und trüben Geister von ihm treiben, dass ihm nicht anders deucht, als wäre er neu geboren. Weil nun diese rohen, unreifen, sauren und herben Geister solche Kraft besitzen, was werden sie tun, wenn sie süß, reif, wohlriechend und wohlschmeckend worden? Ist eben als ein unreifer Apfel, welcher die Zähne stumpft, wenn solcher aber reif, so riecht und schmeckt er lieblich. Denn diese widerwärtigen *Elementa* müssen alle einander verkehren, das Süße muss alle Säure diluieren und das Bittere süß machen, das *Alcali*, das eiserne Salz, das Corrosiv süß machen, wie man am Eisen sieht, wenn solches durch den sauren *Spiritum* ⊕ aufgelöst, ist das *Acidum* vom Eisen getötet und süße.

151. Ich möchte doch noch gerne berichtet sein von Kraft der Elemente, was jedes eigentlich vor Natur, Kraft und Geschmack an sich hat, wie auch ihre Wirkung und Vermögen, wenn diese neue Früchte gebären, wie doch endlich durch die gänzliche Composition, die süße Medizin zu erlangen?

Antwort: Erstlich wird unsere Erde sehr feurig und trocken gemacht, ist das beständige *Astrum, Ignis & Terra*. Diese wird mit ihrem Wasser angefeuchtet, dies tobt und siedet, braust und gibt Blasen, als wenn man *Spiritum* ⊕ auf *Sal Tartari* gießt, denn dieses ist *Agens* und *Patiens*, ob es schon in einem Leibe gewohnt, ist es doch widerwärtiger Natur und wenn es in der Composition diese Zeichen nicht gibt, so ist weder Wasser noch Erde recht zubereitet, diese beide töten einander, das Salz ist trocken und bitter und sehr feurig, das Wasser sauer, innerlich süß, wenn es durch diese Erde läuft, wird daraus das gewünschte *Aqua* ♄, durch diese beiden kommt ein neues Leben und ein neuer Geschmack hervor, unser Lebens ☿, der sich mit seinem Leibe *radicaliter* vereinigt und nun aus dem Wasser oder Meer aufs Trockne geht, da ist keine Schärfe mehr, keine Bittere, sondern das *Manna*, das süße Himmels-Brot, der Baum des Lebens, der das *Humidum Radicale* vermehrt, alle Bäume fruchtbar macht, die grauen Haare auswirft, die Nägel ausstößt

from all pains and illnesses and strengthening its soul and spirit so that it thinks it is in paradise or even in heaven. The patient becomes light of mind and spirit, and all dark and disturbing spirits are driven away from them so that they do not think otherwise than if they were newly born. Now, because these raw, immature, sour, and astringent spirits have such power, what will they do when they have become sweet, ripe, fragrant, and palatable? It is just like an unripe apple, which dulls the teeth, but when it is ripe, it smells and tastes lovely. Because these opposing *elementa* must all transform each other, the sweet must dilute all acidity and make the bitterness sweet; the *alcali* [must make] the iron salt, the corrosive, sweet, as one can see from iron, when it is dissolved by the acidic *spiritum* ⊕, the *acidum* of iron is killed and [becomes] sweet.

151. *I would still like to be told about the power of the elements, what each actually has in nature, power, and taste, as well as their effect and capacity when these give birth to new fruits, and how finally the sweet medicine can be reached through the complete composition?*

Answer: First, our earth is made very fiery and dry, which is the constant *astrum, ignis & terra*. This is moistened with its water, and thereupon it rages and boils, roars and gives bubbles, like when one pours *spiritum* ⊕ on *sal tartari*, because this is *agens* and *patiens*, since although it has been in one body it is nonetheless of a conflicting nature. If it does not show these signs in the composition, then neither water nor earth is properly prepared, these two kill each other, the salt is dry and bitter and very fiery, the water is acid, sweet on the inside. When it runs through this earth it becomes the desired *aqua* ♄, and through these two comes a new life and a new taste, our ☿ of life, which unites with its body *radicaliter* and now goes out of the water or sea onto dry land. There is no longer any sharpness or bitterness, but instead the *manna*, the sweet bread of heaven, the tree of life, which multiplies the *humidum radicale*, makes all trees fertile, casts out the gray hairs,

und sich wie der Eisvogel regeneriert, durch seines Lichtes Strahl das Alter und Tod vertreibt, alle Krankheiten, so in der ganzen menschlichen Natur von innen und außen, müssen weichen und dem Menschen seine erste Jugend geben, die Arznei, so zum langen Leben dient, muss aus der Erde genommen werden, aus diesen centralischen *Polo*.

152. *Wie braucht man denn diesen Lebens ☿?*

Antwort: Man kann ihn auf viele Arten in der Medizin brauchen, in Morschellen, in Zucker, auch kann man ihn in *Spiritus Salis dulcis* solvieren, in einen solvierten Zucker gießen, gelinde zum Salz lassen evaporieren, dies Salz täglich brauchen, ingleichen in dem flüchtigen *Spiritu* ☉ und dies in einem solvierten Zucker zum Salz lassen abrauchen oder man kann ihn in hoch rectificierten *Spiritu Vini* solvieren zu einem wohlschmeckenden und wohlriechenden *Astro Mercurii* übertreiben, bis er alle übergestiegen, auch kann man ihn brauchen , ehe er sich noch zur *Terra Foliata* sublimiert und noch der Kristall des ☿ ist, so kocht man diesen Stein in Öl, Terpentin, Mandel-Öl oder Baum-Öl, mit welchen man nur äußerlich die schmerzhaften Glieder reibt, wer aber in innerlichen Krankheiten nur an diesem Stein leckt, wird gesund, alle Wunden und Schäden, sie sein so arg und böse als sie immer wollen, wenn auch schon alles Fleisch und Geäder verfault, wird es doch durch seines Salzes Kraft alles zurecht bringen, seine Wunder-Kuren sind nicht zu beschreiben, Schwindsucht, Wassersucht, Gicht, Podagra, Schlag, böse Not, Pest und Gift muss davon weichen, denn es erhält den Menschen in steter Gesundheit.

153. *Wenn aber dieser ☿ præcipitiert und fix gemacht, ist er denn hernach noch von größerer Tugend?*

Antwort: Ohne allen Zweifel, denn er bekommt durch das *Aqua Stygia* vielmehr Kraft und Vermögen, denn in diesem Oleo ♄ ist enthalten die ♀ und der Geist der aufgehenden Sonne. Weil aber dieses das rauchende tödliche Wasser, welches niemand trinken konnte, so musste dieser ♌ mit dem Adler streiten, denn das corrosivische Öl solviert die

pushes out the nails, and regenerates itself like the kingfisher. Through its light rays it expels all illnesses, old age, and death, so that the whole human nature, in both its inner and outer aspects, must yield and restore the human being to their first youth. The medicine, which serves for a long life, has to be taken from the earth, from this central *polo*.

152. *How do you use this ☿ of life?*

Answer: You can use it in medicine in many ways, in morsels, in sugar, or you can also dissolve it in *spiritus salis dulcis* and pour it into dissolved sugar. You can let it gently evaporate to salt, and then use this salt daily, or [you can] also [dissolve it] in the volatile *Spiritu* ⊕ and let this smoke off to a salt in a dissolved sugar, or you can dissolve it in highly rectified *spiritu vini*, distilling it to a tasty and fragrant *astro mercurii* until all has distilled over; and you can also use it before it sublimates to *terra foliata* and is the crystal of ☿. Then one boils this stone in oil, turpentine, almond oil, or tree oil, with which one rubs the painful limbs only externally; but whoever only licks this stone in cases of internal illnesses becomes healthy, and all wounds and injuries, no matter how terrible they may be—indeed, even if all flesh and veins are already rotting—will be healed by the power of its salt. Its miraculous cures cannot be described. Phthisis, dropsy, gout, podagra, stroke, cramps, plague, and poison must give way to it, for it keeps people in constant health.

153. *But if this ☿ is precipitated and made fixed, is it afterwards still of greater virtue?*

Answer: Without any doubt, because the *aqua stygia* gives it strength and ability, and also because this *oleo* ♄ contains the ♀ and the spirit of the rising sun. But because this is the smoking, deadly water which no one could drink, this ♌ had to fight with the eagle, for the corrosive

geblätterte Erde, wodurch sein corrosiv und Gift getötet wird, dadurch sie beide ihr Leben lassen, der Adler seine Flügel bei diesem Feuer verbrannt, der ☿ die ♀ beschläft, dann regnet es ☉ aus den Dünsten, so von diesen beiden Leibern aufsteigen und muss die ♀, der grüne ♌, eine harte Transmutation ausstehen, denn sie muss durch diesen Beischlaf in die ☽ verwandelt werden und unser ☿ in das *Aurum horicontale*, ist also freilich dieser Præcipitat weit größerer Tugend, denn jetzt hat der ☿ der ♀ ihren Schatz geraubt und der ♄ bestohlen, seine Sonne, die er in seinem Gefängnis verwahrt, die hat er los gemacht, sich mit ☉ und Silber bekleidet, sind nun beide ohne Gift und ein Leib, die *Dosis* ist sehr klein, die Wirkung mächtig groß und er ist alles in allen der Arzt Raphael, die höchste Arznei, der Schatz des Reichtums und der Gesundheit, hat aber jetzt keinen sehr süßen Geschmack, sondern etwas herbe von der ♀, dem scharfen fliegenden ♄ mit seiner Sense, dadurch er dem ☿ die Flügel abgehauen, jedoch endet sich der Geschmack in einer Süße.

154. *Wie braucht man diesen in der Medizin?*

Antwort: Man kann ihn brauchen in Wein, in Bier oder in *Spiritu Vini*, auch in Konfekt.

155. *Wird denn hier der Spiritus Vini, genannt das Aquavit des Steins, sein Öl, so in Leib, Seele und Geist besteht, auch die Luft genannt wird?*

Antwort: Nein, beileibe nicht, dieses Öl ist Gift und Tod, denn der Stein wird durch dieses Gift sehr hoch gefärbt.

156. *So dient der rote Stein nicht zur Medizin, weil dieses giftige Öl dazu kommt?*

Antwort: Der rote Stein ist die allergrößte und himmlische Medizin und dieses giftige Öl gibt ihm eben die größten Kräfte und muss aus dieser Schlange die größte Medizin werden, die größten Kuren kommen aus dem ärgsten Gift, auch aus dem giftigen Arsenik selber, doch muss dieser Basilisk getötet werden und nicht mehr lebendig sein,

oil dissolves the leafed earth, whereby its corrosiveness and poison are killed. In this conflict they both lose their lives, the wings of the eagle are burned in these fires, the ☿ sleeps with the ♀, then it rains ☉ from the vapors, which rise from these two bodies; and the ♀, the green ♌, has to endure a hard transmutation, because through this cohabitation it has to be transformed into the ☽ and our ☿ in the *aurum horicontale*. This precipitate is of course of far greater virtue, for now the ☿ has robbed the ♀ of its treasure, and has stolen from ♄ his sun, which he [♄] kept in his prison, and which he [☿] has now freed, clothed himself with ☉ and silver, both of which now are without poison and of one body. The *dosis* is very small yet the effect is very great; and all in all it is the highest medicine of Doctor Raphael, the treasure of wealth and health; but now its taste is not very sweet, but rather a bit tart from the ♀, the sharp flying ♄ with his scythe, by which he cuts off the wings of the ☿. But having said this, the taste ends in a sweetness.

154. *How is this used in medicine?*

Answer: You can use it in wine, in beer, or in *spiritu vini*, as well as in confectionery.

155. *Is the spiritus vini mentioned here—the aqua vitæ of the stone, its oil, which consists in body, soul, and spirit—also called the air?*

Answer: No, not at all, this oil is poison and death, for the stone is colored very intensely by this poison.

156. *So the red stone is not used for medicine because this poisonous oil is added to it?*

Answer: The red stone is the greatest and most heavenly medicine of all, and this poisonous oil gives it the greatest powers, for this from serpent the greatest medicine must come. The greatest cures come from the worst poison, even from the poisonous arsenic itself, however this basilisk must be killed and no longer be living. Therefore, a bright

daher wird ihm ein heller Spiegel zugerüstet, dass er sich selbst tötet, dass man diesen Basilisken durch sein giftiges Herz sehen kann. Daher wohl zu merken, dass, wenn der Stein gelb ist, wird er in der Medizin nicht gebraucht. Hier verändert er abermals seinen Geschmack, zuvor war er herbe, jetzt durch das *Oleum* Arsenik wird er aber süße, der Geruch ist eklig, hitzig und trocken, die Farben mehrenteils gelb und himmelblau, bekommt aufs neue den Namen des giftigen Drachens.

157. *Wenn er nun nicht zur Medizin dient, was macht man denn damit?*

Antwort: Man muss diesen giftigen Drachen töten durch ein Feuer, welches die Macht hat, sein Gift zu verändern und in eine heilsame Medizin zu verwandeln, dass dieser giftige Salamander im Feuer leben kann, er verbrennt nicht darinnen, sondern legt nur seinen Gift ab, wenn er getötet wird, so wird er durch sein eigenes Blut lebendig gemacht, wenn sich der Phönix durchs Feuer verbrannt und der Pelikan seine toten Jungen mit Blut besprizt.

158. *Wie tötet man diesen Drachen?*

Antwort: Phöbus schießt seine Pfeile in ihn, sein ganzer Leib ist blau und gelb von Gift, dies sieht man an seinem Oden, wenn er in seiner Höhle liegt und seinen giftigen Dunst von sich haucht, Phöbus aber lauert auf ihn, schießt seine Strahlen, seine feurigen Pfeile in ihn, welches Sonnen-Feuer und Hitze ihn ganz wütend und tobend macht, er will entfliehen, je mehr er flieht, je mehr Phöbus Pfeile in ihn schießt, bis sein ganzer Leib verwundet wird und anhebt zu bluten, in diesem Blut wird er zur Medizin, und muss sein ganzer Leib in lauter Blut schmelzen. Dieses Drachen-Blut ist der Schatz der Welt, der Schatz des Reichtums und der Gesundheit, wer diesen Schlüssel zu des Drachen Höhle findet, dass er die Sonne einlassen kann, dass sie diesen Drachen bestreitet, der hat gewonnen, denn er hat die 2 Feinde des menschlichen Lebens überwunden, die Krankheit und Armut, er ist diesen Feinden entflohen, sein Schiff ist angelandet an dem Land der guten Hoffnung und seinen Feinden entronnen, welche ihn nicht

mirror is prepared for it so that it kills itself, and so that one can see this basilisk through its poisonous heart. Hence it is to be noted that if the stone is yellow it is of no use in medicine. Here it changes its taste once again. Whereas before it was tart, now it is sweetened by the *oleum* of arsenic. The smell is nasty, hot, and dry, and the colors are mostly yellow and sky-blue. It is once again given the name of the poisonous dragon.

157. *If it is not used for medicine, what do you do with it?*

Answer: You have to kill this poisonous dragon by a fire that has the power to change its poison and turn it into a healing medicine, so that this poisonous salamander can live in the fire. It does not burn in it, but only sheds its poison when it is killed; and it will be vivified by its own blood when the phœnix is burned by fire and the pelican sprinkles blood on its dead young.

158. *How does one kill this dragon?*

Answer: Phœbus shoots his arrows into it. Its whole body is blue and yellow with poison, and this can be seen from its breath when it lies in its cave and exhales its poisonous vapor. But Phœbus lies in wait for it and shoots his rays—his fiery arrows—into it. This solar fire and heat make it wrathful and raging, and it wants to escape; but the more it flees the more Phœbus shoots his arrows into it until its whole body is wounded and starts bleeding. In that blood it becomes a medicine, and its whole body must melt into nothing but blood. This dragon blood is the treasure of the world, the treasure of wealth and health, and whoever finds this key to the dragon's cave, which enables him to let in the sun to contest with this dragon, shall triumph, because he overcomes thereby the two enemies of human life: sickness and poverty. He has fled from these enemies and his ship has landed at the port of good hope

verfolgen können, er hat das hermetische Gut gefunden, nun muss er GOTT allein dienen und seines armen Nächsten nicht vergessen, noch sein Herz vor ihm zuschließen.

159. *Was ist doch am meisten bei dieser Arbeit in acht zu nehmen oder was gibt es für sonderbare Geheimnisse bei Verfertigung dieser Arbeit und welches sind die Zeichen seiner Jahre und Reife?*

Antwort: Das erste Geheimnis ist des ☿ seine Reife, dass man den Geburtsort des Steins wohl wissen muss, darauf man Tag und Nacht wachen muss, wer dieses nicht weiß, wird alles verlieren, denn dieser wird von sich selbst brechen und diese Erde wird sich selbst zermalmen und wird dieser Stein schießen, dass er Mehl gibt, sobald als der Stein-Bock geboren, müssen ihn die Jäger erhaschen, sonst ist er nicht zu fangen, er geht fort und lässt einem das ledige Nest.

160. *Wie fängt man ihn aber?*

Antwort: Ich habe in meinem Traktätlein, *Das Mineralische Gluten* genannt, diesem gedacht, welches ich hier mit Verdruss nicht wiederholen will, sondern diese Nachricht zum Überfluss erteilen, wenn jene Zeichen erscheinen, so nehme man etwas davon, werfe es auf ein glühendes Blech, wenn es alles in Rauch weg geht, so sublimiert man es selber und warte nicht bis dieses Erdbeben geschieht, so ist man ohne Gefahr und kann diesem Vogel die rechte Leim-Rute legen.

161. *Was gibt es denn für Zeichen der Reife bei dem weißen Stein?*

Antwort: Das gläserne Meer muss vielmals den glatten Spiegel geben, ehe es ganz erhärtet, es wird oft hart werden, wenn man nun solches aufreißen wollte und gedächte, dieser steinerne Palast wäre erbaut, so würde es sich mit dem Öl nicht mengen, auch würde es das Feuer erlöschen, sondern man muss ihm die rechten Jahre lassen. Es wird vielmals fließen und wieder gestehen, bis es zuletzt seine rechte Härte bekommt, dann springt dies Glas von sich selbst, dass mancher vor Furcht und Warten der Dinge, so noch kommen sollen und vor Schrek-

and escaped his enemies who cannot pursue him. He has found the Hermetic good, now he must serve G o d alone, not forgetting his poor neighbour, nor closing his heart to him.

159. *What is the most important thing to watch out for in this work, or what are the strange secrets in the making of this work, and what are the signs of its years and maturity?*

Answer: The first secret is the maturity of the ☿. One must know the birthplace of the stone, and one must watch day and night for it. Whoever does not know this will lose everything, because it will break apart and this earth will crush itself and the stone will crack so that it gives forth powder; as soon as the steinbock is born, the hunters must catch it, since otherwise it cannot be caught but flees and leaves one with the empty nest.

160. *But how do you catch it?*

Answer: I have mentioned this in my little treatise entitled *The Mineral Gluten*, which I will not repeat here with annoyance, but give this message in abundance: when those signs appear, take some of them and throw them on a red-hot [piece of] tin. If everything disappears in smoke, sublimate it yourself and do not wait until for this earthquake to happen. Then you are safe and can lay out the proper limestick for this bird.

161. *What are the signs of maturity in the white stone?*

Answer: The sea of glass must give the smooth mirror many times; before it hardens completely, it will frequently become hard. If you wanted to rip it open now, thinking that the palace of stone had been built, it would not mix with the oil, and it would also extinguish the fire. Rather, one must let it have the correct years. It will flow many times and will harden again until it finally obtains its proper hardness, at which point this glass will shatter by itself, such that many flee from the chamber in terror, out of fear and expectation for things that are

ken zur Stube hinaus läuft und nicht weiß, wie er daran ist, bis er endlich wiederkommt und findet sein gläsernes Meer in lauter Diamanten zersprungen, grob, klein, eckig, wie grobe gestoßene Kristalle, welches das rechte grobe See-Salz der Natur. Einen guten Freund weiß ich, dem sein Werk in lauter □ Steine gesprungen, als wären sie geschliffen, dieser Diamant muss erweicht werden.

162. *Was hat man für Gefahr oder Furcht bei dem roten Stein?*

Antwort: Dies habe ich in meinem anderen Traktätlein dem *Philosophischen Perl-Baum* schon gedacht, welches gleichfalls nicht wiederholen will, aber zum Überfluss dieses melden: Man muss ihm nicht mehr Feuer geben als er vertragen kann, sonst wird sein ganzer Leib ein Feuer-Geist und geht wieder in sein Chaos, denn er liebt der Sonnen Feuer so sehr und wenn er dadurch zu sehr entzündet wird, kann es leicht kommen, dass er gar verschwindet, dieses habe deswegen anzeigen müssen, weil niemand seine Multiplication so weitläufig ausgeführt und so deutlich als ich beschrieben, damit der Arbeiter nicht aus Unwissenheit in Gefahr läuft, denn dieser Stein hat sehr viel Veränderung, er wird vielmals flüchtig gemacht, wieder fix, wieder flüchtig, auch wieder fix, durch diese vielen Arbeiten werden so viele scharfe Pfeile in ihn getrieben, so viele *Gradationes*, so viele *Illuminationes*, und eben deswegen besitzt er so viele Kräfte.

163. *Ist es denn wahr, dass die rechte Materia in allen Läden überall genug zu bekommen, nicht geachtet, sondern als ein geringes und schlechtes Ding weggeworfen wird?*

Antwort: Ja, vor dieser Zeit ist es in allen Läden zu kaufen gewesen, das Pfund um 6 pf., jetzt aber nur geschmolzen, seine *Minera* nicht, da doch das Geschmolzene untüchtig zum Werk, vor diesem ist die *Minera* so unwert gewesen, dass sie niemand hat gebraucht und dieser Stein nach der Kuh geworfen, da sie aber bekannt werden wollen, hat sie der Feind des menschlichen Geschlechts den Menschen aus den Händen gerissen und ist nach Holland und Italien verkauft

yet to come. Yet they do not know what to do with it until he finally comes back and finds his sea of glass shattered into nothing but diamonds: coarse, small, angular, like roughly crushed crystals, which is the right coarse seasalt of nature. I know a good friend whose work cracked [the glass] into pure ☐ stones as if they had been cut. This diamond must be softened.

162. *What is the danger or fear of the red stone?*

Answer: I have already mentioned this in my other small treatise concerning *The Philosophical Pearl Tree*, which I likewise do not wish to repeat, but in superfluity I will report this: do not give it more fire than it can handle or its whole body will become a fire spirit which will return to its chaos, for it loves the solar fire a great deal, and if it becomes sufficiently ignited by it, it is easy for it to completely disappear. This is why I had to mention this, because no one has explained its multiplication so extensively and so clearly as I, so that the worker does not run into danger out of ignorance, because this stone has very many changes, it is made volatile many times, fixed again, volatile again, fixed again; through all this work so many sharp arrows are driven into it, so many *gradationes*, so many *illuminationes*, and that is why it has so many powers.

163. *Is it then true that the right materia used to be plentiful in every shop everywhere, since it was not highly valued, but was thrown away as a mean and bad thing?*[306]

Answer: Yes, before this time it could be bought in all shops, at the rate of a pound for six *pfennig*, but now it is only available in a melted form and not the hard *minera*. Yet the melted form is unfit for the work. Before this the *minera* was so worthless that no one used it and they threw this stone at the cow, but since it became well known, the enemy of the human race snatched it out of the hands of the people and sold it to

worden, dass man ein *Monopolium* daraus gemacht, nach diesem aber hat man in Deutschland diese *Materia* auch lernen traktieren, doch nur zu einer geringen Lumpen-Sache und gleichfalls ein *Monopolium* daraus gemacht, dass solche fast niemand öffentlich zu kaufen haben kann, da doch das ganze Land voll, in Bergen viele Millionen Zentner stecken, weil sie aber nicht kann verkauft werden, lässt man solche in Bergen stecken, verhindert dadurch so viele fleißige Natur-Forscher, dass solche, weil sie die rechte *Materiam* nicht haben, Geld, Mühe und Zeit müssen vergeblich anwenden, auch verhindert man dadurch der hohen Lands-Obrigkeit ihren Zehnten, die Gewerken, dass sie müssen mit Schaden bauen und arme Berg-Leute, welche ihr Brot haben könnten, welches alles vom bösen Feind angefangen und erregt worden, denn alles Gutes kommt von GOTT, was aber dem menschlichen Geschlecht zum Schaden und Nachteil gereicht, kommt vom bösen Feind.

164. *Weilen aber dieses das wenigste ist, ☉ und Silber durch dieses Werk zu machen, die unedlen Metall in edle zu transmutieren, sondern die Wunder der Magia dadurch auszuführen, so möchte gleich wohl auch gern was davon hören?*

Antwort: Diese Dinge werden nicht öffentlich beschrieben, wer dieses tut, der ist ein Brecher des himmlischen Siegels und wird dem Gerichte GOTTES nicht entrinnen, wird es auch kein Mensch tun, wer diese himmlische Sophia kennt und die Geheimnisse dieser keuschen Braut ausschwatzen und ihr heiliges Ehebette verunreinigen oder beflecken, sonst würde sie ihn verlassen und nimmermehr wieder besuchen, wenn er diese göttlichen Kleinodien würde gemein machen, mit welchen GOTT einen solchen allein beschenkt, dass aber dieser Stein des Reichtums und der Gesundheit gemein gemacht, das ist GOTTES Wille und die Zeit der Offenbarung ist vorhanden, welchen Willen und Trieb GOTTES man nachkommen und erfüllen müssen, denn allezeit haben die unmündigen und der ungelehrte Haufe die Gnade gehabt, dass die GOTTES liebste Freunde gewesen, die er seinen Willen

Holland and Italy so that a *monopolium* was made of it. After this, one has learned to treat the *materia* in Germany, too, but only as a lowly, raggedly thing, and here a *monopolium* was also made out of it, so that almost no one can buy it publicly, although the whole country is full of it and millions of hundredweights of it are stuck in the mountains. But because it cannot be sold, it is left stuck in the mountains, thereby hindering so many diligent natural scientists, who uselessly spend money, time, and effort because they do not have the right *materiam*. This also prevents the high state authorities from receiving their tithes, hinders the tradespeople, who have to build with injuries, as well as the poor miners from being able to earn their bread. All of this was started and stirred up by evil adversaries, because everything good comes from GOD, but what harms and damages the human race comes from the evil adversary.

164. *Since it is the least to make ☉ and silver through this work, to transmute the base metals into noble ones, but [greater] to perform the wonders of magia through it, I would also like to hear something about this.*

Answer: These things are not publicly described, since whoever does this is a breaker of the heavenly seal and will not escape the judgment of GOD. Nor will anyone who knows this heavenly Sophia speak about the secrets of this chaste bride and thereby defile and stain her holy marriage bed, since in that case she would leave him and never visit again when she sees that he would vulgarize these divine jewels which GOD alone bestows upon one. Yet that this stone of wealth and health be made common knowledge is GOD's will, and the time has now come for this revelation to be made in accordance with that will, because the immature and uneducated crowd has always received the grace of God and become GOD's dearest friends to whom he revealed

wissen lassen und ihnen die göttlichen Geheimnisse anvertraut, dies sehen wir aus den Propheten und Aposteln, welche alle ungelehrte Leute gewesen. Auch hat er das weibliche Geschlecht gewürdigt, seine Geheimnisse zu erforschen. Als Mariam, Moses Schwester, die Prophetin. *Item*, die Prophetin Deborah, so Israel gerichtet, auch hat sich der Sohn GOTTES nach seiner Auferstehung am ersten einer Weibes-Person offenbart, damit er von Anfang der Welt gewiesen, dass er alles was von der Welt verachtet, unverständig und ungelehrt gehalten wurde, sich aussuchen und sich nicht an die Welt-Gelehrten binden lassen, sondern sein Geist ist der rechte Lehrer und die hohe Schule, da alle Welt-Gelehrte verstummen müssen bis an das Ende der Welt disputieren und doch nichts ausführen, alle Wissenschaft und Grund nur im Schall der Wörter suchen, den inneren Grund aber fahren lassen, sich um die Schalen beißen, aber den Kern nicht kosten. Dies nehme ein Mensch vor ganz gewiss, wenn er im äußeren Himmel wohl steht und gütig angesehen wird, dass ihm der innere Himmel noch viel günstiger ist und hat Hoffnung durch ein göttliches tugendsames Leben sich GOTT zu nähern und dessen Geheimnisse zu erforschen, er sei Weib oder Mann, in *Christo* sind wir nur einer, alle die Braut *Christi*.

165. Warum sollte aber in der ganzen Natur nur diese einzige Materia sein, daraus dieses hohe Werk zu machen ist, da doch so viele herrliche Mineralia, die lauter güldische ♃ haben und noch wohl gar ☉ halten, welches war flüchtig, so wäre ja diese besser dazu als dieses bleiische Wesen?

Antwort: Die Materialien, so ☉ halten, sind desto schlimmer dazu, was soll das ☉ tun, es hat keinen tingierenden Geist, als wie unsere *Materia*, da auch der rohe Geist tingiert, sein weißer Sublimat, dem er *per se* von sich gibt, welcher das ♀ schön weiß färbt, ingleichen macht solcher das ☿ zum Glas, wenn man sie beide im gelinden Feuer fließen lässt, welches sonst kein mineralischer Körper tut, der die Metalle in kurzer Zeit auf den letzten Glas-Grad bringt und den Metallen ins Herz dringt, was soll nun dieser Vogel tun, wenn er zurück in sein *Centrum* geht, mit solchen sich vereinigt und zum fixen Glas worden.

his will and to whom he entrusted the divine mysteries. This we see from the prophets and apostles, all of whom came from the ranks of common, unlearned people. He also honored the female sex to explore his mysteries, such as Mariam, Moses' sister, the prophetess. Likewise, there was the prophetess Deborah, who judged Israel. After his resurrection, the son of GOD also revealed himself to a woman before anyone else. By doing this he showed, from the beginning of the world, that he sought out everything that was despised by the world, everything held to be ignorant and unlearned, and he did not allow himself to be bound to the educated world. Rather, his spirit is the right teacher and the highest school, where the entire educated world must fall silent. For scholars argue until the end of the world and yet accomplish nothing, seeking all science and reason only in the sound of words, but abandoning the inner reason; they gnaw on the husks, but do not taste the kernel. One can be quite certain of this: if they are well off in the outer heaven and are looked upon favorably, then the inner heaven is even more favorable to them, and they have the hope of approaching GOD through a divine, virtuous life and of investigating his mysteries. This person may be a woman or man; in Christ we are but one, all the bride of Christ.

165. *But why should there be only this single materia in the whole of nature from which this high work is to be made, when there are so many wonderful mineralia that have nothing but gold-bearing ♄, and still probably contain ☉ that was volatile? Would this not be better for [the work] than this lead-like being?*

Answer: The materials that contain ☉ are all the worse. What should the ☉ do? It has no tingeing spirit like our *materia*, for the raw spirit also tinges. The white sublimate which it gives off from itself *per se*, and which colors the ♀ a beautiful white, likewise turns an ☿ into glass if you let them both flow in a gentle fire, which no other mineral body does, and which brings the metals to the last degree of glass in a short time and penetrates the heart of the metals. What should this bird

Zuvor, da er alleine und noch flüchtig und giftig war, färbt er kaum 10 Teile, da er aber wieder neu geboren, tut er es in viel 1000-fältiger Kraft, dies tut kein roher Geist der Metalle, er kann sich nicht also ausbreiten.

166. *Dies wird wohl der tingierende Geist sein, da die Philosophi schreiben: Arsenicum esse Tincturam Albedinis, und nicht Rubedinis, dies ist aber kein gemeiner Arsenik nicht?*

Antwort: Nein, der gemeine Arsenik kommt nicht zu unserem Werk, der gemeine ☍ wird aus dem Cobalt sublimiert, aus dem Wasser-Kies, aus dem rot güldenen Erz, diese haben ein *Acidum* von gemeinem ☿ bei sich, oft sublimieren sie ihn mit ⊖ oder Kalk und anderen Mixturen, unser ☍ aber kommt aus unserem Blei, ist der *Spiritus* ♄, unser Schwefel-Öl, er brennt als *Spiritus Vini*, hat den ☽ ☿ in großer Menge in sich, auch den gelben Riemen um den Leib, denn er färbt den weißen ☿ gelb, in Gold, ist der Drachen Gift und muss er durch die ♀ heilsam gemacht werden, denn er gibt seiner Schwester Honig-Felsen und liebt die *Mineram* ewiglich.

167. *Vorher wurde gemeldet, dass die weiße Tinctur, wenn sie zum gläsernen Meer geworden, von sich selbst springe, das Glas zerbreche, dass es alles kracht, wie geht dies zu?*

Antwort: Was kann es wundern? Man sieht ja solches am *Vitro* ☉. Item, am *Vitro* ♄, wenn solche kalt werden, wie sie springen, dass es knistert und kracht, sich selbst in Splitter teilt, warum nicht auch bei diesem Werk, wenn ihm sein Feuer entgeht?

168. *Mit was für Feuer aber wird dies zum Glas-Grad gebrannt?*

Antwort: Durchs Feuer der Natur, sein eigenes Feuer.

169. *Hat denn der rohe Körper, wie ihn die Natur gibt, ohne einige Arbeit keinen Nutzen in der Medizin?*

now do when it goes back to its *centrum*, unites with it, and becomes a fixed glass? Before, when it was alone and still volatile and poisonous, it hardly stained ten parts. But since it is reborn, it does this with several thousand times the power. No raw spirit of metals does this, it cannot spread like that.

166. *This will probably be the tingeing spirit, since the philosophi write: "Arsenicum esse tincturam albedinis*[307] *and not Rubedinis", but this is not common arsenic, is it?*

Answer: No, the common arsenic does not come into our work, the common ♂ is sublimated from cobalt, from the water pyrite, from the red gold ore; these have an *acidum* of common ♄ with them, and often they sublimate it with ⊖ or lime and other mixtures. But our ♂ comes from our lead, is the *spiritus* ♄, our sulphur oil, it burns like *spiritus vini*, has the ☽ ♄ in it in large quantities, and also the yellow belt around the body because it colors the white ♄ yellow, into gold; it is the poison of the dragon and must be made healing by the ♀, for it gives its sister honey-rocks and loves the *mineram* forever.

167. *Previously it was reported that the white tincture, when it became a sea of glass, shattered itself. The glass breaks, everything cracks. How does this happen?*

Answer: This is not surprising? You can see that in the *vitro* ☉, likewise in the *vitro* ♄ when it becomes cold: how it shatters, how it cracks and crashes and splits itself into splinters. Why not in this work, when its fire escapes it?

168. *But with what fire is this burned to the degree of glass?*

Answer: Through the fire of nature, its own fire.

169. *Has the raw body, as nature gives it, no use in medicine without some work?*

Antwort: Ja, er hat viele und große Tugend, er heilt die Wunden, auch wenn er unter die Pflaster gemacht, in Schwulsten, Milz-Weh, oder Entzündungen, welche von scharfen, sauren, gesalzenen Flüssen ihren Ursprung nehmen, so zieht er diese Säure an sich, als wie ein Magnet das Eisen. Auch hat es die Kraft, dass solches wieder die Zauberei dient, wer es bei sich trägt. Es können auch dadurch Bergwerke gefunden werden, wenn man solches in die Wünschel-Rute tut, ingleichen bei Most-Zeit in den Wein hängt, dass es damit gärt, muss aber überaus klein gerieben sein, so wird es die Säure des Weins brechen, dass es nicht anstößt noch schwer wird, sondern bleibt klar und lauter, dient für viele Krankheiten. Ingleichen wenn man es mit sauren Zitronen-Saft infundiert, etliche Tage in gelinder Wärme hält, solches hernach durch einen destillierten Wein-Essig abrauchen lässt und unter Pillen mischt, so kann die Wassersucht dadurch wegpurgiert werden, man nehme nur nicht andere vehemente und giftige purgierende *Species* dazu, dies ist schon Meisters genug, es greift nicht das Serum allein an, sondern die *Materiam peccantem*, kühlt, stärkt und erfrischt wieder die Natur, dass die zerschmelzenden und fließenden Teile wieder gestehen und erhärten. Auch kann man es mit gleichen Teilen ✷ sublimieren, wenn es recht gemacht, findet man 3 Böden, oben einen hoch zitronengelben Sublimat, mitten einen Boden, welcher weiß, darinnen der fette und ölige ☿, unten ist ein terrestrisches rötliches Wesen, darinnen die roten Stiefeln, das *Acidum* des ✷, steckt, im untersten Teil und kann jedweder Boden voneinander gehoben werden, der Oberste hat den Arsenik in sich, dient nicht wohl zur Arznei, ohne weitern Bereitung, aber wohl in Gradier-Wasser, der Unterste ist der beste zu Medizin, wenn solcher in destilliertem Essig aufgekocht wird, gibt es eine rosenfarbene Solution, diese lässt man geringe evaporieren, dass es in Kristallen anschießt, dieses Salz verändert die Farben, in der Wärme ist es grasgrün, sobald es erkaltet, sieht es himmel-blau, violen-blau und wenn es ganz trocken, rosenfarben. Wenn es aber übergetrieben wird durch die *Cohobation*, bis es alles übersteigt, wird man schon Dinge sehen, welche man sich vorher nicht eingebildet hätte, teils in der

NATURE'S SECRET TREASURE CHAMBER 627

Answer: Yes, it has many and great virtues, it heals wounds, even if it is put under the plaster, in swellings, spleen pains, or inflammations which originate from sharp, acidic, salty fluids, for it draws this acid to itself like a magnet draws iron. It also has the power of defending whoever carries it against magic. Mines can also be found if you put something of it on the dowsing rod. Likewise, by hanging it in the wine at must time, so that it ferments with it; but for this it has to be grated extremely finely so that it will break down the acidity of the wine, so that it does not form disturbances or become heavy, but remains clear and pure; this serves against many diseases. Likewise, if you infuse it with sour lemon juice and keep it mildly warm for a few days, and then let it smoke with a distilled wine vinegar and mix it with pills, the dropsy can be purged away. Just do not add other vehement and toxic purging *species*. This alone is master enough, it does not attack the serum alone, but the *materiam peccantem*. It cools, strengthens and refreshes nature again, so that the melting and flowing parts coagulate and harden again. You can also sublimate it with equal parts of ✶. If done right you will find three levels. The top one is a highly lemon-yellow sublimate, while in the middle there is a level which is white, inside of which is the fatty and oily ☿. As for the lowest level, it is a terrestrial reddish being in which are the red boots. The *acidum* of the ✶ is in the lowest level. Any level can be lifted from the other. The top one has the arsenic in it, and it is not used for medicine without further preparation, but probably in gradating water. The lowest one is the best for medicine, if this is boiled in distilled vinegar it gives a rose-colored solution. This is then allowed to evaporate slightly so that it shoots up into crystals. This salt changes the colors. In the heat it is grass-green, as soon as it cools, it looks sky-blue, violet-blue, and when completely dry, rose-colored. But if it is distilled by *cohobation* until everything is distilled over, one will see things which one would

Vorlage, teils im *Capite Mortuo*, und mir hernach glauben. Wenn das ganze *Destillatum* übergestiegen, bis auf etwas Salz, kann das Grüne von dem Roten separiert werden. Auch kann durch *Spiritum Vini* die Purpur-Farbe davon geschieden werden, gleich der Purpur-Farbe, welche die Maler auf Silber malen und so blutig schimmert, dieses erfreut durchs Gesicht, labt durch den Geruch und erfreut das Herze, was dieses Drachen-Blut und korallenrote Saft für Vermögen hat in der Medizin, wird niemand glauben. Es sollten alle Menschen, welche Profession von Kuren machen, sich dieses lassen angelegen sein, dass sie die Probe erstlich an ihrem eigenen Leib nehmen, auch wenn dieses *Corpus* die *Minera* zum allerkleinsten Staub durch ein Tüchlein geschlagen, in destilliertem Wein-Essig gekocht wird, darinnen ☉, ☿, ○ solviert worden, bis der Essig rosenfarbene, will sich die Solution nicht wohl färben, kocht man es trocken ein, wird aber stets gerührt, röstet es alsdann ein oder zwei Stunden in einer Retorte, dass es dunkel glüht, was übergestiegen und was sich sublimiert, tut man wieder zusammen, kocht es alsdann in Essig, so färbt es sich bald, solches lässt man abrauchen gelinde, bis es will dick als ein Saft werden, dieser sieht grasgrün in der Wärme, sobald es in der Evaporier-Schale will erkalten, sieht es himmel-blau, bald Violen-Farben, wenn es kalt geworden Rosen-Farben, macht man es wieder warm, sieht es wieder grasgrün, hoch himmel-blau, Viol-Farben und Rosen-rot, welches seine letzte beständige Farbe, auch sieht es blass-blau, feil-blau, dunkel und blass, diese Farben kann man einem zeigen, wenn man will, sie haben aber auch bei der Belustigung ihren Nutzen, wenn man es lässt im *Balneo* bis zur Trockne abdünsten, schüttet hoch rectificierten Wein-Geist darauf, je länger solcher darauf steht, je besser er wird, so tingiert er sich und nimmt das subtile in sich, wie sonst, wenn man ☉ durch diese Salien aus solviert, dieses ist abermals solchen Vermögens, welches die Apotheker-Büchsen weit übertrifft.

170. *Sind denn dieses philosophische Arbeiten?*

Antwort: Nein, dies sind keine philosophischen, sondern medizini-

not have imagined before, partly in the receiver, partly in the *capite mortuo*, and afterwards you will believe me. When all the *distillatum* has been passed over, save for a little salt, the green can be separated from the red. The purple color can also be separated from it through *spiritum vini*, like the purple color which the painters paint on silver and which shimmers so bloodily. This is beautiful in appearance, has a pleasant smell, and delights the heart. Nobody will believe what this dragon blood and coral-red juice is capable of in medicine. All people who professionally offer cures should make it their concern that they first take the test on their own bodies, even if this *corpus*, the *minera*, is beaten through a cloth to the tiniest dust, and then boiled in distilled wine vinegar in which ☿, ☽, ☉ were dissolved until the vinegar is rose-colored. If the solution does not want to color well, boil it dry, but keep stirring; and then roast it in a retort for an hour or two so that it glows dimly. What has passed over and what is sublimated is put together, and if it is then boiled in vinegar, it soon colors. This is allowed to smoke off gently until it wants to become thick as a juice. This [juice] looks grass-green in the heat, and as soon as it wants to cool down in the evaporation bowl, it first looks sky-blue, then later violet; and when it has become cold it is rose-colored. If you warm it up again, it looks grass-green again, light sky-blue, violet colored, and rose-red, which is its last and permanent color; it also looks pale blue, violet blue, dark, and pale. You can show these colors if you want, for they are also useful for entertainment. If you leave it in the *balneo* to dry out, pour highly rectified spirit of wine on it; the longer it stands on it, the better it gets, so it tinges itself and takes in the subtle, as usual when one dissolves ☉ through these *salias*, this has again such an ability, that it by far surpasses the apothecary tins.

170. *Are these philosophical works?*

Answer: No, these are not philosophical but medical works, I have

sche Arbeiten, die anderen habe ich schon alle, manche wohl 2 bis 3-mal beschrieben, hier aber schreibe was zur Medizin gehört und nichts philosophisches mehr.

171. *Weil nun dieser rote ♄ so trocken und hitzig, wie kann er zum Glas gebrannt werden?*

Antwort: Dies geschieht durch ihren eigenen Geist und Feuer und kann ihr auch zugleich den Fluss geben, ich will aber hier die *Medicamenta* wiederholen. Man solviert ⊖ im Regen-Wasser, gießt gleich so viel als das Salz gewesen, *Oleum* ⊕ zu, zieht es über, so steigt ein *Spiritus Salis*, zurück bleibt das flüssige *Sal enixum*, dieses 1 Teil und sehr klein gepulverte *Minera* 2 Teile, wohl vermischt und etliche Stunden Feuer geben, die gläserne Retorte muss man zerbrechen, den Sublimat nimmt man klein, sowohl was übergestiegen, das tut man alles wieder aufs *Caput Mortuum*, gießt den *Spiritum Salis*, so von dem *Sale ennixo* übergetrieben darauf, lässt es digerieren, bis es sehr rosenfarben worden, im Feuer ist es allezeit grün, gießt die rosenfarbene Solution ab, destilliert es im *Balneo*, weil etwas gehen will, muss in einem niedrigen Kolben geschehen, hernach wird es destilliert als ein *Spiritus* ⊕, den übergestiegenen *Spiritum* mit dem Sublimat zieht man über, weil er weiß geht, in dem Weißen solviert man das Salz, welches aus dem *Capite Mortuo* mit Regen-Wasser gesotten und wieder zum Salz abgeraucht, lässt es gelinde evaporieren, bis das Salz trocken, darauf gießt man die gefärbten *Spiritus*, hält es im Feuer bis es fix worden und im Glühen nicht raucht, dieser schöne wachsflüssige Stein oder Salz besitzt große Kräfte in der Medizin, es ist fast kein Übel zu arg, in natürlichen und unnatürlichen Krankheiten, man kann die Tinctur mit *Spiritu Vini* extrahieren, auch kann man ihn mit Morschellen vermengen, er wird verbessert wenn man dieses Steins 8 oder 10 Teile nimmt, lässt es im Feuer fließen mit etwas Kohlen-Staub, wirft 2 Teile ☉ darein, gießt es aus, solviert das ganze Wesen, kocht es wieder ein oder übergießt es gleich mit *Spiritu Vini*, die Tinctur zu extrahieren. Man muss Gefäße haben, die nichts einschlucken, sonsten ist es alles verloren, ein Kluger

already described all the others, some probably two or three times, but here I am writing of what belongs to medicine and nothing more philosophical.

171. *Because this red ♃ is so dry and hot, how can it be fired into glass?*

Answer: This happens through its own spirit and fire and can also give it the flux at the same time, but I want to reiterate the *medicamenta* here. One dissolves ⊖ in rainwater, pours in as much *oleum* ⊕ as the salt, and distills it over so that a *spiritus salis* rises and the liquid *sal enixum*[308] remains behind. From this, take one part, and from the very finely powdered *minera* two parts, then mix them well and fire them for several hours. The glass retort has to be broken, and then the sublimate is taken in small amounts, as well as what has gone over. Put everything back onto the *caput mortuum*, pour the *spiritum salis* which has been distilled from the *sal enixo* over it, and let it digest until it becomes quite rose-colored. It is always green in the fire. Pour off the rose-colored solution and distill it in the *balneo*. For if something wants to go, it has to be done in a low flask. Afterwards it is distilled like a *spiritus* ⊕, and the *spiritum* that has passed over is distilled together with the sublimate, because it goes white; and in the white [substance] the salt is dissolved which was boiled out from the *capite mortuo* with rain water and smoked off again down to the salt. It is allowed to evaporate gently until the salt is dry, and then the colored *spiritus* is poured on it. It is kept in the fire until it fixes and does not smoke when glowing. This beautiful, waxy, liquid stone or salt has great powers in medicine and can overcome almost any evil in the form of natural and unnatural diseases. You can extract the tincture with *spiritu vini*, and you can also mix it with the morsels. It will be improved if you take eight or ten parts of this stone, letting it melt in the fire with some coal dust; throw two parts ☉ into it, pour it out, dissolve the whole entity and boil it down again or pour *spiritu vini* over it straight away to extract the tincture. You have to have vessels that do not absorb anything, otherwise it

wird dies schon weiter zu gebrauchen wissen.

172. *Kann man denn durch den ☉ diese Materia nicht auch traktieren?*

Antwort: Ja, man nimmt 3 auch 4 Teile ☉ und diese *Minera* sehr klein gerieben, hält es im Feuer, es entzündet den ☉ nicht, sondern fließt nur mit einer Ebullition, weil die Säure des ☉ bei diesem *Alcali* tötet, es darf nicht starkes Kohlen-Feuer haben, sonst läuft es über, auch dürfen keine Kohlen darein fallen, dass sich der ☉ nicht entzündet, nach 2 oder 3 Stunden, danach man viel oder wenig hat, figiert sich der ☉ dabei und wird ein himmelblauer Stein daraus, diesen kann man mit *Spiritu Vini* extrahieren oder im *Spiritu* von ✶ solvieren, bis es sehr blau worden, den *Spiritum* zieht man über, auf das Zurückbleibende kann man *Spiritum Vini* gießen, den Extract mit dem *Spiritu* vermischen oder kann es in Farben übertreiben, wie es einem gefällt, diese Medizin ist sonderlich vortrefflich in Gehirnkrankheiten, für melancholische oder gar tolle und närrische Menschen, für die böse Not und fallende Sucht, weil hier der ♄ *Lunæ* aufgeschlossen und praedominiert.

173. *Wie wird es aber durch ⊖ traktiert?*

Antwort: Es wird in *Spiritu Salis* solviert, dieser gelinde davon abgezogen, bis es als Öl erscheint, darauf ein starker Wein-Geist gegossen, so gibt sich nach langer Zeit ein Öl oben auf, dies ist gleichfalls eine herrliche Medizin. Es kann aber solche noch besser gemacht werden, wenn man es übertreibt, in dem *Spiritu* sein *Caput Mortuum* solviert, sie zusammensetzt, wieder dephlegmiert, mit *Spiritu Vini* digeriert oder gar in eine Quintessenz bringt.

174. *Kann man es durch ▽ nicht auch traktieren?*

Antwort: Ja, es wird in ▽ solviert, wenn die Solution geschehen und sich das Sediment gesetzt, so gießt man die rosenfarbene Solution ab, solviert ⊖ in so wenig Wasser als es sein kann, gießt das ▽, die Rosen-Farbe darein, zieht solches bis auf das trockne Salz im Sande über, so

is all lost. A clever person will know how to use this further.

172. *Can you also treat this materia with the ☉?*

Answer: Yes, one takes three or four parts of ☉ and [one part] of this *minera*, grated very finely, and keeps it in the fire. It does not ignite the ☉, but only flows with an ebullition, because the acidity of the ☉ kills this *alcali*. It must not have strong coal fire, otherwise it will overflow, and no coals should fall into it lest the ☉ ignite. After two or three hours, depending on whether you have a large or a small [amount], the ☉ fixes and becomes a sky-blue stone. This one you can extract with *spiritu vini* or dissolve in *spiritu* from ✳ until it turns quite blue. Distill the *spiritum* over, pour *spiritum vini* on what is left, mix the extract with *spiritu*, or distill it over in colors as you like. This medicine is particularly excellent in brain diseases, for melancholics, or even for mad and foolish people, as well as for cramps and epilepsy, because here the ♄ *Lunæ* is digested and predominating.

173. *But how is it treated by ☉?*

Answer: It is dissolved in *spiritu salis*. This is gently distilled off until it appears as an oil, at which point a strong spirit of wine is poured on it. After a long time an oil is left on top, and this is also a wonderful medicine. But such things can be made even better if one distills it over, in which *spiritu* one dissolves its *caput mortuum* and then puts them together. Dephlegmate them again, digested with *spiritu vini* or even brings them into a quintessence.

174. *Can one also treat it by ▽?*

Answer: Yes, it is dissolved in ▽, when the dissolution occurred and the sediment settles by pouring off the rose-colored solution and dissolving ☉ in as little water as possible. Then pour the ▽, the rose-colored

bleibt ein rosenfarbenes Salz zurück, dieses kann man mit *Spiritu Vini* digerieren, auch wohl mit etwas infundieren, das Salz muss gar klein gerieben werden.

175. *Ist denn auf dem trockenen Weg keine Medizin daraus zu machen?*

Antwort: Ja, wenn man es mit gleich viel ☿ und *Sale Tartari* schmelzt, es dürfen keine Kohlen darein fallen, sonst reduciert es sich zu einem Könige, halt es 6 Stunden in stetigem Fluss, gießt es in einen gewärmten Mörsel, stößt es, weil es noch warm, sehr klein, gießt starken *Spiritum Vini* darauf, lässt die Tinctur extrahieren, bis sie sehr rot, dieser Tinctur Vermögen ist, das Geblüt zu reinigen, innerliche Geschwüre zu heilen, das *Acidum* in Magen und Därmen zu töten, das Zucken und Fahren in Gliedern zu stillen, denn es schlägt nieder, auch kann man aus diesem *Subjecto*, wenn es *per se* sublimiert, seine weiße Diana oder ✶ genommen werden, mit 4-mal so schwer reinem ☉ vermischt und 2 Stunden zusammen gelinde kochen lassen, bis diese beiden einander figiert, der ☉ nicht mehr in die Höhe steigt, sondern als ein Fett fließt, dies wird ausgegossen, so hat man ein schönes blaues Salz, dies kann man, kleine Messer-Spitzen voll, einnehmen, es hat die Wirkung, dass es allen Schleim hebt, solchen in die Fermentation bringt, davon Stink-Flüsse werden, wird allen Schleim in Adern zerteilen und ausführen, nach der Inclination der Natur, durch Husten, Schnupfen oder wie es die Natur am geschicktesten befindet, es macht weder Erbrechen noch Stuhlgang, man müsste es denn in großer Menge essen, welches aber nicht nötig, auch dient dieses Salz in alle offenen Schäden, welche faulen und stinken, es sei verwahrlost, Rotlaufen, Öl-Schenkel, Krebs und Wolf oder andere die gar keine Heilung annehmen wollen, so ist dieses schon Meisters genug in Geschwulst und hitziger Entzündung; Auch ist es vortrefflich, wenn es unter Salben und Pflaster gemacht wird, für garstige böse Wunden und Aussatz.

176. *Kann man denn diese Materia nicht mit dem ☿ traktieren?*

Antwort: Wenn man dieses *Subjectum* mit 3 mal so schwer corrosivi-

solution into it, and distill this in the sand to the dry salt so that a rose-colored salt remains. This can be digested with *spiritu vini*, and also infused with some of it. The salt has to be ground very finely.

175. *Is it possible to make medicine out of it in the dry way?*

Answer: Yes, if you melt it with the same amount of ☉ and *sale tartari*, no coals must fall into it, otherwise it will reduce to a *regulus*. Keep it flowing steadily for six hours, pour it into a warmed mortar, crush it while it is still warm, very finely, pour strong *spiritum vini* on it, extract the tincture until it is very red; this tincture is able to purify the blood, heal internal ulcers, kill the *acidum* in the stomach and intestines, still tremors and contractions in the limbs, because it precipitates. You can also take your white Diana or ✶ from this *subjecto* if it sublimates *per se* and mix it with pure ☉, four times as heavy, and let them gently boil together for two hours until they both fix each other. The ☉ no longer rises, but flows like a fat. This is poured out, so that you have a nice blue salt. It can be taken in small knife-tips. It has the effect of lifting all mucus, bringing it to fermentation, from which stinking fluxes come. It will break up all mucus in the veins and carry them out according to nature's inclination, through coughing, sniffling, or as nature deems most skillful. It causes neither vomiting nor bowel movements unless one consumes it in large amounts, but this is not necessary. This salt is also used in all open wounds that are rotting and stinking, be it due to neglect, erysipelas, oily thighs, cancer, wolves, or other [injuries] that do not want to accept healing at all. Thus is already sufficient to master swelling and violent inflammation. It is also excellent when it is put under ointments and plasters for ghastly sores and leprosy.

176. *Can you treat this materia with the ☿?*

Answer: If you mix this *subjectum* with corrosive sublimate that is three

schen Sublimat vermischt, gleich in bloßem Feuer treibt, so gibt es einen *Spiritum* und rotes Öl, welches wider die Art der anderen Öle in dem *Spiritu* zu Boden liegt, wenn man solches 1000 Mal schüttelt, wird es doch wieder auf den Grund gehen, in dem *Spiritu* ist die Diana, die Remanenz kann man treiben, bis aller ☿ im Retorten-Hals, welcher weit sein muss, diesen solviert man im Keller zu Öl, das *Caput Mortuum* kocht man in destilliertem Essig und lässt die Rosen-Farbe abrauchen, was diese 4 dienen, mag ein jeder erforschen, ich habe genug verschrieben, man kann den Brei nicht kauen und auch einstreichen, aber hier ist ein Weg der kurz ist, zur Arznei für Menschen und Metalle, doch ist es nicht der große *Universal*-Weg.

177. *Aber wie wird es traktiert und was hat es für Nutzen mit dem gemeinen* ☿?

Antwort: Diese *Minera* wird klein gemacht, mit 2 oder 3 mal ☿ vermischt und übergetrieben, so vermehrt sich der ☿ davon und nimmt den ☽ ♃ davon in sich, je öfter je besser, dies kann man erfahren, wenn man solchen ☿ in ♒︎ solviert, so will er sich nicht gern solvieren, er ist davon fett worden, endlich greift es ihn an und wird himmelblau, diesen ☿ kann man mit der ♀ amalgamieren und hernach solche in ♒︎ solvieren, wenn der ☿ damit digeriert und wieder herübergezogen wird, können auch andere ☉ Kiese und Erze damit reich gemacht werden.

178. *Was ist aber der Liquor Alcahest, davon soviel geschrieben und gezankt wird?*

Antwort: Von dem *Liquore Alcahest*, des *Paracelsi* ist viel disputierens, viele halten ihn für das *Menstruum*, welches aus dem flüchtigen Weinstein-Salz gemacht worden, welches ich aber aus vielen Ursachen nicht glaube, weil dieses nur den ♃ von Metallen scheidet und den ☿ los macht, der *Liquor Alcahest* aber solviert alle metallischen Leiber in ein Salz, solch Saltz kann in Ewigkeit nicht wieder in ein Metall reduciert werden, sondern solviert sich an der Luft zu Wasser, dies Wasser wenn es seine zeitlang putrificiert wird, scheidet sich die Quintessenz oben

times as heavy, and immediately plunge it into the naked fire, you get a *spiritum* and red oil, which, unlike the other oils in the *spiritu*, lies on the ground. When you shake it a thousand times it goes to the bottom again, Diana is in the *spiritu*. The remains can be distilled until all the ☿ is in the retort neck, which must be wide. This is dissolved in oil in the cellar, and when the *caput mortuum* is boiled in distilled vinegar, the rose-colored smoke is let off. Let each seeker explore what these four serve to do. I have prescribed enough. You cannot chew the pulp and rub it in, but this here is a short path to the medicine for humans and metals, but it is it not the great *universal* way.

177. *But how is it treated, and what use is it with the common ☿?*

Answer: This *minera* is made fine, mixed with two or three times ☿ and distilled, so the ☿ of it multiplies and takes in the ☽ ♄ from it, the more often the better. This can be experienced when dissolving such ☿ in ▽, though it does not like to be dissolved, and it has gotten fat from it. Finally it attacks it and becomes sky blue. This ☿ can be amalgamated with the ♀ and thereafter be dissolved in ▽; if the ☿ is digested with it and distilled over again, other ☉ pyrites and ores can also be enriched with it.

178. *But what is the liquor alcahest of which so much is written and squabelled over?*

Answer: There is much debate about the *liquore alcahest* of Paracelsus. Many consider it to be the *menstruum* which is made from the volatile salts of tartar, though for many reasons I do not believe this, because this only separates the ♄ from metals and loosens the ☿, whereas the *liquor alcahest* dissolves all metallic bodies into a salt, and such a salt cannot be reduced back into a metal ever again, but dissolves in the air to form a water. This water, if it is putrefied over a period of time, separates to form a quintessence at the top like an oil, which many

auf als ein Öl, welches viele vor das *Potabile* halten, wenn es mit *Spiritus Vini* vereinigt wird, ich halte ihn für das *Corrosivum specificum*, dieser *Liquor* wird auch das höllische Feuer genannt, solviert alle Steine, Beine, Glas, Sand, bringt alles in ein Salz, alle flüchtigen Körper macht es fix und alle fixen Körper flüchtig, durch die *Cohobation*. Eins zum Exempel: Der ☿ wird dadurch fix, am Geruch wie Bisam, durch die Cohobation geht es in Kristall-Tropfen über, welche wieder in 2 unterschiedene *Liquores* zu bringen und kann solches nimmermehr wieder zu ☿ werden, das *Lapis Philosophorum* kann es nicht bezwingen, die Wunder so durch das *Corrosivum specificum* können getan werden im mineralischen Reiche, auch in den anderen zweien, sind unaussprechlich, aller Leiber ihre Quintessenz kann man mit Freuden erlangen ohne große Mühe, nur dass man solche darinnen zerschmilzt und den *Liquorem* davon zieht, das zurückgebliebene Salz kann durch keine Kunst wieder in ein Metall bracht werden, sondern solviert sich an der Luft zu Wasser, welches Wasser so es putreficiert wird, die Quintessenz als ein Rahm oder Fett oben auf wirft, welches von Wasser zu scheiden, ohne diesen *Liquor* ist kein rechtes *Elixir proprietatis* zu machen oder einige und rechte wahre Arznei aus Rosinen, Harzen, Gummien, alle Hölzer, Wurzeln und Kräuter, was unter dem Himmel ist, so durch kein Corrosiv zu zwingen wird dieses höllische Feuer bemeistern, es wird alles durch dieses höllische Feuer umgewandelt, wohlschmeckend, wohlriechend, bleiben ohne Corrosiv zurück, sonderlich die Metalle als ein süßes wohlriechendes Salz, aus Kristall, Krebs-Augen, Eier-Schalen, wenn sie in dem corrosivischen *Specifico* oder *Alcahest* zerschmolzen, dieser wieder davon gezogen, die *Salia* an der Luft zu Wasser worden, sind solche *Medicamenta* wider den Stein, dergleichen nicht zu finden, auch die rechte und wahre Korallen-Tinctur, wie die groben Korallen und Krebs-Stein, ein Bezoar und Herz bewahrendes Pulver sind, die das ganze Geblüt rectificieren, da sie doch nicht weiter kommen als in den Magen, das *Acidum* brechen und niederschlagen, wie weit besser wird nun ihre Wirkung sein, wenn sie zerschmolzen und ihre groben *Feces* zu Boden geschlagen? Dadurch kann von Perlen ein *Medicament*

hold to be *potabile* when mixed with *spiritus vini*. I consider it to be the *corrosivum specificum*. This *liquor* is also called the hellish fire, and it dissolves all stone, bone, glass, and sand, bringing everything into a salt. It makes all volatile bodies fixed and all fixed bodies volatile through the *cohobation*. An example: the ☿ becomes fixed as a result of the musk-like smell, and through the *cohobation* it turns into crystal drops which can be brought back into two different *liquores* and can never become ☿ again. The *lapis philosophorum* cannot conquer it. The miracles that can be performed through the *corrosivum specificum* in the mineral kingdom, also in the other two, are inexpressible. One can obtain the quintessence of all bodies with joy and without great effort, provided that one melts them inside and draws the *liquorem* away. The salt that remains cannot be brought back into a metal by any art, but dissolves in air to form water. This water is then putrefied, throwing up the quintessence as a cream or fat which can then be separated from water. Without this neither the *liquor* nor the true *elixir proprietatis* can be made, nor any right and true medicines from currants, resins, gums, all woods, roots, herbs, and all that is under the heavens; and what cannot be overcome by any corrosive can be mastered by this infernal fire; everything is transformed by this infernal fire until it becomes palatable and fragrant with no corrosiveness remaining. The metals in particular become like a sweet fragrant salt from crystal, crab-eyes, or egg shells. If they are melted in the corrosive *specifico* or *alcahest* and are distilled off from this again, the *salia* are turned into water in the air. Are such *medicamenta* (contrary to the stone, which is not to be found) also the right and true coral tincture? Just as the coarse corals and crab-stones are a bezoar and heart-preserving powder that rectify the whole blood, and yet they get no further than the stomach, breaking down the *acidum* and precipitating it, how much better will their effect be if they are melted and their coarse feces thrown to the ground? In this way a *medicament* can be made from pearls against the stroke, restoring lost

verfertigt werden, wieder den Schlag, die verlorene Sprache wiederzubringen und das Herz zu stärken; *Summa*, ein jeder arbeite und erfahre es selbst, scheue dies Corrosiv nicht, denn dieses Feuer geht wieder davon und lässt die solvierten Körper in einem herrlichen Zustand zurück und ist dieser *Liquor* nach der Solution so gut als vorher, kann die ganze Haut abziehen und eine neue geben, über Gewächse, Mahle und böse Krankheiten wegbringen, doch dient es in offenen Wunden nicht, wenn mit dem Corrosiv die Haut abgezogen wird, wird der Ort mit dem Balsam von ♃ und *Sale Tartari* gesalbt, die Stücke so zu diesem Corrosiv kommen sind: 1. ✶, dies ist für sich ein Schlüssel, alle Körper zu eröffnen. 2. Der corrosivische Sublimat, dieses ist auch ein Schlüssel, in Alles zu gehen. 3. Das *Oleum* ☿, was dieses für Kräfte besitzt, werden die wissen, so es gemacht, 4. werden diese jetzt gemeldeten durch ein starkes Scheidewasser, so von seinem *Capite Mortuo* abgezogen, solviert und diese flüchtigen Körper in noch kleinere Teilchen gebracht, dass sie alle ein *Spiritus* und Geist werden. Dies ist nun das höllische Feuer, welches alles *in Nihilum* verzehrt.

179. *Wie wird es gemacht?*

Antwort: *Recipe* 🜨 bis auf die gelbe calciniert, ☉ des besten, jedes *ana*, daraus wird ein ℣ destilliert ohne Vorschlagung Wassers, das *Caput Mortuum* solviert man in destilliertem Wein-Essig sein Salz davon, lässt es abrauchen zur Trockne, darauf gießt man das ℣, zieht es stark ab von seinem Salz, dass die Retorte glüht. In dieses abstrahierte ℣ von seinem *Capite Mortuo* in ein Pfund, wirft man 4 Lot corrosivischen Sublimat, und 3 Lot ✶, zieht es über, so ist der ✶ und Sublimat auseinander getrennt und ein herrliches *Menstruum*. In dieses schüttet man gleich soviel *Oleum* ☿, zieht es wieder über, so ist es fertig und immer zu gebrauchen. Welches sehr große Kräfte besitzt, das *Oleum* ☿ wird also gemacht: corrosivischer Sublimat wird in 3 Teilen *Spiritus Salis* solviert und von selbigem wieder abgezogen, bis er durch die *Cohobation* als ein Öl aufsteigt.

speech and strengthening the heart. *Summa*: every single one should work and experience this for themselves. Do not shy away from this corrosive, because this fire goes away again and leaves the dissolved bodies in a wonderful condition, and after the solution this *liquor* is as good as before. The whole skin can be peeled off and a new one is created, it removes growths, marks, and evil diseases, but it is of no use for open wounds. When the skin is peeled off with the corrosive, the place is anointed with the balm of ♄ and *sale tartari*, The components which form this corrosive are: (1) ✶: this is a key to open all the bodies that are before you. (2) The corrosive sublimate: this is also a key to penetrate everything. (3) *Oleum* ☿: those who made it will know what powers it possesses. (4) The above-mentioned [components] are dissolved by a strong separating water from its *capite mortuo*, and these volatile bodies are reduced to still smaller particles, so that they all become one *spiritus*. This is now the infernal fire which consumes everything *in nihilum*.

179. *How is it prepared?*

Answer: *Recipe* ⊕ calcined down to the yellowness, ☉ of the best [quality], each *ana*; some ▽ is distilled from it without water into the receiver, the *caput mortuum* is dissolved in distilled wine vinegar, its salt is left to smoke off until dry, the ▽ is poured onto it. Distill it from its salt so much that the retort glows. Into one pound of this abstracted ▽ from its *capite mortuo* you throw four lots of corrosive sublimate, and three lots of ✶. Distill it over so that the ✶ and the sublimate are separated from one another and [form] a glorious *menstruum*. You pour as much *oleum* ☿ into this at once, then you distill it over again, at which point it is finished and ready to use. This has very great powers. The *oleum* ☿ is made in this way: corrosive sublimate is dissolved in three parts *spiritus salis* and the same is distilled off again until it all rises as an oil through the *cohobation*.

180. *Von dem Corallate des Paracelsi, damit er so viel Wunder-Kuren soll verrichtet haben, möchte ich auch gern hören?*

Antwort: Dies beschreibt er selber also: Dass du nimmst *Liquorem Aquilae* und denselben exsiccierst. Nachdem *Liquorem* ⊕ *Præparati*, denselben imbibierst und wie von den calcinierten *Tartaro* der Process ist, abzuziehen seinen *Spiritum*, dermaßen hier auch, so lange, bis derselbe entzogen wird: Also ist das *Corallium* bereit. Hier meint er durch den *Liquorem aquliæ* den nassen Jungfer ☿, welcher soll eingetrocknet werden und mit dem *Liquore* von ⊕ übergossen. Dasselbe so oft getan, bis er alle darauf bleibt und korallenrot worden, dies können nur die verrichten, welche die Elemente zu scheiden gelernt, im philosophischen Werke, welche den ☿ der Weisen zubereiten können, weil er aber noch einen *Corallat* zu machen gelernt, will ich solchen auch setzen, ist aber eine Beschreibung eines Præcipitats, in *Morbo Gallico*, und anderen garstigen Krankheiten, beschreibt ihn mit diesen Worten: Das Quecksilber soll in einem ▽ calciniert werden, alsdann ein Gradier-▽ 5 oder mehrere Mal davon abgezogen, bis der Præcipitat von Farben gefalle und schön rot ist, alsdann einen rectificierten Branntwein davon gezogen zum 9 mal, alsdann hast du *Præcipitatum diaphoreticum*, &c. Weiter ist noch eine Heimlichkeit zu wissen vom ☿ *præcipitato*, so derselbe nach seiner Calcination und Coloration, mit *aqua sal taberi* abgesüßt, davon destilliert und alle Destillationes mit neuem Wasser so oft und viel getan, bis das Wasser süß davon geht, so ist das Præcipitat süß, dem Zucker oder Honig gleich, in allen *Vulneribus, Ulceribus* und *Morbo Gallico* das größte Haupt-Stück, so erfreut er ja einen jeglichen Alchimisten, denn er das Goldes Augment ist, in das Gold eingeht und bei ihm zu beständigem guten Golde wird, so weit Paracelsus. Weilen aber andere den *Corallat* zu machen gelernt, zu welchem sie das *Elementum Ignis* erfordern, wenn sie nur die rechte ♀ wüssten und eben aus der *Minera* kommt, daraus das *Gluten Aquilæ* geschöpft, weil aber die Künstler, wie schon gedacht, sehr rar, so hat Polemann das *Elementum Ignis* aus der gemeinen ♀ lernen machen, durch den sauren

180. *I would also like to hear about the corallate of Paracelsi, with which he is said to have performed so many miraculous cures?*

Answer: He himself describes it like this: that you take *liquorem aquilæ*[309] and exsiccate the same. After the *liquorem* ☉ *præparati*, you imbibe it, and as with the calcined *tartaro*, where the process is to draw off its *spiritum*, so too here, until it is withdrawn: thus the *Corallium* is ready.[310] By the *liquorem aquilæ* he means here the wet virgin ☿, which should be dried and poured over with the *liquore* from the ☉. The same thing is repeated until it all stays on it and has become coral-red. Only those who have learned to separate the elements in philosophical works can do this, i.e., those who can prepare the ☿ of the wise. But because he still learned to make a *corallate*, I also want to add this, which is but a description of a precipitate. In [a treatise on] *morbo Gallico* and other nasty diseases, he describes it with these words: "The mercury shall be calcined in an ▽⃒, then a gradating ▽⃒ shall be distilled five or more times from it, until the precipitate has fallen from the colors and is beautifully red; then a rectified brandy is drawn from it for the ninth time, then you have *præcipitatum diaphoreticum &c*". Furthermore, there is still a secret to know about ☿ *præcipitato*, which, after its calcination and coloration, is rinsed out with *aqua sal taberi* and distilled from it. All the distillations are performed with new water frequently and repeatedly until the water comes off sweet, and so that the precipitate is also sweet like sugar or honey; the greatest principal part in all *vulneribus, ulceribus*, and *morbo Gallico*. Therefore it pleases every alchemist, because it is the augment of gold which penetrates gold and by this becomes constant fine gold, per Paracelsus. Others have learnt to make the *corallate*, for which the *elementum ignis* is required, but only if they knew the right ♀, which simply comes from the *minera* from which the *gluten aquilæ* was drawn; but because such artists, as already mentioned, are very rare, Polemann[311] has taught how to make the *elementum ignis* out of common ♀, to bring the ♀ into ☉ through the acidic

Spiritum von ✶ das ♀ in 🜨 dadurch zu bringen, weil es aber ein sehr langwieriger, mühseliger und kostbarer Weg, so ist der aus dem gemeinen 🜨 viel besser, weil das ♀ in der Erde durch ein *Sal esurinum* solviert, seinen *Spiritum* noch bei sich hat, ohne große langwierige Kosten und Mühe zu machen, daher mir sicher zu folgen, wie ich es gemacht, wenn der ☿ durch einen *Spiritum Nitri* zum fixen roten Præcipitat gemacht wird, darauf das *Elementum Ignis* aus dem 🜨 gossen und wieder übergezogen zu etlichen Malen, bis er genug gesättigt und dann einen *Spiritum Vini* davon abgezogen, bis der Præcipitat süße, das *Elementum Ignis* aus dem gemeinen blauen 🜨 wird also gemacht: Er wird in der Sonnen zur Gilbe calciniert, diesen destilliert man mit dem größten Feuer, nimmt die Vorlage warm ab, verstopft solche aufs festeste, calciniert das *Caput Mortuum* mit gleich schwer 🜍 bis es alles ausbrennt, solches wirft man, weil es noch warm, in eine gläserne Retorte, gießt den Schwefel-Geist mit seinem *Oleo* darüber, nach und nach, allezeit beide Gefäße zugehalten, dass kein subtiler *Spiritus* verriecht, bis es alles darauf gefüllt, der *Spiritus*, das *Caput Mortuum*, so weit solviert, dass es schwimmt und rot ist, die Retorte legt man in Sand, zieht es in einem Feuer über, gießt den *Spiritum* wieder darauf, zieht ihn wieder ab, dies so oft, bis es alles darauf bleibt und süße wird, wenn die Retorten brechen, muss man neue nehmen, alsdann nimmt man diesen 🜨, destilliert ihn im großen Feuer mit der größten Hitze, man kann auch ein wenig ✶ darunter nehmen, es darf die geringste Feuchtigkeit nicht an sich ziehen, so bekommt man ein blutrotes Öl und Feuer aus der ♀. Wenn es recht gemacht und ein Tropfen Wasser in das Glas fällt, schlägt das Glas in 1000 Stücke und wird dies Feuer verschüttet, aus diesem kann das grüne und süße Öl separiert werden, ist also das *Elementum Ignis* aus dem 🜨 beschrieben.

181. *Das circulierte Salz was ist dieses?*

Antwort: Es ist das doppelte Salz der Weisen und ihr ✶☿ *Philosophorum*, die geblätterte Erde, wer aber dies nicht machen kann, der gebrauche sich der süßen Kristalle aus dem gemeinen Salz, dieses lässt man in

spiritum of ✶; but because it is a very lengthy, laborious, and expensive way, the way proceeding from the common ☉ is much better, because if the ♀ is dissolved in the earth through a *sal esurinum*, it has its *spiritum* still with itself without requiring tedious costs and great effort. Therefore, one can surely follow what I have done, when the ☿ is made into the fixed red precipitate by a *spiritum nitri*, after which the *elementum ignis* is poured out of the ☉ and distilled over again several times until it is sufficiently saturated and then a *spiritum vini* is distilled from it until the precipitate is sweet. The *elementum ignis* from the common blue ☉ is made the same way: it is calcined in the sun to yellowness, and then it is distilled with the greatest fire. The receiver is removed while warm, and is plugged up tightly as possible again. The *caput mortuum* is calcined with an equal weight of ♄ until it all burns out. This is then thrown into a glass retort while it is still warm, at which point the sulphur spirit is gradually poured over it with its *oleo*, always keeping both vessels closed so that no subtle *spiritus* escapes until it is all filled up. The *caput mortuum* is dissolved in the *spiritus* to such a degree that it floats and becomes red. The retort is placed in the sand, distilled in a fire, the *spiritus* is poured on it again and distilled off again; this is repeated frequently until all of it remains on it and becomes sweet. If the retorts break, new ones must be used. This ☉, is then taken and distilled in a large fire with the greatest heat. A little ✶ can also be applied, but it must not absorb the slightest moisture. In this way one receives a blood-red oil and fire from the ♀. If done correctly and a drop of water falls into the glass, the glass breaks into a thousand pieces and this fire is spilled, from which the green and sweet oil can be separated, so the *elementum ignis* from the ☉ was described.

181. *The circulated salt,*[312] *what is this?*

Answer: It is the double salt of the wise and their ✶☿ *philosophorum*, the leafed earth, but if you cannot prepare this, you shall use the sweet crystals from the common salt. This is allowed to flow in a retort which

einer Retorte fließen, welche hinter eine Röhre, darinnen lässt man etliche Tropfen Wasser durch einen Feder-Kiel fallen, alsbald kommen die *Spiritus* mit großer Heftigkeit, damit wird fortgefahren, bis das Salz alles übergestiegen, das dephlegmiert man, darauf gießt man *Spiritum Vini*, zieht ihn ab, cohobiert ihn zu 3 oder 4 Malen oder man solviert das Salz in Rettichen, das destilliert man, solviert es abermals in Rettichen, dies tut man so oft, bis das Salz alles übergestiegen, alsdann dephlegmiert, wenn man solches eine Zeit im Keller stehen lässt, schießt es in Kristallen, solches mit *Spiritu Vini* cohobiert, bis das Salz als ein grünes Öl zurück bleibt, welches viele den grünen ♌ nennen, er ist es aber nicht, denn der grüne ♌ ist aus einem anderen Lande, nämlich das metallische Salz, der Wasser-Stein der Weisen, aus dem Koch-Salz kann auch ein *Sal enixum* gemacht werden, auf diese Weise: ⊖ im Regenwasser solviert, darein gleich soviel *Oleum* ⊕ gegossen, als das Salz gewogen, dieses wird abgezogen, bleibt das *Sal enixum* zurück, damit kann man Gold solvieren, güldische *Crocus*, wenn es wohl damit geflossen, gießt man es aus, extrahiert das rote Salz mit *Spiritus Vini*, dies kann man zu einem güssigen *Sale Alcali* machen, wenn ihm gute tingierende ♃ zugesetzt werden und im Feuer 3 Stunden zur Fixation fließen lässt, aber ohne ♃ muss es mit Kohlen alcalisch gemacht werden, wenn es ☉ allein solvieren soll, auf solche Art kann auch durch Salpeter ein flüssiges Salz gemacht werden, bei diesem geht in der Abziehung ein *Aquafort* über, bei dem Koch-Salz ein *Spiritus Salis*, bei dem Weinstein-Salz und *Oleum* ⊕ ein *Spiritus* wie von Urin, ist also schlechte Kunst, die fixen Salze in *Urinosa* zu verwandeln. In diesen Salien kann man ☿ kochen.

182. *Das Sal Tartari flüchtig zu machen, möchte ich auch gern wissen?*

Antwort: Solches geschieht auf unterschiedliche Art, die erste, welche mir am besten gefällt, geschieht auf diese Weise: Es wird ein sehr feuriges *Sal Tartari* an der Luft solviert, darauf wird gegossen das flüchtige Terpentin-Öl, in einem breiten Zucker-Glas und oft durcheinander gerührt, so nimmt das fette Terpentin-Öl das feurige Salz zu sich und

is located behind a tube. Inside this you insert a few drops of water through a quill, and upon doing so the *spiritus* immediately comes with great intensity. This is continued until the salt is completely sublimated, at which point it is then dephlegmated. Then you pour *spiritum vini* on it and distill it off, cohabating it three or four times. Either this, or you dissolve the salt in radishes,[313] distill it, and dissolve it again in radishes, doing this frequently until the salt is completely sublimated and then dephlegmated. If one leaves such a thing in the cellar for a while, it shoots forth crystals; and these are cohobated with *spiritu vini* until the salt remains as a green oil which many call the green ♌, but it is not, for the green ♌ is from another land, namely the metallic salt, the water stone of the philosophers. A *sal enixum* can also be made from table salt in this way: ☉ is dissolved in rain water and poured into as much *oleum* ⊕ as the weight of the salt. This is distilled, and the *sal enixum* remains. With it one can dissolve gold, gold-bearing *crocus*, if it is flowing well with it, one pours it out and extracts the red salt with *spiritus vini*. This can be made into a pourable *sale alcali*, if well-tinging ♃ is added to it and it is allowed to flow in the fire for three hours for fixation; but without ♃ it must be made alkaline with coals if it is to dissolve ☉ alone. In such a way a liquid salt can also be made with saltpeter, with which an *aquafort* goes over in the distillation, with the cooking salt a *spiritus salis*, with the cream of tartar and *oleum* ⊕ a *spiritus* like from urine. It is therefore bad art to transform the fixed salts into *urinosa*. In these *salia* you can cook ☿.

182. *I would also like to know how to make the sal tartari volatile?*

Answer: This can be done in different ways, the first, which I like best, happens in this way: a very fiery *sal tartari* is dissolved in the air. The volatile turpentine oil is poured onto it in a wide sugar glass, and frequently stirred together so that the fatty turpentine oil absorbs the fiery salt and from these one gets a thick soap. One oil dissolves the other, and when

wird aus beiden eine dicke Seife, ein Öl löst das andere auf, wenn sie nun ein Leib worden und keine Nässe mehr zu sehen, destilliert man es, so steigt das flüchtige *Oleum Terebintini* herüber und weil es sich in der Zeit der 40 Tage mit dem Öl, als dem Weinstein-Salz, in seinem Innersten vereinigt, so reißt es solches mit in die Höhe, dieses flüchtige Weinstein-Salz kann man alle Tage *in infinitum* vermehren, wenn man 4 oder 5 Teile dieses *Spiritus* nimmt, solviert darinnen 1 Teil *Sal Tartari*, zieht es über, so vermehrt er sich. Die 2. Art das Weinstein-Salz flüchtig zu machen, geschieht also: Es wird destillierter Wein-Essig darauf geschüttet, so geht ein ungeschmackvolles Wasser über, damit verfährt man mit Aufgießen des destillierten Wein-Essigs, bis keine Ebullition mehr, sondern der Essig anfängt sauer wieder überzusteigen im *Balneo*, dann destilliert man es mit starkem Feuer, so steigt endlich das *Sal Tartari* in die Höhe, teils wie geblätterter Kalk, solches kann man in *Spiritum* werfen, als ein *Menstruum*, Metalle, Edelgesteine und dergleichen Tincturen zu extrahieren. Der 3. *Modus* ist, dass man den *Spiritum Tartari* über *Sal Tartari* rectificiert, so geht er als Feuer über.

183. *Was hat das flüchtige Weinstein-Salz, der Spiritus, für Nutzen?*

Antwort: Es hat diesen Nutzen, es tötet alle *Corrosiva*, alle Säure, die bei dem Menschen ist, Krankheit, Schmerzen und Reißen machen, welche den Leib zerschmelzen, auch bricht es den Nieren- und Blasen-Stein, in Podagra, Gicht und Zipperlein ist seinesgleichen nicht, welchen *Tartarum* er auflöst und zerschmelzt, absonderlich wenn der *Spiritus* von ungelöschtem Kalk damit einverleibt, auch wider corrosivisches Gift, welchen ein Mensch in Leib bekommen, es sei ein corrosivischer Sublimat oder *Aquafort*, auch wider den Arsenik, wiewohl das Gift des Arseniks nur durch Milchtrinken getötet wird, ingleichen durch Butter, wie ich Exempel weiß, dass Leute dadurch gerettet worden, welche doch viel davon bekommen. Auch ist dieser *Spiritus* der Leber gut und der Milz, wenn solche scirrhosisch und von solcher *Materia* erhärtet, wird es diese steinige *Materia* erweichen und ausführen, auf die Schmerzen des Podagra wird es mit Tüchlein gelegt, welchen

it has now become one body and there is no longer any moisture to be seen, you distill it. The volatile *oleum terebintini* rises over, and because it combines in the period of forty days with the oil, i.e., the tartar salt, in its innermost parts, it draws it up with it. This volatile tartar salt can be multiplied *ad infinitum* every day. If you take four or five parts of this *spiritus*, and dissolve one part *sal tartari* in it, you can distill it over so that it multiplies. The second way of making the tartar volatile is done like this: distilled wine vinegar is poured on it, so that an unpalatable water is distilled over. This is done by pouring the distilled wine vinegar on until there is no more ebullition, but the vinegar begins to increase in acidity again in the *balneo*, because if you distill it with strong fire, then finally the *sal tartari* rises to the top, somewhat like flaked lime. This you can throw in *spiritum* as a *menstruum* to extract metals, precious stones, and similar tinctures. The third *modus* is that the *spiritum tartari* is rectified via *sal tartari*, so it passes over as fire.

183. *What is the use of the volatile tartar salt, the spiritus?*

Answer: It has this use, it kills all the *corrosiva*, all the acids that are in humans, causing sickness, pains, and tearing which melt the body. Also it breaks down the kidney and bladder stone; and in the treatment of podagra, gout, and niggles it is unparalleled. It dissolves and melts the *tartarum*, especially when the *spiritus* of quicklime is incorporated into it. It is also useful against corrosive poison when such substances enter the human body, whether it be a corrosive sublimate or *aquafort*, also against arsenic, although the poison of arsenic is only destroyed by drinking milk, also by butter, as I know of examples whereby people who got plenty of it have been saved by it. This *spiritus* is also good for the liver and the spleen. If this is scirrhous and hardened by such *materia*, it will soften and carry out this stony material. It is placed with a cloth on the pain of the podagra, which it calms, opening all

es stillt, öffnet alle *Obstructiones*, so von zähen schleimigen Geblüt entstehen, wodurch die *Circulation Sanguinis* verhindert wird, doch muss man hierinnen auch nicht zuviel tun. Denn wenn es überflüssig gebraucht, zerstört es die Verdauung, weil es das *Acidum* des Magens tötet, dadurch wird der Magen unkräftig, dass er nicht wohl verdauen kann, wer aber der Säure im Magen zuviel hat, dem ist dieses eine gewisse Kur, wenn man in diesem *Spiritu* einen ♃ auflöst, er sei aus was für einem Metall oder Mineral er wolle, denn dieser Schlüssel ist nach dem *Universal*-Schlüssel der andere geheime Schlüssel im vegetabilischen Reiche und ist ihm kein Schloss so feste, dass er nicht aufschließt. Wenn er nun den ♃ in roten Tropfen übergeführt, hat man dadurch einen roten ♃ Balsam, der in die letzte Verdauung geht und den ganzen menschlichen Körper vor Fäule erhält, welches kein ♃ Öl sonst tun kann, sondern kommt in Magen und Därme, weiter nicht, aber dieser Balsam dringt durch alle Adern und Fleisch, macht den Leib zur lebendigen *Mumia*, sonderlich wenn Zedern-Holz, rote Myrrhen, Aloe und dergleichen Dinge damit gearbeitet werden, denn dieses *Menstruum* nimmt nichts in sich als was Fett und ♃ ist und weil dieses ♃ Öl den flüchtigen und alcalischen Geist bei sich hat und in seinem Innersten damit vereinigt, so führt es ihn ein, dringt durch alle Adern und Nerven, auch in faulen Schäden, wo der natürliche Balsam verloren, wird es den ersetzen und solchem zu Hilfe kommen, von innen und außen, wenngleich schon das Geäder verfault und das Fleisch von Knochen sich abgelöst.

184. *Kann dieses Öl oder flüchtige Weinstein-Spiritus nicht zum Nutzen der Metalle dienen?*

Antwort: Ja allerdings, es tut darinnen lauter *Miracula*, weil es nach dem *Universal*-Schlüssel der andere geheime Schlüssel ist, erstlich im ganzen vegetabilischen Reiche, dass alle Kräfte der Kräuter, Blumen und Wurzeln, Rosinen und alle Species dadurch können in ihre Quintessenz gebracht und der grobe terrestrische Leib dadurch zu Boden geschlagen werden, aber zum metallischen Reiche hat es

obstructiones which are caused by tough, mucilaginous blood, whereby the *circulation sanguinis* is prevented, but one must not do too much here either, for if it is used unnecessarily it destroys the digestive system, because it kills the *acidum* in the stomach. The stomach then becomes weak, so that it cannot digest properly; but if you have too much acid in your stomach, this is a certain cure for you. If you dissolve a ♃ in this *spiritu*, it can be made of whatever metal or mineral one wants, because after the *universal* key, this key is the second secret key in the vegetabilic kingdom and no lock is so strong that it is not unlocked by it. If it now converts the ♃ into red drops, you have a red ♃ balsam, which goes into the last digestion and keeps the whole human body from putrefying. No other ♃ oil can do this, for it only enters the stomach and intestines and goes no further. But this balsam penetrates through all veins and flesh, turns the body into a living *mumia*, especially when cedar wood, red myrrh, aloe, and similar things are worked into it, because this *menstruum* contains nothing but what is fat and ♃ and because this ♃ oil has the volatile and alkaline spirit within it, and unites with it in its innermost being, it introduces it, penetrates through all veins and nerves, even in putefying wounds. Where the natural balsam is lost, it will replace it and come to assist from within and without, even if the veins are already rotting and the flesh is separating from the bones.

184. *Cannot this oil or volatile spirit of tartar be used for the benefit of the metals?*

Answer: Yes, of course, it does nothing but *miracula*, because after the *universal* key it is the other secret key, first in the entire vegetabilic kingdom, so that all the powers of herbs, flowers, roots, raisins, and all species can be brought to their quintessence through it and the coarse terrestrial body are thereby precipitated to the ground. But it has great love and for the metallic kingdom and is also of great use for whoever

große Liebe und bringt auch Nutzen, wer diesen Schlüssel nur wohl anzubringen weiß, er nimmt die ♃ davon, macht die ☿ lebendig und frei. *Item* benimmt den rohen Mineralien ihren Gift, wenn ein guter güldischer ♃ darinnen flüchtig zum blutroten Öl worden, die ☽ damit digeriert wird, kann sie zum wahrhaftigen Golde werden, welches nimmermehr kein Scheidewasser angreifen wird, auch gibt es Nutzen bei dem ☿ und ♄.

185. *Kann denn aus dem großen Vegetabili, als aus dem Weinstock auch ein Stein gemacht werden?*

Antwort: Ja freilich, allerdings, es sind 3 *Universal*-Steine, der mineralische, animalische und vegetabilische, welcher aus dem Wein gemacht wird, von seinem reinen Geist, wenn dieses Salz seinen eigenen ♃ Geist, den *Spiritum Vini* coaguliert, gerinnt und gestehend macht, doch auch durch seinen eigenen Mediatoren, muss er mit seiner Erde vereinigt werden.

186. *Woraus wird denn der animalische Stein gemacht?*

Antwort: Aus dem großen animalischen Tier, dem Menschen, der *Phosphorus* auch, diese haben in der Medizin großen Nutzen und sind viele curiöse Dinge damit zu machen, man kann sie alle beide zum mineralischen Reiche brauchen.

187. *Dienen denn diese Steine nicht zum großen Universal-Stein oder kommen sie nicht dazu?*

Antwort: Man muss keines zum anderen vermengen oder damit vermischen, gleich bleibt seinesgleichen und hält sich zu seinesgleichen, bringt auch seiner Art nach Frucht, wer aber dawider handelt, macht eine Missgeburt und zeugt einen Maul-Esel, als wenn Pferd und Esel sich zusammen gatten, wird dieses unfruchtbare Tier gezeugt. Zum Mineral-Stein darf nichts als ein Mineral kommen, welches alles selber bei sich hat, sich vollkommen zu machen. Wer es anders tut, tut der

knows how to use this key well. It takes the ♄ therefrom and makes the ☿ alive and free. It also removes the poison from the raw minerals. If a good, gold-bearing ♄ within it has turned into a volatile, blood-red oil, when the ☽ is digested with it, it can become real gold, which will never be attacked by any *aquafort*. There are also benefits with the ☿ and ♄.

185. *Can a stone also be made from the great vegetabili, like out of the grapevine?*

Answer: Yes, of course, but there are three *universal* stones: the mineral, the animal, and the vegetabilic, which are made from the wine, from its pure spirit. When this salt coagulates, congeals, and stabilises its own ♄ spirit—the *spiritum vini*—but also through its own mediators, it must be united with its earth.

186. *What is the animal stone made of?*

Answer: From the great animal, the human being, and also from phosphorus. These are of great use in medicine and many curious things can be done with them; both of them can be used in the mineral kingdom.

187. *Do not these stones then serve to form the great universal stone, or do they not come with it?*

Answer: You do not have to mix or mingle one thing with another. Like remains with like and keeps to its own kind, and also bears fruit according to its own kind, but whoever acts against this produces a monstrosity and fathers a mule, that barren animal begotten from the marriage of horse and donkey. Nothing but a mineral should be used for the mineral stone, which has everything within itself to make itself perfect. Whoever does otherwise commits violence against nature and

Natur Gewalt und wird sich am Ende betrogen finden, denn er nicht erhalten wird, was er verlangt und gehofft hat.

188. *Wenn nun in diesen Medicamenten der rote ♃ aus der philosophischen Materia zu Gesicht kommt und zubereitet wird, so kann man ja damit tingieren, wenn man will?*

Antwort: Noch weit gefehlt, dieser rote ♃ tingiert nicht als durch seine fruchtbare Gemahlin, welche er oft umfangen und küssen muss und wenn sie von ihm geschwängert und die solarische Frucht geboren, so muss diese mit Feuer gespeist werden, weil es heißt: Gib dem Feuer Feuer: dem ☿ dem ☿, dann wirst du reich werden. Die philosophische Arbeit ist nicht so leicht getan und ist wohl in meinem *Mineralischen Glutine* und im *Philosophischen Perlen-Baum* als auch hier deutlich genug beschrieben. Wer dies nicht verstehen kann noch will, dem ist nicht zu helfen, wenn mich jemand so unterrichtet hätte, ich wollte ihm mein ganzes Vermögen gegeben haben. Wenn alle Scribenten so klar als ich geschrieben, so wäre es längst Tag worden in der Welt, aber es hat nicht sein sollen, alles hat seine Zeit, die Lilien und Rosen-Zeit auch. Es wird bald Tag werden in der Welt, weil die Morgenröte angeht, worauf das große Uhr-Werk zu Ende laufen wird, da dann die *Ultima Materia* zu der *Prima Materia* kommen wird und das Feuer seine Wunder eröffnen wird, welche seit der Zeit Schöpfung nicht sind offenbart noch erkannt worden, wenn die *Turba* die Tennen fegen wird.

189. *Aus dieser Materia können wohl viele Wunder-Dinge damit gemacht werden, weil es auf vielerhand Art kann traktiert werden?*

Antwort: Es sind an dem *Autorem*, so den *Filum Ariadnes* geschrieben, etliche 70 Wunder von denselben angehängt worden, ich habe derselben durch meine Experienz wohl 1000, glaube ganz gewiss, dass ihrer noch mehr darinnen sind, auch sind sie nicht alle zu erforschen, absonderlich in der Medizin, da die Krankheiten als wie durch eine Zauberei ausgetrieben werden und in seiner Multiplication auch. Wenn es zu anderen mineralischen und metallischen Körpern gesetzt wird,

will find himself cheated in the end, for he will not receive what he has asked and hoped for.

188. *If the red ♃ from the philosophical materia is seen and prepared in these medicamenta, you can tincture with it if you want?*

Answer: Still far from it, this red ♃ tinctures in no other way than through its fertile wife, whom it often has to embrace and kiss, and when she is made pregnant by it and the solar fruit is born, it must be fed with fire, because of the saying: give fire to fire: give ☿ to ☿,[314] then you will become rich. The philosophical work is not done so easily and is described clearly enough in my Mineral Gluten and in the Philosophical Pearl-Tree as well as here. Anyone who cannot or does not want to understand this cannot be helped. If someone had taught me that, I would have given him my entire fortune. If all scribes wrote as clearly as I do, understanding would have come long ago in the world, but this could not have happened in any case since everything has its proper time, for there is a time for lilies and a time for roses too. There will soon come day in the world, because the dawn is rising, whereupon the movement of the great clockwork will run to its end, since the *ultima materia* will come to the *prima materia* and the fire will open its wonders, which have been neither revealed nor recognized since the time of creation, when the *turba* will sweep the threshing floor.

189. *Since it can be treated in many ways, are many things of wonder able to be made from this materia?*

Answer: Some seventy miracles have been appended by the *authorem* who wrote the *Filum Ariadnes*,[315] and I have probably seen a thousand of these miracles through my own experience, and I believe with certainty that there are even more within [this *materia*], especially in medicine, since diseases are cast out as if by magic, and also in its multiplication. But they cannot all be explored. When it is applied to other mineral and metallic bodies so that it can melt and destroy them and

wie es solche kann zerschmelzen, zerstören und seinen inneren Geist durchdringen, klebt an allen, verlässt sie nicht und wenn es Zeit hat, verbessert es solche. Kein Schloss ist ihm zu feste, welches es nicht aufschließt, es nimmt von den Edelgesteinen die Farben, wenn es weiß ist, ist es aber gefärbt, so gibt es ihnen die Farben, auch tingiert er nicht, er werde denn erst selber tingiert, hernach kann er wieder färben.

190. *Was werden denn sonst vor Gläser daraus?*

Antwort: Wie man sie haben will, nach der gemeinen Art mit Pottasche, eine Blaue, mit ☉ aber ein viel schöneres durch Calcination und Extraction ein Glas von Regenbogen-Farben, welches wie brennende feurige Spiegel in der Sonnen und bei Licht leuchtet.

191. *Da die Materia nun so übel zu bekommen, warum haben denn die Philosophi geschrieben, sie wäre allenthalben genug und umsonst zu bekommen, es wäre der Stein am Weg geworfen, nur dass man sich danach bückte und solchen aufhübe?*

Antwort: Ja, zur Zeit, da sie ihre Bücher schrieben, war es also, es lag da am Wege, niemand begehrte es, alle Gruben staken voll von dieser *Materia*, niemand aber wusste es zu gebrauchen, weil nichts daraus zu machen war und lag da am Wege, welches die Bergleute auf die Halde stürzten, war der unwerte und verachtete Stein nach der Kuh geworfen, jetzt aber, da man es hat lernen gebrauchen, ist es nicht mehr also, sondern ist nur in fester Hand.

192. *Wenn nun nur eine Materia ist, welche den Gold-Leim und roten Gummi in sich hat, hat es denn nicht sonderliche Zeichen, dabei man es erkennen kann?*

Antwort: Ob schon nur eine *Materia* ist, so muss man doch die *Signa diagnostica in eligenda Materia* wissen, sonst ist es zum Anfang gefehlt, weil ich aber in meinem Büchlein, dem *Mineralischen Glutine*, schon davon geschrieben, kann es alldorten nachgesehen werden, hier aber zum Überfluss diese Nachricht geben, dass unsere grobe *Materia*,

penetrate their inner spirit, it clings to all of them, does not leave them, and if there is time, it improves them. No lock is so strong that it cannot be opened by this. It takes the colors from the precious stones if it is white, but if it is colored, then it gives them color, and it does not tinge unless it is first tinged itself, thereafter it can color again.

190. *What other glasses are made of it?*

Answer: Whatever you want. According to the common way with potash, a blue glass,[316] but with ☉, through calcination and extraction, a much nicer glass with the colors of the rainbow, which shines like burning fiery mirrors in the sun and in the light.

191. *Since the material is so hard to come by, why did the philosophers write that there was enough of it to get everywhere and for free, that the stone would be thrown in the street, and that one only needed to stoop to pick it up?*

Answer: Yes, at the time they were writing their books it was like that, it was lying there on the street, nobody wanted it, all the pits were full of this *materia*, but nobody knew how to use it because nothing could be made of it and it was lying there on the street, which the miners threw onto the heap, the unworthy and despised stone was thrown at the cow. Now that people have learned how to use it, however, this is no longer the case, and there is simply a tight hold on it.

192. *Now, if there is just one materia that has the gold glue and red gum in it, does it not have special marks by which to recognize it?*

Answer: Although there is only one *materia*, one must know the *signa diagnostica in eligenda materia*, otherwise it will be done wrong from the beginning. But because I have already written about this in my booklet, *The Mineral Gluten*, it can be consulted there, but here I give this additional description: that our coarse *materia*, as it comes from

wie sie aus den Bergen kommt, schon geziert mit den glorwürdigsten Fünklein, weil es die Klarheit aller Metalle in sich hat, welche in unserem *Electro* leuchten, wie die Sterne an diesen Sonnen-Himmel mit Funkeln, im anderen Werk aber, in der anderen Arbeit leuchtet es erst recht, wenn aller Sternen und Planeten Glanz aufgeht, denn unser *Electrum* ist der Himmel, darinnen Sonne und Mond mit allen Planeten laufen müssen, welches unsere *Marcasita plumbea* ist, diese ist nun wieder etlicher Arten, aber nur eine dient dazu und wird sie niemand besser kennen als ich, weilen selber danach gereist und viele Gebirge durchkrochen, wo sie bricht, sonderlich in sächsischen Landen, will aber hier noch ein unfehlbares Kennzeichen geben, wenn sie in ▽ solviert wird und nach der Solution, wenn sich das Sediment gesetzt, das ▽ schön rosenfarben aussieht, ist es die rechte *Materia*, sie ist nicht alle reich, wenn es der besten ist, kann man wohl 12 bis 16 Lot vom Pfund haben, dass zum Werk dient, ist sie aber arm von schönen Farben und schimmert nur weißlich, hat sie weniger, etwa nur 8 oder 6 Lot der schönen Farben in sich, davon im Vertrauen noch viel zu reden wäre.

193. *Wenn man nun nicht das ganze Corpus nehmen darf, wie macht man es dann?*

Antwort: Man muss ein *Menstruum* wissen, solches darinnen zu reinigen, dass das andere nicht zu sich nimmt, sondern nur sein Blut und Leben.

194. *Ist es denn nur auf einerlei Art zu traktieren?*

Antwort: Der ersten Reinigung ist eben nicht nur ein Weg und hier sind viele Verzögerungen, da mancher sich viele Wochen plagt, kann es auch wohl in 8 Tagen und noch kürzer verrichtet werden, in der Nacharbeit ist aber nur ein Weg, in der Composition.

195. *Sie haben aber geschrieben vom feuchten und trockenen Weg?*

Antwort: Er ist freilich erst feucht, hernach aber trocken, es ist *re vera*

the mountains, is already adorned with the most glorious little sparks, because it contains the clarity of all the metals, which shine in our *electro*, sparkling like the stars in this heaven of the sun. Yet in the second work it shines all the more when all the stars and planets rise, because our *electrum* is the sky in which the sun and moon have to run with all the planets, which is our *marcasita plumbea*. This is again only one of several kinds, but only one of them can be used for this; and no one will know it better than I, because I have traveled to it myself and have crawled through many mountains where it fractures, especially in Saxony, but I want to give an unmistakable signpost here: if it is dissolved in ▽ and, after the dissolution, when the sediment settles, the ▽ looks beautifully rose-colored,[317] then this indicates that it is the right *materia*. It is not always rich, but if it is the best, one can probably have twelve to sixteen lots from the pound,[318] which serves for the work; but if the beautiful colors are poor and it only has a whitish shimmer, it has fewer, perhaps only six or eight lots[319] of the beautiful colors within it, of which much could still be said in confidence.

193. *Now, if one cannot use the whole corpus, how then does one do it?*

Answer: One must know a *menstruum* to purify it [the *corpus*] therein, so that the other does not take [the *corpus*] into itself, but only its blood and life.

194. *Is it then only to be treated in one way?*

Answer: The first purification is not just one way and there are many delays, since some people struggle for many weeks. It can also be done in eight days or even less, but in the after-work there is only one way in the composition.

195. *But you wrote about the wet and dry way?*

Answer: It is certainly wet at first, but then it is dry. The *re vera* is only

nur ein Weg. Wie will man doch die *Materia* reinigen in trockener Gestalt? Wie kann sie die *Feces* zu Boden schlagen? Und wenn man es gleich trocken in großem Feuer tun könnte, so ginge doch seine *Anima tingens* verloren und behielte man das ledige Nest, welches eben seinen ☿ veredeln und färben muss, damit er wieder färben kann, muss ihn auch ausbreiten und dünn machen, damit es der Himmel wird, welchen diese Sonne durchlaufen kann, zum Glas machen, zum gläsernen Meer, denn unser metallisches Glas wird nicht mit Kieselsteinen gemacht, sondern mit unserem glasförmigen *Azoth*, welches einen Eingang in die Metalle hat, ist unser Wasser und Salz-Stein der Natur, so metallisch und in das innerste der Metalle eindringt, es haben nur etliche wenige vom trockenen Weg geschrieben, dazu sind sie bewogen worden, wenn sie es mit ♂ und ☉ geschmolzen, aus den Schlacken den ♀ bracht, aufs neue mit dem *Regulis* geschmelzt, bis sie es miteinander vitrificiert, auf *pars cum parte* getragen und einen Wachs an ☉ und ☽ gefunden, haben sie sich gleich eingebildet, es könnte der Stein auf diese Art gemacht werden. Es wird aber dieses Jahr nicht angehen, gehört andere Arbeit und Verstand dazu, wenn es so einfältig und schlecht, hätten es längst alle Unkenbrenner, *item*, wenn sie es im trockenen Weg durch den gemeinen ☿ traktiert, damit animiert und etwas Nutzen bei den anderen Metallen dadurch geschafft, sein sie flugs zu gefahren, ganz gewiss sich eingebildet, ihr Weg wär der rechte und haben große Bücher davon geschrieben, ist aber unmöglich daraus auf solche Weise etwas zu machen, wenn nicht alle Teile voneinander geschieden, dass dieses flüchtige Feuer wieder zurück in sein *Centrum* geht, es anzündet und ins Leben und Wachstum bringt, auch darinnen erstirbt und durch den Tod verherrlicht wird, außer diesem ist alle Mühe vergebens, man sage davon was man will, wer es nicht glauben will, wird es mit seinem Schaden erfahren.

196. *Wie solviert man denn die Perlen durch dieses Menstruum?*

Antwort: Die Perlen werden gestoßen und in der Jungfer-Milch aufgelöst, darauf kann man einen *Spiritum Vini* gießen und digerieren,

one way. How does one want to purify the *materia* in dry form? How can it knock the feces to the ground? And even if one could do it by drying it in a large fire, its *anima tingens* would be lost and one would only be left with the empty nest, which has to refine and color its ☿ so that it can color again. One has to spread it out and make it thin so that it will become the heaven through which this sun can pass; one has to make it into glass, a sea of glass, for our metallic glass is not made with pebble stones, but with our glassy *azoth*, which has an entrance into the metals. This is our water and the salt stone of nature, which is metallic and penetrates into the innermost [depths] of the metals. Only a few have written about the dry way. They were moved to do so when they melted it with ♂ and ☉, brought the ♃ out of the slag, melted it again with the *regulis* until it vitrified it together, carried it *pars cum parte*, and found a wax of ☉ and ☽. They even imagined that the stone could be made in this way, but it will not happen this year. It requires a different work and understanding, and if it were so base and simple-minded, all *Unkenbrenner*[320] or "toad-burners" would have had it long ago. Similarly, if they treated it in the dry way through the common ☿, animated it with the latter, and thereby created something useful for other metals, then they would immediately be lead to imagine that their way was quite certainly the right one. They have written long books about it. But it is impossible to make anything out of it in such a way unless all of the parts are separated from each other so that the volatile fire returns to its *centrum*, ignites it and brings it to life and growth, and then dies therein and is glorified through death. Apart from this, all efforts are in vain. One can say what one wants about it, but whoever does not want to believe it will suffer for attempting it.

196. *How does one dissolve the pearls through this menstruum?*

Answer: The pearls are crushed and dissolved in the virgin milk. One can then pour a *spiritum vini* upon it and digest it, then distill it over to

hernach überdestillieren, so geht eine Perlen-Milch, zuletzt ein gelbes Öl, welche beide im Schlag verlorene Sprache, solche wiederzubringen, ein *Cordiale* desgleichen nach dem *Universal* nicht zu finden.

197. *Wie löst man denn die Metalle darinnen auf?*

Antwort: Diese Zerschmelzung geschieht nicht gewaltsam, sondern allmählich, sie zergehen darinnen und werden zu Öl und Salben, das ☉ in rotem Wasser und die ☽ in weißem, auch können die Metalle erst in Kalk gebracht werden.

198. *Wo bekommt man denn diese Materiam?*

Antwort: Zwischen Frankreich und Spanien, im piemontischen Gebirge, wird *Magnesia pia montana* genannt, etliche ist grauschwarz, das Glas davon wird purpur und amethystenfarben, mit ☽ geschmelzt und ausgekocht, gibt purpurfarbene Solution, verändert die Farben ist grasgrün, wird himmelblau, violenfarben, rosenrot, in Friaul ist eine *Magnesia* ♄, viel reiner und besser als in Piemont, ist graulich, der ☽ zieht eine schöne grüne Farbe heraus, die sich in blau, rot und wieder in grün verwandelt. Zwischen Schwaben und Bayern ist ihr auch, welche durchaus mit einem flüchtigen Blei-Erz gleich durchwachsen, in Böhmen, Sachsen, Saalfeld, Eisfeld, Ungarn und anderen Orten ist ihrer auch, im Kinzinger Tal bei Straßburg, allwo auch eine Farb-Mühle aufgebaut und anderen Orten mehr. Nur dass man die rechte erwählt, sie hat viele Namen, es muss sich aber hier niemand einbilden, weil ich der Farb-Mühlen gedenke, als meinte ich den giftigen schwarzen Cobalt, ich gebe dadurch nur Anleitung der Sache nachzufragen, weil solche gern in Cobaltgängen bricht, auch hat es einen anderen eigenen Gang, welcher Cobalt aber mit Kupfernickel vermischt, allda ist unsere *Materia* nicht anzutreffen, sondern adulteriert mit andern dunklen Geistern, dergleichen auch die Wismut Arten sein, welche mit anderem befleckt.

obtain a pearly milk and finally a yellow oil, both of which can bring back lost speech after a stroke. A *cordiale* like this cannot be found with the exception of the *universal*.

197. *How does one dissolve the metals in it?*

Answer: This melting does not happen violently, but gradually. They melt in it and turn into oil and ointments, the ☉ in red water and the ☽ in white [water]. The metals can also first be made into lime.

198. *Where does one obtain this materia?*

Answer: Between France and Spain, in the Piedmont mountains. It is called *magnesia pia Montana*. Some of it is gray-black, the glass from it becomes purple and amethyst-colored; melted and boiled with ☿, it gives a purple-colored solution, changes its colors, is grass-green, becomes sky-blue, violet-colored, rose-red. In Friuli there is a *magnesia* ♄ which is much purer and better than in Piedmont. It is grayish, the ☿ draws out a beautiful green color that turns blue, red, and then back to green. It also exists between Swabia and Bavaria,[321] where it is thoroughly and uniformly marbled with a volatile lead ore. It is also in Bohemia, Saxony, Saalfeld, Eisfeld, Hungary, and other places. Also, in the Kinzig valley[322] near Strasbourg, wherever a paint mill is built, and in other places still. Only choose the right one. It has many names, but nobody needs to imagine it here because when I think of the color mills, I mean the poisonous black cobalt, and by this I am only giving directions to seek after the subject matter, because it likes to break forth in veins of cobalt. It also has another vein of its own, but here the cobalt is mixed with copper-nickel. Our *materia* is not to be encountered there, for it is adulterated with other dark spirits such as the bismuth species, which are further tainted with others.

199. *Wie versteht man denn dieses, wenn sie sagen, dieser Stein ist alles und ist auch nichts?*

Antwort: Das hat diesen Verstand: Er ist alles, wenn man seine Bereitung weiß und ist auch nichts, wenn man seine Arbeit nicht versteht, ob man gleich die *Autores* wohl versteht und seine ganze Arbeit perfect im Kopfe weiß, wenn man aber die *Praxin* vornimmt, so kommen solche geheime Knoten aufzulösen und werden Dinge gesehen, die einen gleichsam das Gesicht verzaubern, dass man nicht weiß wo aus, es sind gewisse Termine, Zeit und Stunden von der Natur bestimmt, welche man nicht überschreiten darf, welches ich zwar alles und jedes in meinen vorigen Traktätlein und auch hier angezeigt und alles sonnenklar entdeckt, doch wird es noch schwer sein, wenn einer gleichsam mit ungewaschenen Maul zulaufen will und die nackende Diana küssen. Es ist keine Arbeit, wie gesagt, für die Bauern, wie sie Wagen schmieren sollen, es muss einer im Feuer wohl erfahren sein, die Finger verbrannt und manch schönes Buch gelesen haben, sonst wird es ihm einen guten Bauern-Schritt fehlen.

200. *Wie kann aber eine einzige Arznei für alle Krankheiten helfen und allen Menschen ohne Unterschiede?*

Antwort: Hilft doch das Brot allen Menschen für den Hunger und das Wasser allen Menschen für den Durst, so kann auch diese *Universal*-Medizin allen Menschen für alle Krankheit dienen, weil es das Salz der Natur in sich hat, die unsichtbare verborgende Speise des Lebens, aller Planeten Kraft und Influenz, der 4 Elemente Quintessenz, das *Centrum* aus der großen Welt, ein *Astrum* der 7 Metalle, den Schwefel-Balsam, den *Spiritum* des Lebens ☿, welches durch das Natur-Salz zusammen in einem Geist circuliert, so mit unsern Lebens-Geist eine Verwandtnis hat, weil unser Geist sowohl von den *Astris* als dieser. Wenn gleich unser Leib von der Erde, leben wir doch von Gestirn und Elementen, solches sehen wir, wenn die *Elementa* vergiftet, wie die Pest den Menschen wegreißt, so besteht der Mensch auch aus den Elemen-

199. *How is one to understand it when they say that the stone is everything and also nothing?*

Answer: This is the understanding: it is everything if one knows how to prepare it, and it is nothing if one does not understand their work. If one understands the *autores* well and knows all their work perfectly in their head, when one undertakes the *praxin*, then secret knots come unravelled, and things are seen that enchant the vision, so to speak, and one does not know where they come from; certain limits, times, and hours are determined by nature which you must not exceed, all of which I certainly indicated in my previous little treatises as well as in the present one. I have made everything as clear as daylight, but it will still be hard if someone wants to run up to the naked Diana to kiss her with an unwashed mouth, so to speak. As I said, it is no work for the farmers to grease their wagons; one must be experienced in the fire, having burned their fingers and read many a fine book, otherwise one will lack a good peasant's step.[323]

200. *But how can a single medicine help for all diseases and for all people without distinction?*

Answer: If bread helps everyone against hunger and water helps everyone against thirst, this *universal* medicine can also help everyone against all illnesses because it contains the salt of nature, the invisible, hidden food of life, the power and influence of all planets, the quintessence of the four elements, the *centrum* of the macrocosm, the one *astrum* of the seven metals, the balsam of sulphur, the *spiritum* of life ☿, which circulates together with the natural salt in a spirit, and thus has a relationship to our life-spirit, because like this spirit, our spirit is also from the *astris*. Although our body is from the earth, we live from the stars and the elements; and when the *elementa* are poisoned, we see how the plague rips people away. Thus, humans also consist of the elements and have the three *principia* in themselves, and everything

ten, hat die 3 *Principia* in sich, auch alles was in GOTT geschaffen war, besteht aus diesen dreien. Wenn nun bei dem Menschen eins von diesen ins Verderben oder Abnehmen gerät, so ist die Krankheit da und folgt der Tod, ob nun schon viel widersprochen wird, dass eine einzige Arznei für alles helfen könnte, welche 1. alles Unreine aus dem Leibe wegtreiben, 2. das Verdorbene wieder gut machen und 3. verhüten kann, dass die Krankheit nicht wieder kommt, dieses alles sei einer einzigen *Materia* oder Arznei vollkömmlich zu verrichten unmöglich? Weil aber alle Krankheiten daher kommen, wenn den Geist des Lebens die Natur beschwert und verletzt und sie in ihren Verrichtungen verhindert wird, es sei an welchem Ort des Leibes es wolle, von innen oder außen verletzt oder behext, durch böse Luft, schädliche Dämpfe, Gift, verderbte Fermenta, böse Verdauung, der Speisensaft in Magen und Gedärmen, welches das Geblüt alteriert, solches entweder zu bitter, zu sauer, zu sehr gesalzen oder zu wässrig macht. Die meisten Krankheiten stecken im Geblüt, wo der Lebens-Geist seinen seinen Sitz hat, ☉, ♃, ☿, *Humidum Radicale*, so wird denn der Lebens-Geist alteriert und geschwächt, muss derhalben eine Medizin gesucht werden, die die Natur wieder aufrichten kann, welche allein geschickt ist, die Krankheit auszutreiben und nicht der Arzt, die Arznei muss die Natur stärken, den Lebens-Geist ermuntern, erfrischen und zu Hilfe kommen, denselben stärken, verdoppelt, die Lebens-Geister vermehren, die Natur-Drückung erfrischen, aller Verderbung widerstehen und die ganze Natur erquicken und erhöhen, gleich wie nun eine böse Luft einen ganzen Menschen verderben kann und ein wenig Gift einen ganzen Menschen töten, also ist es auch möglich, durch eine einzige kräftige Arznei den ganzen Menschen gesund zu machen und alles Böse auszutreiben, denn diese Quintessenz kann alles Böse, welches die Natur verhindert, wegräumen, die verderbten *Fermenta* wieder zurecht bringen, die Verdauung stärken und die Viscera vor der Verfaulung bewahren. Und ob es gleich an Lunge und Leber fehlt, so kann es durch diese Medizin alles erfrischt werden, weil es alle Adern durchdringt, er restauriert und renoviert alles Geäder, Mark und Bein,

that was created in GOD consists of these three [*principia*]. If one of these now perishes or diminishes in a person, then disease is present and death follows. However it is now highly contested that a single medicine could help for everything; a single medicine which could: (1) drive out everything that is impure from the body, (2) restore what has been corrupted to good condition, and (3) prevent disease from returning. Is all of this impossible to achieve completely with a single *materia* or medicine? For all illnesses arise when the spirit of life, i.e., nature, is encumbered, damaged, and thwarted in its functions, no matter where in the body it may be, damaged or bedevilled from inside or out, whether by malefic air, noxious vapors, poison, corrupted *fermenta*, indigestion, or the gastric juices in the stomach and intestines which alter the blood, making it either too bitter, too acidic, too salty, or too watery. Most diseases are in the blood, where the life-spirit has its seat, ☉, ♃, ☿, the *humidum radicale*. So if the life-spirit is altered and weakened, a medicine must be sought that can restore nature, who is the only one skillful enough to drive out the disease, not the doctor. This medicine must strengthen nature, rouse, refresh, and assist the life-spirit, strengthen it, double it, augment the life-spirits, revive the suppressed nature, resist all corruption, and then quicken and elevate the whole nature. Just as malignant air can ruin a whole person and a little poison can kill a whole person, so too is it possible to make the whole person healthy with a single powerful medicine and to drive out all evil, for this quintessence can clear away all the evils that inhibit nature, set the corrupt *fermenta* straight again, strengthen the digestion, and keep the viscera from putrefaction. And even if the lungs and liver are failing, everything can be revitalized with this medicine because it penetrates all arteries, restores and repairs all bloodvessels, the marrow, and the bones, so that no disunity is created. For then podagra

dass keine Uneinigkeit erfunden wird, denn da weicht Podagra, weil es allen *Tartarum* resolviert und austreibt, die Wassersucht kuriert, das Wasser ausführt, das Natur-Salz stärkt und coaguliert, dass es nicht mehr zerschmelzen und wegfließen kann, *Icterus, Colica, Passio,* &c. Denn ihm weichen alle Dinge, die sich unterstehen, die Natur zu verderben, wie die Sonne die Finsternis vertreibt und die Wärme die Kälte. Also flieht auch die Krankheit vor der Erneuerung der Gesundheit, denn sie hilft für alle Krankheiten, weil in ihr ist das Salz der Natur und der *Spiritus Universalis,* aller Metalle Kraft, das *Astrum* der Sterne, der *Spiritus Mundi,* welches alles dem Natur-Salz einverleibt als einem reinen Geist, welcher es unserem Geist mitteilen kann und weil hierinnen die Quintessenz aller 4 Elemente in gleicher Concordanz stehen, so kann diese Medizin unsere gefallenen Elementa wieder aufrichten, das Salz stärken, dem ☿ *Humidum Radicale* erfrischen, unser trübe Geist und dunkle Luft, welche mit bösen ungesunden Geistern vermischt, clarificieren, die Bösen austreiben und unser Feuer, die natürliche Wärme im Blute anzünden. Denn diese himmlische Arznei ist das reine Paradies, wenn dies in uns aufgerichtet wird, und unsere dunkle finstere Welt durch die Lichtstrahlen des Paradieses erleuchtet, dass unser Leib durch dieses neu generierte *Corpus* auch wieder neu geboren wird, dass wir nicht anders denken, als wären wir im Himmel. Es müsste denn sein, das GOTT einen boshaften Sünder zur Strafe und anderen zum Exempel, oder einem Frommen zur Probe, eine Krankheit aufgelegt oder wollte gar einen Menschen durch den Tod wegnehmen, in solchen Fällen muss man sich dem Willen GOTTES mit willigem Geist ergeben. Aber den anderen wird es allen und jedem helfen, wegen seines Salzes Kraft, welches zugleich heiß und trocken, warm und kalt, die erhitzte Leber wieder abkühlen, die erkaltete erwärmen, die natürliche Feuchtigkeit erfrischen, wenn sie ausgetrocknet. Ist kalte und wässrige Feuchtigkeit im Geblüt, wird sie solches durch den Schweiß austreiben, alle Verstopfungen und Blähung zerteilen und Luft machen, stillt alles Brennen und Drücken, ersetzt des Magens und anderer *Viscerum* veränderte *Fermenta,* macht

gives way because [the medicine] resolves and drives out all *tartarum*, curing dropsy, expelling the water, strengthening and coagulating the natural salt so that it can no longer melt and flow away, *icterus*,[324] *colica*, *passio*,[325] etc. For all things which dare to corrupt nature give way to it like the sun drives out darkness and heat drives out cold. So illness also flees from the regeneration of health, because it helps against all illnesses, because it contains the salt of nature and the *spiritus universalis*, the power of all metals, the *astrum* of the stars, the *spiritus mundi*, all of which are incorporated in the salt of nature as a pure spirit, which it can impart to our spirit. And because the quintessence of all four elements stand in equal concordance, this medicine can correct our fallen *elementa*, strengthen the salt, refresh the ☿ *humidum radicale*, clarify our turbid spirit and dark air which were mingled with malefic, unhealthy spirits, cast out evil, and ignite our fire, the natural heat in the blood. For this heavenly medicine is pure paradise. When it is raised up in us and our dim, dark world is illuminated by the luminous rays of paradise so that our body is reborn again through this newly generated *corpus*, we feel as if we are in heaven. It must then be the case that GOD imposed an illness on a malicious sinner as a punishment and on others as an example, or on a pious person as a test, or even wanted to take a person away through death; in such cases one must submit to the will of GOD with a willing spirit. But [the medicine] will help everyone else because of the power of its salt, which is simultaneously hot and dry, warm and cold, cooling the heated liver again, warming the cold one, refreshing the natural moisture when it has dried up. If there is cold and humidity in the blood, it will drive it out through sweat, breaking up all constipations and flatulence and excess air. It calms all burning and pressure, replacing the *fermenta* of the stomach and other *viscerum* and improving the digestion and

bessere Verdauung und Kochung des Geblüts; In Podagra äußerlich unter Pflaster vermischt und aufgelegt, stillt allen Schmerzen, innerlich dabei gebraucht, wird es solches in der Wurzel wegnehmen, in Gicht, im großen Reißen in Gliedern, wird es alle Schmerzen legen, die *Epilepsia* weicht davon und bleibt weg, in Mutter-Krankheiten, in Fraiß, in Schwindel, in Kopf- und Ohren-Weh, flüssigen Augen, stetigen Herz- und Seiten-Weh, den Tollen und Verwirrten, ingleichen in verstopfter Monats-Zeit, in Unfruchtbarkeit. Auch die Mond-Kinder wird es austreiben, fruchtbar machen, in Apoplexia ist es gewisse Hilfe. Wenn auch gleich das Malum alt worden, wird es doch nach wiederholter Arznei die Gesundheit bringen und den Lebens-Geist in dem verstorbenen Glied wieder einführen, welches sich erstlich mit Kribbeln erzeigen wird in dem Gliede und dann wie heiß Wasser durchlaufen, die Aussätzigen werden davon anfangen zu rauchen wie Brände, bis alle hässlichen Schuppen abfallen, rein und gesund werden, den Wassersüchtigen wird aller Schwulst wegfallen, in der Windsucht und Blähung wird es die Dünste austreiben, den Gängen Luft machen, die Schwindsüchtigen, welche schon den Tod übergeben und vor Mattigkeit weder gehen noch stehen können, solche wird es wieder erfrischen, stärken und gesund machen, dass sie wieder an Geblüt und Leib zunehmen. Die 4-tägigen Fieber werden gleich dadurch gestillt, sollten sie gleich Jahr und Tag, auch noch länger gewährt haben, summa alle Beschwerung in und äußerlich, welche sonst ein Spott der Ärzte, wird doch diese Arznei kurieren, es sei Krebs, Wolf, heißer und kalter Brand, *Sanct Quirins*, oder Johannis-Biß, Öl-Schenkel, Lähme, Contractur, faule Löcher und stinkende Schäden, welche ganz keine Heilung annehmen wollen und ist kein *Malum* so arg, diese Arznei übertrifft sie in ihrer Kraft, denn dieser ☿ oder Præcipitat wirkt nicht wie der gemeine, es sei der ☿ *dulcis*, oder sei præcipitiert gelb, grün, korallenrot, so macht er Salivation, Löcher im Hals, die Zähne los, purgiert heftig, greift das Serum an, ist dem Schwindsüchtigen sonderlich schädlich, welche er hinrichtet, wirkt violent, oft mehr Böses als Gutes, dass er oft *Convulsiones* macht, wenn er ungeschickt gebraucht wird, in

the coction of the blood. In podagra, mixed externally and applied under plasters, it calms all pains; whereas used internally, it will take it away at the root. It will stop all the pains of gout and rheumatism, and *epilepsia* will yield to it and stay away. So too in maternal sicknesses, in gluttony, in dizziness, in headaches and earaches, watery eyes, persistent heart and chest pains, madness and confusion, as well as in the congested menstrual period and infertility. It will also drive out the moon-children, making them fertile; and in apoplexy it is a decisive help. Even if the *malum* is an old one, it will bring health after repeated medication and reintroduce the life-spirit into the deceased limb, which will first engender a tingle in the member and then begin to feel as though hot water was coursing through it. Leprous sores will begin to smoke from it like fires until all the ugly scales fall off, becoming clean and healthy; dropsy will be free of all swelling, in wind afflictions and flatulence it will drive out the fumes, clearing the pathways. It will refresh, strengthen, and heal those suffering from consumption who are almost about to die and can neither walk nor stand because of exhaustion, so improving their condition that they will increase in blood and body again. The four-day fever will be calmed by it immediately, even if it already lasts a year, a day, or even longer. In sum, this medicine will cure all internal and external maladies which normally make a mockery of doctors, be the disease cancer, lupus, hot and cold burns, *Sanct Quirins*, or St. John's bite,[326] swollen thighs, paralysis, contracture, rotting holes, and putrid injuries which resist treatment and refuse to heal. There is no *malum* so severe that this medicine cannot overcome it, because this ☿ or precipitate does not act like the common one, whether the ☿ *dulcis*, or the precipitated yellow, green, or coral-red [☿], which causes salivation, holes in the throat, loose teeth, violent purges, and which attacks the serum. This [common ☿] is particularly harmful to the consumptive, which it executes, and has a violent effect, often more evil than good, frequently causing *convulsiones* when used clumsily in purgations, such that people have

Purganzien, dass Leute darüber dem Tod sind heimgefallen, unser tut dies alles nicht, er sei weiß oder rot, er wirkt nach der Inclination der Natur, ohne Brechen und Purgieren, treibt den Schweiß, macht nicht matt, stärkt die Natur und hilft, derselben unempfindlicher Weise, renoviert, restauriert, vertreibt das Alter, führt die Jugend ein, erhält die Kräfte und treibt durch seines Lichtes-Strahl alles Böse aus.

201. *Sind denn die geheimen Schlüssel der Natur hierinnen alle entdeckt, dass man der ganzen Natur ihre Geheimnisse dadurch in dem mineralischen Reiche erforschen kann?*

Antwort: Ja, so viel mir zum mineralische Reiche von GOTT dem Allerhöchsten offenbart worden und ich durch GOTTES Gnade bekommen, dieses habe meinem Nächsten überreicht, befehle aber doch solche Schlüssel der Göttlichen Majestät, vor welchem hohen Thron ich sie in tiefster Demut niederlege und seiner göttlichen Allmacht heimstelle, wem er sie weiter überreichen und damit beschenken will. Danke ihm mit Herz und Munde für seine große Liebe und Gnade, damit er mich unwürdigen armen Erdwurm beschenkt. Vor allen Dingen aber, dass er mich zu seiner Erkenntnis bracht, dass ich in seinem Lichte, ihn als das wahre Licht erblickt und bitte den hoch-heiligen GOTT inniglich, mich in Liebe und Glauben an ihn beständig zu erhalten, damit ich, wenn mein Geist, Seele und Leib im Tode voneinander geschieden, in der Auferstehung aber wieder vereinigt, zu seinem ewigen Licht eingeführt und mir vollkömmlich sein göttliches Licht, so ich hier im Geiste gesehen, gezeigt werden möge. Amen. Ja, komm, HERR Jesu, Amen.

fallen prey to death. But ours does not do this. It is white or red, it works according to the inclination of nature, without causing vomiting and purging; it drives away the sweat, does not make one faint, strengthens nature and assists her in an impartial manner; it repairs, restores, and drives away old age, introduces youth, preserves strength, and drives out all evil through its luminous radiance.

201. *Have the secret keys of nature all been discovered here, so that the mysteries of all nature can be explored in the mineral kingdom?*

Answer: Yes, as much as has been revealed to me about the mineral kingdom by GOD the Most High, and what I have received by the grace of GOD I have given to my neighbour; but I still command such keys to the divine majesty, before which high throne I lay them down in deepest humility and entrust it to his divine omnipotence to hand it over and bequeath it as a gift to whomever he wills. I thank him with heart and speech for his great love and grace, so that he may bestow it upon me, though I am a poor, unworthy earthworm. Above all, however, I give thanks that he brought me to his realization, and that in his light I see him as the true light, and I pray to the most holy GOD to keep me constantly in love and faith in him, so that I, when my spirit, soul, and body, separated from each other in death, but reunited in the resurrection, are introduced to his eternal light, and that his divine light, which I have seen here in spirit, may be fully revealed to me. Amen. Yes, come, LORD Jesus, amen.

ANHANG

Die *Philosophi* oder *Adepti* schreiben, dass in der Natur eine Blei *Minera* oder metallisches Wesen anzutreffen, welches sich leicht auflösen ließe, wer nun seine Solution und solches mit dem Wasser seiner Art anzufeuchten wüsste, würde ein glückseliger *Medicus* sein, welches *Subjectum* sie *Plumbum Philosophorum* nennen, auch ☿, weil es desselben Weiblein oder das *Lili* der Alten, *Marcasita per excellentiam*. Nun findet sich diese *Minera* in ihrem grauen Kittelchen mit dem Farben der Tauben-Hälse, inwendig mit schönen gelben Rißlein und glorwürdigsten Fünklein, welche leuchten und schimmern als Sterne an diesen Sonnen-Himmel, sie findet sich auch von der Natur, wenn solche alt und über die Zeit gestanden von den *Acido* der Natur zerfressen, welches die Bergleute Cobalt-Blüte, auch Rosen-Blüte nennen, habe auch solcher Steine gesehen, welche dünn mit dieser *Materia* eingesprengt und von der Luft corrodiert, ganz Rosenrot gesehen, weil aber diese corrodierte *Minera* oft mit Cobalt und anderen dunklen fremden Geistern befleckt, so bleibt man lieber bei der *Minera*, weil man versichert, dass solche rein und man die Beste colligieren kann und ob man schon der Besten erwählt, so wird man doch aus einem Pfund nicht 16 Lot erhalten, das zum Werk dient, das andere ist bleiisch und muss man nur die tingierenden Geister davon extrahieren, welches die Blumen ☉, ☽, ♂, ♀ sind, das andere aber fahren lassen. Es ist die Blume und Samen aller Metalle, welches in seiner *Anatomia* zu finden, hat die 3 *Principia*, welches Basilius unter den Nahmen des ☿ auf dem *Triumph-Wagen* eingeführt und wie es soll aufgeschlossen werden in seiner äußeren Reinigung, dass seiner Röte durch den Essig davon genommen wird, hernach durch seinen eigenen Essig eine andere und reinere Extraction zu machen, nach welcher die Scheidung der Elemente erfolgen kann, seine erste Destillation gibt einen gelben *Spiritum*, solcher wird rectificiert im Sand, bis auf ein dunkles Öl, das treibt man mit größerem Feuer über, fängt es allein und bewahrt es vor der Luft, sonst verraucht es alles, wiewohl das Öl noch auf eine geheime Art zu scheiden. In dem

APPENDIX

The *philosophi* or *adepti* write that there exists in nature a lead mineral or metallic being which is easily dissolved. Whoever knows how to moisten this solution and such with the water of its kind, would be a blissful *medicus*. This *subjectum* they call *plumbum philosophorum*, also ♄, because it is the wife of the same or the *Lili* of the old, *marcasita per excellentiam*. Now this *Minera* is found in its little gray tunic which has the colors of pigeon necks, and inside, beautiful little yellow cracks and glorious little sparks, which shine and shimmer like stars in this heaven of the sun. It is also found in nature, when it is old and has stood over time to feed on nature's *acido*, which the miners call cobold-blossom and also rose-blossom.[327] I have also seen such rocks, thinly sprinkled with this *materia* and corroded by the air, all rose-red, but this corroded *minera* is often stained with cobalt and other dark, foreign spirits. One therefore prefers to stay with the *minera*, because one is assured that it is pure and one can colligate the best and although one chooses the best, one will obtain not sixteen loth[328] from a pound that serves for the work. The other is lead-like and you only have to extract the tingeing spirits from it, which are the flowers ☉, ☽, ♂, ♀, but the others can be let go. It is the flower and seed of all metals, which is found in its *anatomia*, the three *principia*, which Basilius introduced under the name of the ♄ in the *Triumphal Chariot*,[329] and how it is to be digested in its outer purification, that its redness is taken away by the vinegar, afterwards to make another and purer extraction with its own vinegar. After this the separation of the elements can take place. Its first distillation gives a yellow *spiritum*, which is then rectified in the sand, except for a dark oil, which one drives over with greater fire, catching it alone and protecting it from the air, otherwise it all smokes up, although the oil can still be separated in a secret way. In the *spiritu* ☿ one

Spiritu ☿ solviert man seinen eigenen 🜨 oder rote Erde, destilliert solches in 40 Tagen zum roten Öl, welches die Scheidung der Elemente ist und kann dieses rote Öl zu keiner Putrefaction oder Extraction gebracht werden, sein ☿, weißes Wasser oder Salz der Natur, ist denn von ihm geschieden, dann nimmt es die Luft, das gelbe Öl, welches auch von Wasser rein geschieden sein muss, in sich, die Probe ist, wenn ein Tüchlein damit genetzt, solches mit verbrennt, als in starkem *Spiritu Vini*, dann ist die Luft wohl rectificiert, welche des roten ♌ Blut in sich nimmt und die Farben eines Regenbogens erscheinen, wenn dieses ☉ in sein erstes Wesen eingeht, wenn nun der rote ♌ und Feuer rein von der Erde geschieden, wird diese Luft, das gelbe ♃ Öl, davon destilliert, bis sich der Helm will anfangen rot zu färben und die Tropfen rötlich fallen, dann wird die Vorlage geändert, das rote ♌ Blut und Salz der Welt allein gefangen, zurück bleibt das ♂ Salz, welches mit dem ♄, weißen, fetten und schlackenreiche ☿ zusammengesetzt wird, dass diese beiden widerwärtigen Fechter mit ihrem Realgar verbrennen und ein vollkommenes Metall daraus wird, welches die beiden Drachen, flüchtig und fix geben, doch sind die *Elementa* noch auf eine andere Art zu scheiden, dass das ganze *Confusum Chaos* in einer Destillation übersteigt und man die *Elementa* in der Rectification scheidet und wenn von des ♌ Blut oder Feuer etwas bei der Erde bleibt, wird es durch *Cohobation* mit übergeführt. Weilen nun die Composition mehr als deutlich beschrieben, so mangelt dem Leser nichts als GOTTES Segen, um welchen er Ihn bitten muss.

dissolves its own 🜨 or red earth, distilling this in forty days to the red oil, which is the separation of the elements. This red oil cannot be used for any putrefaction or extraction. Its ☿, being white water or the salt of nature, is then separated from it. Then it absorbs the air, and the yellow oil, which must also be separated from water. The test is when a cloth wetted with it burns with it as in strong *spiritu vini*, because the air is well rectified, which takes the blood of the red ♌, and the colors of a rainbow appear when this ☉ enters its first essence, when the red ♌ and fire are now separated from the earth. This air, the yellow ♃ oil, is distilled from it, until the helm wants to start turning red and the drops fall reddish; and then the template is changed, and the blood of the red ♌ and the salt of the world are caught alone. The ♂ salt remains, and it is mixed with the ♄, which is placed together with the white, fat, and slag-like ☿ so that these two repelling fencers[330] burn up with their realgar and it becomes a perfect metal, which gives the two dragons, volatile and fixed. But the *elementa* can still be separated in another way, so that the whole *confusum chaos* goes over in one distillation and one separates the *elementa* in the rectification. If something of the ♌ blood or fire remains with the earth, it is also transferred through *cohobation*. Now that the composition is more than clearly described, the reader lacks nothing but GOD's blessing, for which they must ask.

ANNOTATIONS

1 The edition of *Das Mineralische Gluten* in the Wellcome Collection contains two pages of *Errata*. The present edition integrates these corrections. Here, "Plato" is notably amended to "Pluto".

2 The material that was called antimony in the early modern period is what we now call antimony trisulfide. Our element antimony was the *regulus antimoni*, i.e., small king of antimony in Dorothea Juliana Wallich's time.

3 The four alchemical elements are: fire △, air ▲, water ▽, and earth ▼.

4 This is an interpretation of the sentence: *Fac de masculo & foemina circulum rotundum, & de eo extrahe quadrangulum, & quadrangulo triangulum, fac circulum rotundum & habebis lapidem Philosophorum*, first printed in the *Rosarium Philosophorum* from 1550.

5 *Magnesia* is written in the first circle of the cabbalistic figure.

6 *Leo rubeus* (Latin) = "the red lion".

7 The first four lines of this poem can be found in several earlier and later alchemical books, e.g., in: *Hermetischer Rosenkrantz*, Hamburg 1659, p. 1. They are based on a short piece of book 6 of the Latin poem, *Æneis*, by Virgil.

8 The last four lines of this poem can be found in a German translation of Eirenæus Philalethes: *Secrets revealed or An Open Entrance to the Shut Palace*, i.e.: *Eröffneter Eingang zu deß Königs verschlossenen Pallaste Anonymi Philaletæ*, Frankfurt und Hamburg 1673, p. 23. The translated text is based on short Latin sequence from Giovanni Aurelio Augurelli's Latin poem *Chrysopeia*, lib. 2, which was cited by Philalethes.

9   ○ = alchemical symbol for alum, here it stands for a circle in the cabbalistic figure.

10 ⊕ = alchemical symbol for *vitriol*, *vitriol* is written in the second circle of the cabbalistic figure.

11 *Visitando Interiora Terræ Rectificando Invenies Occultum Lapidem Veram Medicinam* = Visit the interior of the earth, and by rectifying you will find the hidden stone which is the true medicine. The first letters of the words of this Latin sentence give the word: VITRIOLVM.

12 This short poem is from: Horn, Caspar, ed.: *Bernhardus Innovatus, das ist deß ... Herrn Bernhardi, Grafen von der Marck und Tervis Chymische Schrifften*, Nürnberg 1643, p. 308.

13 □: not an alchemical symbol, here it stands for the square in the cabbalistic figure.

14 This short poem is also from: Horn, Caspar Horn, ed.: *Bernhardus Innovatus, das ist deß ... Herrn Bernhardi, Grafen von der Marck und Tervis Chymische Schrifften*, Nürnberg 1643, p. 308.

15 △ = alchemical symbol for fire, here it stands for the triangle in the cabbalistic figure.

16 Leafed earth = foliated earth = *terra foliata*.

17 A "sea of glass" is mentioned in Revelation 4:6 and 15:2.

18 ○: alchemical symbol for alum, but here it stands for a circle in the cabbalistic figure.

19 This rhyme, with slightly changing detailed wording, is often used in alchemical context and describes the meaning of the ouroboros in alchemy. It can, e.g., be found in: *Theatrum Chemicum*, vol. 4, Straßburg 1659, p. 258. Here it is part of a longer German language poem by an anonymous author.

20 *Agens*: that which acts, and *patiens*: that which experiences the effect of the action.

21 *Echeneis*: a fish with this name is mentioned already by Pliny. The *Echeneis naucrates* or live sharksucker is a fish which attaches itself temporarily by its modified dorsal fin used as a sucking disc to various hosts such as other marine animals or ships.

22 *Quod Lapidis Pater nunquam concubuerit*: this phrase is used in Agricola, Johann: *Chymische Medicin ...*, part 2, Leipzig 1639, p. 4.

23 Gur is a concept first mentioned by Johann Matthesius in his book, *Sarepta, oder Bergpostill*, Nürnberg 1652. On Gur see: Ana Maria Alfonso-Goldfarb, Marcia H. M. Fernandez: "Gur, Ghur, Guhr, or Bur? The Quest for a Metalliferous Prime Matter in Early Modern Times", *BJHS* 46.1 (2013), 23–37.

24 Theophrasti is Paracelsus, with full name Philippus Aureolus Theophrastus Bombastus von Hohenheim called Paracelsus (c. 1493–1541), an influential German physician, alchemist, theologian, and philosopher.

25 *Aqua fœtida* (Latin) = "stinking water".

26 *Fumus albus* (Latin) = "white smoke".

27 The cupel was a porous pot or vessel used in alchemy in which metals were tried or tested.

28 This German poem is from: Agricola, Johann: *Chymische Medicin ...*, part 2, Leipzig 1639, p. 288.

29 *Venenum tingens* (Latin) = "coloring poison".

30 This is a reference to the Great Bear, the constellation *Ursa Major* in the northern sky.

31 In all three editions there is a misprint in this place: "Ajoht". *Azoth* is an alchemical concept with different meanings used by several authors such as Paracelsus, Basilius Valentinus, Heinrich Kunrath, and others. A glass-shaped *azoth* is mentioned in the *Testamentum* of Ps.-Lullus as "*Azoth vitreus*", see e.g., *Testamentum Raymundi Lulli*, Rouen 1658, p. 22.

32 Lulli refers to Raymundus Lullus or Ramon Lull (c. 1232–c. 1315), a Spanish philosopher and polymath. Alchemical texts published under his name are not from his hand and were all written after his death. Modern research collects them under the authorship of Ps.-Lullus, although these works are from different authors.

33 A "sea of glass mingled with fire" is mentioned in Revelation 15.2.

34 A reference to the German proverb: "Mancher Stein wird nach einer Kuh geworfen, der wertvoller ist als die Kuh" (Sometimes a stone is thrown at a cow that is worth more than the cow).

35 This is roughly what Dorothea Juliana Wallich was doing in her *particular* works for three different German princes.

36 In the original book (first edition) this was wrongly printed as number 47.

37   The German *Scheidewasser* literally means separating water, because with this water, nitric acid or *aqua fortis* (▽), it is possible to separate silver and gold.

38   *Blachmal* usually stands for dross of metal, but also for a black mixture made from copper, silver, and lead sulphides.

39   *Minera Martis solaris Hassiaca* = iron ore from Hesse in Germany, especially from Almerode.

40   In all three editions "○ Kies" was printed, but from the context and what was printed under number 71 it is clear that D.J.W. meant "☉ Kies", so auriferous pyrites instead of aluminous pyrites.

41   In all three editions the word *ÆSustum* was printed, but this has to be two words: *Æs ustum* = "burned copper".

42   In all three editions *æris usti* was printed, but this has to be *æsis usti*.

43   *Verschossen* (in all three editions) does not make sense, maybe this should be *verschlossen* = "closed".

44   *Glut* was wrongly printed in all three editions, but this must be *Glett*, also called *Glätte* or *Bleiglätte*, i.e., lead oxide, PbO.

45   Orpiment and auripigment are both As_2S_3. Maybe for Dorothea Juliana Wallich, orpiment (*Rauschgelb*) was a redder version, whereas auripigment was the usual yellow material (see No. 17 and No. 18).

46   N.B. = *Nota bene* (Latin) = "note well".

47   *Ana*, short for: ana partes æquales (Latin) = of each an equal quantity.

48   This poem is part of a longer poem, which was printed in the book *Güldene Rose* (Golden Rose) by Jacob Rösser (1642–after 1712) printed in Hamburg 1705, pp. 5–13. The part used by Dorothea Juliana Wallich is to be found on pp. 11–12.

49   The German word *Gewächs*, translated here as "growth", also means "plant, vegetation, blossoming, maturing" and is cognate with the English verb "to wax". Its predominant meaning is the "growth" of a plant or tree. Wallich's use of the term to express the development of the "tree" of the three principles is influenced by Jakob Böhme's *De Tribus Principiis, oder Beschreibung der Drey Principien Göttliches Wesens*, 1619. See Weeks: *De Tribus Principiis*, Leiden 2019.

50   The Three Principles of the Divine Being is a concept from the German mystic, Jacob Böhme (1557–1624). See Weeks: *De Tribus Principiis*, Leiden 2019.

51	The three worlds, the outer, the inner, and the angelic world, is also a concept of Jacob Böhme. See Weeks: *Boehme*, Albany 1991, p. 93ff.
52	Kings 8:27.
53	In all three editions verse "6" is printed, but this is verse 5.
54	In all three editions verse "2" is printed, but this is verse 11.
55	In all three editions verse "26" is printed, but this is verse 25.
56	The concept of the seven source-spirits was developed by Jacob Böhme.
57	*Schiedligkeit* is an ingenious term from Jacob Böhme. It has been translated as "divisiveness". See Zuber: *Böhme and Alchemy*, 2019.
58	Revelation 21:1.
59	Revelation 21:5.
60	2 Peter 3:13.
61	Psalms 102:26.
62	*Item* (Latin) = "also".
63	5 Moses 33:13.
64	Isaiah 65:17 and 2 Peter 3:13.
65	1 Moses 13:14–15.
66	*Inqualieren* is an ingenious term from Jacob Böhme. Weeks translated it as: "infuse", "imbue", or "unite with". See: Weeks: *Aurora*, Leiden, 2013, p. 42.
67	*Ruach* is a Hebrew word which is usually translated as the "spirit" of God.
68	Genesis 1:3–5.
69	Genesis 1:6–8.
70	Genesis 1:9.
71	Genesis 1:14.
72	Genesis 1:20.
73	Genesis 1:24.
74	In all three editions the count jumps from 7 to 10; 8 and 9 are missing. This has been corrected in this new edition.
75	*Wasserstein der Weisen* = "Water Stone of the Wise", title of an anonymous alchemical book: *Wasserstein der Weysen*, Frankfurt am Main, 1619.
76	*Urim* and *Thumim* are are divinatory stones in the Hebrew bible.

77 Last lines from the German church song, *In dulci jubilo*.
78 This phrase is also cited in Wallichs third book under Morienus, question 68.
79 On theriac made from the Syrian serpent Tyrus see: Jonathan Rubin: "The use of the 'Jericho Tyrus' in theriac: A case study in the history of the exchanges of medical knowledge between Western Europe and the Realm of Islam in the middle ages", *Medium Ævum* 83 (2014) 234–253.
80 This German poem was printed in: Vigilantium de Monte Cubiti: *Dreyfaches hermetisches Kleeblatt*, Nürnberg 1667, pp. 110–111.
81 1 Samuel 9.
82 1 Samuel 28.
83 2 Maccabees 1:18–23
84 The seven seals are described in Revelation 6.
85 Psalms 139: 8–10.
86 In all three editions ☉ was printed, but from the context it is ☾.
87 On the number x, see Wallich's third book, question 133.
88 Isaiah 1:16–18.
89 Matthew 18:9.
90 Matthew 18:8.
91 1 John 5:4.
92 Ephesians 5:14.
93 Revelation 20:6.
94 Luke 18:13.
95 Matthew 9:2.
96 Luke 23:42.
97 Luke 23:43.
98 In all three editions ☉ was printed, but from the context it is ☾.
99 Isaiah 65:17.
100 2 Peter 3:13.
101 Revelation 21:5.
102 Revelation 21:3.

ANNOTATIONS

103 1 Moses 13:14.
104 1 Moses 13:15.
105 Joel 2: 28–29.
106 Hosea 13:14.
107 *Sanctus* (Latin) = "holy". "Three times *Sanctus*", i.e., "*Sanctus, Sanctus, Sanctus*" is a hymn in Christian liturgy.
108 Luke 23:43.
109 2 Kings 6:15.
110 2 Kings 6:17.
111 1 Moses 2:17.
112 1 Moses 3:4.
113 In all three editions *Höhe* = "height" was printed, but this must be *Hölle* = "hell".
114 A sidereal spirit, meaning the astral spirit or spirit of the stars, was a concept of Jacob Böhme's. He considered the sidereal spirit to be the soul of the great world (macrocosm) coming from the sun and also taking its light and life from the sun.
115 1 Corinthians 15:54.
116 1 Thessalonians 5:23.
117 E.g., Matthew 3:11.
118 Psalms 51:10.
119 Hebrews 6. 4–6.
120 Psalms 51:11.
121 John 16:7.
122 John 16:13.
123 On Naaman see: 2 Kings 5.
124 1 Moses 3:9.
125 1 Moses 6:3.
126 According to Baro Urbigerus, *circulatum majus* and *circulatum minus* are alternative denominations of the grand elixir and the vegetable elixir, respectively. See: Urbigerus: *Aphorismi Urbigerani*, 1690.
127 1 Corinthians 15:41.

128 John 14:2.
129 "Where are those joys?"
130 "New songs".
131 "In the King's court".
132 Text of the third stanza of the German church song, *In dulci jubilo*, which contains alternating Latin and German lines.
133 Ezekiel 47:8–9.
134 Ezekiel 47:12.
135 In the Christian tradition, Raphael is one of the archangels (a high ranked angel), his name can be translated from the Hebrew as "God heals".
136 Matthew 6:20.
137 Matthew 6:21.
138 Matthew 6:10.
139 Matthew 6:13.
140 Revelation 12:10.
141 Revelation 7:9.
142 Revelation 15:2–4.
143 *Sapienti sat* (Latin) = "Enough for the wise".
144 Luke 8:10.
145 Book of Wisdom 9:4.
146 Revelation 1:11.
147 Revelation 1:12–13.
148 Revelation 1:14–16.
149 Revelation 1:17–18.
150 Hosea 13:14.
151 Revelation 4:1.
152 Revelation 4:2.
153 Revelation 4:3.
154 Revelation 4:4.
155 Revelation 4:5–8.

156 Revelation 4:9.
157 Revelation 4:10–11.
158 Castor and Pollux are twin half-brothers in Greek and Roman mythology.
159 Revelation 5:1.
160 Revelation 5:3.
161 Revelation 5:6.
162 Revelation 5:7–8.
163 Revelation 5:9.
164 Revelation 10:1–7 (with some omissions).
165 Revelation 10:8–9 (with some omissions).
166 Revelation 10:10–11 (with some omissions).
167 Revelation 11:1–2 (with some omissions).
168 Revelation 12:1–5 (with some omissions).
169 Revelation 12:14–17 (with some omissions).
170 Revelation 15:1–4 (with some omissions).
171 That is, illness and poverty.
172 1 Moses 19:1–28.
173 2 Kings 1:9–10.
174 2 Moses 8:19.
175 Psalms 18:10.
176 *Sal esurinum* (Latin) = "hungry salt".
177 In all three editions, *Spiritus Vini* was printed but from the context it should be *Spiritus Salis*, i.e., hydrochloric acid (see also question 10).
178 *Nodum Gordium* (Latin) = "Gordian knot".
179 Book title of: Anon.: *Wasserstein der Weysen*, Frankfurt am Main, 1619.
180 Cyllenius is a surname of Hermes because he was born on Mont Cyllene in Arcadia.
181 These twelve persons were taken by Wallich from the *Viridarium Chymicum* from 1624. They first appeared 1617 in Michael Maier's *Symbola Auræ Mensæ*. With this question (65) begins the section of the book which refers to the *Viridarium Chymicum*.

182 Hermes Trismegistus, the thrice great Hermes, is a legendary person from ancient Egypt, the purported author of the *Corpus Hermeticum*, which was written around 200 CE.

183 *Mariæ Hebræ* or *Maria Hebræa* = "Mary the Jewess" is a legendary female alchemist from ancient Alexandria in Egypt living around 200 CE.

184 Democritus was an ancient Greek philosopher living around 400 BC. Alchemical writings attributed to Democritus are pseudonymous.

185 Morienus Romanus is a legendary seventh-century Christian hermit and alchemist who is said to have lived near Jerusalem. Alchemical writings attributed to him are pseudonymous.

186 Avicenna, an important Persian physician and philosopher from around 1000 CE, was also the author of alchemical texts.

187 Albertus Magnus was a German theologian and philosopher from the thirteenthth century. Alchemical writings attributed to him are mostly pseudonymous.

188 Arnaldus de Villa Nova was a French or Spanish physician living around 1300 CE. Most alchemical writings attributed to him are pseudonymous.

189 Thomas Aquinas (1225–1274) was an Italian theologian and philosopher. Alchemical writings attributed to him are pseudonymous.

190 Raymundus Lullus or Roman Lull, a Spanish theologian and philosopher who lived around 1300 CE. Alchemical writings attributed to him are pseudonymous.

191 Roger Bacon was a thirteenth-century philosopher who was also active as an alchemist.

192 Melchior Cibinensis was a sixteenth-century alchemical writer from Hungary.

193 Michael Sendivogius (1566–1636) was a highly influential Polish alchemist.

194 This question may refer to Figura XXXI in *Viridarium Chymicum*, also displayed on p. 155 of *Septimana Philosophica*.

195 *Aries* (Latin) = "ram".

196 This refers to the book: Daniel Stolcius von Stoltzenberg: *Viridarium Chymicum*, Frankfurt am Main, 1624.

197 This is the description of Figura XXXIV in *Viridarium Chymicum* and

also displayed on p. 96 of *Philosophia Reformata*.

198 This is the description of Figura XXXV in *Viridarium Chymicum* and also displayed on p. 96 of *Philosophia Reformata*.

199 This is the description of Figura XXXVI in *Viridarium Chymicum* and also displayed on p. 96 of *Philosophia Reformata*.

200 This is the description of Figura XXXVII in *Viridarium Chymicum* and also displayed on p. 107 of *Philosophia Reformata*.

201 This is the description of Figura XXXVIII in *Viridarium Chymicum* and also displayed on p. 107 of *Philosophia Reformata*.

202 This is the description of Figura XXXIX in *Viridarium Chymicum* and also displayed on p. 107 of *Philosophia Reformata*.

203 This is the description of Figura XL in *Viridarium Chymicum* and also displayed on p. 107 of *Philosophia Reformata*.

204 This is the description of Figura XLI in *Viridarium Chymicum* and also displayed on p. 117 of *Philosophia Reformata*.

205 This is the description of Figura XLIII in *Viridarium Chymicum* (Figura XLII was not included by Wallich). This figure was also displayed on p. 117 of *Philosophia Reformata*.

206 This is the description of Figura XLIV in *Viridarium Chymicum* and also displayed on p. 117 of *Philosophia Reformata*.

207 This is the description of Figura XLV in *Viridarium Chymicum* and also displayed on p. 126 of *Philosophia Reformata*.

208 This is the description of Figura XLVI in *Viridarium Chymicum* and also displayed on p. 126 of *Philosophia Reformata*.

209 This is the description of Figura XLVII in *Viridarium Chymicum* and also displayed on p. 126 of *Philosophia Reformata*.

210 This is the description of Figura XLVIII in *Viridarium Chymicum* and also displayed on p. 126 of *Philosophia Reformata*.

211 This is the description of Figura XLIX in *Viridarium Chymicum* and also displayed on p. 167 of *Philosophia Reformata*.

212 This is the description of Figura L in *Viridarium Chymicum* and also displayed on p. 167 of *Philosophia Reformata*.

213 This is the description of Figura LI in *Viridarium Chymicum* and also displayed on p. 167 of *Philosophia Reformata*.

214 This is the description of Figura LII in *Viridarium Chymicum* and also displayed on p. 167 of *Philosophia Reformata*.
215 This is the description of Figura LIII in *Viridarium Chymicum* and also displayed on p. 190 of *Philosophia Reformata*
216 This is the description of Figura LIV in *Viridarium Chymicum* and also displayed on p. 190 of *Philosophia Reformata*.
217 This is the description of Figura LV in *Viridarium Chymicum* and also displayed on p. 190 of *Philosophia Reformata*.
218 This is the description of Figura LVI in *Viridarium Chymicum* and also displayed on p. 190 of *Philosophia Reformata*.
219 *Conjunctio* (Latin) = "conjunction", in alchemy the process of combining different substances.
220 This is the description of figure LVII in *Viridarium Chymicum* and also displayed on p. 216 of *Philosophia Reformata*.
221 *Augmentatio* (Latin) = "augmentation", another alchemical term for multiplication.
222 This is the description of Figura LVIII in *Viridarium Chymicum* and also displayed on p. 216 of *Philosophia Reformata*.
223 This is the description of Figura LIX in *Viridarium Chymicum* and also displayed on p. 216 of *Philosophia Reformata*.
224 *Cibatio* (Latin) = "cibation", i.e., in alchemy the process of feeding the alchemical crucible with fresh material during the course of an operation.
225 This is the description of Figura LX in *Viridarium Chymicum* and also displayed on p. 216 of *Philosophia Reformata*.
226 In all three editions *Wunden* (= "wounds") was erroneously printed, but from the context, and after *Viridarium Chymicum*, it must be *Wunder* (= "wonders").
227 This is the description of Figura LXI in *Viridarium Chymicum* and also displayed on p. 224 of *Philosophia Reformata*.
228 *Aqua Vitæ* (Latin) = "water of life", a common archaic name for a mixture of water and ethanol with an ethanol content higher than in wine, beer, or the like, produced by distillation from water-ethanol mixtures with a lower ethanol content.
229 This is the description of Figura LXII in *Viridarium Chymicum* and also

displayed on p. 224 of *Philosophia Reformata*.

230 This is the description of Figura LXIII in *Viridarium Chymicum* and also displayed on p. 224 of *Philosophia Reformata*.

231 This is the description of Figura LXIV in *Viridarium Chymicum* and also displayed on p. 224 of *Philosophia Reformata*.

232 This is the description of Figura LXV in *Viridarium Chymicum* and also displayed on p. 243 of *Philosophia Reformata*.

233 In Wallich's counting, Figure 33 follows Figure 31 in all three editions. In this new edition we use the correct numbering. This is the description of Figura LXVII in *Viridarium Chymicum* (Figura LXVI was not described by Wallich). This figure is also displayed on p. 243 of *Philosophia Reformata*.

234 *Animæ extractio sive Imprægnatio* (Latin) = "the extraction or impregnation of the souls".

235 This is the description of Figura LXVIII in *Viridarium Chymicum* and also displayed on p. 243 of *Philosophia Reformata*.

236 *Ablutio vel mundificatio* (Latin) = "ablutions or cleansing".

237 This is the description of Figura LXIX in *Viridarium Chymicum* and also displayed on p. 262 of *Philosophia Reformata*.

238 *Animæ Jubilatio, sive sublimatio* (Latin) = "the jubilation or sublimation of the soul".

239 This is the description of Figura LXX in *Viridarium Chymicum* and also displayed on p. 262 of *Philosophia Reformata*.

240 *Germinatio* (Latin) = "growth".

241 This is the description of Figura LXXI in *Viridarium Chymicum* and also displayed on p. 262 of *Philosophia Reformata*.

242 *Fermentatio* (Latin) = "fermentation".

243 This is the description of Figura LXXII in *Viridarium Chymicum* and also displayed on p. 262 of *Philosophia Reformata*.

244 *Illuminatio* (Latin) = "illumination".

245 This is the description of Figura LXXIII in *Viridarium Chymicum* and also displayed on p. 281 of *Philosophia Reformata*.

246 *Nutrimentum* (Latin) = "nourishment".

247 This is the description of Figura LXXIV in *Viridarium Chymicum* and

also displayed on p. 281 of *Philosophia Reformata*.
248 *Fixatio* (Latin) = "fixation".
249 This is the description of Figura LXXV in *Viridarium Chymicum* and also displayed on p. 281 of *Philosophia Reformata*.
250 *Multiplicatio* (Latin) = "multiplication".
251 This is the description of Figura LXXVI in *Viridarium Chymicum* and also displayed on p. 281 of *Philosophia Reformata*.
252 *Revificatio* (Latin) = "revival, revivification".
253 This is the description of Figura LXXVII in *Viridarium Chymicum* and also displayed on p. 300 of *Philosophia Reformata*.
254 *Perfectio* (Latin) = "perfection".
255 This is the description of Figura LXXVIII in *Viridarium Chymicum* and also displayed on p. 300 of *Philosophia Reformata*.
256 *Leo Viridis* (Latin) = "the green lion".
257 This is the description of Figura LXXIX in *Viridarium Chymicum* and also displayed on p. 300 of *Philosophia Reformata*.
258 *Aurum nostrum* (Latin) = "our gold".
259 This is the description of Figura LXXX in *Viridarium Chymicum* and also displayed on p. 300 of *Philosophia Reformata*.
260 *Resuscitatio Regis* (Latin) = "the awakening or resuscitation of the king".
261 This is the description of Figura LXXXI in *Viridarium Chymicum* and also displayed on p. 224 of *Philosophia Reformata*.
262 *Tres fontes perennis aquæ* (Latin) = "three wells of perennial water".
263 This is the description of Figura LXXXIII in *Viridarium Chymicum* (The description of Figura LXXXII was not included by Wallich). This figure is also displayed on p. 354 of *Philosophia Reformata*.
264 This is the description of Figura LXXXIV in *Viridarium Chymicum* and also displayed on p. 354 of *Philosophia Reformata*.
265 This is the description of Figura LXXXV in *Viridarium Chymicum* and also displayed on p. 354 of *Philosophia Reformata*.
266 This is the description of Figura LXXXVI in *Viridarium Chymicum* and also displayed on p. 359 of *Philosophia Reformata*.
267 *Generatio* (Latin) = "the birth".

268 This is the description of Figura LXXXVII in *Viridarium Chymicum* and also displayed on p. 359 of *Philosophia Reformata*.
269 *Conjunctio* (Latin) = "the conjunction".
270 This is the description of Figura LXXXVIII in *Viridarium Chymicum* and also displayed on p. 359 of *Philosophia Reformata*.
271 *Mortificatio* (Latin) = "the death".
272 This is the description of Figura LXXXIX in *Viridarium Chymicum* and also displayed on p. 359 of *Philosophia Reformata*.
273 *Putrefactio* (Latin) = "putrefaction".
274 This is the description of Figura XC in *Viridarium Chymicum* and also displayed on p. 361 of *Philosophia Reformata*.
275 *Albificatio* (Latin) = "the whitening".
276 This is the description of Figura XCI in *Viridarium Chymicum* and also displayed on p. 361 of *Philosophia Reformata*.
277 *Rubificatio* (Latin) = "the reddening".
278 This is the description of Figura XCII in *Viridarium Chymicum* and also displayed on p. 361 of *Philosophia Reformata*.
279 This is the description of Figura XCIII in *Viridarium Chymicum* and also displayed on p. 361 of *Philosophia Reformata*.
280 This is the description of Figura XCVI in *Viridarium Chymicum* (Figura XCIV and XCV were not used by Wallich) and also displayed on p. 53 of *Occulta Philosophia*.
281 This is the description of Figura XCVII in *Viridarium Chymicum* and also displayed on p. 54 of *Occulta Philosophia*.
282 This is the description of Figura XCVIII in *Viridarium Chymicum* and also displayed on p. 56 of *Occulta Philosophia*.
283 This is the description of Figura XCIX in *Viridarium Chymicum* and also displayed on p. 59 of *Occulta Philosophia*.
284 This is the description of Figura C in *Viridarium Chymicum* and also displayed on p. 60 of *Occulta Philosophia*.
285 This is the description of Figura CI in *Viridarium Chymicum* and also displayed on p. 61 of *Occulta Philosophia*.
286 This is the description of Figura CII in *Viridarium Chymicum* and also displayed on p. 63 of *Occulta Philosophia*.

287 This is the description of Figura CIII in *Viridarium Chymicum* and also displayed on p. 65 of *Occulta Philosophia*.

288 This is the description of Figura CIV in *Viridarium Chymicum* and also displayed on p. 67 of *Occulta Philosophia*.

289 This is the description of Figura CV in *Viridarium Chymicum* and also displayed on p. 70 of *Occulta Philosophia*.

290 This is the description of Figura CVI in *Viridarium Chymicum* and also displayed on p. 75 of *Occulta Philosophia*.

291 This is the description of Figura CVII, the last figure in *Viridarium Chymicum* and also displayed on p. 72 of *Occulta Philosophia*.

292 This is the description of Figura XIII in *Viridarium Chymicum* and also displayed on p. 67 of Maier: *Tripus Aureus*.

293 In the first edition *Saltz* = "salt" was printed, in the second and third editions, the correct *Schatz* = "treasure".

294 This is a famous sentence from the *Tabula Smaragdina*.

295 *Vas, fornax, ignis, pondus, cum tempore latent* (Latin) = "vessel, furnace, fire, weight, and time are hidden". This saying can, e.g., be found in: Johannes Tilemann: *Aphorismi Hippocratei*, 1665, p. 90.

296 The word *Vater* = "father" is missing in all three editions, so that this sentence reads as if Saturnius were the son of the goddess Vesta. But according to Roman mythology, Saturnius is the husband of Opis and the father of Vesta. Therefore, this missing word was added.

297 This statement can be found (not literally but in essence) in *Atalanta Fugiens*, p. 109. The original Latin text is: *Draco mulierem, & haec illum interimit, simulquesanguine perfunduntur*.

298 *Aqua Stygia* is another name for *Aqua Regia*.

299 Cælus was a Roman god of the sky.

300 Crœsus (around 550 BCE) was the fabulously rich king of Lydia in what is now Western Turkey.

301 *Aqua ardens* (Latin) = "burning water", *aqua vitæ* (Latin) = "water of life".

302 The Cosmopolitan, or Cosmopolite, was a synonym for the Polish alchemist Michael Sendivogius. This synonym was mainly used for the French editions of his works. Question and answer 132 refer to Sendivogius' *Parable or Philosophical Riddle*, an *addendum* to his first book, *Twelve Treatises about the Philosopher's Stone* (first Latin edition: Prague 1604, first

ANNOTATIONS 695

German translation, Straßburg 1606, first English translation, London 1650).

303 Adfar of Alexandria is a legendary Arab alchemist who supposedly lived around 1050 in Egypt and was the teacher of Morienus.

304 Nicolas Flamellus (ca. 1330–ca. 1413) was a French writer. After his death he was stylized into a successful alchemist. Alchemical texts published under his name are pseudonymous. The most famous of these texts is the *Livre des figures hiéroglypiques*, Paris, 1612.

305 Johannes de Padua or Giovanni Padovani or John of Padua was a medieval Italian alchemist and possible author of *Consummata Sapienta seu Philosophia Sacra, Praxis de Lapide Minerali*, Magdeburg, 1602.

306 With this question starts the part of Wallich's third book in which her mostly real experiments and experiences with her secret *minera* are described. These are the questions: 163, 169, 171–177, 189–195, and 198.

307 *Arsenicum esse Tincturam Albedinis* (Latin) = "arsenic is the white tincture".

308 *Sal enixum* = potassium bisulfate $KHSO_4$

309 *Liquorem aquilæ* (Latin) = "the eagle's liquor".

310 This description of the preparation of Paracelsus' *corallate* was printed in: Paracelsus: *De Morbo Gallico*, Straßburg 1578, pp. 207–208.

311 Joachim Polemann (ca. 1620–ca. 1672) was a German alchemist. In his book, *Novum Lumen Medicum*, Amsterdam, 1659, p. 68, he describes how to produce the *Elementum Ignis* from copper (= ♀).

312 The circulated salt or *Sal circulatum* was first mentioned by Paracelsus in: Paracelsus: *De Renovatione & restauratione*, in: Johann Huser (ed.): *Bücher und Schrifften des ... Paracelsi*, vol. 6, pp. 100–114, pp. 106, 113.

313 Dissolving a salt in a radish root was mentioned earlier by Paracelsus, see: Gerhard Dorn: *Theophrastische Practica*, 1619, pp. 73, 77.

314 Michael Maier: *Atalanta Fugiens*, Frankfurt am Main, 1617, pp. 48–49: "Gib Feuer zum Feuer, Mercurium zum Mercurio" and "*Da ignem igni, Mercurium Mercurio*". Somewhat similar before already in the *Turba Philosophorum*.

315 Anon. (August Hauptmann): "Neun und Siebenzig grosse und sonderbahre Wunder, so bey einem Special angegebenen Subiecto theils von der Natur, theils aber in der angeführten Arbeit sich befunden", as appendix to Heinrich von Batsdorff: *Filum Ariadnes. Das ist Neuer Chy-*

mischer Discurs von den grausamen verführerischen Irrwegen der Alchymisten, Leipzig and Gotha, 1690.

316   This describes the blue smalt production from cobalt ore, potash, and sand.

317   This describes the preparation of an aqueous cobalt(II) nitrate solution by the reaction of nitric acid with a cobalt-containing ore.

318   12–16 lot from 1 pound is in today's units about 233.8 to 311.7 g from 467.7 g (1 pound = 24 lot = 467.7 g).

319   6–8 lot: 116.9 to 155.9 g.

320   *Unkenbrenner* was a legendary fraudulent alchemist from Southern Germany around 1430. His name became synonymous with an uneducated fraudulent alchemist.

321   These locations: Piedmont, Friuli, and between Bavaria and Swabia, and the details of the corresponding descriptions are taken from: Johann Rudolph Glauber: *Theutschlandes Wohlfahrt*, volume 3, Amsterdam, 1659, pp. 91–94.

322   Kinzig valley in the Black Forest. On the opposite side of Strasbourg, France, the Kinzig river flows into the Rhine.

323   A peasant step is a slow step.

324   In all three editions, *Iclerus* was printed, but this must be *Icterus* (Latin) = "jaundice".

325   *Colica passio* (Latin) = "intestinal pain".

326   *Sanct Quirins Buss* (= St. Quirinus' penance) or *Sanct Johannis Buss* (= St. John's penance): Old German terms for a disease, a nasty rash associated with fever.

327   *Kobaltblüte* = literally "Cobalt blossom", a red cobalt arsenate mineral, in modern mineralogy denoted as erythrite or cobalt red, with a peach blossom red color.

328   1 pound = 24 lot = 467.7 g; 16 lot = 311.8 g.

329   This refers to the book (Ps.-)Basilius Valentinus: *Triumph Wagen Antimonii*, Leipzig, 1604.

330   In all three editions, *Fehler* = "mistake" was printed, but this should be *Fechter* = "fencer".

BIBLIOGRAPHY

AGRICOLA, Johann: *Ander Theil Joannis Agricolæ P. & M.D. Commentarium, Notarum, Observationum & Animadversionum in Johannis Poppii Leipzig 1639.* This is part 2 ("ander Theil") of the first, two-volume edition of Johann Agricolas *Chymical Medicine.*

ALFONSO-GOLDFARB, Ana Maria, Marcia H.M. Fernandez: "Gur, Ghur, Guhr or Bur? The Quest for a Metalliferous Prime Matter in Early Modern Times", *BJHS* 46 (2013) 23–37.

BATSDORFF, Heinrich von: *Filum Ariadnes. Das ist Neuer Chymischer Discurs von den grausamen verführerischen Irrwegen der Alchymisten*, Leipzig and Gotha 1690.

DORN, Gerhard: *Theophrastische Practica, das ist Auszerlesene Theophrastische Medicamenta, beneben eigentlicher Beschreibung derer Præparation: Auch richtige Nutz und Gebrauch,* 1619.

GLAUBER, Johann Rudolph: *Theutschlandes Wohlfahrt,* volume 3, Amsterdam 1659.

Hermetischer Rosenkrantz, Das ist: Vier schöne, außerlesene Chymische Tractätlein ..., Hamburg 1659.

HORN, Caspar (Ed.): *Bernhardus innovatus das ist, deß hocherfahrnen vortrefflichen und waaren Philosophi Chemici Herrn Bernhardi Grafen von der Marck und Tervis Chemische Schrifften von der hermetischen Philosophia Oder Vom gebenedeiten Stein der der Weißen,* Nürnberg 1643.

KRAFT, Alexander: "Dorothea Juliana Wallich, geb. Fischer (1657-1725), eine Alchemistin aus Thüringen". *Genealogie. Deutsche Zeitschrift für Familienkunde* 33 (2017) 539–555.

KRAFT, Alexander: "Dorothea Juliana Wallich (1657–1725) and Her Contributions to the Chymical Knowledge about the Element Cobalt". in: Annette Lykknes, Brigitte van Tiggelen (eds.): *Women in Their Element. Selected Women's Contributions to the Periodic System*, Singapore 2019, pp. 57–69.

LULLUS, Raymundus: *Testamentum Raymundi Lulli*, Rouen 1658.

MAIER, Michael: *Atalanta Fugiens*, Frankfurt am Main 1617.

MAIER, Michael: *Symbola Aureæ Mensæ Duodecim Nationum*, Frankfurt am Main 1617.

MAIER, Michael: *Tripus Aureus hoc est Tres Tractatus Chymici Selectissimi*, Frankfurt am Main, 1618.

MAIER, Michael: *Septimana Philosophica: Qua Ænigmata Aureola De Omni Naturæ Genere*, Frankfurt am Main 1620.

MATTHESIUS, Johann: *Sarepta, oder Bergpostill*, Nürnberg 1652.

MONTE CUBITI, Vigilantium de: *Dreyfaches hermetisches Kleeblattin welchem begriffen dreyer vornehmen Philosophorum herrliche Tractätlein ...*, Nürnberg 1667.

MYLIUS, Johann Daniel: *Philosophia Reformata*, Frankfurt am Main 1622.

Occulta Philosophia von den verborgenen Philosophischen Geheimnüssen der heimlichen Goldblumen, Frankfurt am Main 1613.

PARACELSUS: *De Renovatione & restauratione*, in: Johann Huser (ed.): *Bücher und Schrifften des ... Paracelsi*, vol. 6, 1569.

PARACELSUS: *De Morbo Gallico*, Straßburg 1578.

PHILALETHES, Eirenaeus: *Eröffneter Eingang zu deß Königs verschlossenen Pallaste Anonymi Philaletae*, Frankfurt und Hamburg 1673.

POLEMANN, Joachim: *Novum Lumen Medicum*, Amsterdam 1659.

Rosarium Philosophorum, Frankfurt am Main 1550.

J.R.V.M.D. [RÖSSER, Jacob]: *Güldene Rose, d.i. Einfältige Beschreibung des allergrössesten von dem Allmächtigen Schöpffer Himmels und Erden Jehovah, In die Natur gelegten, und dessen Freunden und Außerwehlten zugetheilten Geheimnisses, als Spiegels der Göttlichen und natürlichen Weißheit*, Hamburg 1705.

RUBIN, Jonathan: "The Use of the 'Jericho Tyrus' in Theriac: A Case Study in the History of the Exchanges of Medical Knowledge Between Western Europe and the Realm of Islam in the Middle Ages", *Medium Ævum* 83 (2014) 234–253.

SENDIVOGIUS, Michael: *Von dem Rechten wahren Philosophischen Stein Zwölff Tractätlin in einem Wercklin verfasst ...*, Straßburg 1606.

STOLCIUS VON STOLTZENBERG, Daniel: *Viridarium Chymicum Figuris Cupro Incisis Adornatum, et Poeticis PicturisIllustratum*, Frankfurt am Main 1624.

The Holy Bible Containing the Old and New Testaments (King James Version), London 1991.

TILEMANN, Johannes: *Aphorismi Hippocratei*, 1665.

URBIGERUS, Baro: *Aphorismi Urbigerani, or Certain Rules, Clearly Demonstrating the Three Infallible Ways of Preparing the Grand Elixir or Circulatum Majus of the Philosophers, Discovering the Secrets of Secrets*, London 1690.

VALENTINUS, Basilius: *Triumph Wagen Antimonii*, Leipzig 1604.

WALLICH, Dorothea Juliana [= D.I.W.]: *Das Mineralische Gluten, Doppelter Schlangen-Stab, Mercurius Philosophorum, Langer und kurtzer Weg zur Universal-Tinctur*, Leipzig 1705.

WALLICH, Dorothea Juliana [= D.I.W.]: *Der Philosophische Perl-Baum, das Gewächse der drey Principien, zu Deutlicher Erklärung des Steins der Weisen, wie er mit seinen Wurtzeln in der äusern und finstern Welt, mit seiner Blüthe aber in der Paradisischen- und Licht-Welt, und mit seiner reiffen Frucht in der Englischen und Himmlischen Welt stehet und wachset*, Leipzig 1705.

WALLICH, Dorothea Juliana [= D.I.W.]: *Schlüssel zu dem Cabinett der geheimen Schatzkammer der Natur, zur Such- und Findung des Steins der Weisen, durch Fragen und Antwort gestellet*, Leipzig 1706.

Wasserstein der Weysen: das ist ein chymisch Tractätlein darin der Weg gezeiget, die Materia genennet und der Proceß beschrieben wird zu dem ... Geheymnuß der Universal Tinctur zukommen, Frankfurt am Main 1619.

WEEKS, Andrew: *Boehme. An Intellectual Biography of the Seventeenth-Century Philosopher and Mystic*, Albany, 1991.

WEEKS, Andrew: *Aurora (Morgen Röte im auffgang, 1612) and Ein gründlicher Bericht or A Fundamental Report (Mysterium Pansophicum, 1620) by Jacob Boehme*, Leiden 2013.

WEEKS, Andrew: *De Tribus Principiis, oder Beschreibung der Drey Principien Göttliches Wesens, Of the Three Principles of Divine Being, 1619, by Jacob Boehme*, Leiden 2019.

WILLARD, Thomas: "Beya and Gabricus: Erotic Imagery in German Alchemy", *Mediævistik* 28 (2015) 269–281.

ZUBER, Mike: "Jacob Böhme and Alchemy: A Transmutation in Three Steps", in: Bo Andersson, Lucinda Martin, Leigh Penman, Andrew Weeks (Ed.): *Jacob Böhme and his World*, Leiden 2018, pp. 262–285.

ABOUT THE EDITOR

DR. ALEXANDER KRAFT is a chemist and historian of chemistry from Germany. He studied chemistry at the Humboldt University in Berlin where he received his doctorate in 1994 in the field of semiconductor electrochemistry. Since 1995 he has worked as an employee, founder, or consultant for various start-up companies in Germany, the USA, Sweden, and Spain in the fields of electrochemical water treatment and dynamic smart windows (electrochromism, thermochromism). Since 2007, Kraft has also been working as a historian of alchemy and chemistry, with a focus on the history of Prussian blue and the history of chemistry and alchemy in Berlin. He has published three German-language books in this field. These include works on the history of chemistry in Berlin (*Chemie in Berlin: Geschichte, Spuren, Persönlichkeiten*, 2012), the history of Prussian blue (*Berliner Blau: Vom frühneuzeitlichen Pigment zum modernen Hightech-Material*, 2019) and the history of alchemy in Saxe-Meiningen around the year 1700 (*Zwei Herzöge im Goldrausch: Die Alchemie am Fürstenhof von Sachsen-Meiningen von 1680 bis 1724*, 2023). Since 2016, Kraft has been researching the biography and writings of the alchemist Dorothea Juliana Wallich. Together with colleagues in the Alchemy Network at the Gotha Research Center (Erfurt University), selected experiments by Wallich are being recreated. The ligand exchange thermochromism she discovered is still the basis for developmental work towards dynamic glazing in modern construction.

Rubedo Press

HERMETIC AND ALCHEMICAL SOURCETEXTS
Bilingual Series

Hermetic Recreations, Including the Scholium (anonymous manuscript, late eighteenth–early nineteenth century). French/English. Edited, translated, and annotated by Christer Böke, John Koopmans, Stanislas Klossowski de Rola, and Aaron Cheak, 2017.

The Basilian Aphorisms, or the Hermetic Canons of the Spirit, Soul, and Body of the Major and Minor World (1608). Latin/English. Edited, translated, and annotated by Mirco A. Mannucci and Aaron Cheak, 2018.

The Key to the Hermetic Sanctum (anonymous manuscript, eighteenth century). French/English. Edited, translated, and annotated by Christer Böke, John Koopmans, Juan Duc Perez, and Aaron Cheak, 2020.

The Reign of Saturn Transformed into an Age of Gold, by Huginus à Barma (1657). Latin/English. Translated by Michael A. Putman. Edited and Annotated by Aaron Cheak and Mirco A. Mannucci, 2020.

Pathways to the Universal Tincture: Collected Alchemical Writings, by Dorothea Juliana Wallich. German/English. Edited and annotated by Alexander Kraft. Translated by Alexander Kraft with Michael A. Putman and Aaron Cheak, 2025.

HERMETIC AND ALCHEMICAL STUDIES

Becoming Gold: Zosimos of Panopolis and the Alchemical Arts in Roman Egypt, by Shannon Grimes, 2018.

The Path to the New Hermopolis: The History, Philosophy, and Future of the City of Hermes, by Mervat Nasser, 2019.

Alchemical Traditions: From Antiquity to the Avant-Garde. Edited by Aaron Cheak. Revised and expanded edition. Forthcoming.

※

www.ingramcontent.com/pod-product-compliance
Lightning Source LLC
Chambersburg PA
CBHW030335010526
44119CB00047B/504